Modern Residential Wiring

by

Harvey N. Holzman

Master Electrician and Member IAEI
Halcott Center, New York

South Holland, Illinois
THE GOODHEART-WILLCOX COMPANY, INC.
Publishers

Copyright 1986

by

THE GOODHEART-WILLCOX COMPANY, INC.

All rights reserved. No part of this book may be reproduced, stored in a retrieval system, or transmitted in any form or by any means, electronic, mechanical, photocopying, recording, or otherwise, without the prior written permission of The Goodheart-Willcox Company, Inc. Manufactured in the United States of America.

Library of Congress Catalog Card Number 85-27259
International Standard Book Number 0-87006-482-7

23456789-86-09876

Library of Congress Cataloging in Publication Data

Holzman, Harvey N.
  Modern residential wiring.

  Includes index.
  1. Electric wiring, Interior.   I. Title
  TK3285.H65 1986   621.319'24   85-27259
  ISBN 0-87006-482-7

INTRODUCTION

MODERN RESIDENTIAL WIRING provides you with a solid background of electrical principles and practices, as well as a thorough understanding of electrical *Code* requirements. Once having mastered the information given here, you will be well equipped to design and install modern, safe residential wiring systems that meet the electrical power demands of modern dwellings.

MODERN RESIDENTIAL WIRING covers not only the "how" but the "why" of safe electrical wiring practice. Although the content is concerned primarily with residential installations, many of the same concepts and principles may be applied to commercial and light industrial electrical construction.

Chapters are arranged in a logical sequence. The order of instruction follows the normal order in which the installation would be made. However, each chapter is designed to stand alone and may be studied independently to suit individual need.

MODERN RESIDENTIAL WIRING makes the study of electrical wiring easy. Even the most complicated procedures are simply explained and easy to understand. Procedures are explained step-by-step while the many illustrations are fully integrated into the easy-to-read text. The illustrations should be carefully examined as they will often clarify and explain the more difficult principles of electricity and the requirements of the National Electrical Code.

Harvey N. Holzman

Persons using this book are advised to familiarize themselves with the National Electrical Code.® Certain tables in this book (credited to NEC® or National Fire Protection Association) are the property of that professional association and may not be reproduced without their permission.

They are reprinted for your convenience with permission from NFPA 70-1984, *National Electrical Code,* Copyright©1983, National Fire Protection Association, Quincy, MA 02269. This reprinted material is not the complete and official position of the NFPA on the referenced subject which is represented only by the standard in its entirety.

CONTENTS

1. ELECTRICAL ENERGY FUNDAMENTALS 7
2. ELECTRICAL CIRCUIT THEORY 19
3. ELECTRICAL CIRCUIT COMPONENTS 25
4. TOOLS FOR THE ELECTRICIAN 32
5. SAFETY AND GROUNDING ESSENTIALS 40
6. WIRING SYSTEMS 52
7. BOXES, FITTINGS, AND COVERS 59
8. INSTALLING BOXES AND CONDUCTORS 69
9. DEVICE WIRING 85
10. PLANNING BRANCH CIRCUITS 101
11. READING BLUEPRINTS AND WIRING CIRCUITS 114
12. THE SERVICE ENTRANCE 131
13. APPLIANCE WIRING AND SPECIAL OUTLETS 153
14. LIGHT COMMERCIAL WIRING 161
15. FARM WIRING 165
16. MOBILE HOME WIRING 172
17. LOW-VOLTAGE CIRCUITS 179
18. ELECTRICAL REMODELING 187
19. ELECTRICAL METERS 200
20. ELECTRICAL TROUBLESHOOTING 207
21. SPECIALIZED WIRING 212
22. MOTORS AND MOTOR CIRCUITS 227
23. ELECTRICAL CAREERS 240
24. MATH REVIEW 247
25. TECHNICAL INFORMATION 250
26. DICTIONARY OF TERMS 258

Residential wiring is a satisfying career for those who can master the principles of electrical wiring and who familiarize themselves with the Code requirements. These electric company workers are installing the pad transformer which will supply electrical power to the residence shown. The dwelling, itself, is wired by a qualified electrician.   (New York State Electric and Gas Corp.)

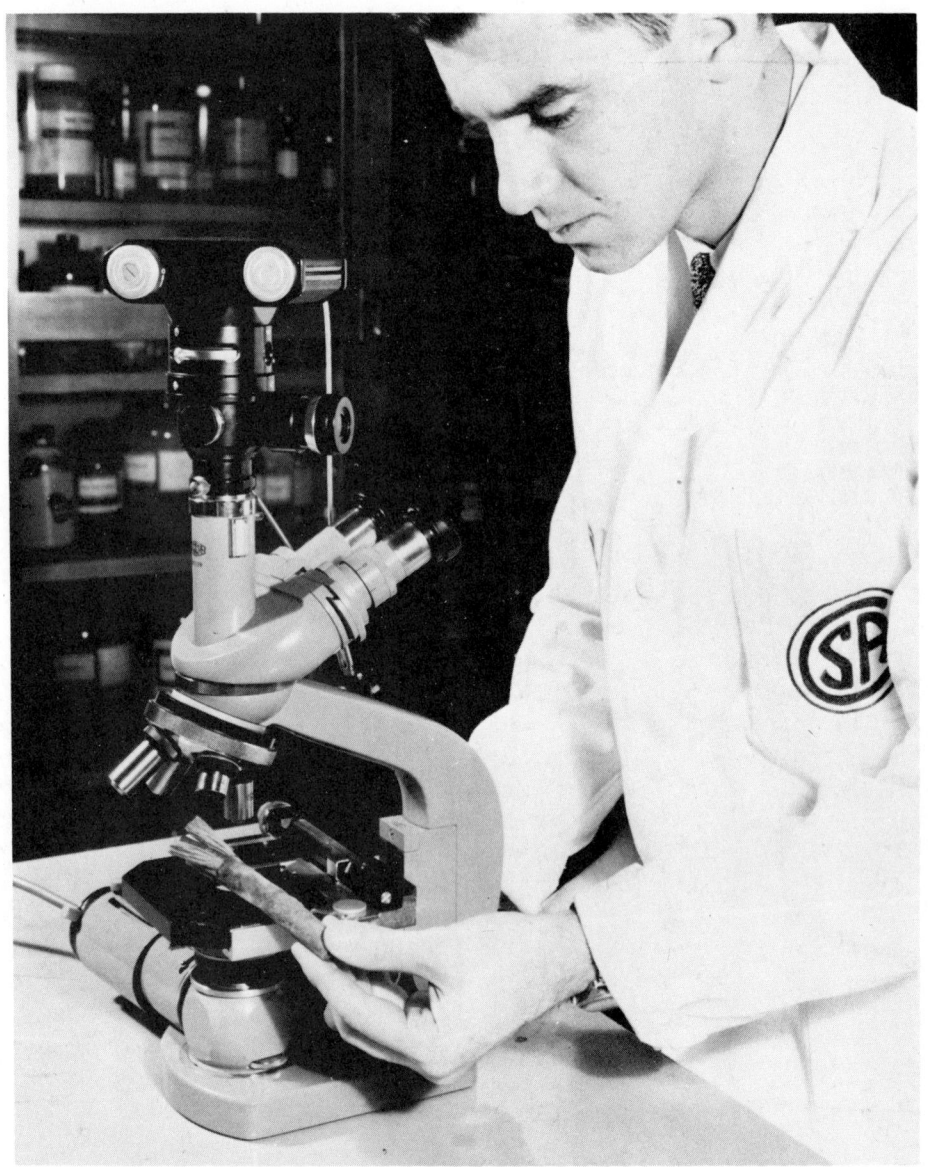

CSA technician examines cable used in barns after it has been subjected to fungas attack in controlled laboratory tests. (Canadian Standards Assoc.)

Chapter 1
ELECTRICAL ENERGY FUNDAMENTALS

In this chapter you will learn what electricity is, how it is produced, transmitted, and measured. This information will provide a background that will make your study of wiring practices and principles more easily grasped.

Electricity is a form of energy. Like other forms of energy, it can do work or it can be changed into another form which will operate appliances or produce light and heat.

ELECTRON THEORY

Electricity is closely related to matter. That is, it is made up of particles. According to the ELECTRON THEORY all matter is made up of atoms.

Each of these atoms is made up of a nucleus (center) consisting of positively charged particles called PROTONS and neutral (noncharged) particles called NEUTRONS. Negatively charged particles called ELECTRONS, are attracted to the positively charged protons in the nucleus and whirl rapidly around it much like the earth moves around the sun. A typical atom is shown in Fig. 1-1.

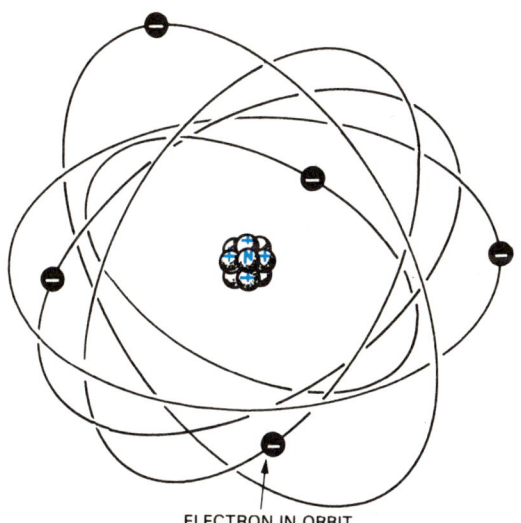

Fig. 1-1. Basis of electricity is the atom which is the foundation of all matter. In some elements, such as metals, orbiting electrons can easily be made to leave the atom and travel.

There can be one or more electrons traveling around in any one atom and they can be at different distances from the nucleus in their circular path.

Electrons have the ability to travel. They may be made to leave one atom and move to another.

FREE ELECTRONS

In some types of matter, electrons are tightly bound to the nucleus and can leave only with great difficulty. In others the bond is so relaxed that moving is easy. These electrons are called "loosely held" or "free."

ELECTRON TRAVEL

Electrons (negative charges) pass readily out of the atoms of some substances. In other substances the electrons can travel only with much effort. Materials through which electrons pass easily are called CONDUCTORS. Those where passage is difficult are called INSULATORS.

Pure metals, carbon and most liquids are excellent conductors. Among good insulators are dry gases, glass, rubber, mica, silk, and cork.

According to the commonly held electron theory, metals conduct electricity well because they have a great number of loosely held or free electrons. In insulators there are few, if any, free electrons; thus, little or no electricity is conducted.

EQUILIBRIUM

If a negatively charged body is connected to a positively charged body, by a conductor, such as a wire, the electrons will be attracted to the positive particles. Nature prefers everything to be neutral—equal numbers of positive and negative charges. As a result, electrons will flow from the negatively charged body to the positive one. See Fig. 1-2. This flow is called CURRENT. Even if two negatively charged bodies are connected in this manner, free electrons will flow from the body with the most electrons to the one with fewer. Flow will continue until EQUILIBRIUM (balance) is reached.

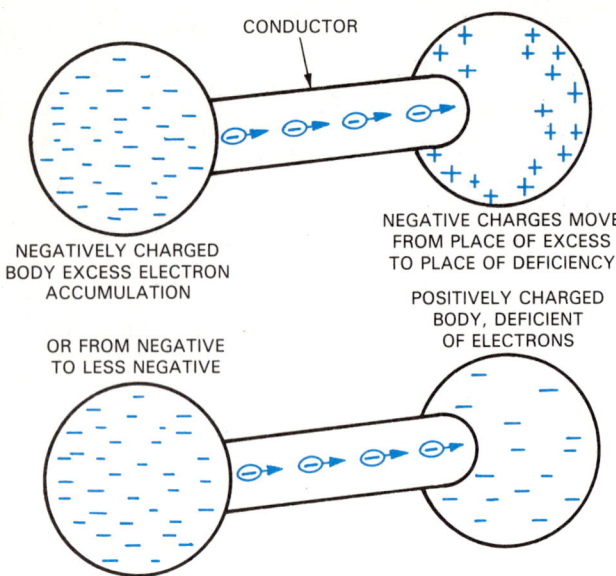

Fig. 1-2. Electrons will move from a body with an excess of electrons to a body with fewer or no electrons provided there is a path known as a conductor. This is also known as the Law of Static Charges.

Bodies with different concentrations of free electrons are said to be at DIFFERENT POTENTIALS. Therefore, electrical potential is the ability to provide free electrons. In electricity, differences in potential are measured in VOLTS (V).

TYPES OF ELECTRIC CURRENT

There are two ways or methods of having electric current flow. One way is called *direct current* (dc), the other, *alternating current* (ac).

In direct current the electricity flows in *one direction only*. Alternating current *reverses its direction* usually 60 times each second and is, thus, often called *60-cycle electricity* or 60 hertz (60 Hz).

A *cycle* is one complete electrical wave or vibration. We will look at electrical waves in more detail later.

Campers, automobiles, electric toys, flashlights, and the like, employ dc electricity, for the most part. House lights, appliances, electric tools, and other devices, normally operating from household electricity, operate on ac.

SOURCES OF ELECTRICAL ENERGY

To maintain a flow of electrons or electricity it is necessary to have a source which is always at a greater electrical potential. That is, it always has an excess of free electrons which are ready to move along a conductor to where the electrical potential is less. The energy may be produced chemically, as with a battery, or mechanically as with a generator.

BATTERY POWER

A basic chemical device for providing electrical power is the cell. Fig. 1-3 shows a simplified cell made up of two metal rods suspended in a mild acid solution. The rods react chemically to the acid. Electrons on the surface of the copper rod are stripped away and travel to the zinc rod. Then the copper rod has fewer electrons and is positively charged. It becomes known as the positive pole or ANODE. The zinc rod has an excess of electrons and is known as the negative pole or CATHODE.

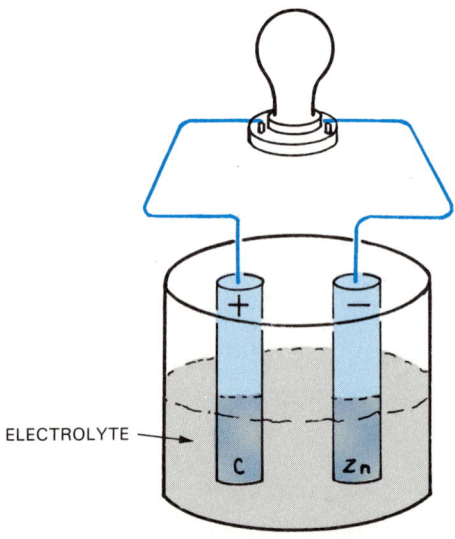

Fig. 1-3. A simple wet cell can store electricity as chemical solution. Conductors (wires) and light make this a simple circuit. Wet cell is the forerunner of today's flashlight batteries. Electrodes are two different metals, copper and zinc.

A common practice is to group a number of cells and use them as a source of current rather than just a single cell. See Fig. 1-4. This group of cells is known as a BATTERY. Either a liquid chemical or a more or less dry chemical paste can be used in the battery. The first type is known as a WET cell battery, the latter as a DRY cell battery.

GENERATOR POWER

Batteries are just one of the ways to generate or drive an electrical current. A second method is with a GENERATOR. Driven by some mechanical force, this device creates differences in electrical potential between two electrical poles. It moves electrons from one terminal and deposits them on the other terminal. This is accomplished through electromagnetic induction. This process will be explained later as we discuss generators and alternators.

Electrical Energy Fundamentals

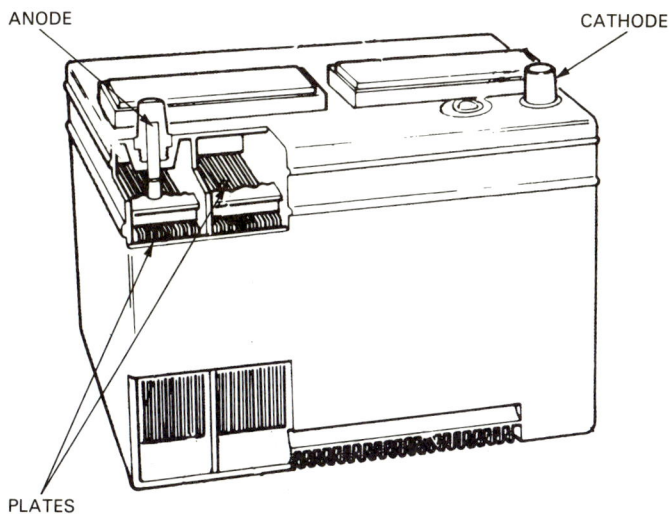

Fig. 1-4. Group of wet cells connected electrically and suspended in electrolyte is called a battery. It can supply a steady current of electricity.

CIRCUIT (ELECTRICAL PATH)

No electron flow will occur unless there is a pathway over which the electrons can move. In a water system the pathway is the piping which moves the water from storage to where it is used. In electrical wiring the pathway is called a circuit, Fig. 1-5. The circuit allows the electrons to flow from the negative terminal to the positive terminal. The circuit is made of conducting materials or wires.

A simple circuit will consist of:
1. A power source. For a residence, the power source could be considered the electrical generating stations. However, primary sources include small generators and batteries.
2. Conductors (wiring) to provide a path for the current so it can travel from one point to another.
3. A load or loads. These are devices through which the electricity produces work.
4. A device or devices for controlling current. Included would be switches and fuses or circuit breakers.

TYPES OF CIRCUITS

The two basic electrical circuits are the SERIES and PARALLEL. A third, called a complex circuit, is a combination of the other two.

Series circuit

A series circuit is one in which only one path is provided for the current. The electricity must always flow through every device in the circuit. If one device is burned out the circuit will not function. Fig. 1-6 is a simple diagram of a series circuit.

Except for switches and fuses or circuit breakers which are used to control electricity flow, series circuity is not practical for residential wiring. A simple example will explain why. Older Christmas tree lights were often wired in series. Current had to pass through every light bulb for the string of lights to work. If one bulb burned out all the lights went out. Finding which bulb was defective was a trial and error operation. Every bulb had to be removed and tested. Residential circuits are set up so that a nonfunctioning load will not stop electrical current in the remainder of the circuit.

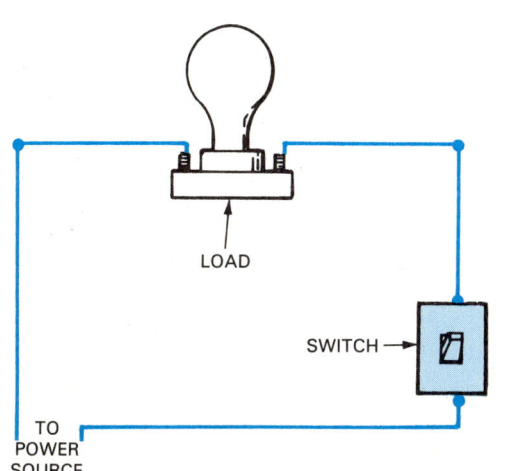

Fig. 1-5. Simple electrical circuit includes a pathway (conductors), power source, load (light) and a switch for controlling the current.

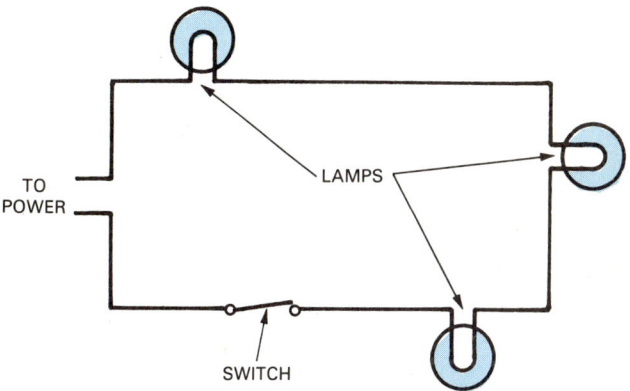

Fig. 1-6. Simple diagram of series circuit. The three lights are hooked into the circuit so that if one were not working the circuit would be open and all lights would go out.

Parallel circuit

Like the series circuit, the parallel circuit has a complete path for the current to follow. The difference in a parallel circuit is that there is more than

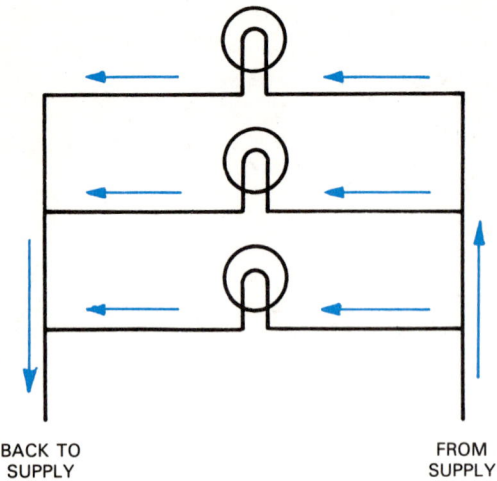

Fig. 1-7. Parallel circuit diagram. Note that each lamp is on its own electrical path. Were one to burn out when circuit is closed, the others would remain lit.

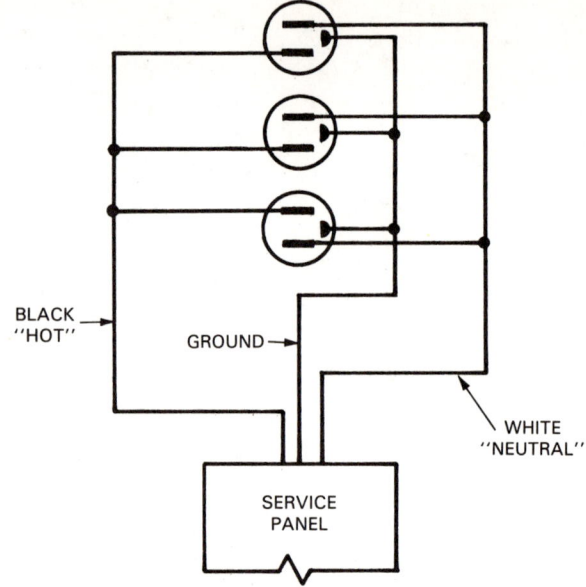

Fig. 1-8. Typical parallel circuit for 120 V housewiring for electrical outlets. Each outlet is connected to ground as well as to "hot" and "neutral" wires.

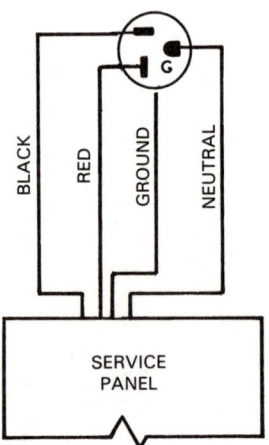

Fig. 1-9. A housewiring circuit supplying 240 V. The black and red conductors each supply 120 V to the outlet.

one path for the current, Fig. 1-7. Each path goes to a load which can operate independently of the other branches and/or loads in the circuit. If one load, such as a lamp, burns out, the other branches would continue to operate since a path still exists from one supply terminal through the circuit to the other supply terminal.

Under normal conditions, another advantage of parallel circuits is that the current draw of each branch affects only that branch. It depends only on the resistance of the load in that branch. (In a series circuit, resistance in one load affects the rest of the loads on the circuit. Loads such as lamps, heaters, or motors would not operate properly. In fact, some loads, such as motors, could be severely damaged by voltage drop.)

Except for switches and fuses or circuit breakers, all residential wiring is done in parallel. Figs. 1-8 and 1-9 are examples of parallel circuits found typically in housewiring.

LOADS

In electrical wiring, a load is any device that uses an electric current and converts the energy to another form. The devices include:
1. Lamps or light bulbs.
2. Resistances such as heating elements.
3. Electric motors such as are used to drive tools or operate appliances.
4. Radios, televisions and other electronic devices.

Each load is designed to operate at definite voltages. A voltage rating is given with every device. It is important that the device be operated at that rating. Otherwise, it can be damaged or destroyed. In some cases the entire circuit can be damaged. This is an important reason why series circuits are not used in residential wiring. House circuits are designed to handle a number of loads. Series wiring could result in drastic voltage drops while electricity moved from one load to the next.

ELECTROMAGNETIC INDUCTION

In a battery, chemical action displaces the electrons and produces the electron flow or current through a conductor. There is a mechanical method of doing the same thing.

Two men, Michael Faraday and Joseph Henry, working independently, discovered ELECTROMAGNETIC INDUCTION. By passing a wire conductor through a magnetic field, Fig. 1-10, they could produce a flow of electrons in the conductor. They had induced the first electrical current using magnetism.

Electron flow occurs because they have been dislodged from their orbits around particles of matter (atoms) and are attracted to where there are fewer electrons. If you were to fashion the conductor into a loop (or coil) and rotated it between two magnetic poles you would have a crude generator. See Fig. 1-11.

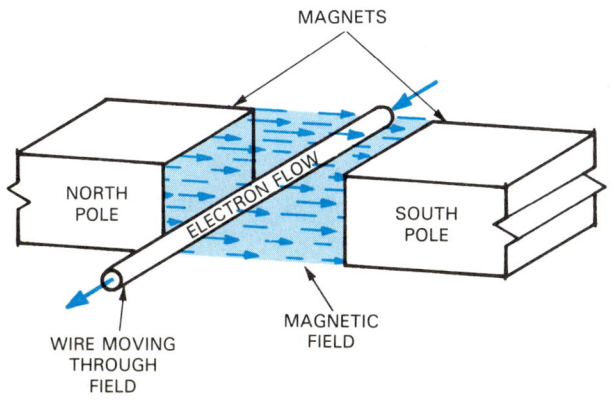

Fig. 1-10. Passing a conductor between the poles of a magnet will displace electrons in the conducting wire and cause them to move through the conductor. This event is called electromagnetic induction.

The strength or force of electron flow is called electromotive force or voltage. Generator voltage is determined by four factors:
1. Number of turns of wire making up the conductor coil or loop which is cutting through the magnetic field between the magnet poles.
2. The speed at which the conducting loop is rotating.
3. The strength of the magnetic field. This depends on the strength of the magnet.
4. The angle at which the conducting coil cuts through the magnetic field. Voltage is at its greatest when the coil is moving at right angles to the magnetic field lines of force. See Fig. 1-12.

GENERATORS AND ALTERNATORS

As you have just learned, generators and alternators are rotating machines driven by some mechanical force. They use the electromagnetic induction principle to convert the mechanical force to electrical energy. There are ac and dc generators. The ac generator is usually called an alternator.

DIRECTION OF INDUCED CURRENT

Current from a spinning generator will always move in a direction determined by what is called the LEFT HAND RULE for generators, Fig. 1-13. The rule says: If the thumb, forefinger, and middle finger of the left hand are held at right angles to each other, and the thumb is pointing in the direction the wire is moving, while the forefinger is pointing in the direction of the magnetic field (north to south) then the middle finger will point in the direction of the induced electron flow. This statement is also known as Fleming's Rule.

Alternating current generator

The ac generator, or alternator, consists of several distinct parts. Each has a specific function:
1. A coil or armature. This part rotates. It has the conducting wire which cuts across a magnetic field. The electron flow begins in this part.
2. Nonmoving poles or opposite ends of field magnets. They create the magnetic field. In some generators, the poles are magnetized by a portion of the electrical current generated.
3. Slip rings. These metal parts are always in contact with one of the terminals of the coil or armature. They transfer the current to the brushes.

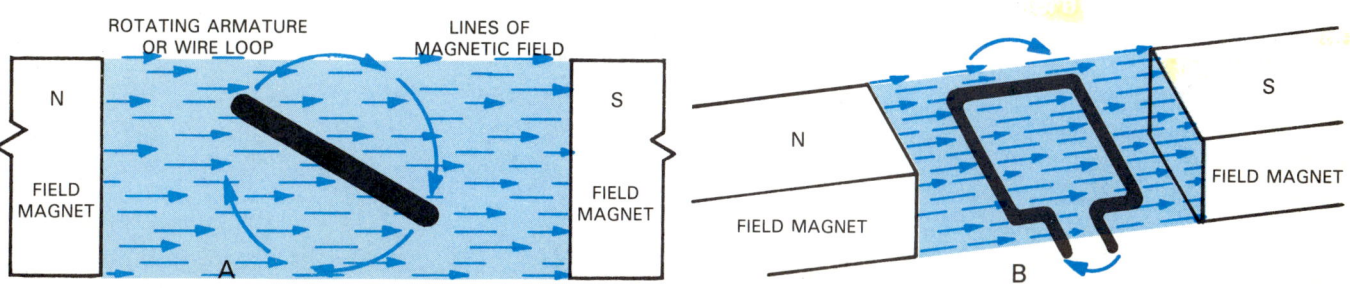

Fig. 1-11. A simple generator of electricity. Rotating coil of wire has current induced in it as loop cuts a magnetic field created between the poles of the magnet. A—Straight-on view of generator. Voltage would be low with loop in this position. Wire would be cutting field at sloping angle. B—Perspective view showing loop cutting magnetic field at right angle. Voltage would be at its highest.

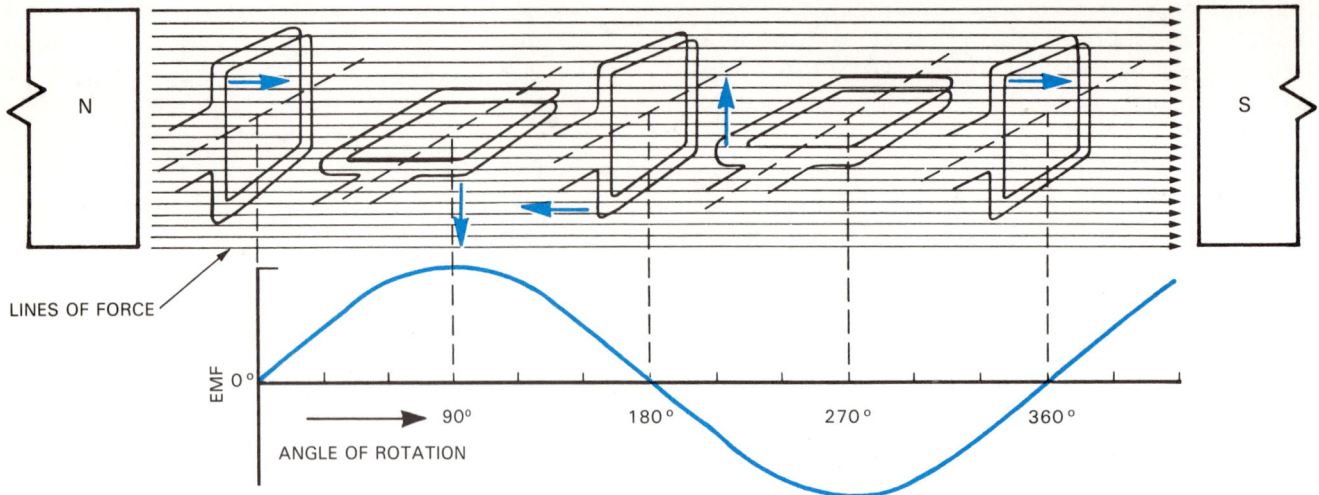

Fig. 1-12. Voltage produced by generator varies. Compare position of coil above with sine wave below. Peaks in wave represent points of highest voltage during one complete rotation.

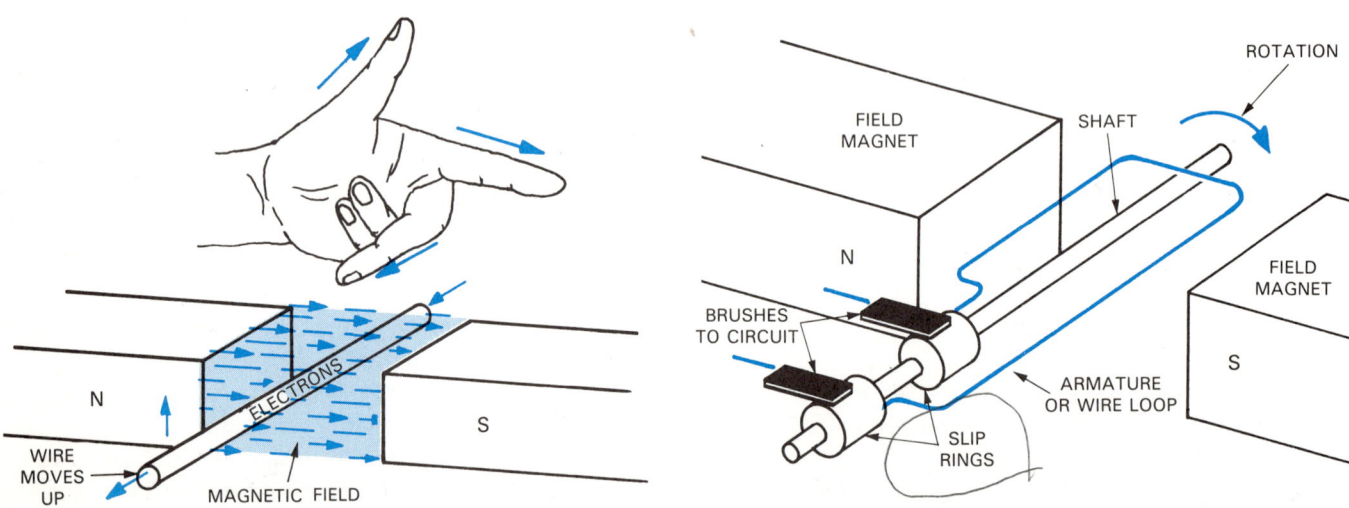

Fig. 1-13. Left hand rule for generators. If wire is moving in direction of thumb and the magnetic field is moving in the direction of the forefinger (north pole to south pole), then electric current is in direction of the middle finger.

Fig. 1-14. Basic elements of an alternator. Wire loop carries induced current. Electrons flow out one brush, through circuit and back in through other brush.

4. Brushes. Two brushes transfer current from the slip rings to the external circuit. One is in contact with each slip ring.

How ac generators work

As the armature is turned, it cuts through the magnetic field created by the field magnets. Electrons flow through the wires of the armature. The electrons move into one of the slip rings, then to one of the brushes and, finally, into the external circuit. See Figs. 1-14 and 1-15.

Every half-turn, the electron flow reverses, following the generator rule. Then the electrons flow out to the external circuit through the opposite slip

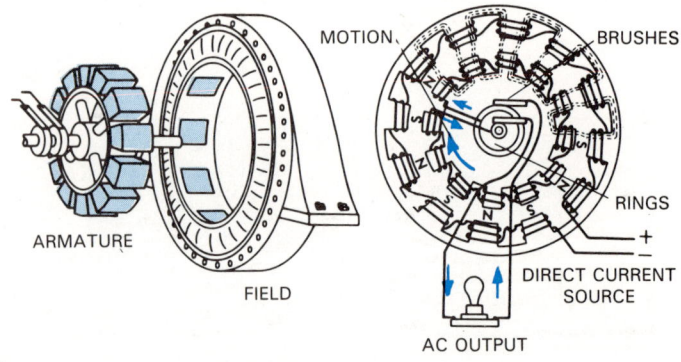

Fig. 1-15. A commercial ac generator (alternator) looks much different from the simple sketches you have seen. This unit uses some of the current it produces to magnetize the field.

Electrical Energy Fundamentals

ring and brush. This creates the reversal of electron flow which accounts for alternating current.

The speed at which the generator is turned determines the frequency of the current. One complete turn is called a cycle, Fig. 1-16. It produces 1 hertz (cycle) of current. In the United States, alternating current is 60 hertz (Hz). It produces 120 changes of current direction per second. This happens so rapidly that the changes cannot be detected as flicker in an electric light.

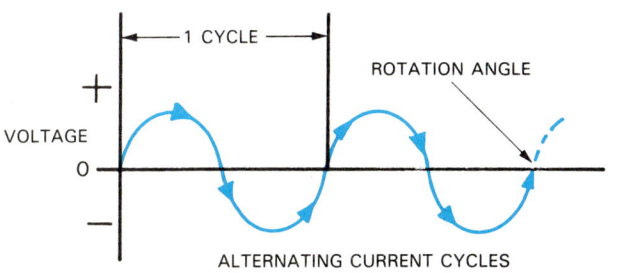

Fig. 1-16. A cycle of alternating current is one complete revolution of the alternator. It is represented on the sine wave as the wavy line shown. The curve above the centerline represents voltage in one direction while the curve below the line represents voltage in the opposite direction. Voltage falls to zero momentarily just as the conducting coil is moving parallel to the magnetic field.

THE DIRECT CURRENT GENERATOR

The dc generator is similar in construction to the alternator except that the two slip rings are replaced by one split ring or COMMUTATOR. The two segments of the commutator are connected to the two ends of the armature, Fig. 1-17.

As the armature rotates, an alternating current is set up in the coil exactly as in the alternator. However, because the segments of the commutator change brushes every half-turn, the ac of the coil is changed into a pulsating dc in the external circuit, Fig. 1-18.

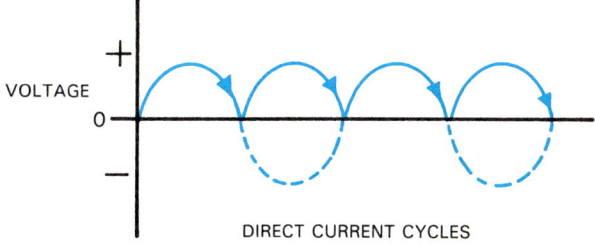

Fig. 1-18. Typical dc current pulsates as shown in the wave pattern. Dotted line shows where the sine curve would go if it were alternating current.

Single and three phase electricity

An ac generator may produce either single phase or three phase electric power. A generator with a single armature coil produces single phase voltage. To produce three phase voltage the generator must have three armature coils on its stator. (The stator is the part of a generator which does not rotate.)

Coils of a three phase generator are located exactly 120 degrees apart. As the rotor spins, voltage is created at three different intervals during each turn of the rotor. The separate currents are 120 electrical degrees apart in time.

Household power is produced as single phase current. Three phase current is used in industry where heavy use is made of electricity. It is very necessary

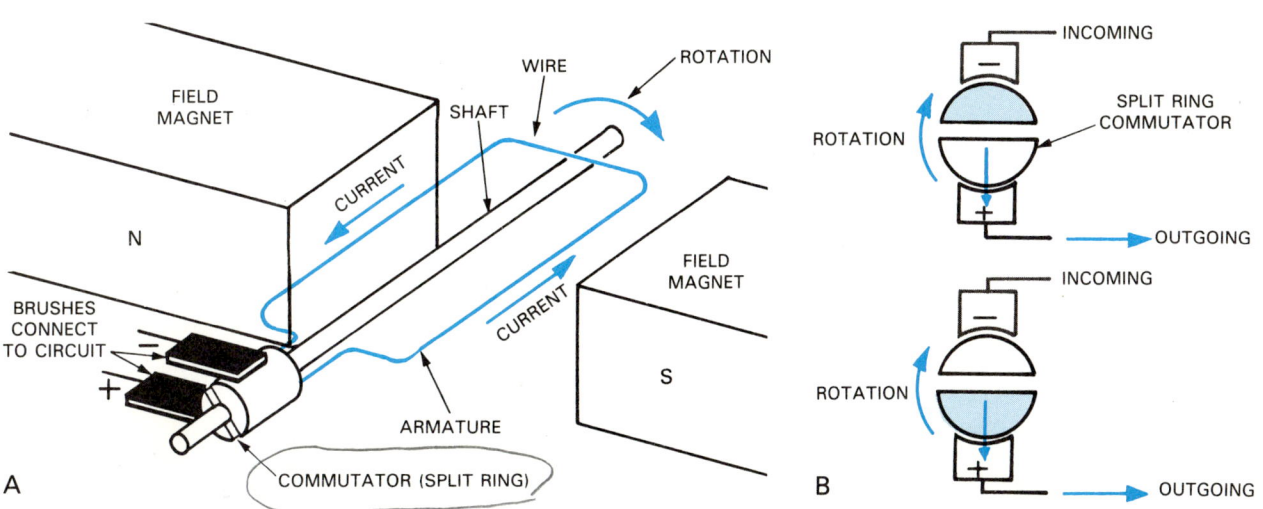

Fig. 1-17. If you look at simple dc generator you can see why current always flows in same direction. A—Each end of armature coil connects to different half of split-ring commutator. B—As current reverses in the armature, opposite sides of split-ring commutator contact the brushes so current flows through brushes in same direction as before.

where there are many motors. Three phase motors are simpler, less expensive, and more powerful than single phase motors. Power plants like to generate three phase power because it is easier to transmit over long distances.

TRANSMISSION OF ELECTRIC POWER

Electrical power is generated by large power stations and sent out over long, high voltage conductors to be used in factories, schools, offices and homes. See Fig. 1-19. So that it can be transmitted more easily and economically, voltage of generated power is stepped up to a higher voltage as it leaves the power station. It must be carried to different points, redistributed, and stepped down to lower voltages so it can be used. This requires transformers and substations.

Most electrical power is stepped up to 132,000, 238,000, or 750,000 volts at the power station. It is sent out to substations over high voltage cable. Here step-down transformers reduce it to lower voltage for further distribution. Before being brought into households, it is further reduced to 120 and 240 V, Fig. 1-20.

TRANSFORMER PRINCIPLES

A transformer is a device which transfers electrical energy from one circuit to another without benefit of a direct mechanical link. It uses electromagnetic induction principles to increase or decrease voltage between an electrical power source and its load.

Basically, a transformer is made up of two coils of wire wrapped around an iron core, Fig. 1-21. There is no connection between the two coils of wire. Each is linked to a conductor or different circuit.

Fig. 1-20. A 240/120 V step-down transformer. This type is used to reduce voltage to levels that can be used in households.

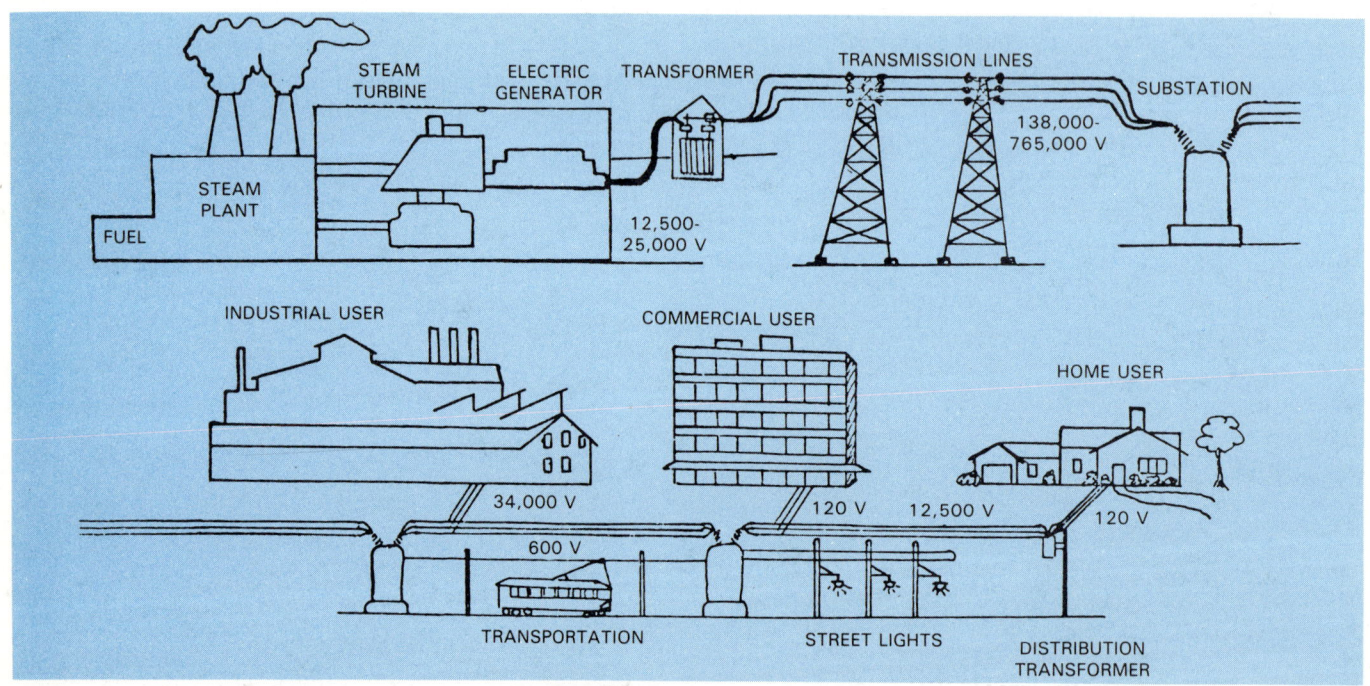

Fig. 1-19. Electric power used in residences is generated in power stations and transmitted over long distances through high voltage lines. Before it is brought into a home voltage is reduced to 240 or 120 volts. (NSTI)

Electrical Energy Fundamentals

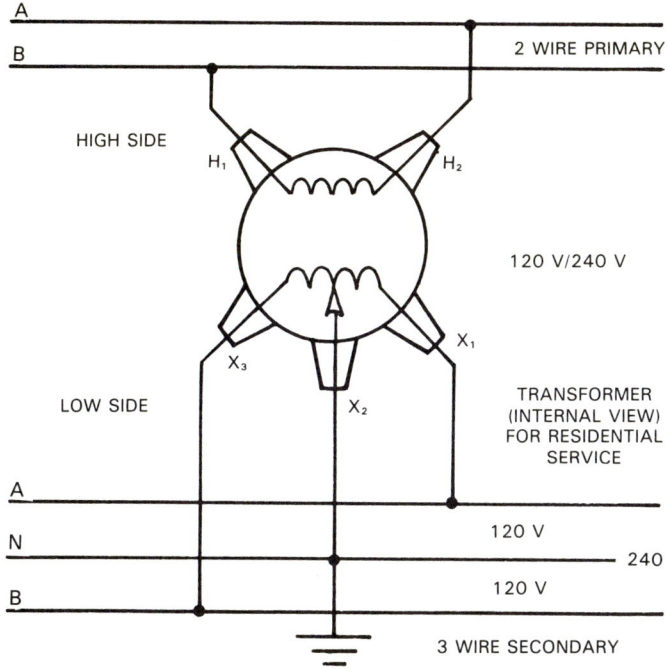

Fig. 1-21. Simple diagram of a transformer. Coil of wire from primary side creates a magnetic field. A current is induced in the secondary coil at a different voltage.

The coil which is part of the circuit hooked up to the power source is called the PRIMARY coil. The coil linked to the other circuit is called the SECONDARY coil. The core of the transformer is made up usually of laminations of iron. The coils are held close together so that a magnetic field is set up between them. This magnetism is induced by the current flowing through the primary coil.

How transformer coils change voltage

Voltage into and out of a transformer is related to the number of turns of wire in the coils. If the number of turns is greater in the secondary coil than in the primary, the voltage out will be greater than the voltage in. Such a transformer would step up (increase) voltage coming from the power source. Suppose that the incoming electrical power were at 110 volts on a transformer which has 100 turns on the primary coil and 600 turns on the secondary coil. The secondary has six times as many turns as the primary. Therefore, the output voltage will be six times greater. Six times 110 V = 660 V.

In a STEP-DOWN transformer the secondary (output) coil has fewer turns than the primary (input) coil. Therefore, the input voltage is greater than the output voltage.

ELECTRIC MOTORS

Electric motors, Fig. 1-22, are like generators. They transform one kind of energy to another. Whereas generators convert mechanical energy to electrical energy, electric motors change electrical energy to rotating mechanical energy.

A dc motor is constructed exactly like a dc generator. Rotation takes place when the polarity of the armature is in opposition to the poles of the field magnet. In other words, LIKE magnetic poles in the motor repel each other to start the rotation. Constant shifting of the magnetic polarity keeps the motor spinning.

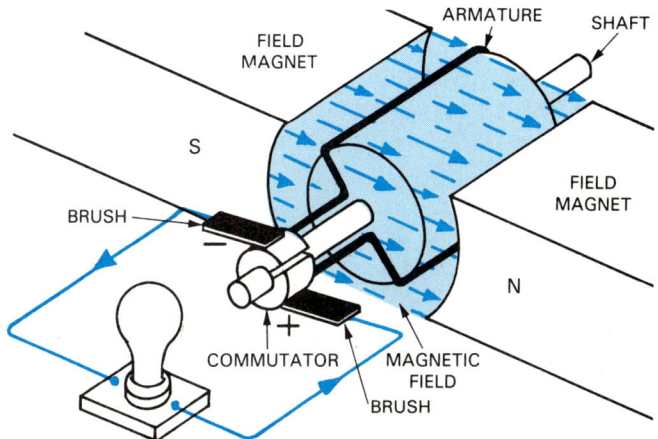

Fig. 1-22. A dc motor is constructed in the same way as a dc generator.

Motors which operate on ac are somewhat different in construction but their operation is similar. They will be discussed in a later chapter.

MEASURING ELECTRICITY

Measuring electricity is essentially finding out how much, how fast, and with what force the electricity goes through something. The terms or units used to measure electricity are *amperage, voltage, resistance,* and *wattage.*

The full definition of these terms is vital not only to measuring electricity but to understanding its very nature.

AMPERAGE

The rate at which electricity flows is called the amperage. Technically, the amount of electrons passing a given point in an electric conductor is directly related to the amperage. One ampere, the basic unit of amperage, is equivalent to 6,240,000,000,000,000,000 electrons going past a given point in one second, Fig. 1-23.

However, as already stated, we may consider amperage as the rate of electron flow. The units of amperage are simply called amperes or "amps," abbreviated A or I.

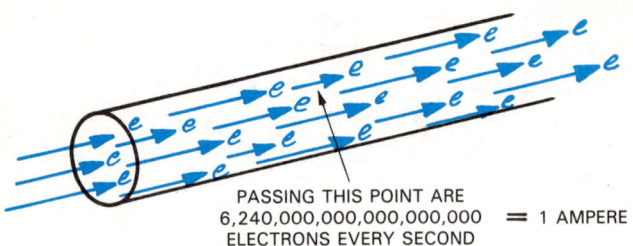

Fig. 1-23. Measuring one ampere.

VOLTAGE

Electrical pressure or force by which electrons are moved through a conductor is termed voltage. The unit of measuring electrical pressure is the volt (V). Most home wiring is rated 120 and 240 volts of electrical pressure.

RESISTANCE/OHMS

The opposition to the flow of electrons through a conductor is called resistance and is similar to friction. This electrical resistance is measured in units of OHMS (Ω). Like other forms of friction, it is responsible for creating heat and consequent loss of power. All devices which utilize electric power are forms of resistances.

WATTAGE

Wattage refers to the amount of power derived from an electrical device or system. It is, in a sense, the most important product obtained from electrical energy. It is this power that we wish to put into use. It is for this reason that electrical energy is sold to us by the power company in amounts measured in watts (W), watt-hours (Wh), or kilowatt-hours (kWh), the units of electrical power. To determine the wattage of any given device or electrical system, we must know the amperage and voltage, since WATTS or VOLT AMPERES = VOLTS × AMPS.

Wattage and horsepower (hp) are also related. One horsepower or 550 foot-pounds/sec., is equivalent to 746 watts.

ELECTRICAL CODES AND SAFETY STANDARDS

The most informative and authoritative body of information concerning electrical wiring installation in the United States and perhaps the world, is the *National Electrical Code®*. It establishes a set of rules, regulations, and criteria for the installation of electrical equipment. Compliance with these methods will result in a safe installation.

It is drafted by a team of experts assembled for this purpose by the National Fire Protection Association (NFPA). This team is formally called the *National Electrical Code* committee. They revise and update the NEC every three years. It is imperative that any person wishing to do electrical wiring obtain and study the NEC. For a copy of the *Code,* write to: NFPA, Batterymarch Park, Quincy, MA 02269.

Throughout this text we will refer to articles and sections of the *Code.* We quote certain portions, tables and examples directly from its text. The terms we use—NEC and *Code* refer to this body of information.

STATE AND LOCAL CODES

Although the NEC, itself, has no legal basis, it is often made mandatory under local or state rulings. In such cases it becomes a legal document.

CODE ENFORCEMENT

Section 90-4 of the *Code* grants full power to the local inspection authority to interpret and/or modify meanings and intentions of the NEC.

For this reason, the local inspector should be consulted whenever a question of methods or materials arises.

Almost every State, region or locality, has some sort of electrical code. Many use the NEC in whole or in part. Further, some communities add regulations beyond those outlined in the NEC.

An electrician must be familiar with these codes also. Failure to do so will most probably result in violations and an inspection failure.

INSPECTION, PERMITS, AND LICENSING

In many areas, permits and licensing are required to do electrical work. In such areas, the utility company requires that permits be obtained and the work be done by or under a licensed electrician *before* they will furnish power. Contact the electrical inspector or power company supervisor for information pertaining to permits and licensing.

EQUIPMENT TESTING AGENCIES

There are many recognized testing agencies throughout the U.S. and Canada. These agencies test materials and equipment submitted to them by electrical material manufacturers, Fig. 1-24. If the materials and equipment submitted measure up to the testing agencies' expectations, then they are listed as suitable for electrical installation. In addition, the product will be labeled. It will bear a recognized emblem or sticker.

Almost all reputable manufacturers of electrical materials and equipment submit their products for

National Electrical Code® and NEC® are Registered Trademarks of the National Fire Protection Association, Inc., Quincy, MA.

Electrical Energy Fundamentals

Fig. 1-24. United Laboratories engineer prepares an electric motor for testing. (Underwriters Laboratories, Inc.)

Fig. 1-25. UL label insures minimum safety requirements have been met.

Fig. 1-26. CSA certification mark is applied to products which meet the Association's testing standards. (Canadian Standards Assoc.)

testing. Products which are not listed, should be avoided. In fact it violates the *Code* to use an unlisted product when there are listed ones available.

The most widely known testing agencies are: Underwriters Laboratories (UL) and Canadian Standards Association (CSA). Products listed by these and other testing agencies are sure to be well constructed and safe. See Figs. 1-25 and 1-26.

OCCUPATIONAL SAFETY & HEALTH ACT (OSHA)

First enacted in 1970, this law permits the Secretary of Labor to create and enforce health and safety standards to protect persons in all occupations within the U.S.

Everyone should become familiar with the various requirements, rules and regulations under this Act. The regulations can be obtained by contacting the Office of Information Services, OSHA, U.S. Dept. of Labor, Washington, D.C. These safety regulations are often more detailed than those of the NEC and, in some instances, supercede the *Code* rulings. When in doubt, consult the inspecting authority.

New OSHA safety standards entitled "Design Safety Standards for Electrical Systems" became effective in the spring of 1981.

These new standards update the previous OSHA standards on how electrical systems should be installed. The overall object of these rules is to ensure the safe use of electrical equipment by employees. In this respect, many of the new standards cover only those portions of direct access and use by personnel.

Consistent with previous OSHA requirements, the new rules follow closely those of the *National Electrical Code*.

All persons involved in the electrical trade should become familiar with the new OSHA electrical standards, which specifically address the following areas:

1. Hazardous locations.
2. Wiring methods, components & equipment.
3. Special systems.
4. Specific purpose equipment.
5. Wiring design & protection.
6. Electric utilization systems.
7. Definitions.
8. General requirements.

REVIEW QUESTIONS — CHAPTER 1

1. Electricity is a form of _____.
2. Substances which allow electricity to pass through them easily are called _____, whereas those which block the electron flow are called _____.

3. The two types of electric current we use are called _____ current and _____ current.
4. A grouping of chemical cells, called a _____, is used as a convenient source of electric current and supplies only (alternating, direct) current.
5. A _____ is a pathway for electric current.
6. In a _____ circuit current must go through every device in the circuit or it will not work.
7. The circuit diagram shown is a (series, parallel) circuit.

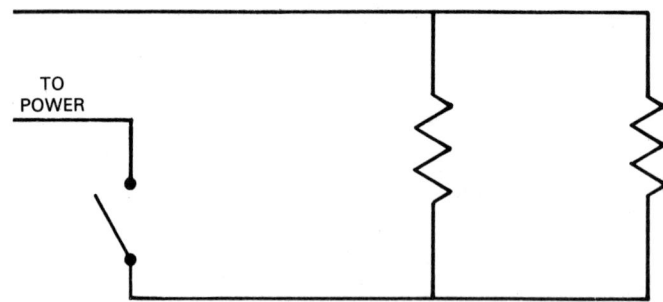

8. (Indicate the most correct answer) An electrical load is
   a. An electric light bulb.
   b. Resistances such as a heat element.
   c. An electrical appliance.
   d. A radio or television.
   e. Any device that uses electric current.
9. What did Faraday and Henry discover?
10. In what important way do ac and dc generators differ?
11. Explain Fleming's Rule.
12. A _Transformer_ uses _magnetic induction_ principles to increase or decrease voltage between an electrical power source and its load.
13. An electric motor is much like a generator. (True) or False?
14. Define each of the following terms used for measuring electricity:
    a. Amperage.
    b. Voltage.
    c. Resistance.
    d. Wattage.
15. Watts = Volts × Amps. True or False?
16. Why is the *National Electrical Code* important to an electrician? Most states have adopted it as law

Chapter 2

ELECTRICAL CIRCUIT THEORY

Particularly crucial to understanding circuitry is resistance. Without resistance the circuit would be for nothing. Without it, electric current could create neither heat nor light. This facet of electricity was first explored by Georg Simon Ohm in 1827. He found that there was a definite relationship between current, voltage, and resistance. He expressed his findings as follows: *The current in a circuit is directly proportional to the voltage applied to the circuit and inversely proportional to the resistance of the circuit.* Put another way, current increases as voltage increases. But, as resistance increases, voltage decreases. This statement has become so basic to electrical circuitry that it is called Ohm's law. This law can be applied to the entire circuit as well as any of its parts.

ELECTRICAL RESISTANCE

Mathematically, the law takes on the following form: $I = \frac{E}{R}$. I is the current measured in amperes. E is voltage or EMF, (electromotive force) measured in volts. R is the resistance in ohms.

The resistances of electrical conductors vary widely, but, in general, depend upon four factors:
1. The nature of the material.
2. The length of the conductor.
3. Their cross-sectional area (thickness).
4. The temperature.

Nature of the material

As we learned in Chapter 1, good conductors have a large number of free electrons that will travel. Generally, metals are better conductors than nonmetals. The best conductors are pure metals such as silver, copper and aluminum. The last two are widely used in electrical wiring. Alloys such as tungsten, nicrome, German silver, brass, and bronze have higher resistance. They heat up quickly as electrons flow and are often used in devices designed to produce heat.

Substances such as cotton, rubber, and plastic are extremely poor conductors and strongly resist the flow of electrons through them. Poor conductors such as these make excellent electrical insulators and prevent electrons from straying from their appointed paths. This is why we often find rubber and plastic used as coverings for electrical wiring.

Length and thickness

All other factors being equal, the longer the electrical conductor, the greater its resistance to current. On the other hand, if electrons have a greater area through which to pass, they will come up against less obstruction and, therefore, less resistance. Generally, the thicker the wire, the less the resistance, Fig. 2-1.

In most electrical applications, the area of a conductor is measured in circular mils, (CM).

The *circular mil* is the cross-sectional area of a wire 1 mil or .001 in. in diameter.

To compute the area of a wire in circular mils, we must change its diameter to mils. For example, a wire with a diameter of .250 in. would equal 250 mils (.250 ÷ .001). Then, we can find the wire area in circular mils by squaring the diameter:

Area, in circular mils = (Diameter in mils)² or
A = 250²
A = 62.500 CM or 62.5 MCM (one thousand circular mills).

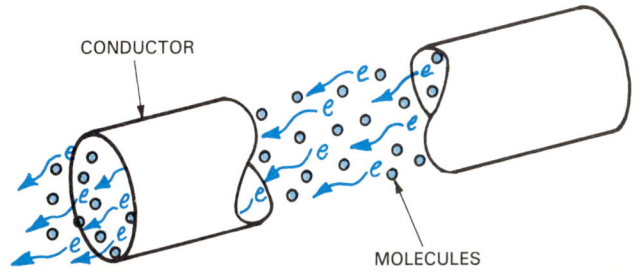

Fig. 2-1. It is much easier for free electrons to weave their way around molecules of a large conductor.

TEMPERATURE

Because of their molecular structure different materials react differently to changes in temperature. As the temperature of metals rises, the resistance to current increases. Heat creates greater activity among the molecules and, thus, more friction or resistance. Therefore, as a wire heats up, its resistance increases greatly. In nonmetals, the reverse is true.

CIRCUIT FUNDAMENTALS

Electric circuits, as you learned in Chapter 1 consist of a source of emf (electromotive force, voltage) and one or more complete pathways of electron flow. Both dc and ac require a definite pathway in order to be of any use. To keep the text sequence as simple as possible, we will restrict our present discussion of circuitry to that used primarily with direct current and the very simplest of alternating current circuits. Later we will expand our discussion to more complex ac circuits.

THE SERIES CIRCUIT

A series circuit, you will recall, has only one loop or path. When devices such as resistors are in series, they are placed one after another so that the current flows through all of them in succession. The three resistors illustrated in Fig. 2-2 are in series arrangement.

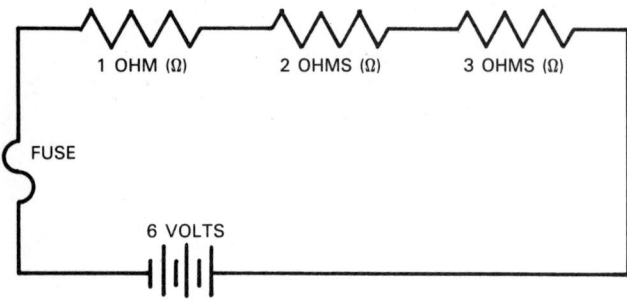

Fig. 2-2. A simple series circuit. It has three loads or resistors hooked up to a single loop.

Series circuits follow certain rules which may be summarized as follows:
1. *The current (amperage) of a series circuit is the same throughout.* By applying Ohm's law, we can find the current in Fig. 2-2:

   $I = \dfrac{E}{R}$

   $I = \dfrac{6\ V}{6\ \Omega}$

   $I = 1\ A$

Therefore, 1 Amp is not only the total amperage, but the amperage flowing through each of the resistances.

2. *Total resistance is the sum of the individual resistances.* Therefore, the total resistance of the series circuit, in Fig. 2-2, is 1 ohm + 2 ohm + 3 ohm = 6 ohms.
3. *The voltage at the source is equal to the sum of the voltages at each of the resistances.* Using Ohm's law:

   Voltage of Resistor A = Amperage × Resistance
   = 1 A = 1 Ω
   = 1 V
   Voltage of Resistor B = Amperage × Resistance
   = 1 A × 2 Ω
   = 2 V
   Voltage of Resistor C = Amperage × Resistance
   = 1 A × 3 Ω
   = 3 V
   Total voltage = 1 V + 2 V + 3 V
   = 6 V

4. *A break anywhere in the circuit stops the electron flow in the entire circuit. This is the main disadvantage of series circuits.*
5. *Ohm's law applies to any part of the entire series circuit.*

Example: Using the foregoing rules along with Ohm's law, you can easily master series circuits. Fig. 2-3 is another series circuit. Find:
a. The total resistance R_t.
b. The amperage at the source, I_s.
c. The voltages of resistors V_1 and V_3.

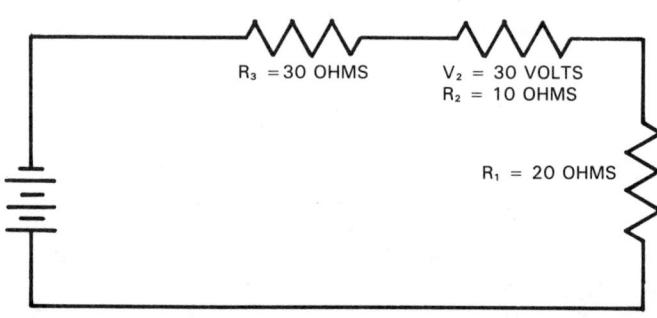

Fig. 2-3. In this series circuit, find total resistance, amperage and voltage at the battery.

Solution:

a. $R_t = R_1 + R_2 + R_3$
   $= 20\ \Omega + 10\ \Omega + 30\ \Omega$
   $R_t = 60\ \Omega$

b. $I_t = \dfrac{E\ total}{R\ total} = I_2 = I_3$

(Total amperage is designated I_t or I_s)

$$I_t = I_2$$
$$= \frac{E_2}{R_2}$$
$$= \frac{30 \text{ V}}{10 \text{ }\Omega}$$
$$I_t = 3 \text{ A}$$
c. $V_1 = I_1 \times R_1$
$$= 3 \text{ A} \times 20 \text{ }\Omega$$
$$= 60 \text{ V}$$
$$V_2 = 30 \text{ V (given)}$$
$$V_3 = I_3 \times R_3$$
$$= 3 \text{ A} \times 30 \text{ }\Omega = 90 \text{ V}$$

Check to see if the total voltage is *equal* to the sum of the individual voltages:
$$V_t = I_t \times R_t$$
$$= 3 \text{ A} \times 60 \text{ }\Omega$$
$$= 180 \text{ V}$$
$$V_1 + V_2 + V_3 = 180 \text{ V and}$$
$$60 \text{ V} + 30 \text{ V} + 90 \text{ V} = 180 \text{ V}$$

PARALLEL CIRCUITS

As you learned in Chapter 1, a parallel circuit is an electrical circuit having two or more different conducting pathways. The term parallel, in electricity, does not necessarily mean physically or geometrically parallel, but simply "alternate routes." These parallel or alternate routes are best called *branches.* Parallel circuits are more commonly used in electrical circuitry for reasons which will become clear as we go on. Fig. 2-4 shows a simple parallel circuit containing three resistances (20, 30, and 60 Ω). Arrows indicate the electron flow.

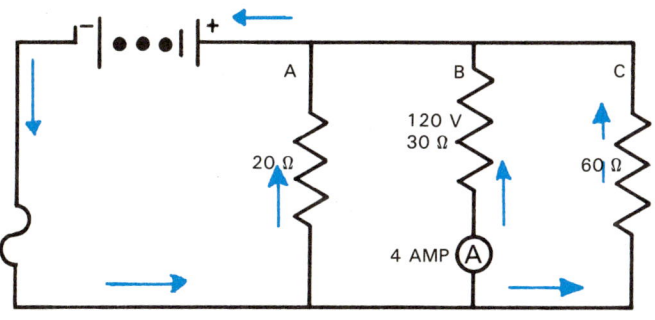

Fig. 2-4. Parallel circuits provide numerous alternate routes for the electrons to follow. These routes, in residential wiring, are called branches.

PARALLEL CIRCUIT RULES

Like series circuits, parallel circuits follow certain rules:
1. *The voltage across all branches, loops, of a parallel circuit are the same and are equal to the voltage at the source.* As in Fig. 2-4, the voltage in each branch is 120 V. This is shown in branch B. Likewise, the voltage at the source is 120 V. Mathematically: source voltage = branch A = branch B = branch C.
$$V_s = V_a = V_b = V_c$$
2. *The total current (amperage) is equal to the sum of the currents flowing through each of the branches.* Therefore, total amperage = branch A + branch B + branch C.
$$I_t = I_a + I_b + I_c$$
Using Ohm's law, the current in the branch A is $\frac{120 \text{ V}}{20 \text{ V}} = 6$ A. The current in branch C is $\frac{120 \text{ V}}{60 \text{ }\Omega} = 2$ A. When this is added to the 4 A of current (given) of branch B, total current is 12 A.
3. *The total resistance in a parallel circuit is the reciprocal of the sum of the reciprocals of the separate resistances in parallel.* Now, before throwing up our hands, let us look at this mathematically:
$$\frac{1}{R \text{ total}} = \frac{1}{R \text{ branch A}} + \frac{1}{R \text{ branch B}} + \frac{1}{R \text{ branch C}}$$
Now use the resistance information from the circuit in Fig. 2-4:
$$\frac{1}{R_t} = \frac{1}{20 \text{ }\Omega} + \frac{1}{30 \text{ }\Omega} + \frac{1}{60 \text{ }\Omega}$$
Find a common denominator:
$$\frac{1}{R_t} = \frac{3}{60 \text{ }\Omega} + \frac{2}{60 \text{ }\Omega} + \frac{1}{60 \text{ }\Omega}$$
$$\frac{1}{R_t} = \frac{6}{60 \text{ }\Omega} \text{ and finally, } \frac{R_t}{1} = \frac{60 \text{ }\Omega}{6} = 10 \text{ }\Omega$$
4. *Ohm's law applies equally well to the total circuit or any of the loops or branches.*
$$E = I \times R$$
5. *A break or opening in any branch of a parallel circuit does not stop the flow of electrons to the remaining branches.*

SERIES PARALLEL COMBINATION CIRCUITS

Practical circuits, for the most part, are far more complex than those just outlined. Quite often, resistances in a particular circuit are designed both in series and in parallel. When resistances are arranged both ways in a single circuit, the circuit is called a *network*.

Simple networks consist of a single source of emf. Some resistors are in series, and some are in parallel. A very simple network of this type is shown in Fig. 2-5. We will use this network to show the steps taken to simplify any network and thus reduce it to a rather simple circuit.

In the network shown, the emf is 12 V. The resistance of each component resistor is also given.

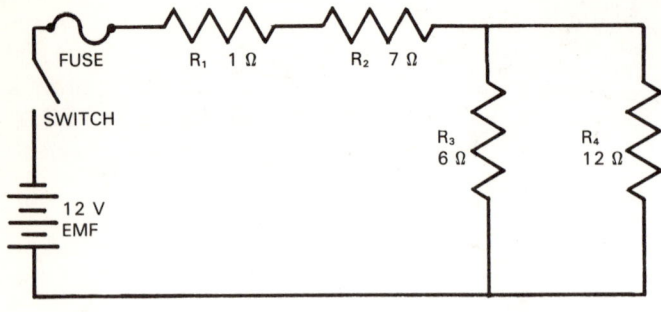

Fig. 2-5. A network, or combination series - parallel circuit, is a common type of electrical pathway employing the advantages of both.

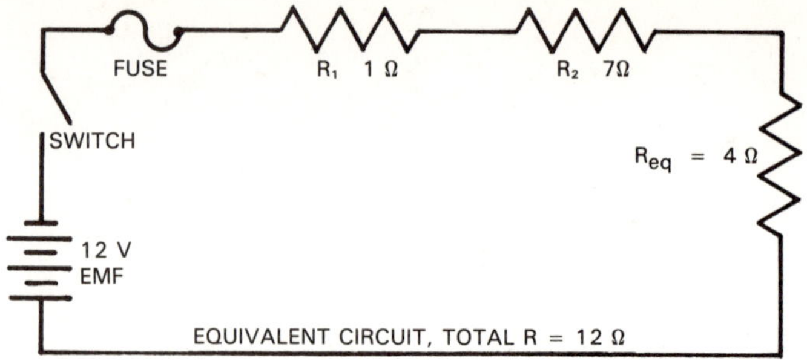

Fig. 2-6. An equivalent circuit of Fig. 2-5. R_3 and R_4 have been "reduced" to one resistance, "R equivalent."

First, we need to find the total resistance. To do this we must reduce the parallel resistances into an *equivalent single resistance*, (R_{eq}), and add this resistance to the resistances arranged in series. Equivalent Resistance from parallel portion:

$$\frac{1}{R_{eq}} = \frac{1}{R_3} + \frac{1}{R_4},$$

$$= \frac{1}{6\,\Omega} + \frac{1}{12\,\Omega}$$

$$= \frac{2}{12\,\Omega} + \frac{1}{12\,\Omega}$$

$$\frac{1}{R_{eq}} = \frac{3}{12\,\Omega}; \quad \frac{R_{eq}}{1} = \frac{12\,\Omega}{3}$$

Therefore, $R_{eq} = 4\,\Omega$

Now add this resistance to R_1 and R_2 = 12 Ω = total Resistance

Next, we may draw a simpler *equivalent circuit* showing the equivalent resistance *in series* with R_1 and R_2. See Fig. 2-6.

Having found the total resistance, and given the emf, we can find the total current through the circuit using Ohm's law:

$$\frac{E}{R} = I = \frac{12\,V}{12\,\Omega} = 1\,A$$

Next, find the voltages across each of the resistors:

$V_1 = I_t \times R_1 = 1\,A \times 1\,\Omega = 1\,V$
$V_2 = I_t \times R_2 = 1\,A \times 7\,\Omega = 7\,V$
$V_{eq} = I_t \times R_{eq} = 1\,A \times 4\,\Omega = 4\,V$

This is verified by the fact that the emf or total voltage is equal to the sum of the various resistors in a series-arranged circuit.

All that remains is to find the portion of the total current (1 ampere) that goes through each of the parallel-arranged resistors. (Remember, voltage is the same through both R_3 and R_4.)

$$I_3 = \frac{V_{eq}}{R_3} = \frac{4\,V}{6\,\Omega} = \frac{2}{3}\,A$$

$$I_4 = \frac{V_{eq}}{R_4} = \frac{4\,V}{12\,\Omega} = \frac{1}{3}\,A$$

Again, verification comes from the fact that the total current (1 ampere) is equal to the sum of the amperages through each branch of the parallel-arranged resistors. Thus, 1 ampere = $\frac{2}{3} + \frac{1}{3}$ ampere.

Considered as a whole, electrical circuits are mostly of the network type. The line or source voltage (supplied by electrical conductors entering the service meter and service entrance panel) is carried by means of wires across every device, appliance and branch circuit. These are connected in parallel.

Each of the parallel-arranged devices may be in series with others (but not necessarily so). This will become more apparent when later chapters discuss wiring of various devices.

CIRCUITS SUMMARY

Fig. 2-7 is a summary of the various circuits discussed in this chapter. It outlines the major characteristics of each type. Persons planning to become professional electricians should familiarize themselves with this basic knowledge of circuitry. Anybody could do *certain* electrical wiring without this information, but, in the long run, knowing how circuits act is of great benefit. Further, this information is essential to good electrical planning.

Ohm's law applies to any and all parts of a circuit. Fig. 2-8 is a simple aid to remembering the law. Simply cover the part you wish to find and you discover what to multiply or divide by. Thus, to find E we multiply I and R; to find R, we divide E by I; to find I we simply divide E by R.

ENERGY MEASUREMENT

Electrical energy can be measured in units called *joules* (J). The unit is named after the English physicist, James Prescott Joule, who found the rela-

Electrical Circuit Theory

SERIES CIRCUIT-PARALLEL CIRCUIT NETWORK

	Series Circuits	Parallel Circuits	Network
Resistance (R) Unit: Ohm Symbol: Ω	$R_t = R_1 + R_2 + R_3$ Sum of individual resistances	$\frac{1}{R_t} = \frac{1}{R_1} + \frac{1}{R_2} + \frac{1}{R_3}$	Total resistance equals resistance of parallel portion and sum of series resistors
Current (I) Unit: Ampere Symbol: A	$I_t = I_1 = I_2 = I_3$ The same throughout entire circuit	$I_t = I_1 + I_2 + I_3$ Sum of individual currents	Series rules apply to series portion of circuit. Parallel rules apply to parallel part of circuit
Voltage (E) Electromotive Force (emf) Unit: Volt Symbol: V, E	$E_t = E_1 + E_2 + E_3$ Total voltage and branch voltage are the same	$E_t = E_1 = E_2 = E_3$ Total voltage and branch voltage are the same	Total voltage is sum of voltage drops across each series resistor and each of branches of parallel portion

Fig. 2-7. Characteristics of electrical circuits can be summarized as shown.

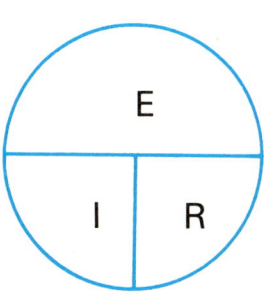

Fig. 2-8. This drawing is an easy way to remember the mathematical relationships of Ohm's law. It will tell you when to multiply or divide.

tionship of electrical energy to heat production. This relationship is known as *Joule's Law: Electrical Energy (joules) = E × I × t*. E is electromotive force or volts; I is current or amperage, and t is time in seconds. One joule equals one watt per second or 0.7376 foot-pounds or 0.24 calories.

POWER

Power is the rate of doing work. In electricity, this rate is measured in watts (W) or volt-amperes (VA). To determine the amount of electrical power for a simple circuit, multiply the voltage by the current. Power equals volts × amperes.

Electrical power and mechanical power are closely related. This makes sense since it is easy to transform one into the other. In fact, 1 horsepower (hp) is the same as 746 W of electrical power.

Because watts are relatively small units, it is sometimes simpler to speak of kilowatts or megawatts (kilo means 1000 and mega means 1,000,000). A kilowatt (kW) is 1000 W. One megawatt (MW) is 1,000,000 W.

Electrical energy and power are measured in power-time units. The *kilowatt-hour* (kW-hr) is conventionally the unit used. A kilowatt-hour is equivalent to 1000 W being used for a 1 hour period.

The cost of a kilowatt-hour varies, but averages around 5.5 cents. The rate is "stepped" so that the more you use, the less the cost per kilowatt-hour. Power, amperage, resistance and voltage can be related mathematically using the formulas in Fig. 2-9. Fig. 2-10 shows typical power ratings for common household electrical devices.

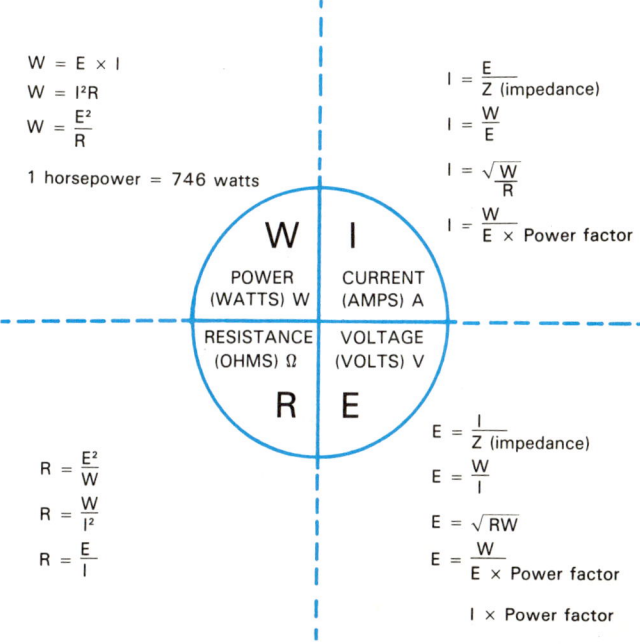

Fig. 2-9. Mathematical relationships of power, amperage, resistance and voltage are shown in this memory device.

UNIT	POWER IN WATTS
Air Conditioner	1200
Clock	3
Dishwasher	2000
Freezer	500
Fryer, deep fat	1200
Iron	1000
Projector	500
Range, electric	15,000
Rotisserie	1600
Sewing Machine	100
Television	350
Vacuum	600
Washing Machine	700
Water Heater	4000
Waffle Iron	800

Fig. 2-10. Power requirements of different domestic appliances.

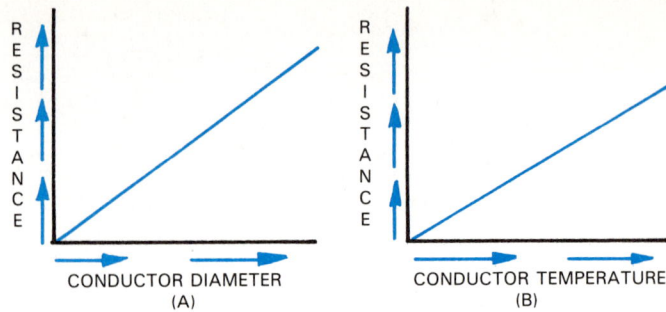

REVIEW QUESTIONS — CHAPTER 2

1. Ohm's law relates resistance, _voltage_ and _current_ of electrical circuits.
2. Certain insulators, such as _rubber_, and _plastic_ are used as wire coverings because of their high _resistance_ to electron flow.
3. A certain series circuit has a current of 2 amperes. What is the source voltage, if the total resistance is 6 ohms? _12_ volts.
4. Which of the following graphs would best express the correct relationship of resistance to other conductor characteristics:
5. What is the total amperage of a parallel circuit having a total resistance of 12 ohms and a source voltage of 24 volts?
6. Ohm's law applies to series, parallel and network circuits. True or False?
7. Power is the rate of doing work. True or False?
8. Electrical energy is measured in units called:
   a. Joules.
   b. Watts.
   c. Horsepower.
   d. Kilowatt-hours.
9. A kilowatt, kW, is the same as:
   a. 10 watts.
   b. 100 watts.
   c. 1000 watts.
   d. 10,000 watts.
   e. None of these.
10. Which of the following is the largest consumer of electrical power?
    a. A clock.
    b. A television.
    c. A sewing machine.
    d. An electric range.

Chapter 3
ELECTRICAL CIRCUIT COMPONENTS

As you learned in Chapter 1, almost all metals may be drawn into wire and theoretically will make suitable conductors of electricity. However, for all practical purposes, only copper and aluminum are used in residential and commercial wiring, Fig. 3-1. They are low cost and offer low resistance to current. For information on the resistance of copper wire in various gauges from 4/0 down to 40, refer to Fig. 25-3 on page 251.

Copper is the most common of conductor materials for electrical wiring because of its low cost and desirable characteristics. It is strong and resists oxidation (corrosion). Aluminum also is cheap and has good conductance. However, it is subject to problems as a result of oxidation and expansion and must be properly prepared.

Before using aluminum you should consult the NEC and any local codes. Questions about its use should also be directed to the local inspector. Local codes may have restrictions against its use.

WIRE SIZES

The size and type of wires to be used vary greatly. It depends on the amount of electricity they must conduct, how far they are to conduct it, and in what type of situation (location) they are to be placed. In fact, whether or not a wire is to be insulated or covered with a nonconducting substance depends upon its application. Except in rare instances, wire carrying alternating current is insulated.

Discussion of wire sizes is limited to copper wire. For aluminum, always figure two sizes larger.

Fig. 3-2 shows the diameters of the most commonly used copper wires. Sizes are designated by number, Fig. 3-3. The larger the wire number, the smaller its diameter. Wire Nos. 8 and up are either solid or stranded. Wires Nos. 6, 4, 2 and down are always stranded. The standard for wire sizes is called the American Wire Gauge (AWG). Wire size is important for two reasons:

1. *Ampacity.* This is safe current carrying

Fig. 3-1. Copper and aluminum are the only practical metals for electrical conductors.

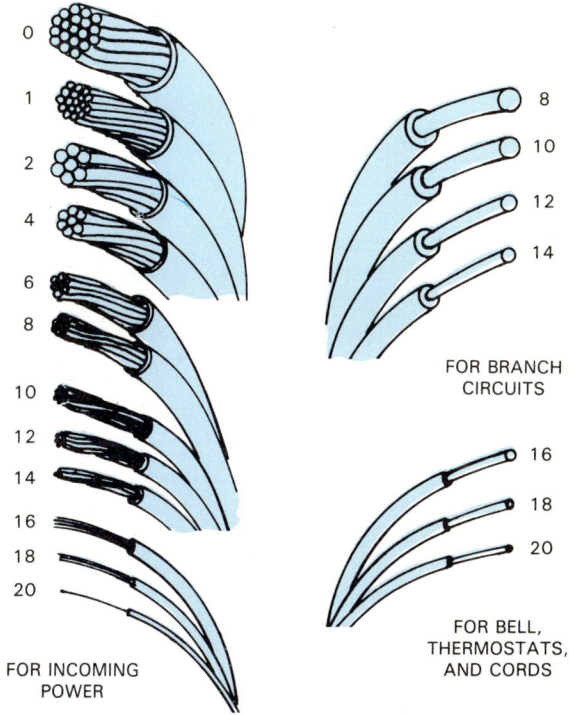

Fig. 3-2. Copper wire sizes shown with insulation. Where alternating current is used, wires are usually insulated. (GE Wiring Devices Dept.)

Modern Residential Wiring

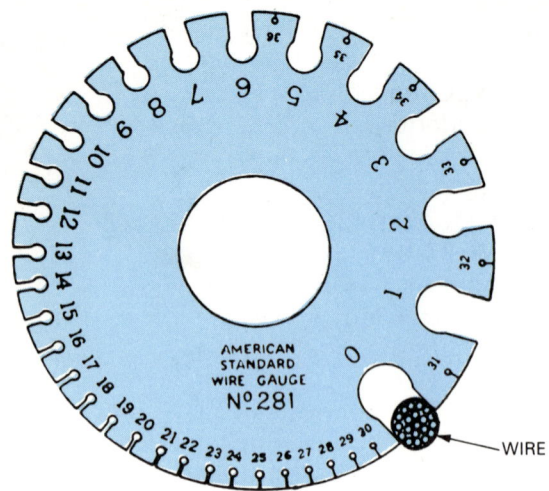

Fig. 3-3. Sizes of wires can be determined by use of a wire gage such as this. Wire size is marked on the insulation for most electrical wire.

WIRE COVERINGS (INSULATION)

Bare or uninsulated wire is rarely used in wiring applications. In almost every instance, insulated or covered wire is needed. Usually type THW, THHN or TW within cable or conduit is used for housewiring, Fig. 3-4. These conductor coverings are suitable in either wet or dry locations.

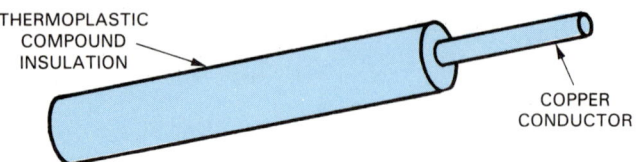

Fig. 3-4. Type TW wire is a common electrical conductor.

The most common coverings for wire used in electrical wiring are plastic or rubber. They are excellent nonconductors of electricity. In addition, these insulated wires are most often covered or protected by substances such as braiding, jute, paper, more plastic, or metal. See Figs. 3-5 and 3-6.

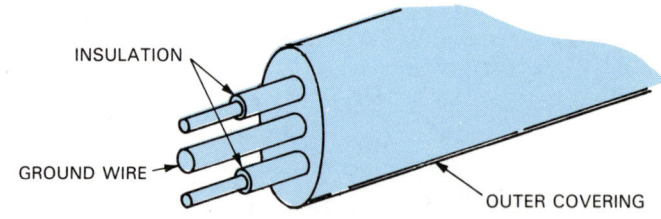

Fig. 3-5. Type TW wires embedded in a moisture and fungus resistant plastic outer jacket, form a durable electrical cable.

capacity of the wire and its consequent heat dissipation (shedding) ability.

2. *Voltage drop.* This is the loss of electrical pressure (voltage) due to resistance when trying to force the electrons long distances or through too small a cross-sectional area.

FIGURING VOLTAGE DROP

To compute voltage drop, use the following equation:

$$VD = I \times \frac{KL}{CM}$$

Where VD = voltage drop in volts.
 I = amperage in amperes.
 K = a constant of 10.8 = Ω per mil-foot copper = 10.8
 L = Length of copper conductor in feet (one way)
 CM = Circular mills (area of conductor). See the *National Electrical Code*, Table 8, Conductor Properties.

For example, the voltage drop of 100 ft. of AWG 10 copper wire carrying 30 A would be:

$$VD = I \times \frac{KL}{CM}$$

$$= \frac{30 \times 10.8 \times 100}{10380}$$

= 3.12 volts (approx. 2.6 percent of 120 volts)

Thus, the actual voltage of this conductor is 120 − 3.12 or roughly 116.88 V. In this respect, not only is wire size (cross section) important, but its *length*, as well.

Except for service wires, most branch circuit wires range from No. 14 AWG to No. 10 AWG. Of the insulated types No. 12 AWG is the most common.

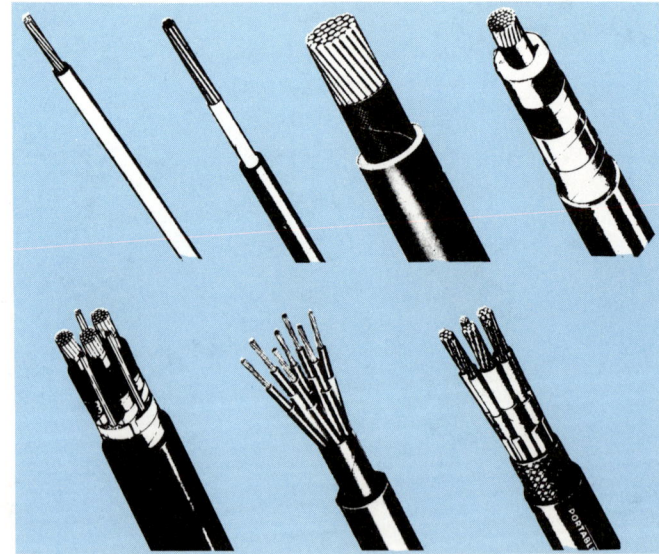

Fig. 3-6. There are numerous conductor types and coverings used in the trade today.

Electrical Circuit Components

A list of the various types of conductor insulation is provided in Fig. 3-7. Once again, the type of wire used depends upon the situation and/or environment where the wire is to be placed. Fig. 3-8 is a short list of insulated conductor types and some conditions under which they may be used.

The NEC, a set of carefully drawn up rules and regulations concerning the types of materials and methods used in installing electrical equipment, is very specific as to which types of wires are to be used under specified conditions.

See NEC Table 310-13 for more details.

INSULATED WIRE COVERINGS

Covering Types	Letter Designation
Rubber	RH, RHH, RHW, RUH, RUW
Thermoplastic Compound	T, TW, THW, TBS, THHN
Thermoplastic and Asbestos	TA
Silicone and Asbestos	SA
Asbestos	A
Varnished Cambric	V
Asbestos and Varnished Cambric	AVA, AVL, AVB

Fig. 3-7. Type of covering on conductors is often indicated by its letter designation.

CONDUCTOR APPLICATION

Wire Type	Apply Where	Temperature Maximum °C	(°F)	
RH	Dry	75	(167)	
RHH	Dry	90	(194)	
RHW	Dry/Wet	75	(167)	
RUH	Dry	75	(167)	
RUW	Dry/Wet	60	(140)	
T	Dry	60	(140)	
TW	Dry/Wet	60	(140)	
THHN	Dry	90	(194)	
THW	Dry/Wet	90	(194)	Under Special Conditions
THWN	Dry/Wet	75	(167)	
XHHW	Dry	90	(194)	
	Wet	75	(167)	
MI	Dry	85	(185)	Under Special Conditions
	Wet	250	(482)	
SA	Dry	90	(194)	For Special Applications
		125	(257)	
FEP	Dry	90	(194)	
V	Dry	85	(185)	
AVA	Dry	110	(230)	
AVL	Dry/Wet	110	(230)	
AVB	Dry	90	(194)	

Fig. 3-8. An insulated conductor is made for every type of temperature and humidity condition where electrical wiring is needed.

CONDUCTOR MARKINGS

The NEC requires that all electrical conductors be suitably identified. This is accomplished by marking the conductor covering. The marking, Fig. 3-9, will indicate:

For all conductors:
1. AWG size.
2. Insulation type.
3. Voltage rating.
4. Testing agency (UL, CSA, etc.) approval.

For cable:
5. Number of conductors.
6. Outer finish or covering.

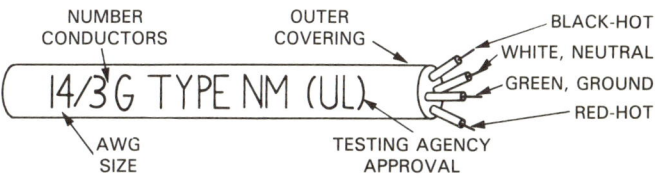

Fig. 3-9. Cable type is marked on the covering.

INSULATION COLOR

Color coding of insulation on conductors is also used to identify wiring, Fig. 3-10. The general coloring scheme is:
1. For current carrying conductors, use black, red, blue, and yellow usually in that order.
2. For neutral wire, white or natural gray.
3. The grounding conductor or ground wire may be bare, green, or green with yellow stripes.

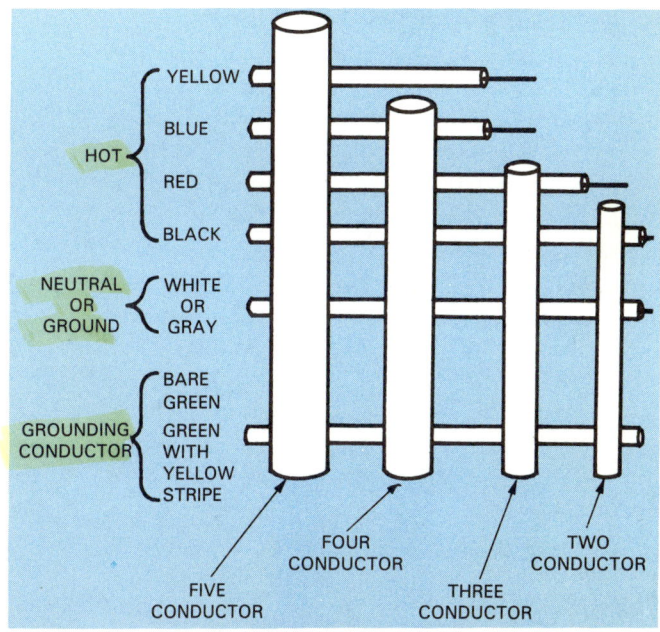

Fig. 3-10. Color coding of insulation indicates for what purpose each conductor is to be used.

Modern Residential Wiring

PROTECTIVE DEVICES

All electrical circuits must have some sort of safety device. The purpose of the safety device is to prevent the circuit from being stressed or overloaded. Fig. 3-11 shows allowable maximum amperages for wires sizes from No. 14 to No. 4/0. Electricians should refer to NEC Table 310-16 for details.

Overcurrent in a conductor could cause serious damage to the covering. The conductor could then become a serious hazard to anything or anybody coming in contact with it.

MAXIMUM AMPACITY ALLOWANCES OF INSULATED COPPER WIRE

Wire Size (AWG)	60°C Types: T, TW, RUW	75°C Types: RH, RHW, RUH, THW, THWN, XHHW
14	15[1]	15[1]
12	20[1]	20[1]
10	30[1]	30[1]
8	40	50
6	55	65
*4	70	85
*2	95	115
*1	110	130
*1/0	125	150
*2/0	145	175
3/0	165	200
4/0	195	230

[1]Load current rating and overcurrent protection shall NOT exceed these figures.
*For 3-wire, single-phase residential services, the allowable ampacity of RH, RHH, RHW, THW and XHHW Copper Conductors shall be for sizes no. 4-100 amp., no. 2-125 amp., no. 1-150 amp., no. 1/0-175 amp., and no. 2/0-200 amp.

Fig. 3-11. Ampacity is the current load a conductor is allowed to carry. NEC sets safe load-carrying ability of different size conductors.

Overcurrent protection devices include fuses and breakers, Fig. 3-12. Both are manufactured in various shapes and sizes, but are all designed to do the same thing: stop the flow of current should it exceed safe limit.

All fuses and breakers are rated in amperes. The rating must not be greater than the overall capacity of the circuit being protected.

CIRCUIT PROTECTION

Electrical circuits are often prey to destructive and dangerous currents. This can be caused by OVERLOADS and SHORT CIRCUITS, brought on by any number of factors which can damage the circuit. Protection of the circuit against such damage is essential.

OVERLOADS AND SHORT CIRCUITS

Overcurrents in the form of overloads and/or short circuits are defined as currents in excess of the normal current for a given circuit. Overloads are overcurrents which can range between twice and ten times the normal current, but are confined and contained within the normal electrical pathways.

Short circuits are relatively large overcurrents which may exceed the normal current by hundreds of times and flow between the circuits and other objects outside the circuit. Any overcurrent is potentially destructive to equipment and dangerous.

PROTECTIVE DEVICE RATING

Protective devices—fuses or circuit breakers—must be adequate to withstand overcurrents, par-

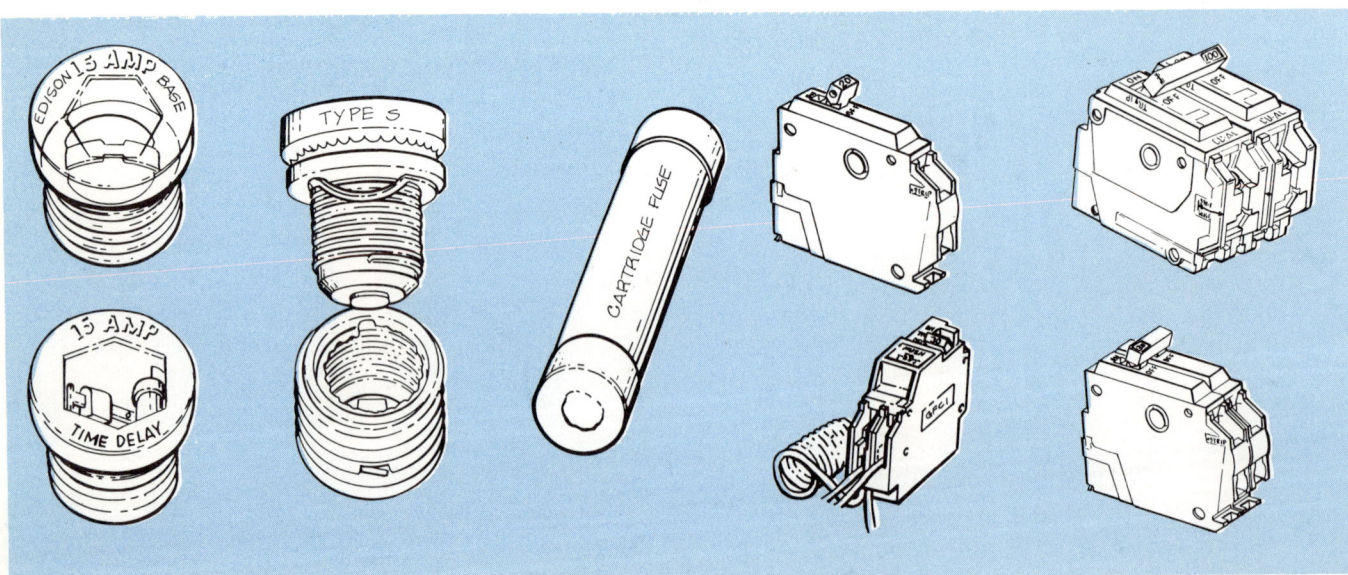

Fig. 3-12. Various designs of fuses and circuit breakers are available for overcurrent protection. (GE and Leviton Mfg. Co., Inc.)

Electrical Circuit Components

ticularly those created by short circuits. Each device has a limit at which it will stop current. This is called the *interrupting rating.* The interrupting rating of most circuit breakers ranges between 10,000 and 20,000 AIC. Fuse ratings are generally around 200,000 AIC. (The abbreviation AIC means "amperage interrupting capacity.")

Naturally, this is much larger than the normal current rating of the fuse or breaker. Normal rating is the current it is designed to carry under normal operating conditions (15, 20, 30 or 60 A etc.). All overcurrent protective devices are labeled as to their normal current rating and maximum or interrupting rating.

Guides for properly sizing fuses are found in Fig. 3-13 and in Article 240 of the NEC.

1.16 Guide For Sizing Fuses

General guidlines are given for selecting fuse ampere ratings for most circuits. Some specific applications may warrant other fuse sizing; in these cases the load characteristics and appropriate N.E.C. sections should be considered. The selections shown here are not, in all cases, the maximum or minimum ampere ratings permitted by the N.E.C. Demand factors as permitted per the N.E.C. are not included in these guidelines.

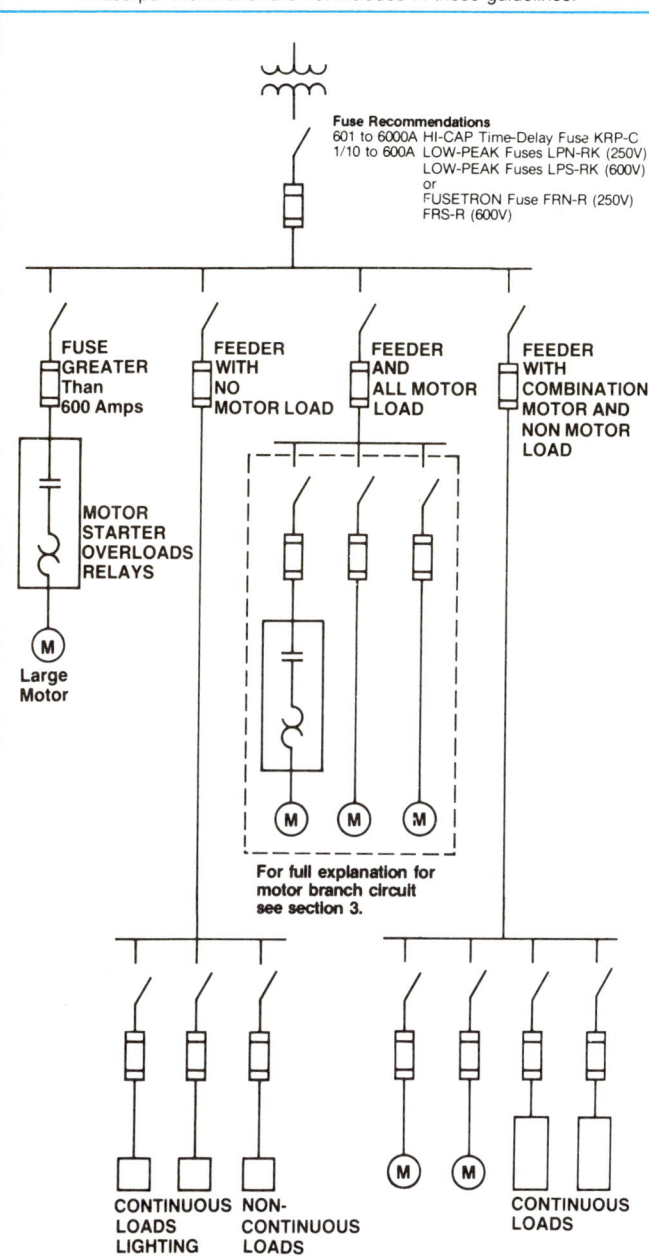

Dual-Element Time-Delay Fuses
(LOW-PEAK or FUSETRON).

1 Main Service. Size fuse according to method in **4**.

2 Feeder Circuit With No Motor Loads. The fuse size must be at least **125%** of the continuous load† plus **100%** of the non-continuous load. Do not size larger than ampacity of conductor*.

3 Feeder Circuit With All Motor Loads. Size the fuse at **150%** of the full load current of the largest motor plus the full-load current of all motors.

4 Feeder Circuit With Mixed Loads. Size fuse at sum of
a. **150%** of the full-load current of the largest motor plus
b. **100%** of the full-load current of all other motors plus
c. **125%** of the continuous, non-motor load† plus
d. **100%** of the non-continuous, non-motor load.

5 Branch Circuit With No Motor Load. The fuse size must be at least **125%** of the continuous load† plus **100%** of the non-continuous load. Do not size larger than ampacity of conductor*.

6 Motor Branch Circuit With Overload Relays. Where overload relays are sized for motor running overload protection, the following provide backup, ground fault, and short-circuit protection:
a. **Motor 1.15 service factor or 40°C rise:** size fuse at **125%** of motor full-load current or next higher standard size.
b. **Motor less than 1.15 service factor or over 40°C rise:** size the fuse at **115%** of the motor full-load current or the next higher standard fuse size.

7 Motor Branch Circuit With Fuse Protection Only. Where the fuse is the only motor protection, the following fuses provide motor running overload protection and short-circuit protection:
a. **Motor 1.15 service factor or 40°C rise:** size the fuse at **110%** to **125%** of the motor full-load current.
b. **Motor less than 1.15 service factor or over 40°C rise:** size fuse at **100%** to **115%** of motor full load current.

8 Large Motor Branch Circuit—Fuse larger than 600 amps. For large motors, size KRP-C HI-CAP time-delay Fuse at **150%** to **225%** of the motor full load current, depending on the starting method; i.e. part-winding starting, reduced voltage starting, etc.

Non-Time-Delay Fuses
(LIMITRON and T-TRON, typically)

1 Main service. Size fuse according to method in **4**.

2 Feeder Circuit With No Motor Loads. The fuse size must be at least **125%** of the continuous load† plus **100%** of the non-continuous load. Do not size larger than the ampacity of the wire.

3 Feeder Circuit With No Motor Loads. Size the fuse at **300%** of the full-load current of the largest motor plus the full-load current of all other motors.

4 Feeder Circuit With Mixed Loads. Size fuse at sum of:
a. **300%** of the full-load current of the largest motor plus
b. **100%** of the full-load current of all other motors plus
c. **125%** of the continuous, non-motor load† plus
d. **100%** of the non-continuous, non-motor load.

5 Branch Circuit With No Motor Load. The fuse size must be at least **125%** of the continuous load† plus **100%** of the non continuous load. Do not size larger than the ampacity of conductor*.

6 Motor Branch Circuit With Overload Relays. Size the fuse as close to but not exceeding **300%** of the motor running full load current. Provides ground fault and short-circuit protection only.

7 Motor Branch Circuit With Fuse Protection Only. Non-time-delay fuses **cannot** be sized close enough to proivde motor running overload protection. If sized for motor overload protection, non-time-delay fuses would open due to motor starting current. Use dual-element fuses.

† 100% of the continuous load can be used rather than 125% when the switch and fuse are listed for continuous operation at 100% of rating. Most bolted pressure switches and high pressure contact switches 400A to 6000A with Class L fuses are listed for 100% continuous operation.
* Where conductor ampacity does not correspond to a standard fuse rating, next higher rating fuse is permitted when 800 amperes or less (240-3. Exc. 1).
♦ In many motor feeder applications dual element fuses can be sized at ampacity of feeder conductors.

Fig. 3-13. Manufacturers provide information for matching fuses to load. (Bussmann Mfg. Div. McGraw-Edison Co.)

When fuses blow or breakers trip, the cause should be checked out before the fuse is replaced or breaker reset. Never replace fuses or breakers with protective devices of a higher amperage rating than the one intended for that circuit. If you do, you defeat the purpose of the protection device. We will look at fuses and breakers in more detail later in Chapter 12.

CIRCUIT CONTROLLERS (SWITCHES)

Although fuses and breakers protect the circuit and have the ability to turn the electric current "on" or "off", they are not a convenient control unit. For this reason, SWITCHES are used to operate all, or sections (branches), of any circuit. Again, there are numerous sizes, shapes, and styles of switches. Their purpose, however, is the same, see Fig. 3-14 and Fig. 3-15.

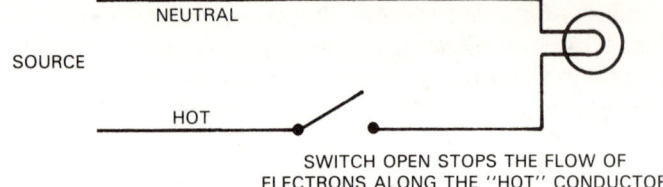

Fig. 3-15. Switch operates by interrupting the current through the conductor. "Hot" wires are connected to the screw terminals as indicated in the schematic.

Switches	Purposes
Single-pole, single-throw switch	Controls light or outlet from single location.
Three-way switch	Controls light or outlet from two locations.
Four-way switch	Controls light or outlet from other locations in between pair of three-way switches.
Dimmer switch	Like single-pole, single-throw switch, but also contains rheostat or voltage regulating device which allows all or only portion of electrical energy to outlet or light fixture.
Pilot lighted switches	Used to control light or outlet which is not in sight. It indicates, by use of pilot light, whether power to device is on or off.
Time-delay switch	Used where delayed shut-off is desirable.
Others: Toggle switch, Pushbotton switch, Pull Chain switch, Photoelectric, Knife switch, Tap switch, Plate switch, Locking switch.	

Fig. 3-16. Table of switch types and where they are designed to be used.

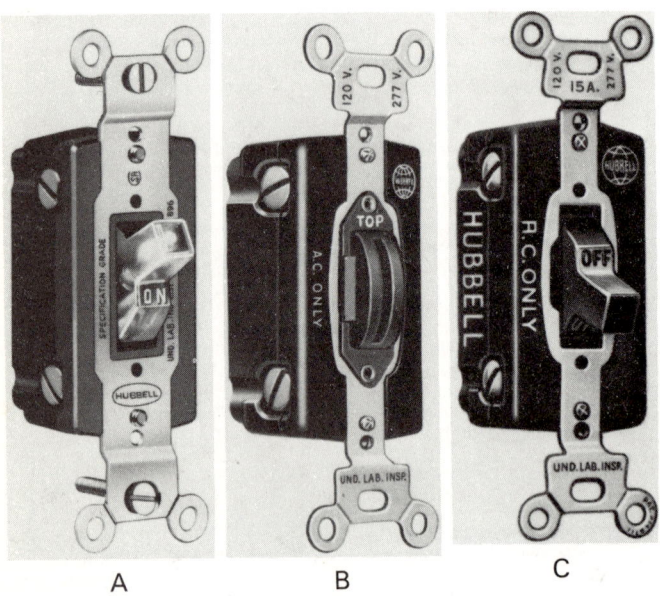

Fig. 3-14. Switches act as circuit controllers. A—Lighted handle switch. B—Key operated switch. C—Standard toggle switch. (Harvey Hubbell Inc.)

Like fuses and breakers, each switch is labeled for its maximum amperage and voltage rating. Do not place any switch in a circuit which can deliver more than the maximum permitted by the switch. It will damage the switch and could become dangerous. Also, switches are placed in "hot" conductors—NEVER in the neutral. Fig. 3-16 is a list of different types of switches and, where appropriate, their special functions.

ELECTRICAL ENERGY CONSUMERS

The purpose of the electrical circuit is to distribute electrical energy to one or more uses within the circuit. All the uses, termed *loads*, are essentially types of resistors. They impede the flow of electricity while, at the same time, consuming the energy that is provided.

Except when heat is needed, electrical resistance is to be minimized wherever possible. In toasters, waffle irons, electric broilers, and heaters, the resistance is the workhorse.

However, keep in mind that all users or loads must have some resistance or they would not be able to operate. For example, an ordinary light bulb's filament offers a fair amount of resistance so that it can, because of friction, get hot and glow to create light. Other fixtures, such as fluorescent lights and electric clocks, offer less resistance.

Resistance is also present in other components of an electric circuit: sources, conductors, controllers, and protectors. Each offers some resistance.

As mentioned before, resistance in an electric circuit is measured in OHMS. This unit of measure-

Electrical Circuit Components

ment is defined in terms of current and voltage. That is, 1 ohm is the amount of resistance which will permit the flow of 1 ampere of current when 1 volt of electrical force or pressure is applied.

REVIEW QUESTIONS — CHAPTER 3

1. What are the two best metal conductors? *copper, alum*
2. The safe current carrying capacity of a wire is its:
   a. Voltage drop.
   b. Voltage.
   c. Ampacity.
   d. Resistance.
3. The electrical pressure of a circuit is its:
   a. Voltage drop.
   b. Voltage.
   c. Ampacity.
   d. Resistance.
4. As the AWG number becomes smaller, the wire size becomes (larger, smaller).
5. The consumers or uses of electrical energy are more properly called _loads_.
6. The Greek letter Ω, omega, is the symbol or abbreviation for the _ohm_ which signifies resistance in an electrical circuit.
7. A switch is always placed in the (hot, neutral) conductor of a circuit.

Chapter 4

TOOLS FOR THE ELECTRICIAN

In the course of installing electrical wiring, an electrician will use a number of tools, Fig. 4-1. It is important that the tool pouch be kept orderly with a stock of good tools. The tools must be kept in good repair so the job can be done quickly and well. If tools are not available, cannot be found, or are in poor condition, time is wasted and good work is not possible.

ESSENTIAL TOOLS

Certain tools are essential. Without them the electrician cannot and should not attempt to wire anything. The list includes:

Wire cutter: This could be a lineman's pliers, side cutting pliers like the needlenose. It will be used for small wire.

Cable ripper: Various types are available for NM or UF cable.

Wire stripper: Either a manual or automatic type is satisfactory to remove conductor jacket from conductor. A knife is acceptable IF the electrician has developed a practical technique for using it.

Cable cutters: This is needed to cut larger cable like types SE, SEC, or SEU in sizes 6 to 4/0 or larger. A hacksaw is acceptable as a substitute.

Neon test light: This is satisfactory as minimal equipment for testing whether conductor is energized. Preferred is a voltage meter, VOM or "amp clamp" type for more specific information.

Screwdrivers: Assorted sizes and types including both Phillips head and straight blade.

Allen wrench set: This is necessary to tighten main lugs on certain load centers, panels, etc.

Pump pliers (expandable jaw): These are needed to tighten weatherproof connectors as used on service entrances installed with cable.

Hammers: Several types are needed (a) sledge or pump for driving ground rods; (b) electrician's preferred for general electrical work.

Tubing cutter: Same type as used by plumbers; used to cut EMT (IMC and rigid conduit are rarely used in residential wiring).

BASIC TOOL LIST

Striking Tools
Claw hammer
Lineman's or electrician's hammer

Drilling tools
Electric drill, 1/2 in. chuck
Electric drill, 3/8 in. chuck
Electric drill, 1/4 in. chuck
Drill bits, various sizes
 auger
 wood twist
 metal
 masonry
 expansive
 bit extenders

Soldering & Wire-Joining Tools
Soldering iron
Soldering gun
Propane torch
Solder, rosin core
Soldering paste
Blow torch
Crimping tool

Fastening Tools
Standard screwdriver
Phillips screwdriver
Offset screwdriver
Torque head screwdriver
Adjustable wrench
Allen wrenches
Socket/ratchet wrenches
Box end wrenches

Measuring Tools
Folding ruler
Carpenters extension ruler
Steel tape
12 inch ruler or meter stick
Wire gage
VOM

Cutting and Sawing Tools
Files
Crosscut saw
Keyhole saw
Hacksaw
Circular saw, 7 in.
Reciprocating saw
Pocketknife
 or electrician's knife
Cable cutters
Chisel, wood
Wire strippers
Cable strippers

Pliers
Slip joint pliers
Lineman's pliers
Side cutting pliers
Diagonal pliers
Long nose pliers
End cutting pliers
Curved jaw pliers

Special & Miscellaneous Tools
Fish tape wire puller
Wire pulling lubricant
Conduit or pipe cutter
Reamer
Conduit bender (hickey)
Fuse puller
Tape and die set
Flashlight
Plumb bob
Test light, continuity tester
Level
Conduit threader
Trouble light
Gas generator, about 1500 W
Portable space heater
Assorted wood or fiberglass ladders
Wire grips
Chalk line
Tool pouch

Fig. 4-1. A well stocked tool pouch is important if the electrician is to do his/her work well and without wasted time for lack of the right tool for the task.

EMT conduit bender: Absolutely necessary to work with EMT.

Screw starter or scratch awl: Electrician can do without it but installation of screws is so much easier and faster that all electricians should have them.

Level: Most equipment must be installed plumb.

Electrician's locking blade knife: Needed for all-purpose cutting.

Plaster knife: Wide and flexible type to repair damaged plaster when remodeling old work.

Other necessary tools: Set of socket wrenches; trouble light; tool pouch; drill and drill bits; adjustable wrench; ladders.

Knives, drill bits, chisels, saws, and other tools meant for cutting should be kept sharp. A dull tool does poor work. It may also cause injury to a worker or damage the electrical materials or the structure itself.

Gripping tools or turning tools such as pliers and screwdrivers should be inspected frequently. Plier jaws with rounded teeth will slip and cause damage. If reconditioning of any tool is impossible, the tool should be replaced. Inspect screwdriver blades for rounding and other signs of damage and wear. Recondition or replace as necessary.

Handles of striking tools should be inspected frequently for looseness and weakness. Wooden handles can crack and splinter from misuse or misdirected blows. Loose wooden handles can be repaired with new wedges. Broken ones should be replaced. Proper use and care of tools should be a part of every electrician's training. There are excellent pamphlets on the market for this purpose, Fig. 4-2.

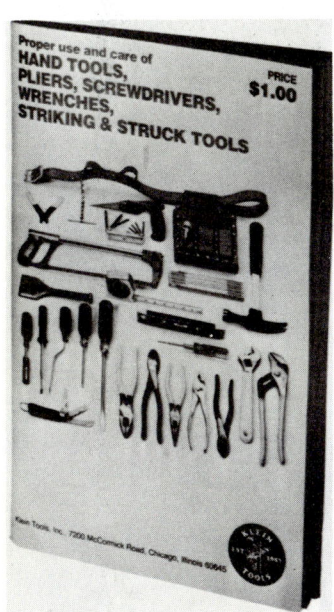

Fig. 4-2. Good tools require proper use and maintenance. (Klein Tools, Inc.)

STRIKING TOOLS

A hammer is used for driving nails and staples, for attaching hangers and electrical boxes, and for driving chisels. Many electricians prefer the electrician's hammer. It has an extra long neck for driving nails through bottoms of electrical boxes. The claw which is straight rather than curved, is used to remove nails or to loosen structural materials such as plaster during electrical remodeling. See Fig. 4-3.

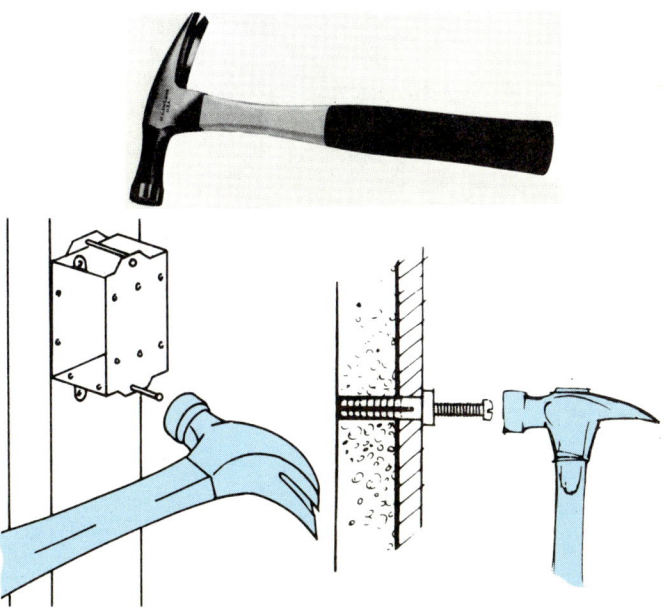

Fig. 4-3. Hammers are essential for attaching electrical components and installing hangers.

General rules regarding the proper use of striking tools follow:
1. Always choose the right size and weight hammer or striking tools for the intended use.
2. Never strike tools together unless they are designed for the purpose.
3. Striking tools should strike a surface squarely; never at an angle.
4. Always wear safety glasses when using striking tools.

CUTTING AND SAWING TOOLS

The handsaw is needed for notching or cutting studs. For heavier cuts, a power circular saw may sometimes be used. When cutting in tight quarters, a keyhole saw is preferred, Fig. 4-4. This saw is capable of cutting circles and irregular shapes. It can also start a cut through a small hole drilled into the material.

A reciprocating saw is handy wherever a keyhole saw is used. It is faster and will handle thicker material.

Modern Residential Wiring

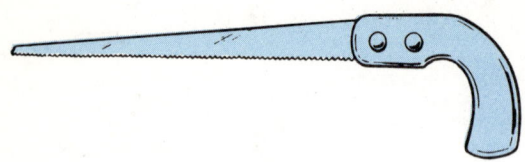

Fig. 4-4. Keyhole saw. It can make cuts where a handsaw cannot be used for lack of space. (GE Wiring Devices Dept.)

The wood chisel will trim away small amounts of wood in framing members to make room for mounting of fixtures or boxes. It is preferred for notching studs, joists, plaster, flooring and old-style lath.

Cutting tools require careful use and frequent maintenance. For their proper use and care:
1. Wear gloves when using cutting tools.
2. Use the correct size and type of cutting tool for the purpose intended.
3. Do not force cutting tools past reasonable hand pressure.
4. Oil pivot points of the tool frequently.
5. Keep cutting edges sharp.
6. Tighten nuts and bolts of pivot properly when applicable.

A hacksaw, Fig. 4-5, will cut conduit and cable. It may also be used in place of a handsaw for cutting away plaster and lath.

Fig. 4-5. Hacksaw is used for cutting conduit, cable, plaster, and, sometimes, lath. (Parker Mfg. Co.)

WIRE AND CABLE CUTTERS

A number of tools can be used for clamping and cutting wire conductors, Fig. 4-6. Multipurpose tools will cut, strip, and crimp, Fig. 4-7. Wire strippers are good for quick removal of insulation without damaging the wire conductor material. A pocketknife or electrician's knife is suitable for removing insulation.

Cable cutters are designed to cut large diameter copper and aluminum cable. Handles 15 to 17 in. long provide sufficient leverage to the jaws to cut cable 350 MCM (.350 in. or about 3/8 in.) thick. The jaws can be replaced if they become worn or damaged, Fig. 4-8.

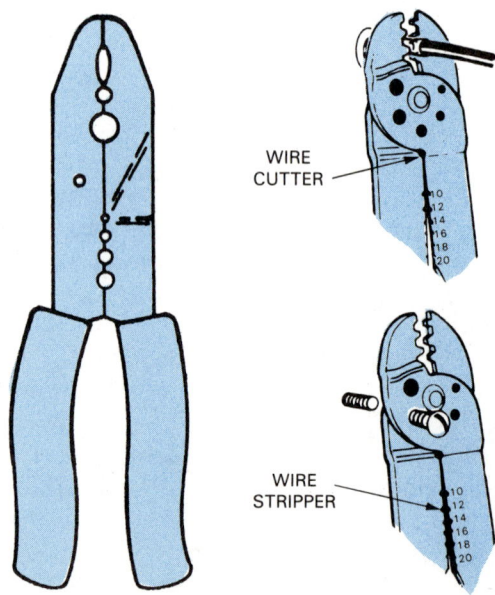

Fig. 4-7. A crimping tool also serves as a cutter, wire stripper, and bolt cutter. It is an all-around tool favored by many electricians. (Vaco Products Co.)

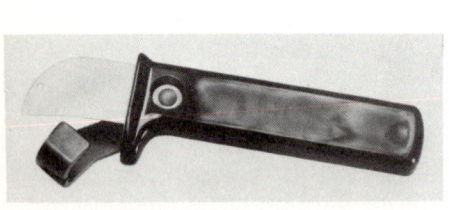
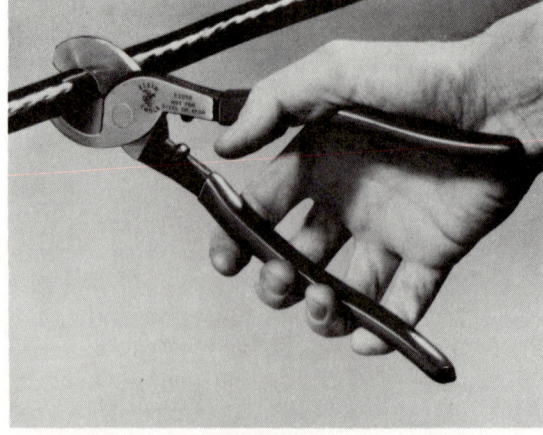
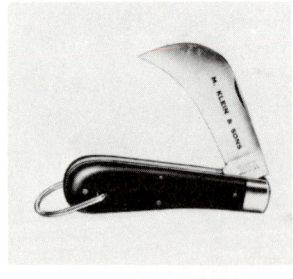

Fig. 4-6. Left. Cable ripper strips away insulation without damaging the conductor. Hook beneath blade provides leverage, requires less hand pressure. Center. Cable cutter. Insulated handles are important on many electrician's tools because they protect the electrician from potential shock. Right. Pocketknife designed for removing insulation. (Klein Tools, Inc.)

A cable stripper is used on larger conductors from 9/16 in. to 1 7/16 in. in diameter. It is clamped onto the cable and rotated. The blade can strip away the insulation without damaging the conductor wire. See Fig. 4-9.

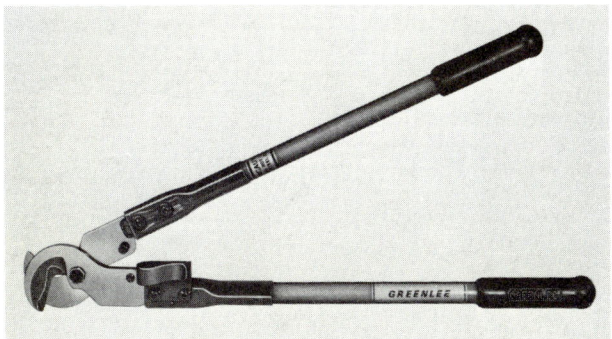

Fig. 4-8. Large conductors are cut with the cable cutter shown. Tool length is nearly 2 ft. long. (Greenlee Tool Div., Ex-Cello Corp.)

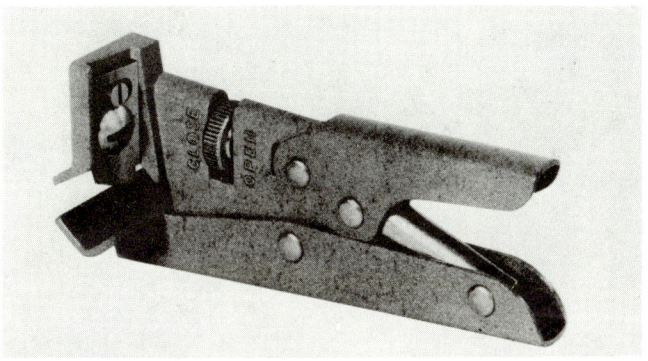

Fig. 4-9. Insulation stripper. Tool self-feeds as it is rotated around cable. Cable is held in V-shaped trough.

PLIERS

Pliers, as shown in Fig. 4-10, may be used for holding, shaping and cutting. The basic tools are the slip joint pliers and the lineman's pliers or side cutters. Both can be used for holding parts such as locknuts and for gripping wires that are being pulled. The lineman's pliers can also be used for cutting wires. Sometimes it will be used as a crimper for connectors used on Romex.

Diagonal pliers have high leverage and are needed for cutting wires where it is difficult or impossible to use side cutting pliers. One instance of their use is for cutting conductors in BX after the metal armor has been cut.

They are sometimes used for shaping loops in the wires for screw terminals. However, if available, a long nose pliers makes better loops.

A good all-around tool is the lever-jaw wrench

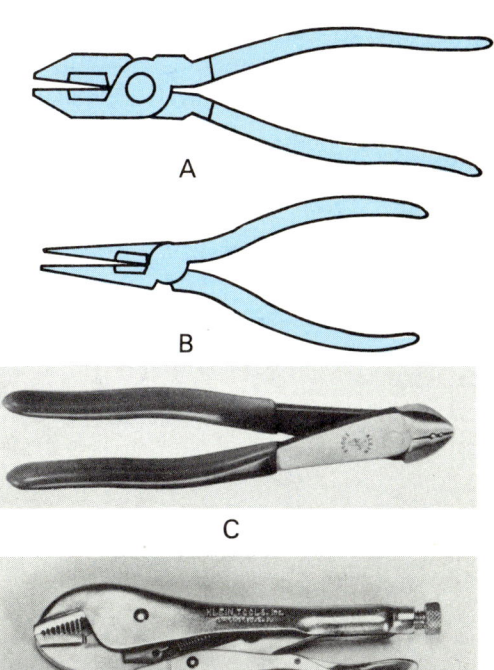

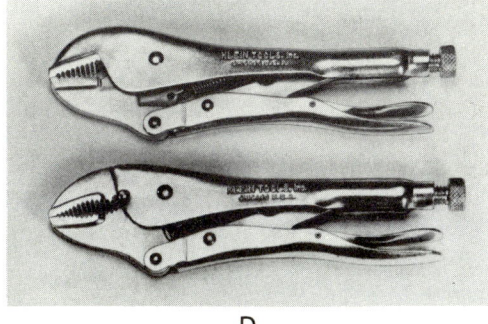

Fig. 4-10. Types of pliers. A—Lineman's pliers is designed for pulling wires. B—Long nose pliers is best for forming terminal loops on wires. (GE Wiring Devices Dept.) C—Diagonal pliers with stripping holes. D—"Vise grips" or lever-jaw wrench will grip parts and leave hands free to work. Either curved jaw or straight jaw models are available. (Klein Tools, Inc.)

usually called a "vise grips." It serves as a pliers, lock wrench, open end wrench or a pipe wrench. As you can see, pliers are quite versatile. These tools can grip, clamp, tighten, turn, pull and cut objects. Certain rules should be followed to prolong their usefulness and make them safe:

1. Never use pliers to strike another object.
2. Do not use for cutting, unless designed to do so.
3. Keep properly oiled.
4. For electrical work, use plier-type tools having properly insulated handles. They will help protect against electrical shock.

FASTENING TOOLS

The most-used fastening tool in the electrician's tool box is the standard screwdriver. It is needed primarily to tighten terminal screws, attach switches and outlets to their boxes, install switch and outlet covers, and install fixtures. It is also a handy tool for tightening locknuts. Phillips screwdrivers are needed for some screw heads used in electrical work. Several sizes are desirable and are, therefore, recommended, Fig. 4-11. In addition, open end wrenches, box end wrenches, socket wrenches and nut drivers are needed frequently and should be kept handy. See Fig. 4-12. For proper use and care of fastening tools:

1. Never use them for striking other tools.
2. Do not use make-shift extension handles, sometimes called "cheaters," for increasing leverage.
3. Avoid overtightening or overtorquing when using tightening tools.
4. Again, use the right type and size of the tool for its intended purpose. Proper fit is essential for good performance.

DRILLING TOOLS

Whether using a hand drill, push-pull drill, brace and bit or electric powered drill, the common function is to make a "clean" hole. In order to accomplish this, the tool must be in good working condition. The bit must be sharp. It must also be of the proper type and size for the hole to be drilled and the material you are drilling. Fig. 4-13 illustrates an ordinary brace and a power drill. Fig. 4-14 and Fig. 4-15 show a variety of drill bits, accessories, and hole saws.

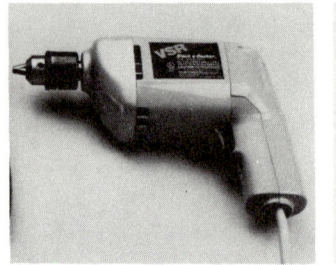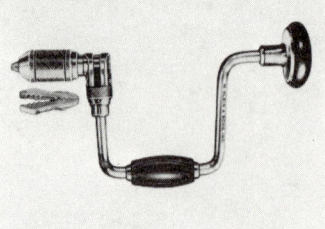

Fig. 4-13. Two drilling tools. Left. Electric drill can be used with suitable bits to drill any size hole in wood or metal. (Black & Decker). Right. Hand operated drill is called a brace. It is still one of the most versatile boring or drilling devices.

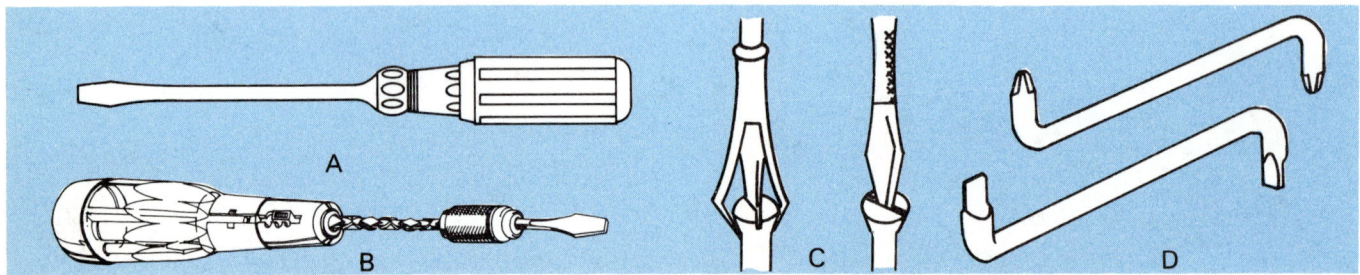

Fig. 4-11. Tools for fastening. A—Standard screwdriver. (Klein Tools, Inc.) B—Spiral ratchet screwdriver with reversing control. (Stanley Tools) C—Self-gripping and adjustable screwdrivers. D—Offset screwdrivers.

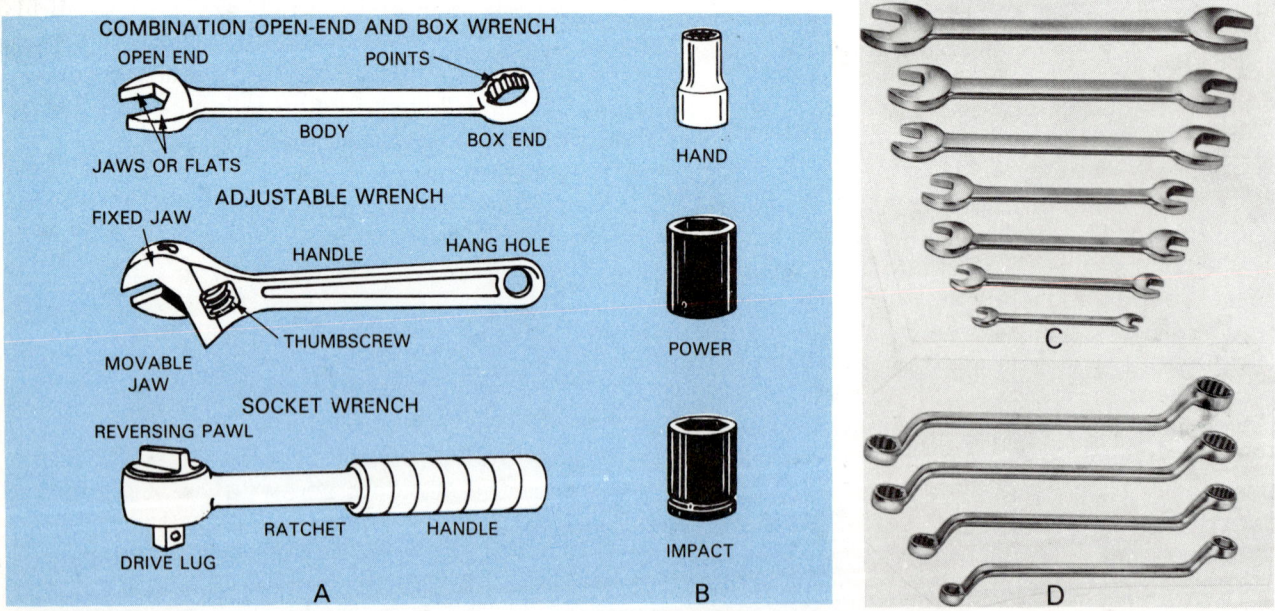

Fig. 4-12. Other fastening tools. A—Parts of wrenches. (Klein Tools, Inc.) B—Sockets. C—Open end wrenches. D—Box end wrenches. (Duro Chrome Tools)

Tools of the Trade

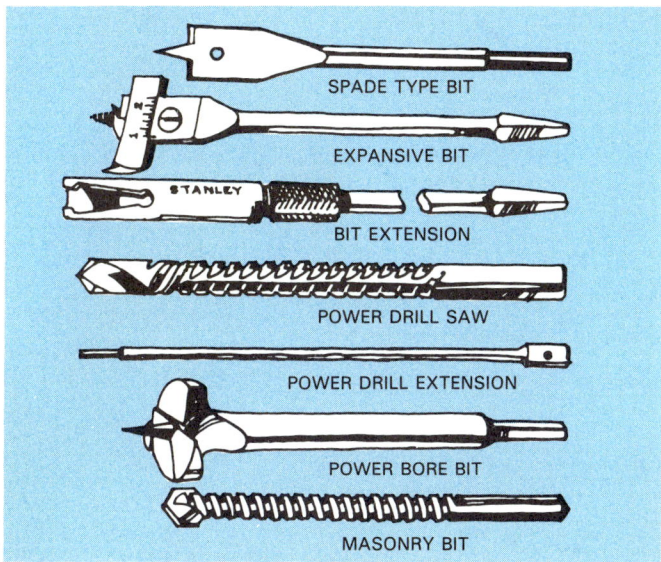

Fig. 4-14. The electrician will use a variety of bits and extenders in the course of installing wiring in a house. (Stanley Tools)

Fig. 4-15. For boring large holes, the heavy duty hole saw is a useful tool. Diameters range from 3/16 in. to 4 1/8 in. (Black & Decker)

SOLDERING TOOLS

The development of various types of wire connectors has all but eliminated the once common technique of joining wires by soldering. Still, soldering may occasionally be necessary. Upon such rare occasions, a soldering gun or propane torch like those illustrated in Fig. 4-16, will fill the bill.

Such tools require almost no maintenance except for occasional tinning or cleaning of the soldering tip. Complete instructions provided by the manufacturer are adequate for the care and use of these tools.

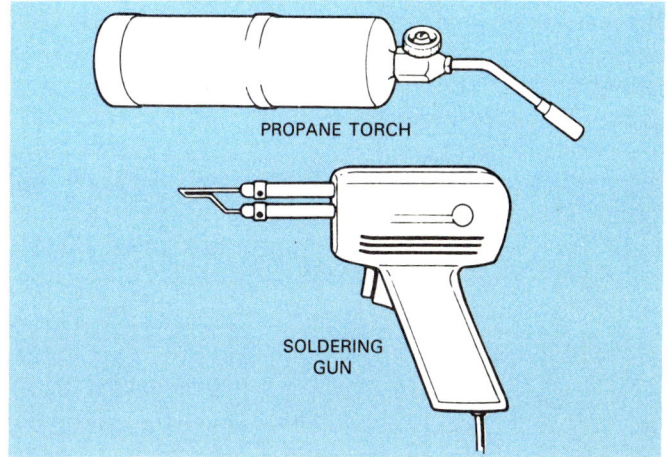

Fig. 4-16. Tools for soldering.

MEASURING TOOLS

Two of the most important measuring devices needed by electricians and other trades people are the folding rule and steel tape, Fig. 4-17. The folding rule requires almost no maintenance, except a drop of oil at each hinge once a year or so. Typical rules are 6 ft., 8 ft., and 10 ft. long with hinges for folding at 6 in. intervals. Those having both conventional U.S. and metric scales are best.

Fig. 4-17. Electricians should keep more than one type of rule. Left. Steel tape. Right. Folding rule. (Stanley Tools)

MISCELLANEOUS TOOLS

Throughout this text we will see illustrations and references to various tools used in the electrical trade. Being specific in their use, they defy a general category. Nonetheless, this group comprises the largest group of devices used by the electrician.

The circuit tester can be used to determine if there is current in a circuit and if it is properly grounded. Continuity testers contain their own power source and are needed for checking continuity (complete circuit) in unenergized circuits.

Trouble lights serve a two-fold purpose. They bring light into out-of-the way places and, at the same time, provide an electric power outlet for drills, saws, and other electrical tools. Be sure to select one with a grounded outlet.

A pipecutter is useful for cutting pipe and conduit. While a hacksaw serves the same purpose, the cutter is preferred as it produces a neater cut. The pipe reamer will remove sharp and ragged edges from pipe that could abrade insulation. See Fig. 4-18 for various tools.

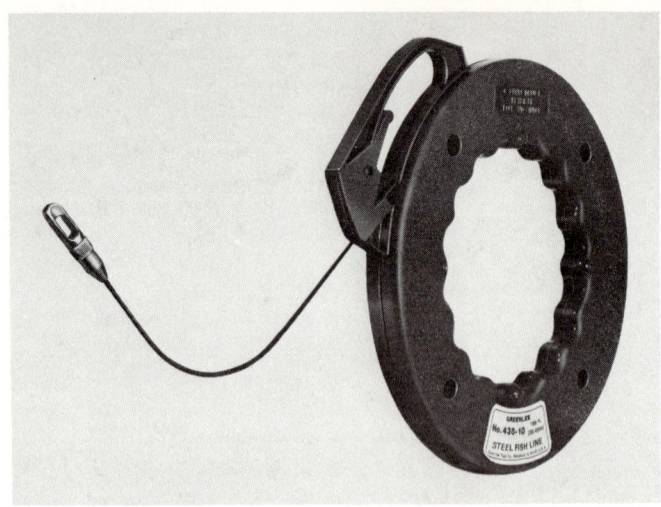

Fig. 4-19. Steel fish tape is used to pull wires through wall spaces and conduit runs. (Greenlee Tool Div., Ex-Cello Corp.)

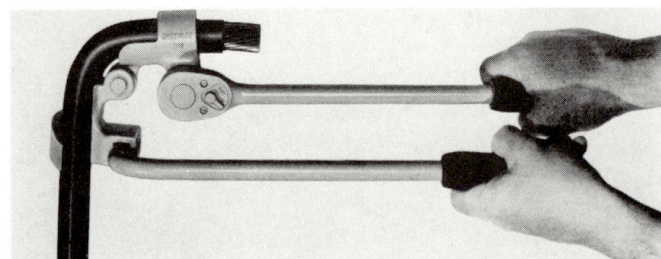

Fig. 4-20. Manual conductor bender makes easy work out of bending large, usually stubborn, conductors.

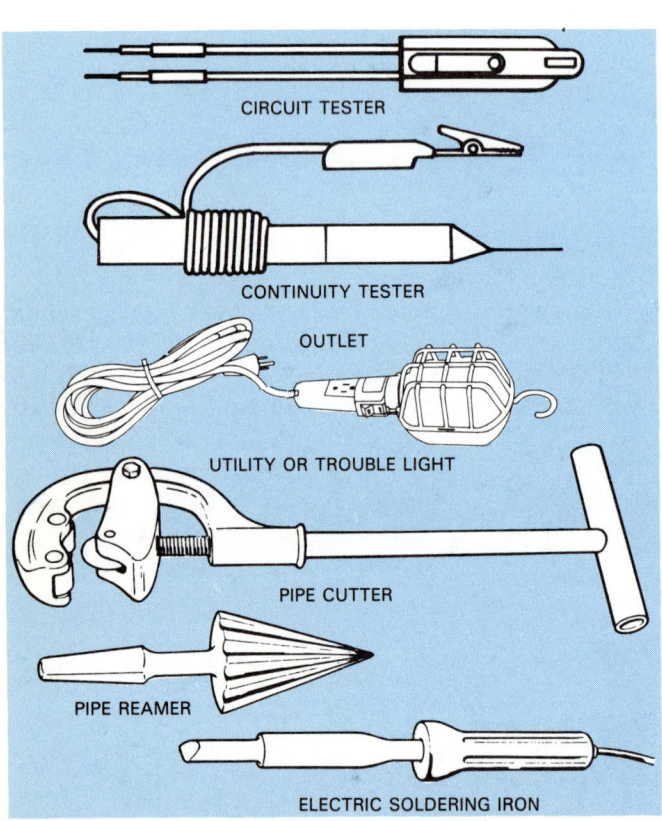

Fig. 4-18. Miscellaneous tools such as these are frequently useful to the electrician. (GE Wiring Devices Dept.)

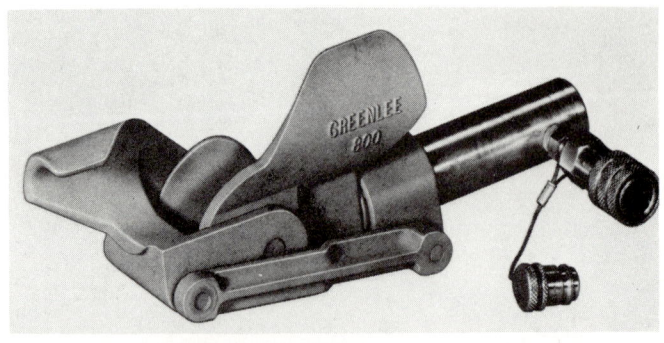

Fig. 4-21. Hydraulic conductor bender. This tool is a "must" when wire bending space is limited. (Greenlee Tool Div., Ex-Cello Corp.)

Fig. 4-19 through Fig. 4-23 illustrate a few more essential tools. No doubt, you will come across dozens of others when involved in actual wiring procedures.

Electrical contractors who hire on electricians to work for them expect the worker to have his or her own tools. While not all of the tools need be purchased at one time, the list presented is necessary for the well stocked electrician's tool kit. Since tools are expensive to replace, it is important to follow the instructions for their use and care. Refer again to Fig. 4-1 for a checklist of the tools you need to get started.

For additional information concerning the proper use of hand tools, contact the Hand Tool Institute, 331 Madison Avenue, New York, NY 10017.

Tools of the Trade

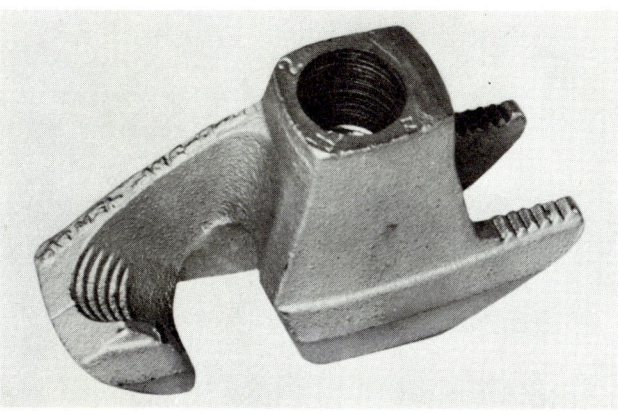

Fig. 4-22. Conduit hickey. This bender is often used to bend larger sizes of rigid conduit. (Appleton Electric Co.)

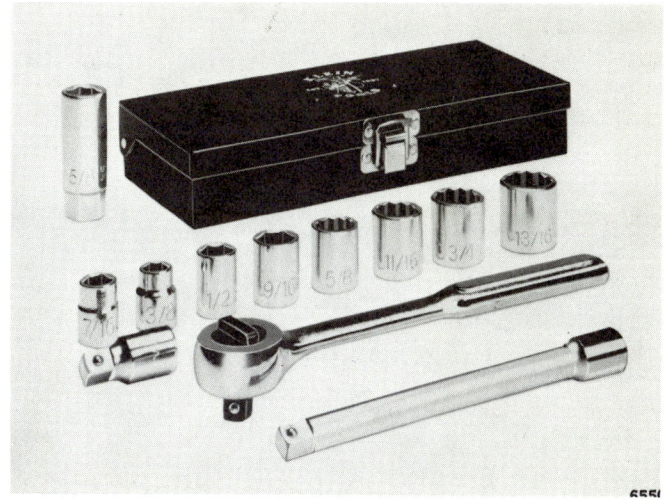

Fig. 4-23. Socket set will be used frequently by the electrician as he or she installs service entrance and entrance panels. (Klein Tools, Inc.)

REVIEW QUESTIONS — CHAPTER 4

1. List those tools which are essential for the electrician to have on the job.
2. It is better to buy cheap tools and replace them often. True or False?
3. How is an electrician's hammer different from a standard claw hammer?
4. Which of the following statements is true? When using cutting tools such as cable cutters and side cutters:
   a. Never use more than reasonable hand pressure.
   b. Use a handle extender if the material does not cut easily.
   c. It is alright to use a hammer to make them cut the material.
5. Pliers are used for _____, _____ and _____.
6. The most-used fastening tool in the electrician's tool pouch is the _____ _____.
7. Insulated handles are important on many electrician's tools because _____ _____.
8. Where can you get additional information on proper use of hand tools?

Chapter 5

SAFETY AND GROUNDING ESSENTIALS

Safety when working with electrical systems, equipment, and tools should be foremost in the electrician's mind. Safety in electrical wiring means avoiding accidents by proper work habits and by maintaining good housekeeping in the areas where work is underway.

Another important aspect of wiring safety is the installation of a well-grounded system which follows *Code* and which will be serviceable and safe for years to come. More will be said about grounding later.

SAFETY DURING INSTALLATION

Certain safety rules should be learned and followed faithfully while installing electrical wiring. Only if the rules are followed can the electrician hope to avoid accidents and injury. The basic rules are:

1. Remain alert always. THINK before you act.
2. Avoid quick movements. Work slowly and deliberately. Plan each step of your work carefully.
3. Never work on live wires unless absolutely necessary. Disconnect the circuit from main source, if possible.
4. Never assume that the circuit is de-energized (dead). Test it with a test light before you start to work.
5. Use tools correctly and for their intended purpose. Keep them in good working order. Tools used for work on live circuits, especially, should have insulated handles.
6. Do not wear metal objects when working on electrical devices. Remove jewelry, rings, watches, and any other metallic apparel from the body.
7. Make sure that all electrical equipment is properly grounded.
8. Keep work areas dry and free of debris. If necessary, clear the area of loose material or hanging objects. Cover wet floors with wooden planking. Wear rubber boots or rubber soled shoes.
9. Wear gloves when possible.
10. Place a rubber barrier or other nonconductive shield around exposed conductors or equipment near the work area.
11. Lift objects carefully. Use the leg muscles. Keep back straight while lifting.
12. Wear clothing that is neither floppy nor too tight. Loose clothing will catch on corners and rough surfaces causing unsafe motion. Clothing that binds is uncomfortable and maybe just as unsafe as clothing that is too loose. See Fig. 5-1.
13. Wear safety glasses when using striking tools.

Fig. 5-1. Safety in electrical wiring may save your life. Move cautiously around electrical equipment that is live. Observe proper clearances. Wear clothing that is neither too loose (floppy) nor too tight. Keep floors clear of debris. (OSHA)

LADDER SAFETY

Ladders used by the electrician should be made of wood or fiberglass. They should be set at a safe angle and tied off to prevent a slip or fall.

Avoid placing a ladder at too steep an angle. As a rule of thumb, the horizontal distance from the base of the ladder to the structure should be one-quarter the length of the ladder, Fig. 5-2.

Make certain the base of the ladder has firm support. Ground or floor should be level. Use extreme care if surface is wet or icy. Special blocking may be needed to prevent slipping.

Stepladders should be set level. They must be fully opened so the hinges will lock. Keep joints tight so the ladder does not wobble or lean. Discard step ladders in poor condition.

When working on the ladder always face it. Never use the top rung or step for support. Check straight ladders and extension ladders often for step rot or other structural damage.

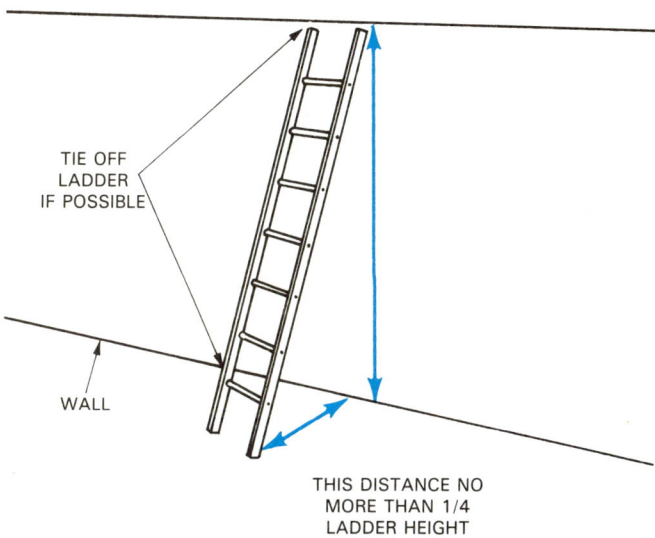

Fig. 5-2. Use only wooden or fiberglass ladders for overhead electrical work. Do not set ladder at too steep an angle. Distance between base and structure should be at least a quarter of the ladder height.

SCAFFOLDING

Wooden scaffolding is preferred for electrical work. If a metal scaffolding must be used, it should be grounded. Handrails and toe boards should be secured to the top level. If the scaffold has wheels, they should be locked to prevent unwanted movement. Tie the scaffold off as well, especially if it is several platforms high. Always use a ladder to get onto a scaffold. NEVER climb the braces.

Both scaffolding and ladders should be handled carefully to prevent injury or property damage. Be particularly careful not to contact overhead wires or other exposed wires.

TEMPORARY WIRING

Temporary wiring is done to provide electrical power only during construction, remodeling, demolition, repair, or large-scale maintenance of a building. It is meant to be used for a short period and then removed. Article 305 of the *National Electrical Code* covers this subject.

Although temporary wiring safety requirements may be less exacting than permanent wiring, most of the same rules apply. However, there are several important differences.

Temporary wiring can be used on a construction site during the entire term of construction, regardless of the type of construction. In addition, temporary power can be brought in for emergency situations, for testing, and for experimentation usage.

Time limitations are set on other uses. For example, temporary Christmas lighting or carnival power must be removed within 90 days.

The key factors for installation of all temporary wiring are:
1. All temporary wiring, devices, and equipment should be located in a safe place and should be as neat as possible. It should be kept overhead as much as possible, Fig. 5-3.

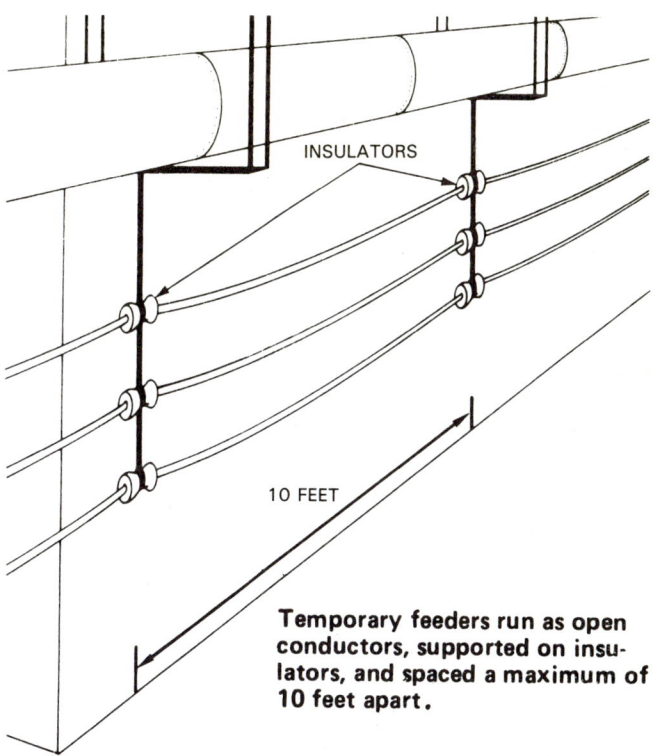

Fig. 5-3. Safe temporary power lines should be well-supported and should be neatly strung from the power drop to the temporary power panel. (OSHA)

2. Protect temporary wiring from physical damage.
3. Place ground-fault type receptacles in outlet boxes for personnel protection. See Fig. 5-4.
4. Grounding must be in compliance with Article 250 of the *National Electrical Code.*
5. All lamps should have protective covers or caging. They should be mounted at heights of 7 ft. or more.
6. All circuits must originate from approved panels or power outlets and must have overcurrent protection.
7. Receptacle circuits and lighting circuits are to be separated.
8. Weatherproof devices and housings should be used wherever there is dampness or exposure to weather.
9. Locking type plugs and connectors should be used on power ends.
10. Temporary wiring should be inspected frequently for damage.

EFFECTS OF ELECTRICAL SHOCK

Alternating current electrical shock is dangerous and often fatal. Likelihood and severity of shock depends on many variables. Degree of electrical shock will be different if:
1. Moisture is present. If floor or clothing is wet or if person is perspiring, shock will be more severe than if conditions are dry.
2. Floor is metal, wood, or concrete. The better the floor conducts, the more severe the shock. Wood and concrete are insulators. Metal is a conductor.
3. The circuit is grounded.

Voltage does not have as much to do with severity of shock as the amount of current. A shock of 10,000 V is no more deadly than 120 V.

On the other hand, a shock approaching 10 milliamperes (0.01 A) is painful. Possibly severe effects will be experienced. Refer to Fig. 5-5. The values listed in the chart and the bodily effects are only an average. Conditions of size, skin moisture, food and liquids in the body, surrounding humidity, and clothing would tend to change the severity of the effect.

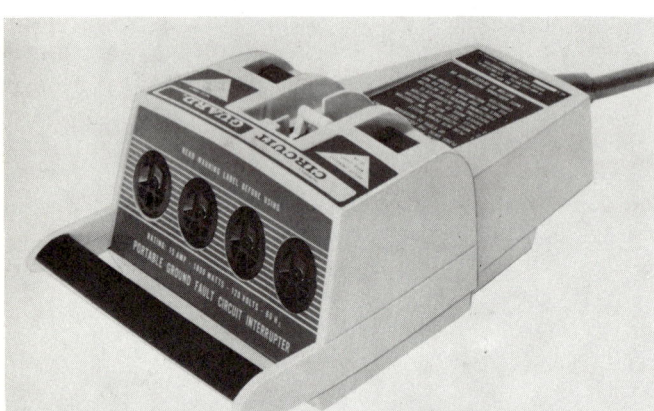

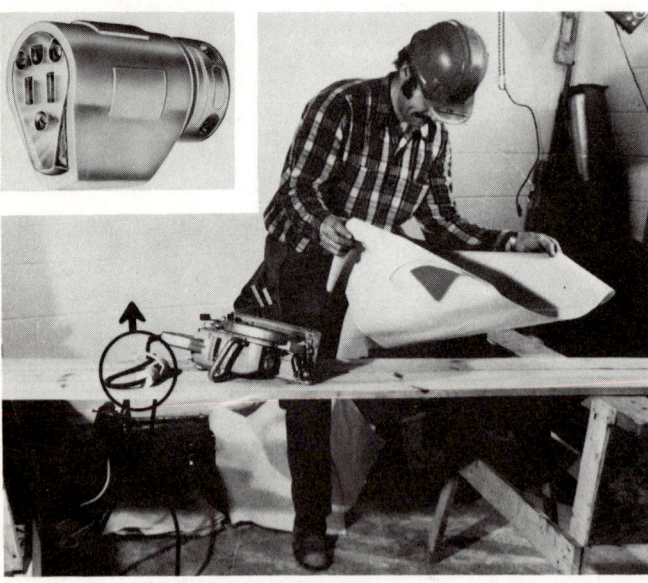

Fig. 5-4. To protect personnel on construction sites, devices should be used to guard against electrical shock. Top. Portable ground-fault circuit interrupter. (Hubbell Wiring Device Div.) Bottom. A ground continuity monitor checks constantly that an extension cord is properly grounded. (Daniel Woodhead Co.)

AVERAGE EFFECTS OF ELECTRIC CURRENT ON BODY

Amount of current (in amperes)	Effects on body
0.001 1 milliampere	Felt slightly as mere tingle
0.001 to 0.01 1 mA to 10 mA	Would probably cause muscles to contract freezing individual preventing him/her from releasing object
0.01 to 0.1 10 mA to 100 mA	Fatal after several seconds duration
0.1 or more 100 mA or more	Almost always fatal

Fig. 5-5. Even very small amounts of current passing through the body can be lethal. Severity is affected by many factors such as moisture on the body or clothing, amount of food or liquids in the body, etc.

HELPING A SHOCK VICTIM

Immediate action must be taken to help a shock victim. This is particularly true if the victim is still in contact with the source of electrical current. If you cannot shut off the power IMMEDIATELY, try instead, to move the person from contact with the electricity.

Use great care that you do not receive a shock yourself. Use a stick or insulated material to move the conductor or the victim. Do not take a chance on injury from shock yourself.

Call a doctor or paramedics to the scene. If the victim is not breathing, perform artificial respiration or mouth-to-mouth resuscitation until medical assistance arrives.

GROUNDING

The most important element in wiring safety is grounding. This refers to the connection of all parts of a wiring installation to ground or to other systems that are well grounded. This deliberate and purposeful connecting of the system and the earth provides for the safety and protection of:
1. Persons using the electricity.
2. The electrical system itself.
3. Persons using power tools or appliances.

GROUNDING THEORY

Grounding protects by limiting the possibility of damage to electrical equipment and conductors and by preventing shock to persons contacting electrical equipment.

Electrical systems could receive extremely high voltage from several sources:
1. Lightning striking electrical wiring or service wiring, Fig. 5-6.
2. Insulation breaking down and allowing high voltage across the supply (step-down) transformer.
3. Accidental contact between the service conductors and high tension wires.

In any of these mishaps, very high voltages could enter the electrical system. Voltage could be in excess of several thousand volts. In a system designed to carry no more than several hundred volts, the damage would be substantial. Without proper grounding the high current related to the high voltage would create so much heat that conductor insulation would melt. Surrounding structural materials could ignite and burn.

With a system properly grounded, this excess voltage and current is directed to earth rapidly. In reality, the excess would travel through the grounding conductor and not affect the electrical system at all. See Fig. 5-7.

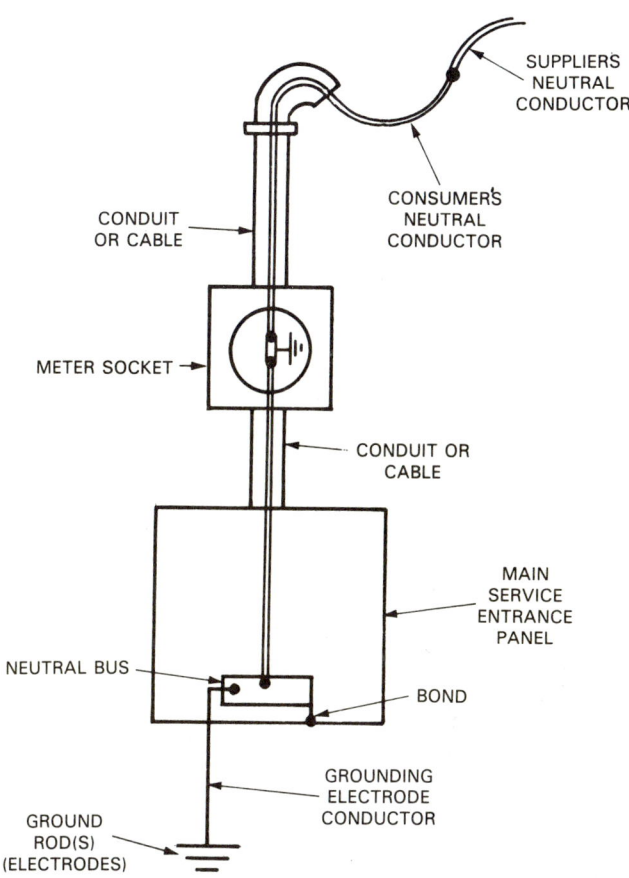

Fig. 5-7. In system grounding the service entrance is connected to earth so that excessive voltage and accompanying overcurrent would be directed to ground. Note in schematic how grounded conductor is grounded.

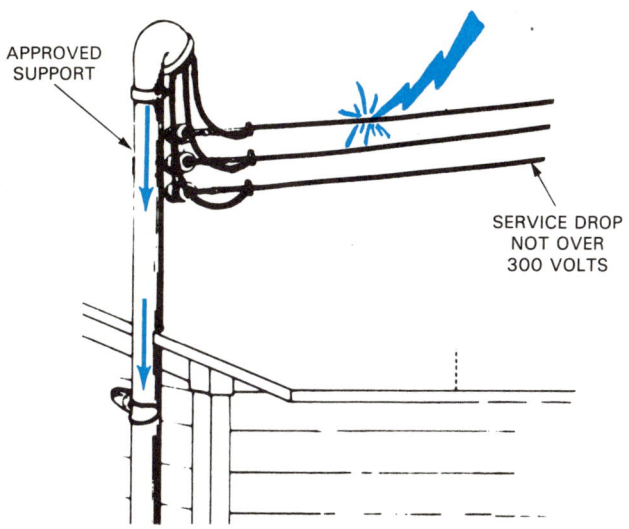

Fig. 5-6. Lightning striking supply wiring could send high voltage and overcurrent into an electrical system if it is not grounded. (U.S. Labor Dept.)

Other problems can occur in the electrical system's equipment or conductors. These include SHORT-CIRCUITS and GROUND FAULTS.

Short-circuits are conducting connections, accidental or intentional, between any of the conductors of an electrical system. This connection may be from line to line or line to the neutral (grounded) conductor.

A ground fault is a conducting connection, intentional or accidental, between any conductors of an electrical system and the normally noncurrent-carrying conducting material enclosing the conductors. (Such enclosures would include metal conduit, electrical boxes, metal cabinets of appliances, etc.)

In either case, without proper grounding, an unwanted electric current could be induced in the system over its exposed portions. Proper equipment grounding would carry this unwanted current back to the service panel. Here it would cause the overcurrent device (fuse or breaker) to open the circuit and de-energize (shut down) the circuit. Excess current would go to ground. See Fig. 5-8.

There are basically two kinds of grounding for electrical wiring. One is SYSTEM GROUNDING and the other is called EQUIPMENT GROUNDING. They have different purposes and involve two different electrical paths.

SYSTEM GROUNDING

System grounding is the intentional connection of one conductor of the electrical system to the earth. This ground connection is usually to the neutral conductor if one is available in the system. However, it can be a line conductor as in a three-phase, three wire delta system or a two-wire, 120 V single-phase system.

The neutral or line conductor selected to be grounded is connected to a grounding electrode. This may be:
1. A driven ground rod. The rod reaches 8-10 ft. deep into the ground for contact with moist earth. Usually, it is connected to one or more driven ground rods for better grounding.
2. A metallic water piping system of 10 ft. or more in length.
3. A buried (metal) ground plate.

Other grounding means are outlined in Article 250 of the *National Electrical Code.* Any one of these methods may be used to ground the system. A combination of several is also acceptable.

Grounding of electrical systems is required to:
1. Limit the voltage entering the system from surges in power from the supply transformer or from lightning.
2. Hold all parts of the electrical system (including noncurrent carrying enclosures) at zero potential to ground.

System grounding serves to drain off high voltage that may accidentally enter the system from:
1. Lightning.
2. Breakdown of insulation in the supply transformer.
3. Chance contact between high voltage supply lines and incoming service wires.

Fig. 5-8 is a simple example of a system ground. The grounded conductor (neutral or white wire) forms an unbroken path from all the various circuits to the service entrance panel. All the neutral conductors meet in the service panel and connect to a common bus bar called the NEUTRAL BUS. This bus bar is in turn, connected to the ground bus by

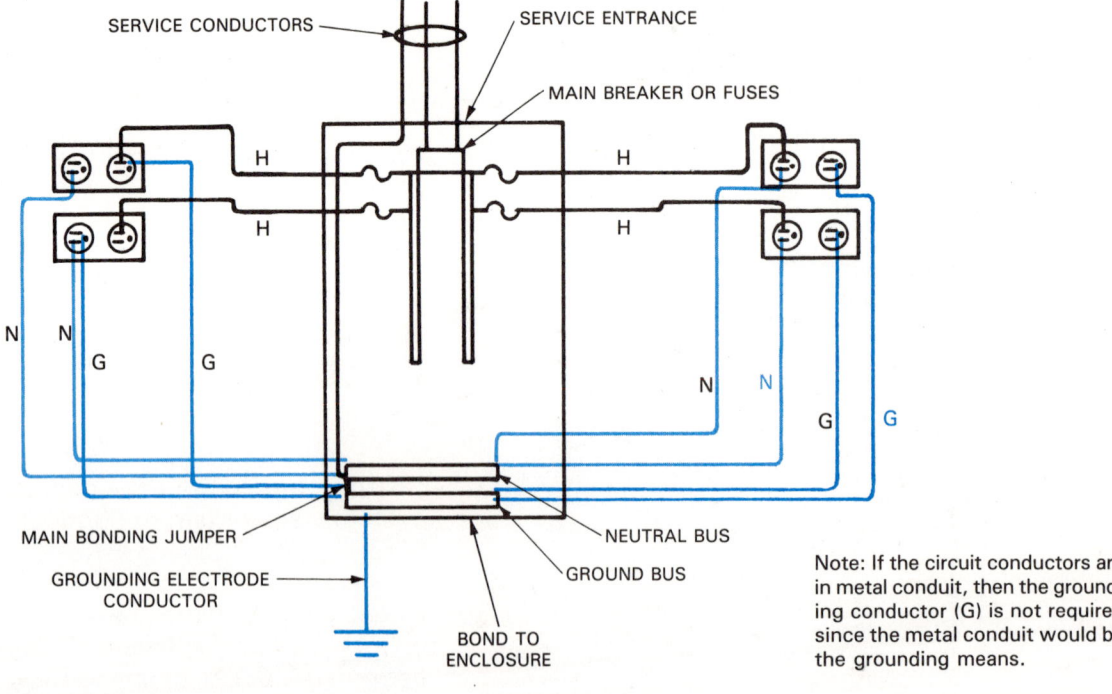

Note: If the circuit conductors are in metal conduit, then the grounding conductor (G) is not required since the metal conduit would be the grounding means.

Fig. 5-8. Schematic of a system ground. All neutral conductors in the system are connected to ground at service panel. Grounding conductor is continuous from the service panel ground bus to the grounding electrode.

Safety and Grounding Essentials

the main bonding jumper. The ground bus is connected to the grounding electrode. Should overcurrent accidentally reach the system, it will be drained off harmlessly into the ground before it can damage any part of the system.

The neutral wire in an electrical system, also known as the grounded conductor, should never have its continuity interrupted by switches, fuses, or circuit breakers. This would cancel its value as a path to ground back at the service panel. Devices which interrupt the circuit can be used only on hot wires (ungrounded conductors).

The grounding of the neutral wire is done at both the building's service panel and at the power supply pole. See Fig. 5-9. A commonly accepted method of system grounding is to connect the neutral conductor to one or more ground rods. The rods must be driven to a depth of at least 8 ft., Fig. 5-10. The depth is important as the rods must be in contact with moist earth for good grounding.

When more than one rod is used they should be spaced a minimum of 6 ft. apart. Technically the ground rods are known as GROUNDING ELECTRODES. They are connected to the neutral conductor by a stranded copper conductor no smaller

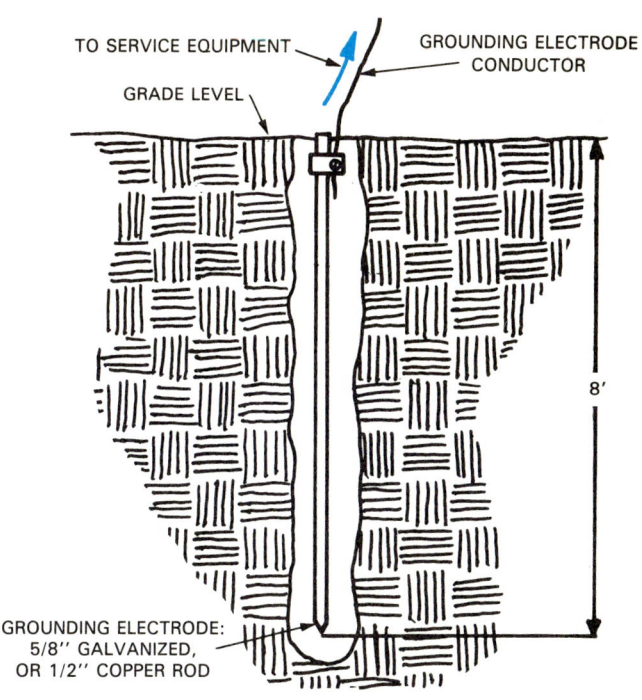

Fig. 5-10. A ground rod. Also called grounding electrode, this rod is driven to a depth of 8 ft. This assures a good connection of the neutral service conductor with the earth.

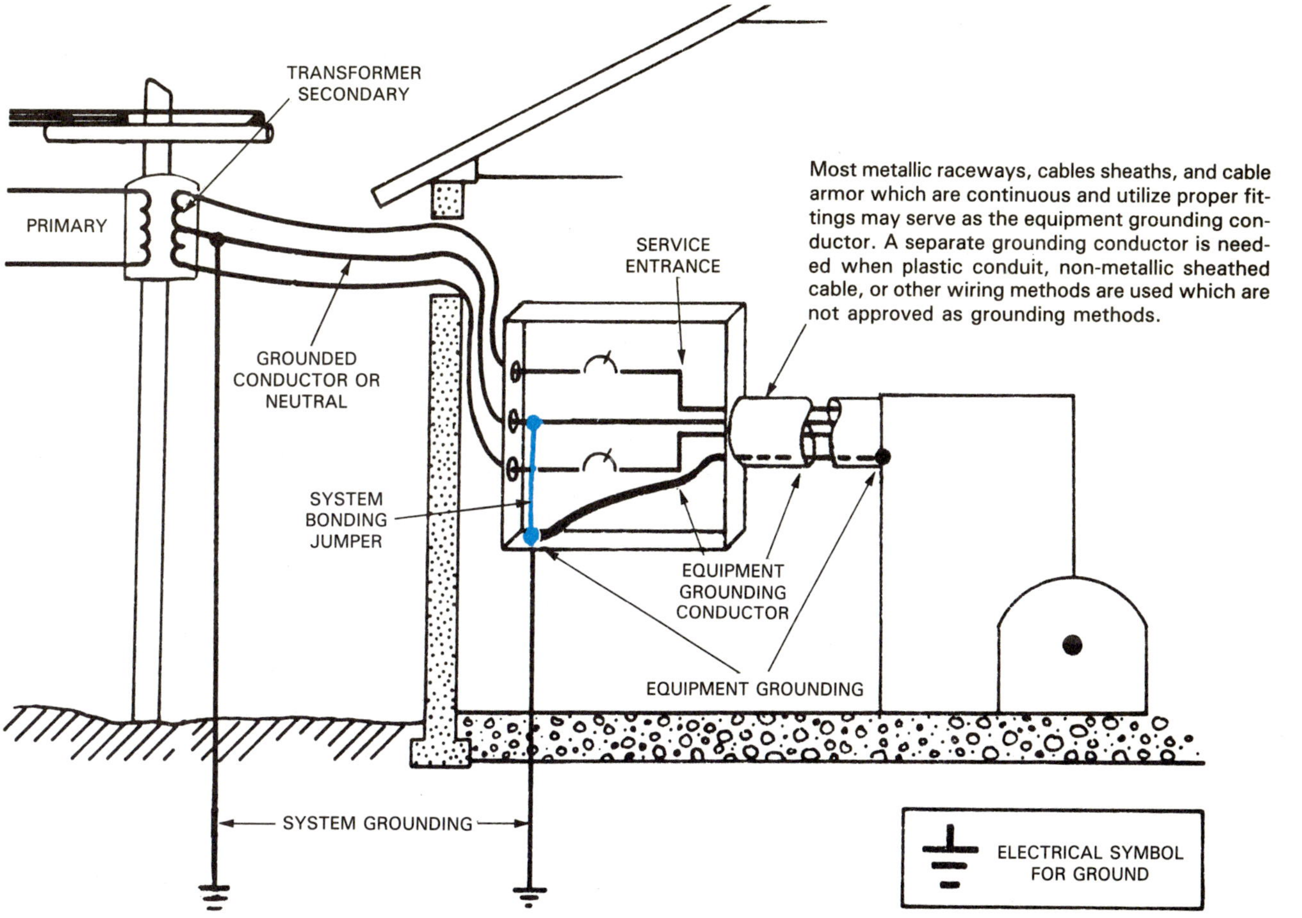

Fig. 5-9. A schematic illustration of both system and equipment grounding. System must be grounded at both the service pole and at the service panel. (OSHA)

than No. 8 AWG. This bare wire is known as the GROUNDING ELECTRODE CONDUCTOR. See Fig. 5-11 for proper sizing of the grounding conductor.

Another method of system grounding is to connect the neutral wire to the metal piping of the water supply system in the building, Fig. 5-12. It is a com-

Table 250-94
Grounding Electrode Conductor for AC Systems

Size of Largest Service-Entrance Conductor or Equivalent Area for Parallel Conductors		Size of Grounding Electrode Conductor	
Copper	Aluminum or Copper-Clad Aluminum	Copper	*Aluminum or Copper-Clad Aluminum
2 or smaller	0 or smaller	8	6
1 or 0	2/0 or 3/0	6	4
2/0 or 3/0	4/0 or 250 MCM	4	2
Over 3/0 thru 350 MCM	Over 250 MCM thru 500 MCM	2	0
Over 350 MCM thru 600 MCM	Over 500 MCM thru 900 MCM	0	3/0
Over 600 MCM thru 1100 MCM	Over 900 MCM thru 1750 MCM	2/0	4/0
Over 1100 MCM	Over 1750 MCM	3/0	250 MCM

Where there are no service-entrance conductors, the grounding electrode conductor size shall be determined by the equivalent size of the largest service-entrance conductor required for the load to be served.
*See installation restrictions in Section 250-92(a).
See Section 250-23(b).

Fig. 5-11. *Code* specifies size of grounding electrode used for system ground.   (National Fire Protection Assoc.)

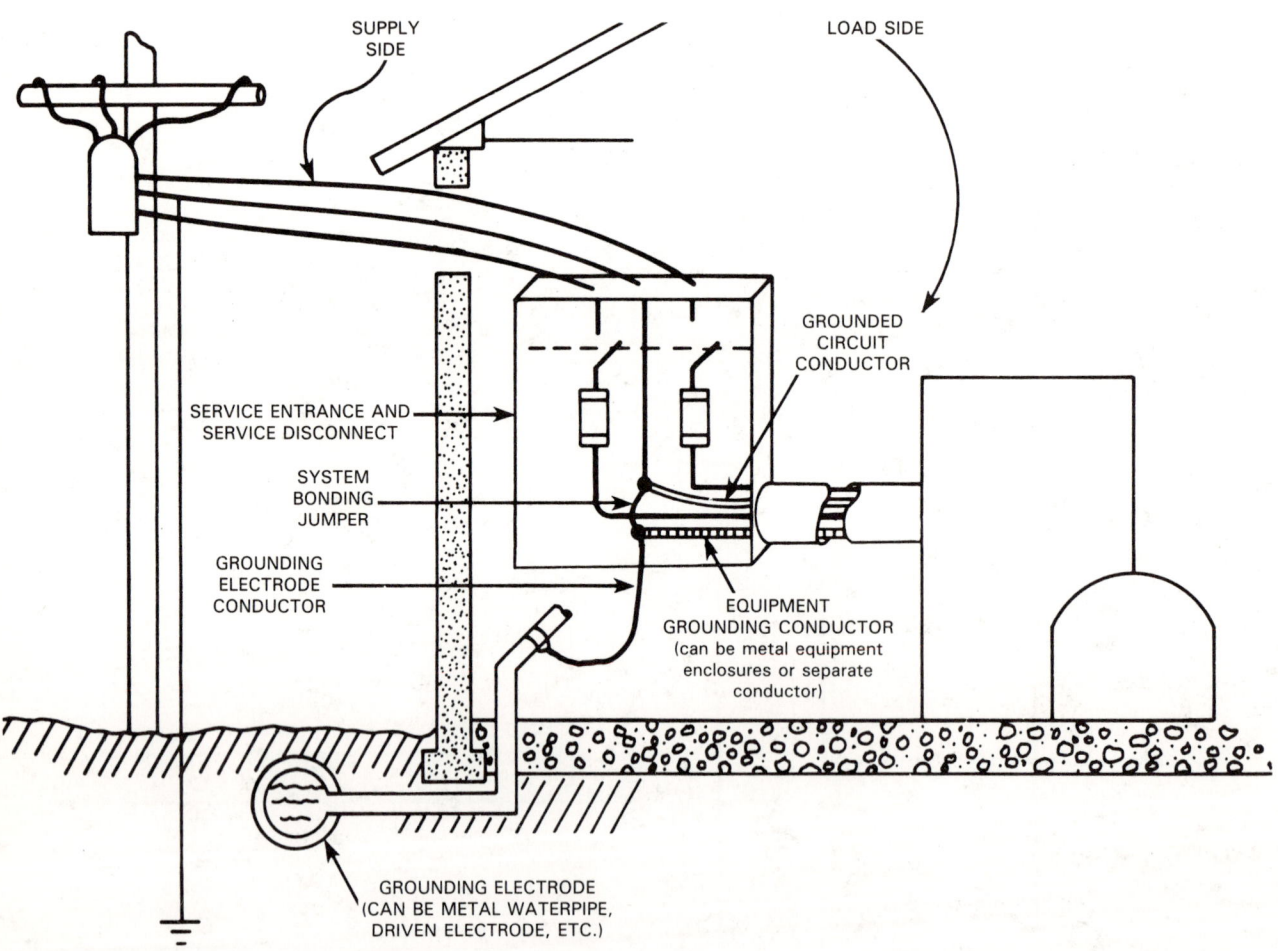

Fig. 5-12. Grounding connections for system grounding using a water pipe as grounding electrode.
(U.S. Labor Dept.)

Safety and Grounding Essentials

mon practice to use both grounding methods together. It is necessary that the two grounds be interconnected or bonded.

BONDING

Bonding means joining all metal parts of the wiring system—boxes, cabinets, enclosures, and conduit. It ensures having good, continuous metallic connections throughout the grounding system. Refer to Article 100 of the *National Electrical Code.* Bonding is required at:
1. All conduit connections in electrical service equipment.
2. Points where a nonconducting substance is used that might impair continuity. That is, where current may not be able to flow past the nonconductor. Such bonding is done by connecting a jumper around the nonconducting substance.
3. All service equipment enclosures whether inside or outside the building.

If a grounding conductor is properly bonded to the metal service equipment, a fault from line to ground (from an ungrounded [black] conductor to grounded equipment such as enclosures, raceways, cables, boxes, etc.,) on the load side of an overcurrent device will follow a low-resistance path to the grounded conductor at the transformer. This would allow the circuit protection device (fuse or circuit breaker) to open rapidly.

Circuits having conductors enclosed in nonmetallic sheaths or plastic conduit must have an additional grounding wire. This wire must be interconnected between outlets. Fig. 5-13 indicates the size of the grounding wire, 15 to 400 A. This wire assures a continuous equipment ground.

When metal conduit is used, the conduit itself is the grounding means *if the conduit provides an uninterrupted path back to the service entrance.* If it does not, an additional grounding conductor must be run in the conduit, Fig. 5-14.

SIZING GROUNDING CONDUCTORS

RATED AMPERAGE OR AMPERAGE SETTING OF AUTOMATIC CURRENT DEVICES PLACED AHEAD OF EQUIPMENT, CONDUIT, ETC.	AWG SIZE	
	COPPER	ALUMINUM AND COPPER-CLAD ALUMINUM
NOT EXCEEDING 15 A	14	12
NOT EXCEEDING 20 A	12	10
NOT EXCEEDING 30 A	10	8
NOT EXCEEDING 40 A	10	8
NOT EXCEEDING 60 A	10	8
NOT EXCEEDING 100 A	8	6
NOT EXCEEDING 200 A	6	4
NOT EXCEEDING 300 A	4	2
NOT EXCEEDING 400 A	3	1

Fig. 5-13. Sizes of grounding conductors to use with various sizes of overcurrent devices. Refer to NEC Table 250-95 for additional details.

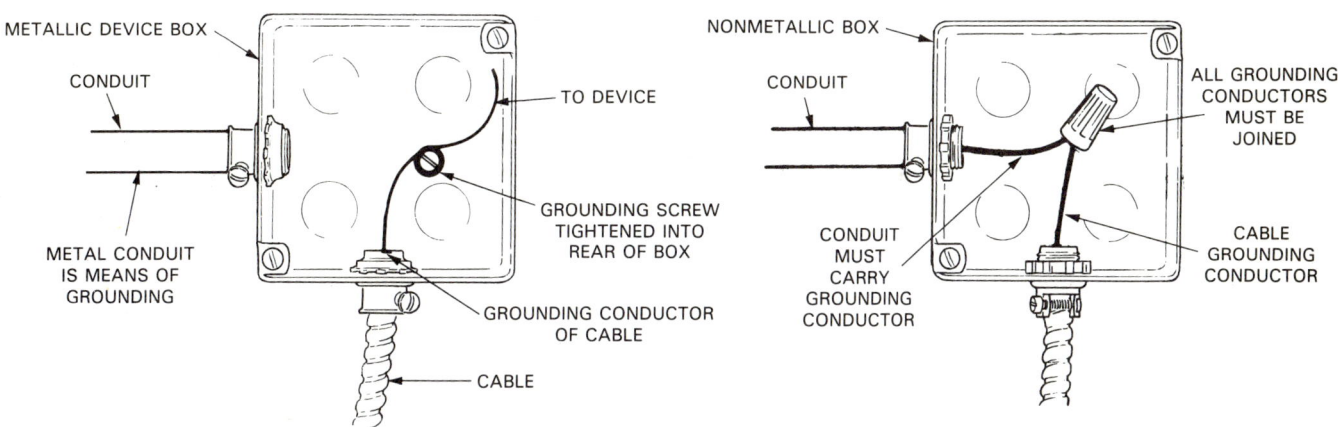

Fig. 5-14. The proper method of bonding electrical devices to the electrical box and, in turn, the electrical system.

HOW THE SYSTEM GROUND WORKS

In theory, with the neutral wire grounded at the supplier's pole (transformer) or at the building, no harm would come from touching the neutral wire. Any current present would be running to the earth. It would be the same as touching the earth.

Since every circuit is connected to the main service entrance, the circuit and all the loads on it are also grounded.

Article 250 of the *National Electrical Code* deals with grounding details and should be understood by anyone intending to do electrical work. In fact, a thorough understanding of both system grounding and equipment grounding is essential to completing an electrical installation which is safe to both persons and property. This is, in fact, the fundamental basis for the creation of the *National Electrical Code*.

EQUIPMENT GROUNDING

Equipment grounding is the method which bonds the grounding conductor to equipment enclosures (motor frames devices) and metallic noncurrent carrying equipment, Fig. 5-15. A permanent and unbroken path is formed back to the service neutral and the grounding electrode. This path is thus continuous with the grounded, neutral conductor both on the supply and load side of the service. If a system does not have a grounded conductor or neutral, one of the line or phase conductors may be intentionally grounded. This is common when there is a three-phase, three-wire delta system or a two-wire, 120 V, single phase system. One conductor of the system is grounded by connecting it to the grounding electrode.

This ground protects the electrical equipment from damage should a live conductor make contact with the equipment housing or frame. This is known as a GROUND FAULT. It occurs when conductor insulation fails or when a wire comes loose from its terminal point. Not only is this damaging to the electrical equipment but it exposes the equipment user to danger of serious shock. For this reason all metal electrical enclosures, frames, wireways, and conduits must have good ground continuity.

The relationship between the system ground and the equipment ground is shown in Figs. 5-16 and 5-17. Because of the misunderstanding of grounding theory, the subtle but important differences are often overlooked even by experienced journeymen. It may well be the most common *Code* violation.

GROUND FAULT CIRCUIT INTERRUPTERS (GFCI)

The importance of a good grounding system in electrical wiring cannot be overemphasized. However, in certain situations good grounding is not enough. Too many individuals have lost their lives from improperly maintained or imperfect grounding of electrical equipment. Such a situation exists whenever defective, worn, or misused equipment is handled by the operator, especially in damp or wet areas.

Although properly rated fuses and/or circuit breakers normally provide adequate protection of equipment against overloading, faults, and short-circuits, they do not entirely protect people using such equipment. For this reason, the NEC now requires that GFCI protected circuits be used at the following locations of a residence:

1. All 125 V, single-phase, 15 and 20 A receptacles in bathrooms and outdoors where there is direct grade access. This requirement also applies to bathrooms in motel guest rooms.

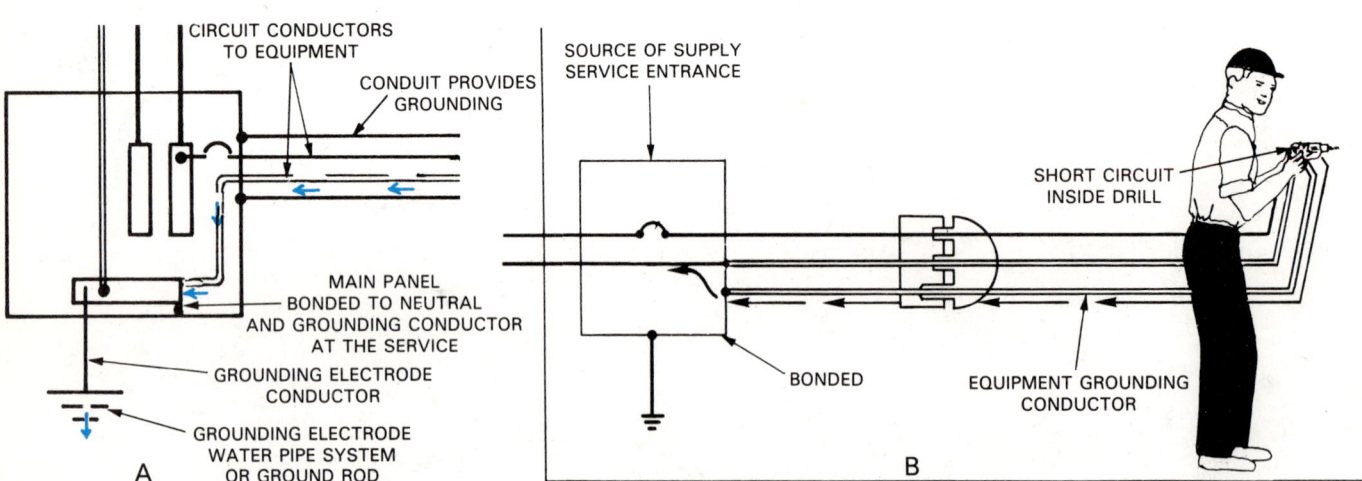

Fig. 5-15. Schematic examples of how an equipment ground works. A—Path current will take during a ground fault when there is a proper equipment ground. B—How equipment grounding protects tool user. Current follows grounding conductor back to ground. Otherwise current could pass through worker's body to ground. (OSHA)

Safety and Grounding Essentials

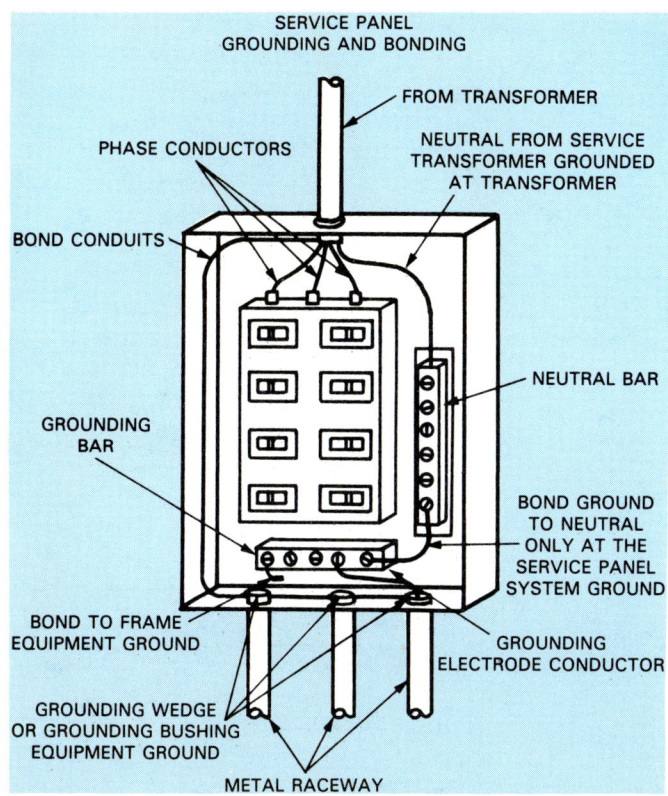

Fig. 5-16. Service panel sketch shows methods of bonding and grounding. Note differences in equipment and system grounding.

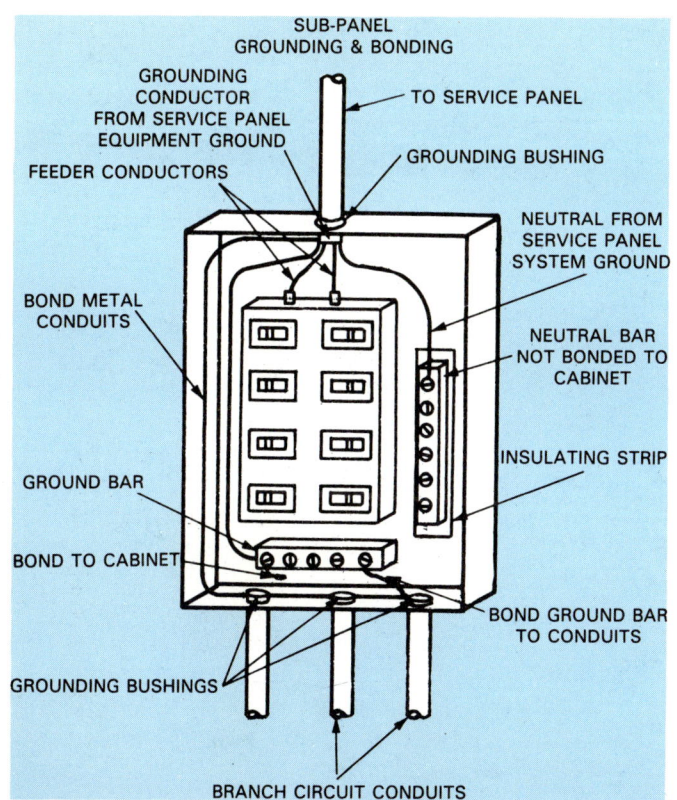

Fig. 5-17. Subpanel grounding and bonding. Relationship of system grounding and equipment grounding is shown. Refer to callouts.

2. All 125 V, single-phase, 15 and 20 A receptacles in garages unless:
   a. Not readily accessible.
   b. Dedicated for a cord-and-plug-connected fixed appliance.
3. All 120 V, single-phase, 15 and 20 A receptacles which are not part of permanent wiring (such as those used for temporary power during construction).

These requirements may be met by installing GFCI breakers or receptacles on the circuits. See 5-18.

Fig. 5-18. Ground-fault circuit interrupters are good for extra protection. They are available both as breakers and receptacles. The device shown here is a GFCI receptacle.

HOW THEY WORK

GFCIs open the circuit if a current leakage or fault (to ground) exceeds 0.006 A, (6 mA). Note, a person can be electrocuted by a current of only 2 mA.

Normally, the current going to the appliance or device along the black ("hot") wire and the current returning to the source along the white or grounded neutral conductor are equal. See Fig. 5-19.

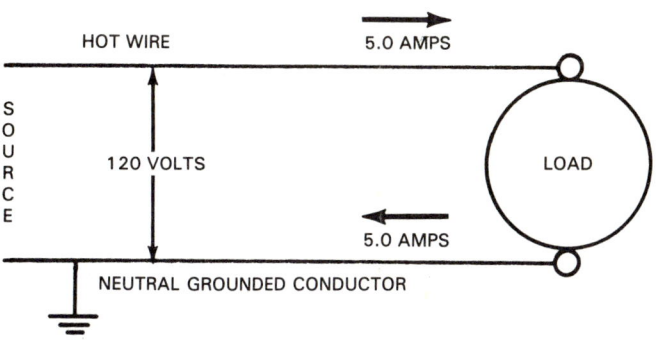

Fig. 5-19. As shown in the schematic, a balance normally exists between the current to and from the load.

Modern Residential Wiring

During a ground fault or short circuit, however, an imbalance occurs and the difference represents a dangerous shock hazard. See Fig. 5-20.

A GFCI, monitoring the circuit, will sense *any* imbalance greater than 6 mA and will immediately open the circuit, stopping the dangerous current flow. Fig. 5-21 shows schematic of a GFCI (breaker) installed. Any fault occurring in this circuit will activate the GFCI and it will immediately open, or deenergize, it.

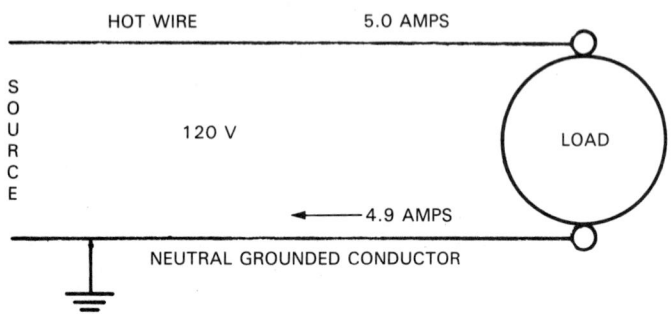

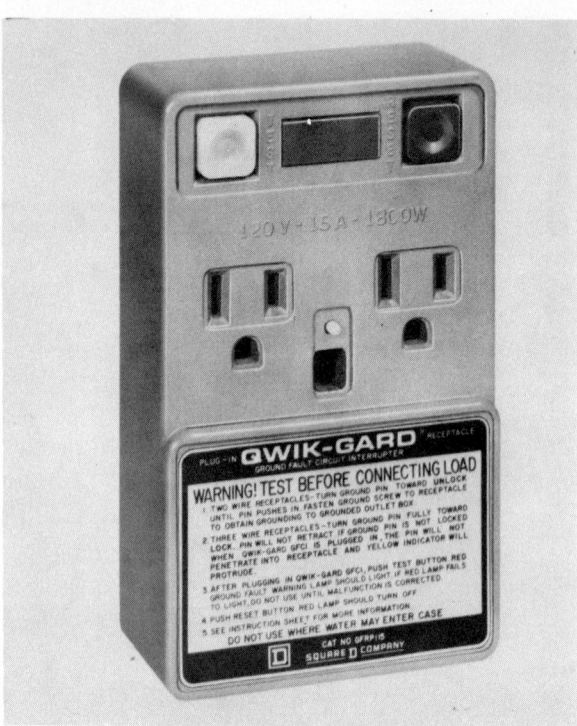

Fig. 5-20. When an imbalance occurs between the hot and the grounded conductor, a serious danger, called a fault or current leak, is present. Contact with such a fault could be fatal.

A portable ground fault circuit interrupter. Unit is plugged into either a two-wire or three-wire receptacle. (Square D. Co.)

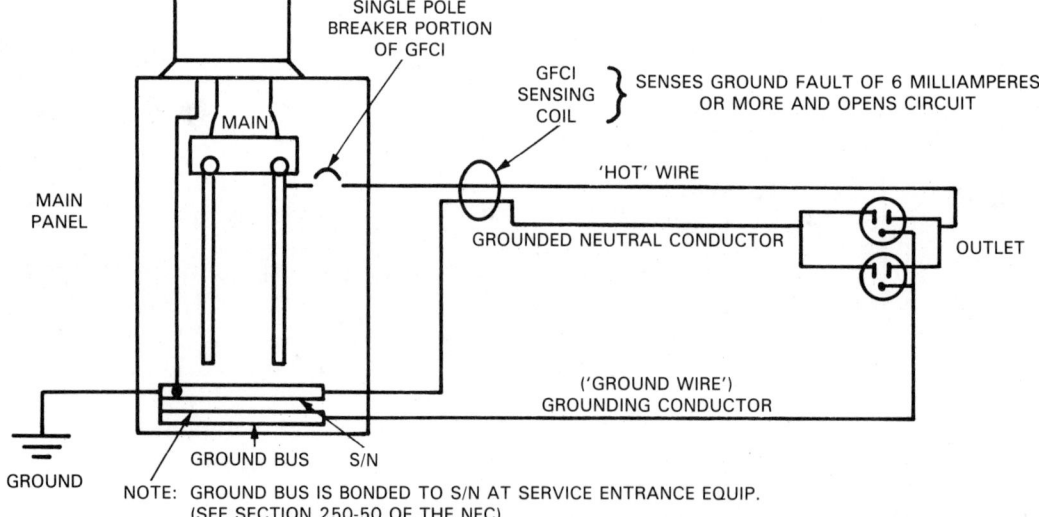

Fig. 5-21. Schematic of a breaker type GFCI. It protects the entire circuit by "tripping" should it sense current leak of 0.006 A (6 mA).

REVIEW QUESTIONS — CHAPTER 5

1. List eight safety rules an electrician ought to observe when working with electrical equipment and tools.
2. An aluminum ladder should not be used when working around electrical wiring. True or False?
3. Grounding of temporary wiring must be in compliance with Article _305_ of the *National Electrical Code*.
4. The amount of current passing through the body is an important factor to remember in electrical shock. A current approaching _.010_ is painful and possibly severe effects will be experienced.
5. To aid a victim of electrical shock list which of the following you would NOT do and explain why:
   a. Quickly grasp the victim and drag him/her

away from the current carrying material.
b. Shut off the power immediately.
c. Do not move the victim but immediately give artificial respiration until medical help arrives.
d. If shutoff switch is too far away, use a wooden stick to remove victim from current carrying equipment.
e. When there is no danger of shock, check if the patient is breathing. If not, give mouth-to-mouth resuscitation until medical help arrives.
f. Have someone call a doctor or paramedics.
6. A GFCI breaker or receptacle will open a circuit where there is a current leakage to ground fault of:
a. .06 A
(b.) .006 A
c. .6 A
d. .2 A
e. Any of the above.
7. List three locations where GFCIs are required.
8. Explain what is meant by:
a. Ground fault.
b. Short circuit.

mil = .001

9. _bonding_ refers to the connection of all parts of a wiring installation to ground or to other systems that are well grounded.

Match up the terms in column A with the definitions in column B:

10. _f_ Grounding electrode conductor.
11. _d_ Grounding.
12. _c_ Ungrounded conductor.
13. _a_ Grounding electrodes.
14. _e_ Grounded conductor.

a. Driven ground rods, water pipes or metal framework of structure.
b. Senses current leakage or ground fault.
c. "Hot" wire.
d. Connection of electrical system with the earth.
e. Neutral or white wire of an electrical system.
f. Conductor which connects electrodes to neutral conductor.

15. What is the minimum size grounding conductor permitted for grounding raceways of equipment of a 60 A circuit?

Chapter 6

WIRING SYSTEMS

Conducting material for carrying electricity is commonly called wiring. A wiring system includes the wire, its insulating cover, its protective cover, and connectors which fasten it to a junction or other type of box. See Fig. 6-1.

In some systems the protective covering is separate from the conducting wire and the two are assembled by the electrician on the job. In other cases the protective covering is attached to the conductors during manufacture. You will study and learn to recognize both types of wiring systems in this chapter.

For homes and outbuildings such as garages, sheds, or other structures, you may use combinations of several wiring systems. There are many from which to choose. This chapter will describe the various systems manufactured and where to use them.

Regardless of the system of wiring chosen or required, it is important to have a continuous ground throughout every part of the system and every circuit. Refer to Chapter 5 for a complete discussion of grounding requirements.

TYPES OF CONDUCTORS

As you know from earlier chapters, conductors are the wires which carry current from one point to another in a circuit. A common 120 V ac circuit will always have two, and sometimes, three or more wires in it. One, known as the hot wire, carries current. Another, called a neutral wire, completes the circuit. When used, a third wire is called the ground wire, as you learned in Chapter 5.

Conductors are not only different in size but in the way they are insulated and/or protected. The wiring systems most used today are:
1. Rigid metal conduit.
2. Intermediate metal conduit (IMC).
3. Thin-walled conduit known as electrical metallic tubing (EMT).
4. Flexible metal conduit (Greenfield).
5. Liquid-tight flexible metal conduit (LTFMC).
6. Armored cable (BX or AC).

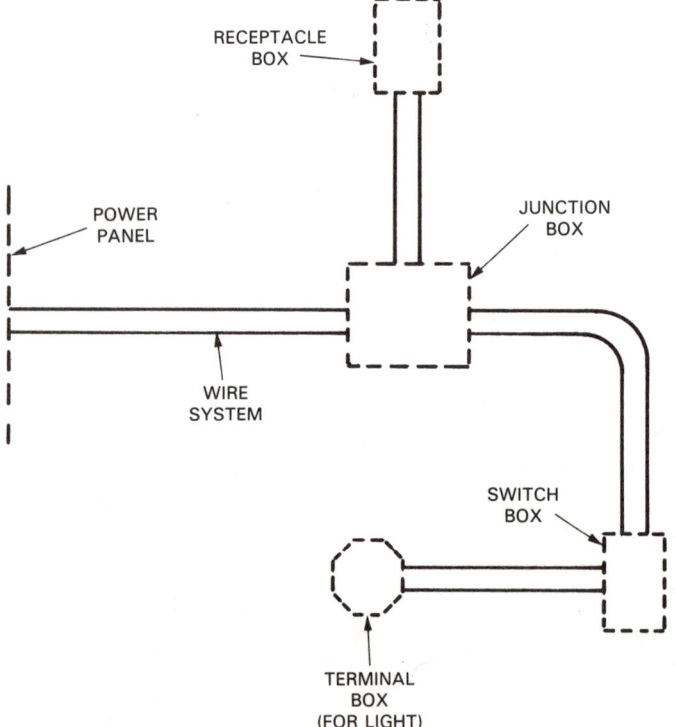

Fig. 6-1. Simple sketch of wiring system. It is the part of the circuit that carries current from source through boxes to load.

7. Nonmetallic sheathed cable (NM, NMC, UF and USE).
8. Nonmetallic conduit (PVC).

All these systems, Fig. 6-2, are equally adequate for most normal installations. Some cannot be used where there are unusual hazards such as extreme moisture, explosive gases, etc.

The system selected depends upon:
1. Type of dwelling (style).
2. Materials and type of construction used (log, concrete, post and beam, platform, etc.).
3. Surroundings of dwelling (hot, cold, wet, or dry).
4. Cost of the electrical materials.
5. Contractor's preference.
6. Local, county, state and NEC requirements
7. Preference of owner (client).

Wiring Systems

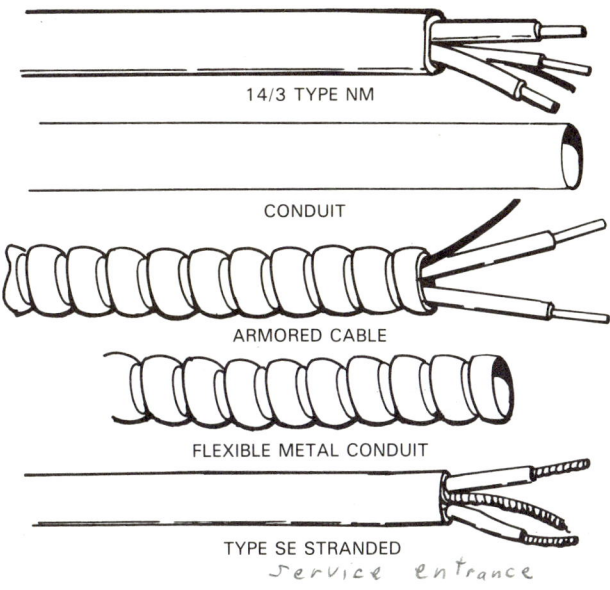

Fig. 6-2. Many wiring systems are available. Each is safe when used for purpose designed. National and local codes should always be consulted.

RIGID CONDUIT

Rigid conduit is galvanized and somewhat similar to water pipe. It is used for both indoor and outdoor applications. The pipe is smooth inside and out and must be threaded, reamed, and bent as necessary for proper installation, Fig. 6-3.

Rigid conduit may be cut with a hacksaw and threaded with dies similar to water pipe dies. It must be reamed to eliminate rough edges that could damage wires as they are pulled through.

To bend the conduit, we use a tool called a hickey or a conduit bender. Bending procedures for this and other types of conduit are found in Chapter 8.

Rigid conduit is available in 10 ft. lengths, and diameters of 1/2 in. to 6 in. It usually bears the Underwriters Laboratories Inc. label (UL listed) or other bonafide testing laboratory label.

Conduit is anchored to electrical outlet boxes, panels, etc., with a threaded locknut and bushing. All fittings should keep a continuous ground throughout the system. See Fig. 6-4. All connections must be tight.

In addition to being permitted indoors and out, rigid conduit may be buried directly in the earth or concrete. It can also be placed in cinder fill or beneath it unless there is a permanent moisture condition.

For more information on the use and restrictions of rigid metal conduit refer to Article 346 of the *National Electrical Code.*

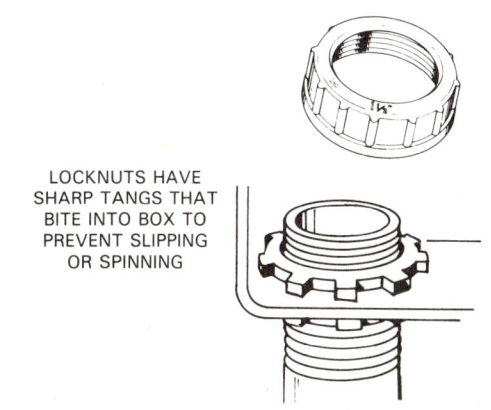

Fig. 6-4. Connecting rigid conduit requires the use of a locknut and bushing firmly tightened onto the pipe thread. (RACO Inc.)

INTERMEDIATE METAL CONDUIT

Intermediate conduit (IMC) is permitted for use in all atmospheric conditions and in all types of occupancies. Like its rigid counterpart, it is available in 10 ft. lengths and diameters from 1/2 in. to 6 in. Again, each length, or part of a length must be properly connected and joined to other lengths or enclosures using the right fittings. Both threaded and no-thread types are available. They are shown in Chapter 7, Boxes, Covers, and Fittings.

EMT OR THIN-WALL CONDUIT

The technical name for thin-wall is electrical metallic tubing or EMT. It is almost identical to rigid conduit or IMC except that the pipe walls are thinner. This makes it lighter and extremely difficult to thread. Therefore, EMT is not threaded but is connected length-to-length or to electrical boxes with suitable pressure couplings and connectors.

EMT is cut, bent and reamed the same as rigid conduit. EMT is manufactured in trade sizes 1/2 to 4 in. with 1/4 in. increments from 1/2 to 1 1/2 in. sizes. Increments of 1/2 in. are produced in 1 1/2 in. to 4 in. conduit. The 4 in. size is the maximum permitted.

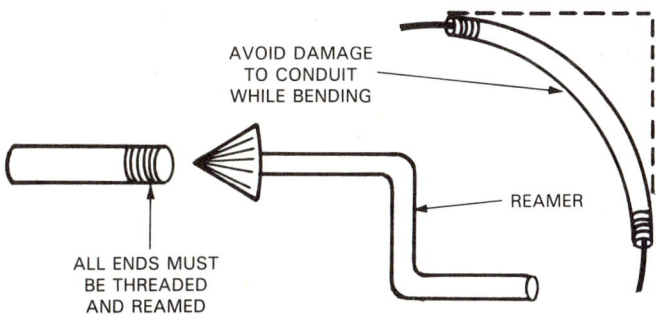

Fig. 6-3. Rigid conduit must be cut, threaded and reamed properly before it is installed. NEC specifies that conduit must not be injured inside or outside during bending or cutting.

Modern Residential Wiring

CONDUIT FILL

Everybody involved in doing electrical wiring must be familiar with the subject of conduit fill. It is a basic *Code* requirement that when three or more conductors are used in conduit, the fill for the most part, must not exceed 40 percent of a conduit's cross-sectional area. See Fig. 6-5. The percentage applies to both new and old work and to all conduit: rigid, IMC, EMT, and plastic. Refer to Table 1, Chapter 9 of the *National Electrical Code*.

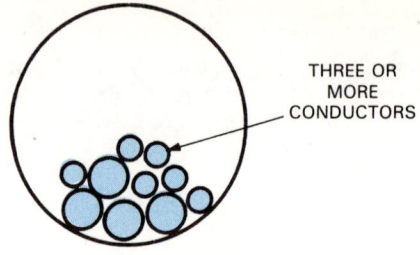

Fig. 6-6. *Code* allows only 40 percent of conduit's cross-sectional area to be filled. Check tables in Chapter 9 of *Code*.

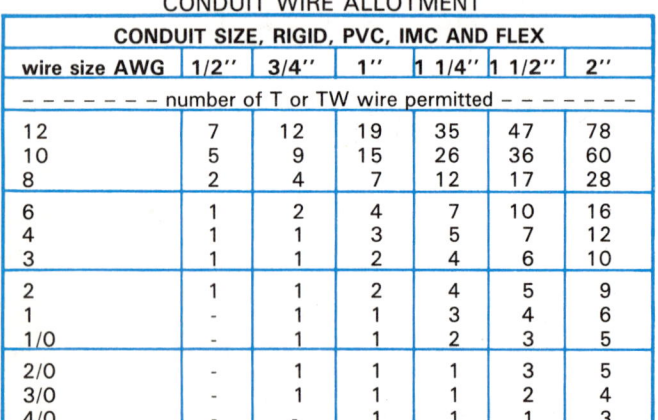

| CONDUIT WIRE ALLOTMENT |||||||
| CONDUIT SIZE, RIGID, PVC, IMC AND FLEX |||||||
wire size AWG	1/2"	3/4"	1"	1 1/4"	1 1/2"	2"
— — — — —	number of T or TW wire permitted — — — — —					
12	7	12	19	35	47	78
10	5	9	15	26	36	60
8	2	4	7	12	17	28
6	1	2	4	7	10	16
4	1	1	3	5	7	12
3	1	1	2	4	6	10
2	1	1	2	4	5	9
1	-	1	1	3	4	6
1/0	-	1	1	2	3	5
2/0	-	1	1	1	3	5
3/0	-	1	1	1	2	4
4/0	-	-	1	1	1	3

Fig. 6-5. Code sets the number of wires which can safely be installed in all types of conduit. Rule guards against damage to wires and to their insulation while they are being pulled through the conduit. Note that the chart refers to T and TW wires. Refer also to NEC Table 3A.

FINDING CONDUIT SIZE

Wires of the same size. When all conductors are the same size, finding the conduit size is easy. Tables 3A, 3B, and 3C in Chapter 9 of the *Code,* list the allowable fill for 1/2 to 6 in. conduit with conductors up to 750 MCM (thousand circular mils).

For example, if eight No. 12 (AWG) THHN conductors are to be run in a conduit. Table 3B shows that 1/2 in. conduit is permitted. Suppose that you wanted to use three No. 6 RHH conductors? Table 3C indicates that a 1 1/4 in. conduit would meet the requirement.

Wires of different size. When three or more conductors of different sizes are to be pulled into a conduit, the conduit size can be determined with the aid of *National Electrical Code* Tables 4-8, Chapter 9. The overall guide, as before, is the "40 percent fill rule."

As an example, suppose you needed to know the conduit size necessary for enclosing four No. 10 THHN and six No. 12 TW conductors, as shown in Fig. 6-6. You would follow these steps:

1. Determine the cross-sectional areas of the conductors from Table 5.
   No. 10 THHN = 0.0184 sq. in.
   No. 12 TW   = 0.0172 sq. in.
2. Multiply by the number of conductors of each size to find the total area.
   4 No. 10 THHN = 4 × 0.0184 = 0.0736
   6 No. 12 TW   = 6 × 0.0127 = 0.1032
                                   Total = 0.1768
3. Refer to Table 4. Under the column "over two conductors 40%" you will find that 0.21 sq. in. is the 40 percent fill of a 3/4 in. conduit. Since 0.0768 is less than that figure, you will select that size.

PULLING WIRES

To pull wires through conduit you will need an electrician's fish tape. See Chapter 4, Tools of the Trade.

Wires are only pulled through after all conduit is:
1. Connected to all outlet boxes.
2. Thoroughly tightened.
3. Checked for burrs which could damage the insulation on the wire.

FLEXIBLE METAL CONDUIT (GREENFIELD)

Actually, this type of conduit is the same as flexible armored cable described later. However, there are no wires in it. Also, it does not protect the conductors from weather conditions as well as the nonflexible conduits. It is only used indoors and in very dry locations.

Greenfield is routed and attached as if it were nonflexible conduit; but it has the advantage of bending without using tools such as conduit bender. There are special couplings, locknuts and bushings for attaching Greenfield. The wire is fished through as with other conduit.

One important requirement should be noted when using Greenfield. A grounding wire—green insulated or bare—must run from outlet box to outlet box and back to the service entrance.

Frequently Greenfield is used with EMT or rigid conduit. It is substituted where bends are needed. Instead of making complicated bends and angles in EMT or rigid conduit, a section of Greenfield can be inserted. There are special bushings, locknuts, and couplings to connect Greenfield to the other forms of conduit.

LIQUID-TIGHT FLEXIBLE METAL CONDUIT (LTFMC)

This type of conduit is similar to Greenfield, but is covered with a continuous plastic sheath, Fig. 6-7. This system of wiring is becoming more and more popular.

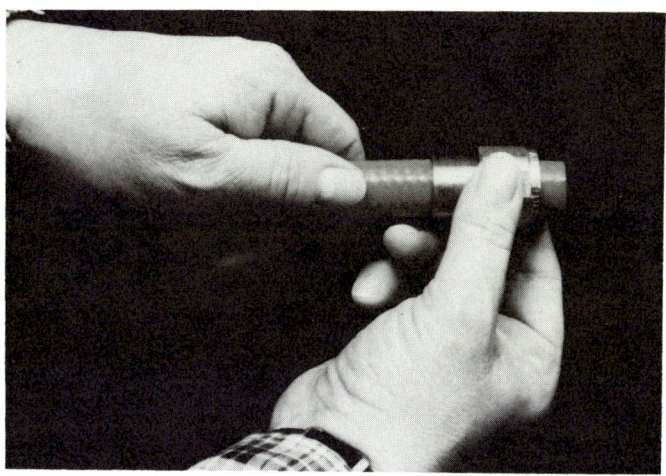

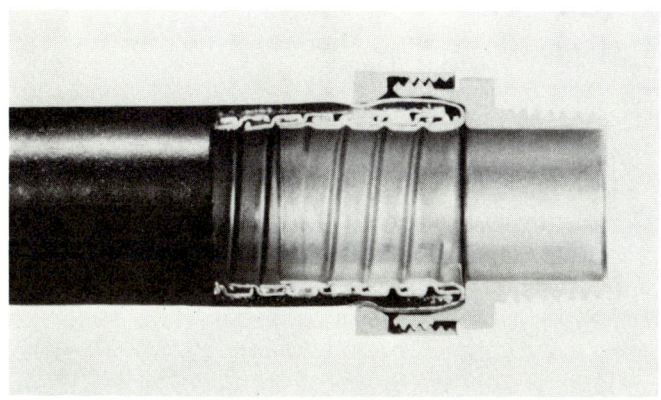

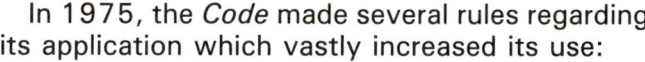

Fig. 6-7. Liquid-tight flexible metal conduit with section cut away to show interior. An inner flexible metal wall is surrounded by outer plastic coating. This makes conduit leakproof. (Appleton Electric Co.)

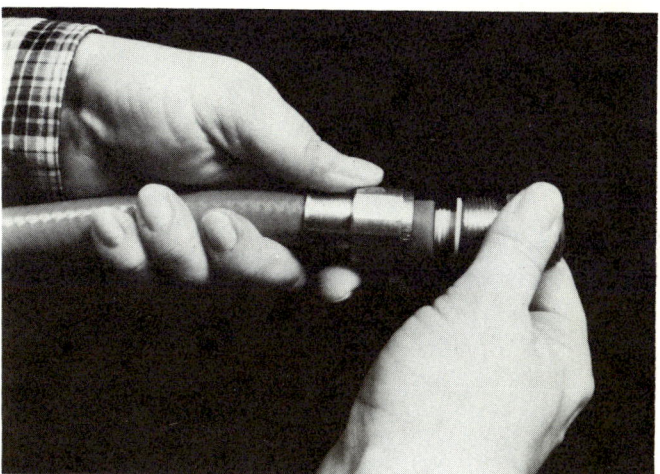

Fig. 6-8. Installing the LTFMC end fitting (connector) onto the conduit. Top. Slide compression nut with captive floating sleeve over conduit. Bottom. Use connector body to help thread nylon ferrule into inner wall.

In 1975, the *Code* made several rules regarding its application which vastly increased its use:
1. It can be used where hazardous conditions exist.
2. It can be concealed or left exposed, particularly where operation and maintenance of the system requires flexiblity and protection from liquid or gaseous fumes.
3. Sizes from 1/2 in. through 4 in. are permitted.
4. The number of conductors allowed in a single conduit must follow the percentage of fill outlined in Table 1, Chapter 9, of the *National Electrical Code.*
5. It must be secured every 4 1/2 ft. and within 12 in. of every box and fitting.

Although not used for general wiring, LTFMC has some distinct advantages. Because of it extreme flexibility, it can be used to connect machinery which is portable and/or which vibrates during normal operation.

Use of LTFMC requires use of some special fittings and connecting procedures. See Figs. 6-8 through Fig. 6-14.

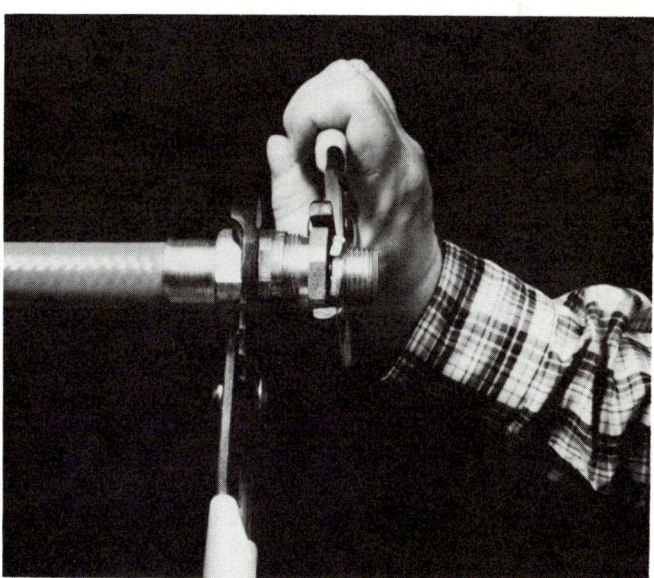

Fig. 6-9. Secure connector tightly to conduit. (Appleton Electric Co.)

55

Modern Residential Wiring

Fig. 6-10. Sealing gasket assures a liquid-tight connection. It seals out oil, dirt, dust, and chemicals. (Appleton Electric Co.)

Fig. 6-11. Insulated throat of this connector provides an excellent buffer against wire damage. This is especially helpful when electrical connections are made to vibrating units. (Appleton Electric Co.)

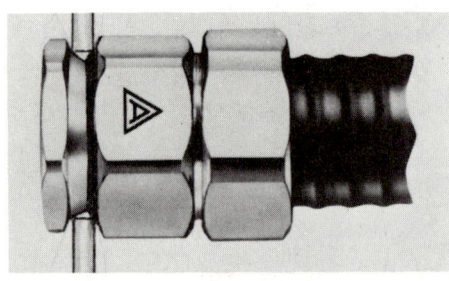

Fig. 6-12. The insulated throat connector from Fig. 6-13, as it appears when installation is complete.

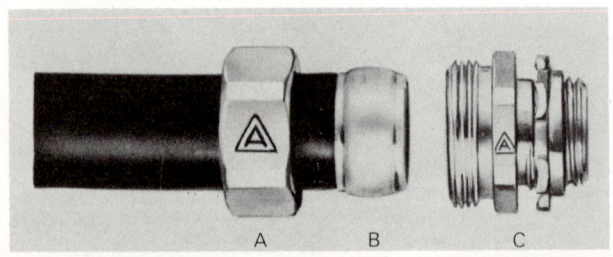

Fig. 6-13. After cutting conduit to length, install compression nut A first. Then thread ferrule B into the conduit. Finish preparing connector body C. (Appleton Electric Co.)

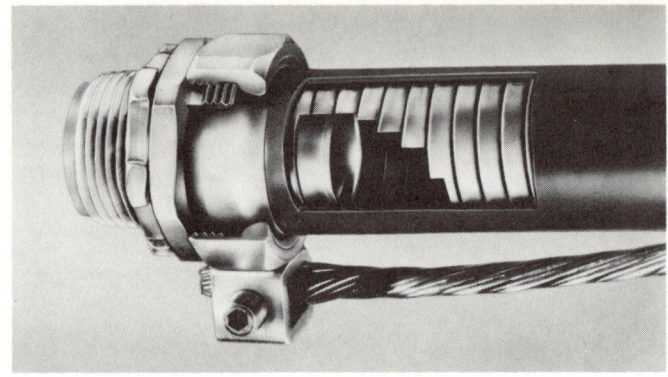

Fig. 6-14. Liquid-tight connector with grounding lug. Lug provides external grounding which can easily be checked and maintained.

ARMORED CABLE

Armored cable (BX) is like flexible conduit. Its outer covering is the same. The difference between them is that two, three or more type T or TW wires are already installed in BX. Each wire is wrapped with a strong paper to protect it.

In addition, a bare copper strip is found between this paper and the outer steel jacket. This strip provides a continuous ground when installed correctly.

Fig. 6-15, shows BX cable fitting installed on box. Fig. 6-16 illustrates the grounding strip firmly attached to the connector set screw, insuring a good ground continuity.

BX was once a popular method of electrical wiring. However, it must be used only in dry locations, since its cover is not weatherproof. For specific limitations check the *Code* and consult the local inspector.

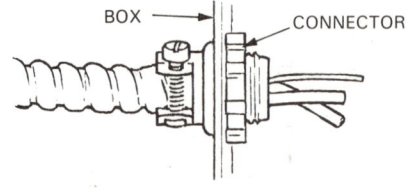

Fig. 6-15. Method of attaching BX or armored cable connectors.

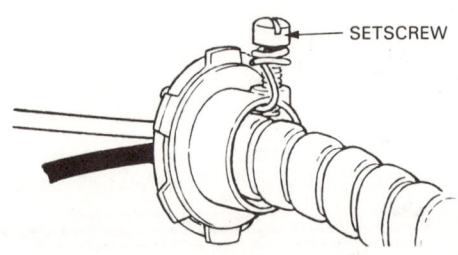

Fig. 6-16. When the setscrew is tightened down, cable will be bonded to BX and box.

Some additional uses of BX cable:
1. It can be fished or run through masonry block.
2. It can be used as underplaster extensions.
3. It can be concealed behind walls.
4. It can be used as exposed, surface extensions.

Armored cable is expressly forbidden for use in:
1. Wet or damp locations.
2. Hazardous locations.
3. Areas where there are corrosive agents, fumes, or vapors.
4. Commercial garages.

NONMETALLIC SHEATHED CABLE

Probably the most widely used conductor system employed today in residences is nonmetallic sheathed cable. This type of cable has two, three or more insulated conductors wrapped in a strong plastic or braided outer sheath. Often included is a bare copper ground wire, Fig. 6-17.

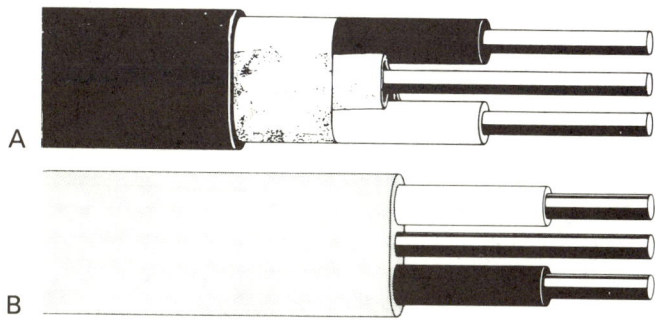

Fig. 6-17. A—Type NM cable with grounding conductor is often called Romex. It is very similar to NMC and is a very popular method of wiring residences. B—Type UF cable is used indoors or out. It is moisture and corrosion resistant and is suitable for direct burial underground.

TYPES

Type NM cable may be used only in dry locations. If it becomes damp or wet the outer plastic braid or jute may be destroyed.

Type NMC is used in either dry or wet locations. Its tough outer plastic cover is waterproof. However, type NMC may not be in direct contact with the ground.

Type UF nonmetallic cable can be buried directly below grade. For this reason, it is more properly classified as underground feeder cable. In addition, it can be used for any purpose suitable for NM or NMC. Thus, it is suitable as a branch circuit cable.

Type USE (underground service entrance) cable is used to connect the house service with the power company's pole. It may be used indoors, out-doors, above or below ground. It is extremely tough and durable and, therefore, the most versatile of the wiring cables.

RIGID NONMETALLIC CONDUIT

Nonmetallic conduit, at first glance, looks like metallic conduit. However, looks are about the only similarity. PVC (polyvinyl chloride), as it is commonly called, is corrosion-proof and may be placed underground as well as above grade, indoors and out, Fig. 6-18. It weighs much less than metal pipe. It is designed to be used everywhere that other types of conduit are used. It is only restricted in certain hazardous locations and where it could be physically damaged. Check to see if local code permits its use. Do not use it to support fixtures.

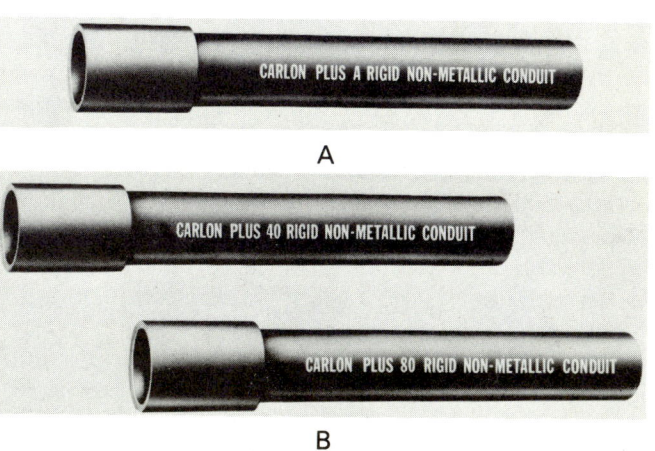

Fig. 6-18. Various types of rigid nonmetallic conduit. A—Type A is used primarily for underground installations. B—Schedule 40 PVC and Schedule 80 PVC can be used either above ground or below ground. (Carlon, an Indian Head Co.)

PVC is easily worked. It can be bent using special bending boxes, which consist of heating elements, to warm the conduit. See Fig. 6-19. This allows it to be easily bent by hand, using a simple bending "jig." In addition, preformed bends may be obtained.

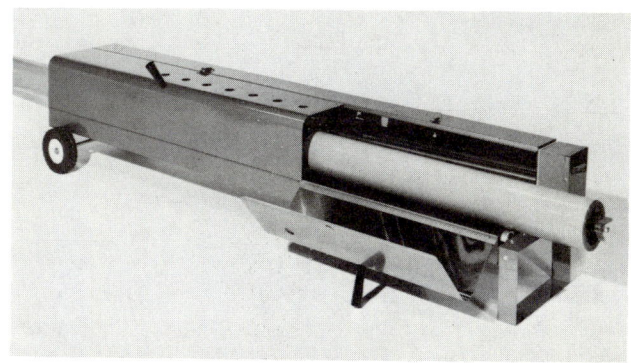

Fig. 6-19. A "hot box" is used to heat nonmetallic conduit to make it more pliable and ready to bend. (Carlon, an Indian Head Co.)

CAPACITY

It can hold the same amount of wires as metallic conduit. Refer to NEC Tables 3A and 3B, Chapter 9 of the *Code*.

To install PVC, all that is needed, besides the bending apparatus, are the proper connectors and fittings along with the solvent-cement used to join the fittings to the pipe itself.

SPECIAL NOTES

You must run grounding wire along with the conductors and bond these to the devices, metal boxes, etc., all the way back to the main panel grounding bus. Also, PVC must be supported more frequently that metallic conduit due to its "flexibility," Refer to Table 347-8 of the *National Electrical Code*.

PVC, has some excellent advantages due to its corrosion resistant properties and can be used with nonmetallic boxes and fixtures, creating a virtually moisture proof electrical system, a real plus in damp areas like farm buildings. However, PVC is a toxic chemical when heated and this factor, may, in time, limit its use in certain locations.

More information on the use of PVC conduit, can be obtained through your local power company, electrical inspector and code enforcement agency. Article 347, of the *National Electrical Code,* provides more insight into its use and should be reviewed prior to installing nonmetallic conduit.

REVIEW QUESTIONS — CHAPTER 6

1. Name five key factors which determine the wiring system used for a particular installation.
2. When wiring with rigid metal conduit a ground wire is required. True or False?
3. Rigid conduit is cut, threaded and bent to execute a proper and neat installation. It is also important that it be _____ to remove any sharp burrs which could damage the _____ during wire installation.
4. Explain how rigid conduit is anchored to electrical boxes.
5. Intermediate metal conduit is commonly sold in the following lengths.
   3 ft.   6 ft.   8 ft.   10 ft.   12 ft.
6. The number of wires that a conduit can carry is governed by (select most correct answer):
   a. Number that can be easily pulled through the conduit.
   b. Never more than four (two hot wires, neutral wire, ground wire).
   c. *National Electrical Code* sets limits for size of conduit.
7. Flexible metal conduit is also called _____, and is used only in very dry locations.
8. LTFMC stands for _____.
9. Another name for BX is _____.
10. List four places where BX cannot be used.
11. Using Fig. 6-7, calculate the number of No. 12 TW wires permitted in 1/2 in. conduit.
12. Which article of the NEC outlines rules regarding rigid metal conduit?
13. The following are types of nonmetallic sheathed cable. Indicate where each may be used:
    NM   NMC   UF   USE
14. Nonmetallic conduit, also known as _____, is corrosion proof and may be placed _____ as well as _____ grade.
15. Article _____ of the NEC provides more information on the use of nonmetallic conduit.

National Electrical Code® and NEC® are Registered Trademarks of the National Fire Protection Association, Inc., Quincy, MA.

Chapter 7

BOXES, FITTINGS, AND COVERS

Boxes and covers are necessary electrical wiring containers. They house and protect electrical conductors and electrical devices. Boxes are designed to make tight connections with conduit or cable assemblies and keep exposed metal conductors away from surrounding building materials that might be combustible. The box also protects people against accidental shock from exposed conducting metal.

The NEC requires that all joints, connections, and splices be housed inside approved enclosures. Fig. 7-1. Always refer to Article 370 of the *Code* for questions regarding boxes or fittings. Specific references to the *Code* will be made as discussion of boxes progresses.

BOX CONSTRUCTION

Boxes may be constructed of metal, plastic, or fiberglass. They must be strong enough to withstand the stresses put on them during installation and resist bending or twisting when holding fixtures or devices.

BOX SHAPES

There are four common box shapes in use today, Fig. 7-2:
1. Square.
2. Octagonal (eight-sided).
3. Rectangular.
4. Circular.

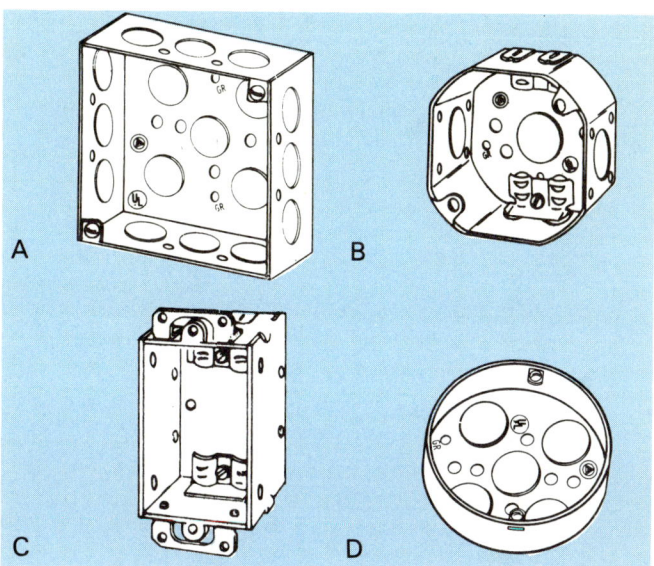

Fig. 7-2. There are four basic shapes for electrical boxes. A—Square, generally used as switch or junction box. B—Octagonal, generally used for ceiling boxes. C—Rectangular, preferred for wall receptacles and switches. D—Round, used only for fixtures such as ceiling lights. (Appleton Electric Co.)

Fig. 7-1. Boxes are required by *Code* to house points where wires are spliced or where they connect to fixtures or devices.

Each of these shapes is made in various widths, depths, and knockout arrangements. Metal boxes usually have a galvanized (zinc coating) finish. They can be used for either concealed or open work where there are no explosive or flammable vapors present. Special types are made weathertight for outdoor applications. Others are produced especially for hazardous locations.

TYPES AND USES

Some electricians separate types of boxes by usage:
1. Ceiling boxes intended for ceiling fixtures and usually mounted on or between beams or joists. Boxes may be square, octagonal or round.
2. Wall boxes for housing switches and receptacles. Usually rectangular, these boxes are generally mounted on studs in frame buildings or set into masonry walls. Special masonry boxes are shown in Fig. 7-3. Common boxes range in size from 1 1/2 to 3 1/2 in. deep, 2 in. wide, and 3 in. high.

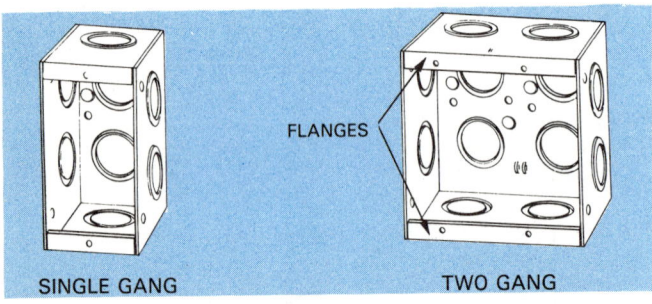

Fig. 7-3. Masonry boxes have special flanges across face for attaching devices.

3. Junction or pull boxes can be either of the foregoing types. These are installed wherever conductor splices must be made in a location not appropriate for a switch, receptacle, or fixture. In most cases, splices can be planned where an electrical device will also be installed. It is less expensive to install a larger box for the extra wiring than to install a separate box for the junction. Electrical boxes used strictly as junction boxes must be covered with a solid plate of the same material as the box. See Fig. 7-4.

PULL BOXES

Pull boxes are boxes installed with conduit whose sole purpose is to provide a place from which to pull wires. A fish tape is inserted at the box. Any type or shape of box is suitable as a pull box.

HANDY BOX

When surface mounting is necessary or required, a modified form of box is used. Modified boxes are seamless and have rounded corners to prevent injuries. Sometimes these modified boxes are referred to as "utility boxes" or "handy boxes." Refer to Fig. 7-5.

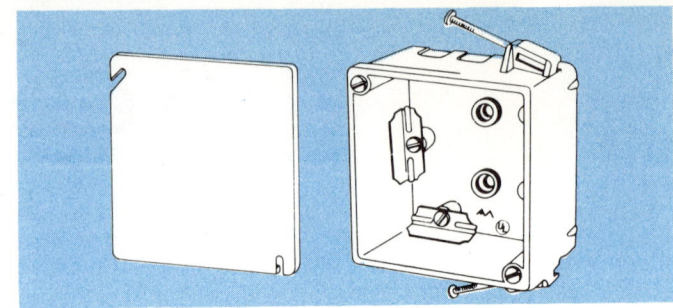

Fig. 7-4. When a box is used to house only a splice or splices it should be closed up with a solid cover of the same material. It must also be accessible.
(Allied Moulded Products Inc.)

KNOCKOUTS AND PRYOUTS

Boxes are designed with great flexibility. Wires can be brought in from any side. Some ceiling boxes also contain a knockout which can be removed to receive a lug for mount fixtures.

There are two basic styles of knockouts. One is made by scoring the material in such a way that only a thin layer holds the knockout in place. A sharp tap with a hammer will remove it or break it loose so it can be grasped and removed with a pliers. In the second type the knockout is cut all the way through except for one or two spots. These knockouts can be removed with a screwdriver and pliers. Some have a slot for inserting the blade of a screwdriver. A sharp twist will remove them. This type is called a pryout. See Fig. 7-6.

Knockouts may be removed only to provide an opening for a cable, conduit, or fitting. No other openings are permitted in electrical boxes. Snap in seals or special plates are available for covering unused knockout openings, Fig. 7-7.

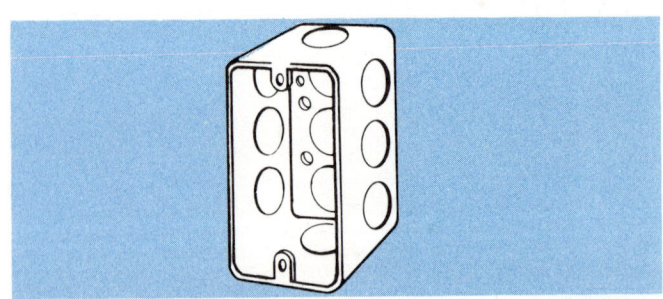

Fig. 7-5. Handy box, also known as utility box, is designed for surface mounting. Corners are rounded for better appearance and to avoid injury. (Appleton Electric Co.)

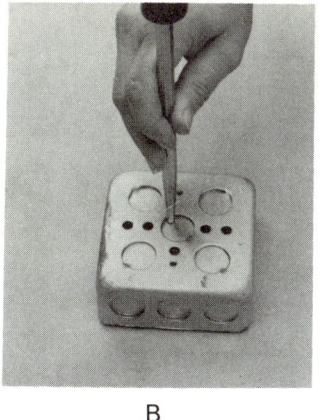

is not handy. Nonmetallic boxes are one piece and cannot be ganged. Larger sizes must be used when several switches or outlets are to be mounted at one location. Fig. 7-8B shows a multigang box design.

BOX MOUNTING SYSTEMS

Boxes must be securely fastened to a structural member of the wall, ceiling or floor of the dwelling. Wall boxes are normally mounted to studs with nails, brackets, or both. Ceiling boxes are fastened

Fig. 7-6. Removing knockout. A—Tapping with hammer and screwdriver. B—Prying out with twist of screwdriver blade inserted in slot.

Fig. 7-7. Knock out seal. These are installed in unused openings of boxes where knockouts have been removed. (Electroline Mfg. Co.)

GANGING BOXES

As shown in Fig. 7-8A, some switch boxes are designed to be ganged (joined together) to form double, triple, or more device accommodations. Boxes with removable sides are the only ones which can be used for this purpose.

To set up a ganged box, remove adjoining sides from the two boxes by removing the small screw holding the side plate. Join the boxes and fix them together by reinstalling the screw, Fig. 7-9. Double, triple or even larger boxes are available making ganging unnecessary. However, most codes allow ganging and it may be done when a large box

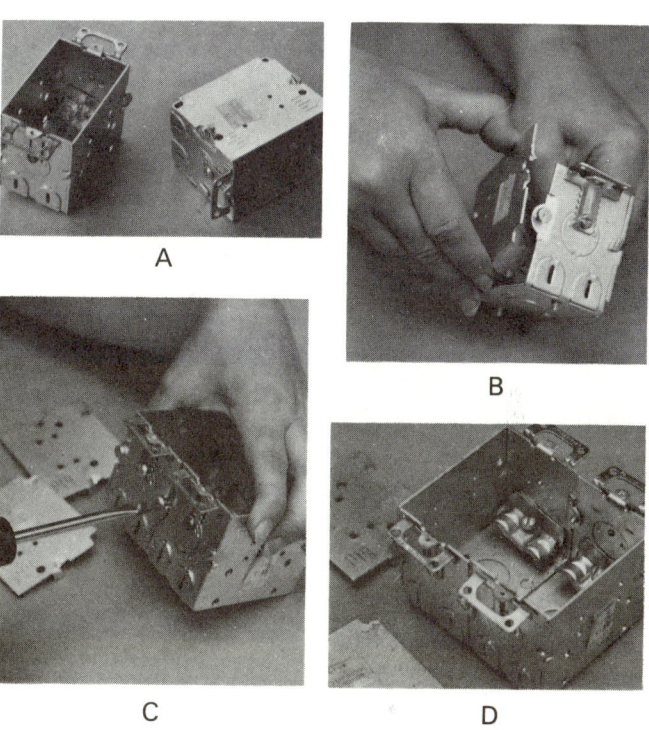

Fig. 7-9. Steps for ganging up two or more single metal electrical boxes. A—Two boxes to be joined. Note that in this type, screws and tangs hold sides. B—Remove one side of both boxes by backing off screw. C—Attach boxes and tighten screw over slotted edge at each end. D—Ganging of boxes is completed. Extra sides are discarded.

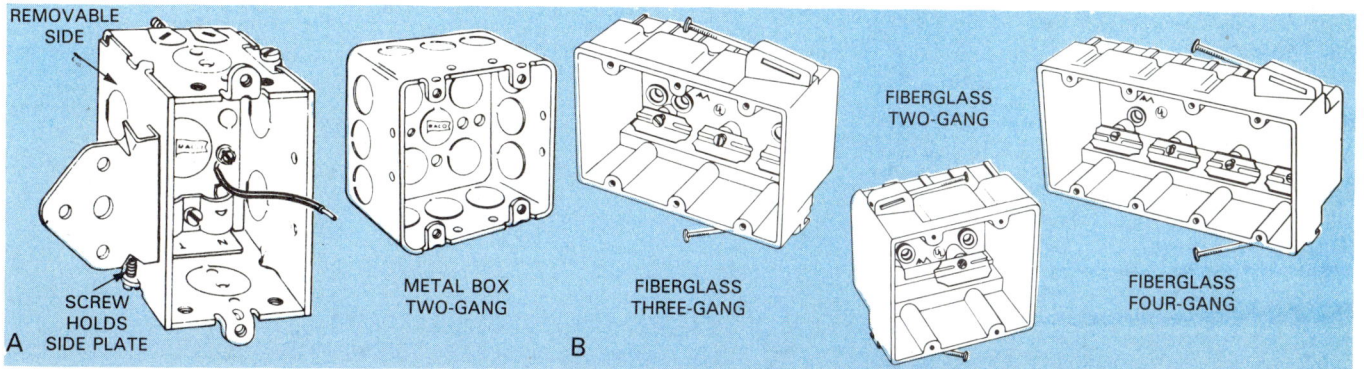

Fig. 7-8. A—Some boxes have removable sides so several boxes can be ganged (joined together). Remove screw and discard side. B—Multigang boxes are also manufactured. (RACO Inc., and Allied Moulded Products Inc.)

Modern Residential Wiring

directly to joists or to wood or metal brackets which bridge the joists. Some brackets have built-in fasteners which bite into the wood frame when struck with a hammer. Fig. 7-10 and Fig. 7-11

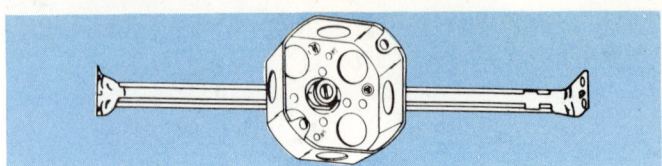

Fig. 7-11. Bar hanger with octagonal box attached. This type of bracket is used to mount boxes between ceiling joists. Similar type is shown in Fig. 7-12. (Appleton Electric Co.)

show different arrangements of brackets. Another bracket design for plastic boxes is shown in Fig. 7-12 below.

NONMETALLIC BOXES

Plastic, polyvinyl chloride (PVC), and fiberglass boxes are particularly popular. This type of box, shown in Fig. 7-12, is used only with nonmetallic cable and conduit.

Plastic and fiberglass boxes are rugged, lightweight and resist corrosion. Further, their brackets make it easy to position and mount them to wood or metal structural members. Since cable clamps are

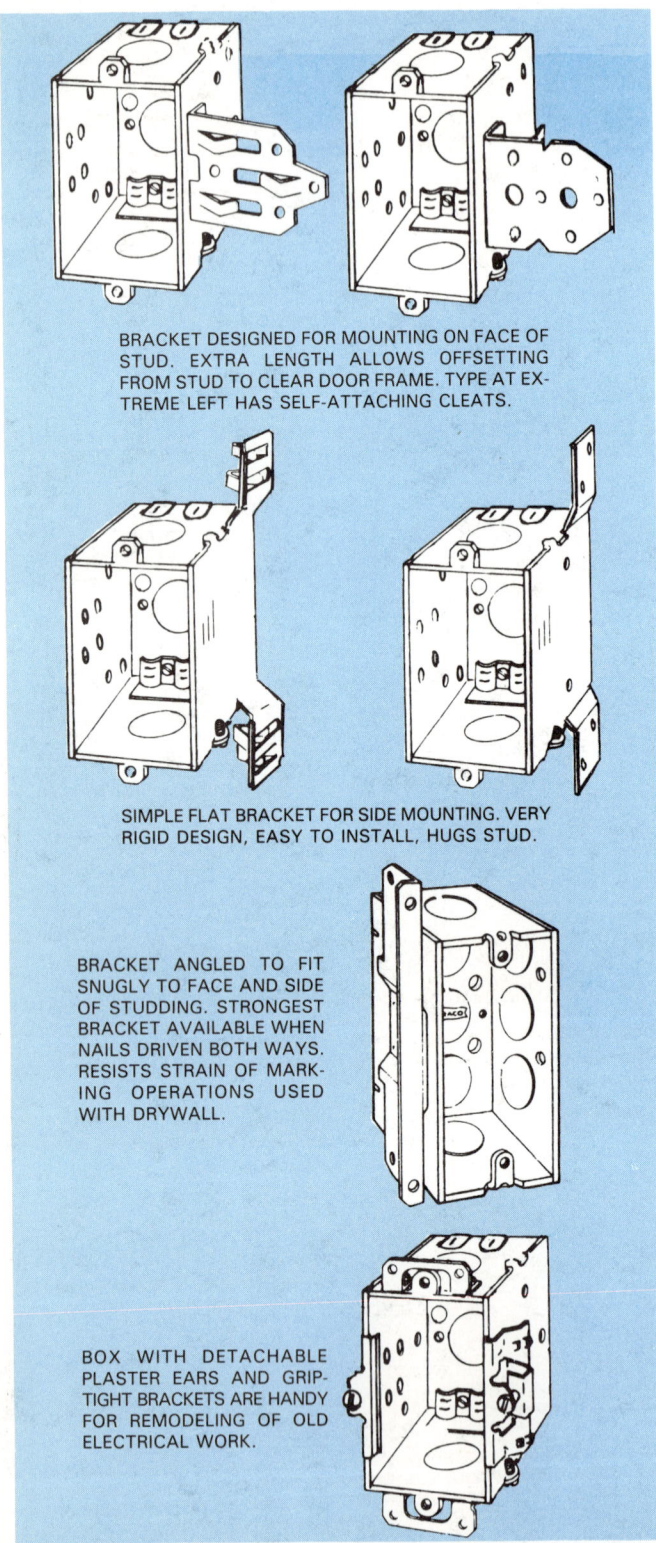

Fig. 7-10. Various brackets are designed to make boxes easy to attach to building frame. Some have their own fasteners. (RACO Inc.)

Fig. 7-12. Popularity of nonmetallic boxes is growing. A—Fiberglass box using captive nail attaching system. B—Knockout of fiberglass box is thin portion of box and is broken out with a sharp tap from a screwdriver. (Allied Moulded Products, Inc.)

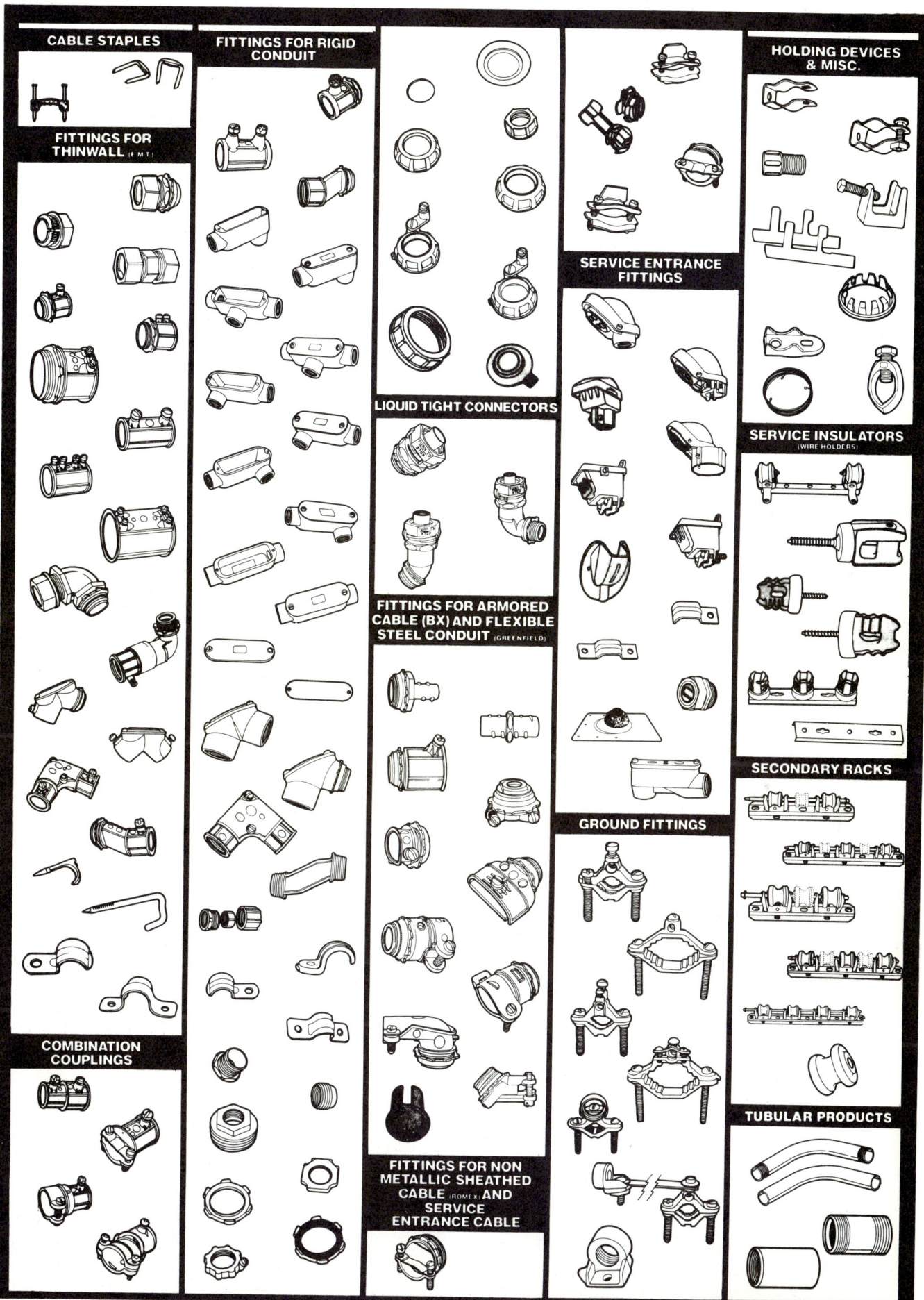

Fig. 7-13. Fittings for electrical wiring. Many are used with or on electrical boxes. (Electroline Mfg. Co.)

not required in the single gang box, an additional conductor can be accommodated. Perhaps the greatest advantage is that no grounding connection is needed to the box, itself, as with metal boxes.

FITTINGS

Fittings are parts of a wiring system which are designed to interconnect conduit, conductors, or boxes. They include accessories such as box extensions, clamps, ground clips, connectors, bushings, locknuts, nipples, couplings, conduit bodies, and holding devices. Many of these products are shown in Fig. 7-13. Refer also to Chapter 6 for information on conduit and cable connectors.

BOX EXTENSION RINGS

Extension rings are actually boxes without bottoms. They are used to:
1. Provide additional space when extra depth is needed such as to add more wires.
2. Bring a box flush to the wall or ceiling surface when the box has been mounted too deep.
3. Assist in remodeling job when the electrician wants to add surface wiring.

Flanges on the extension ring provide a method of fastening it to the box. The ring must be of the same size and shape as the box. See Fig. 7-14.

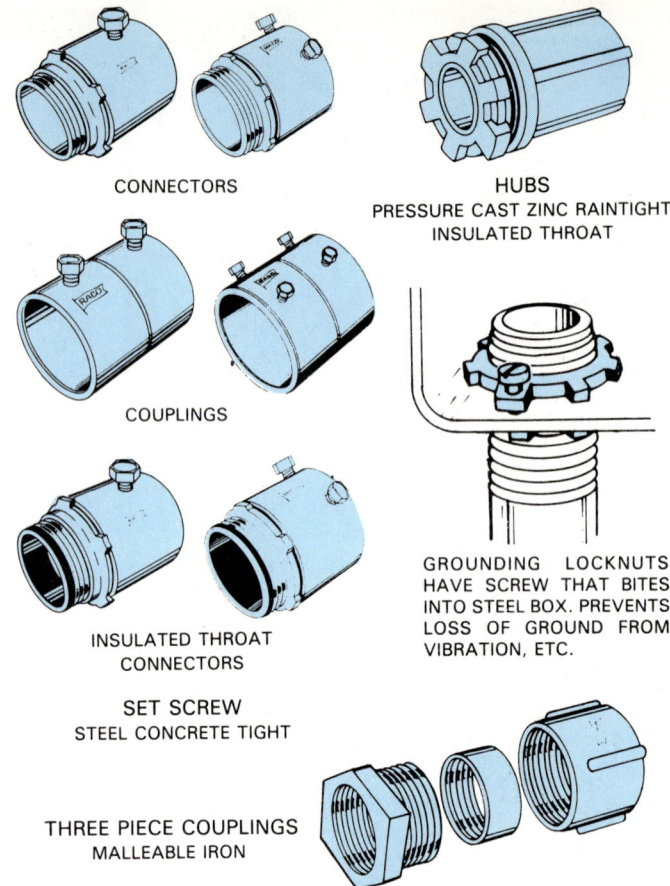

Fig. 7-15. Fittings for intermediate conduit (IMC) may be either threaded or no-thread type. (RACO Inc.)

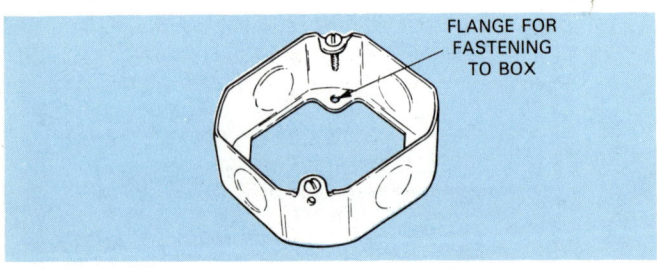

Fig. 7-14. Box extension ring adapts boxes for greater depth and capacity. It mounts to boxes and is fastened through flanges and tapped holes.

CONNECTORS AND CLAMPS

Connectors and clamps are fittings which secure conductors to an electrical box. A number of connectors are shown in Fig. 7-15 through Fig. 7-18. These fittings are available in straight, 45°, and 90° angles. Some have a screw which can be turned down to apply holding pressure on metal coverings or conduit. Others use a clamping device which will not damage the fabric of softer cable coverings. The clamping action prevents the cable conductors from being separated or pulled from the electrical box.

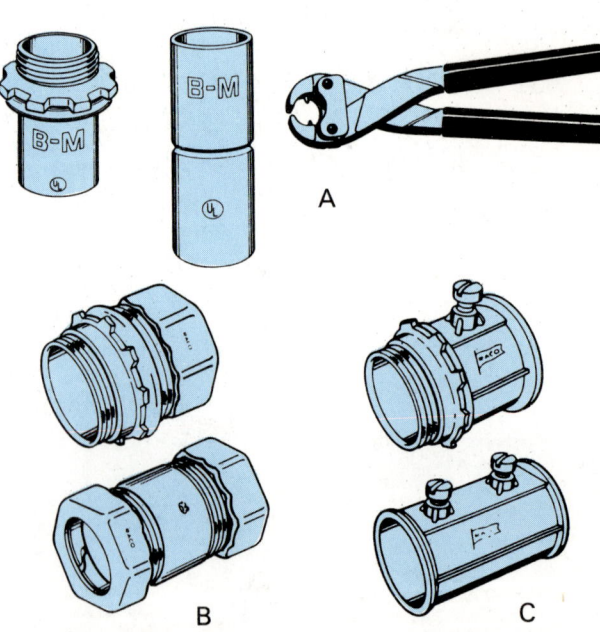

Fig. 7-16. EMT (thinwall conduit) may be extended or connected to boxes by various types of fittings. For one type a special tool called an indenter is needed. A—Indenter type fittings and special tool to install them. B—Compression type fittings tighten onto conduit as nut is turned on. C—Setscrew type fittings.

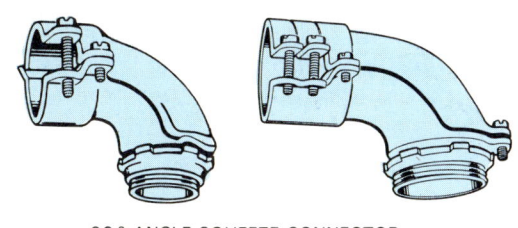

90° ANGLE SQUEEZE CONNECTOR

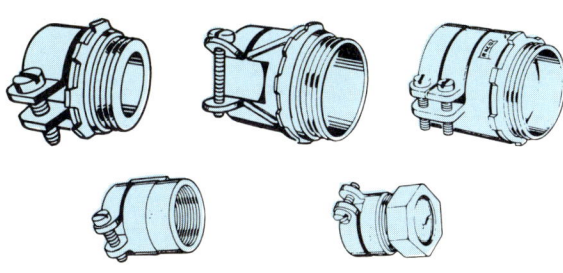

BOX CONNECTORS
SCREW-IN SQUEEZE TYPE

Fig. 7-17. Greenfield (flexible conduit) connectors. Two 90° types are shown at left. Note that insides are smooth to prevent damage to wires as they are pulled through fittings. (RACO Inc.)

Fig. 7-18. BX or armored cable connectors commonly use setscrews to assure a tight connection between fitting and cable covering. (RACO Inc.)

Box connectors used with metal clad cable are designed to provide a continuation of the ground between the box and the metal covering. Many boxes have built-in cable clamps. When such boxes are used, no other connector is needed. See Fig. 7-19 and Fig. 7-20. A clamp intended for nonmetallic cable is shown in Fig. 7-21. It is attached to the box through one of the knockouts and is fastened with a locknut. Note installation instructions in the illustration.

RIGID CONDUIT BODIES

Certain of the conduit fittings are necessary wherever conduit is being run on the job. Connectors and couplings are in this category. Another group of fittings called "bodies" is occasionally handy to have when the electrician is working in areas where space is restricted. The second column of Fig. 7-13 shows a number of different conduit bodies, each designed for a special purpose. The

Fig. 7-19. Some boxes have built-in clamps (arrow) to hold armored cable or nonmetallic cable. (GE Wiring Devices Dept.)

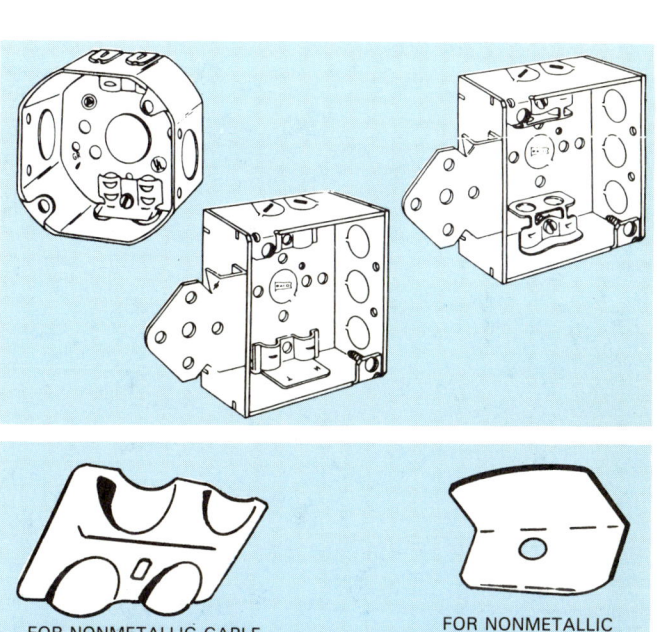

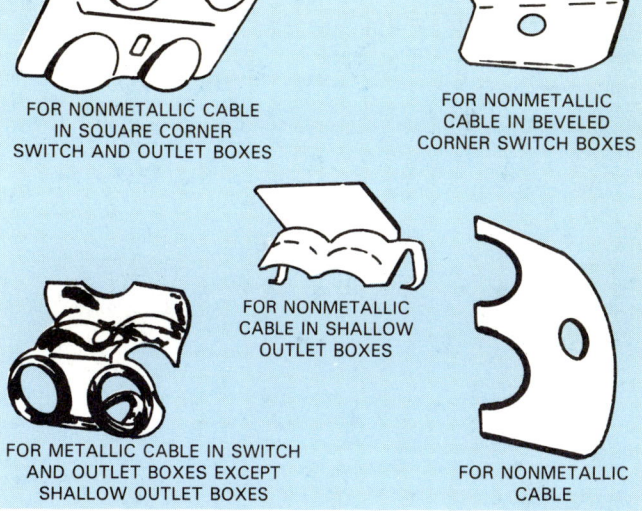

Fig. 7-20. Clamping devices are used to prevent strain on wire connections made to outlets or switches. Top. Boxes with clamping devices. (RACO Inc.) Bottom. Different clamp designs are suited to different types of cable. (Electroline Mfg. Co.)

oval conduit body shown in Fig. 7-22 is used in place of a 90° bend. Connections may be threaded or compression type.

GROUND CLIP OR SCREW

A ground clip or ground screw serves to attach a ground wire from a nonmetallic cable to the electrical box. The grounding is required only on metal boxes. The clip is a spring device that holds the wire in tight electrical contact with the box when it is hooked over the edge of the box. The ground screw threads into a drilled and tapped hole in the base or side of the box. Fig. 7-23 shows both types.

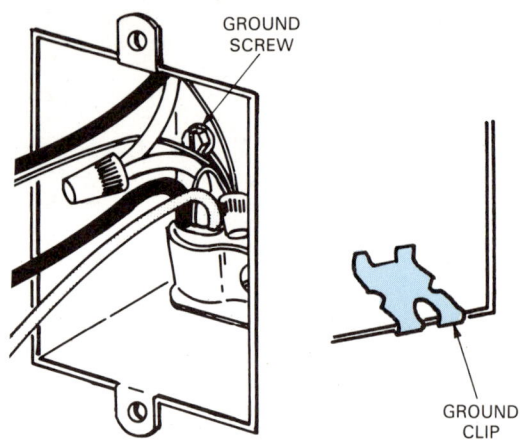

Fig. 7-23. Nonmetallic cable must have a grounding wire. When used with a metal box, the ground wire must be attached to the box by some means. Left. Attachment by ground screw (arrow). (GE Wiring Devices Dept.) Right. Ground clip.

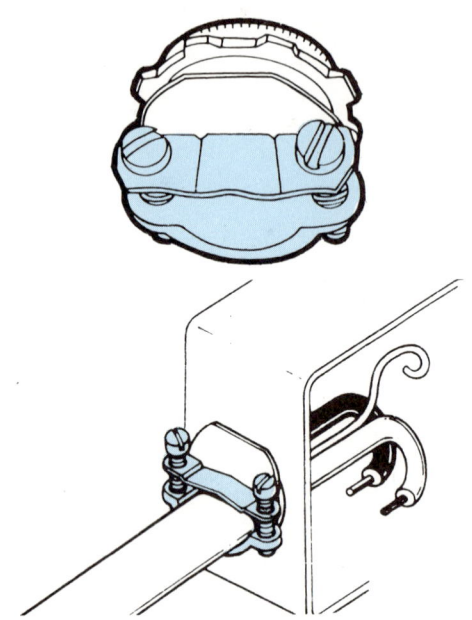

1. Remove locknut and push fitting into knockout hole in box. Screw on locknut and tighten by tapping with screwdriver.
2. Cut cable to desired length with wire cutter.
3. Pull cable through fitting into the box. Cable leads should extend approximately 6″ to 8″ into box.
4. Tighten screw or screws on fitting to secure the cable.

Fig. 7-21. Clamp for nonmetallic cable and how to install it. (Electroline Mfg. Co.)

BUSHINGS

NEC requires a grounding bushing wherever conduit enters a nonthreaded opening on a box. The conduit is secured on the one side by a locknut and on the other by a locknut and grounding bushing. Some grounding bushings have a grounding lug where a grounding wire can be attached. Fig. 7-24 shows several types of ground bushings.

FILL ALLOTMENT

Fill allotment refers to the number of conductors the *Code* will allow in certain sizes of boxes. Article 370 of the *Code* outlines how outlet, switch, and junction boxes are to be used. It is very specific about the number of conductors permitted. Fig. 7-25 summarizes the NEC requirements of NEC Table 370-6a.

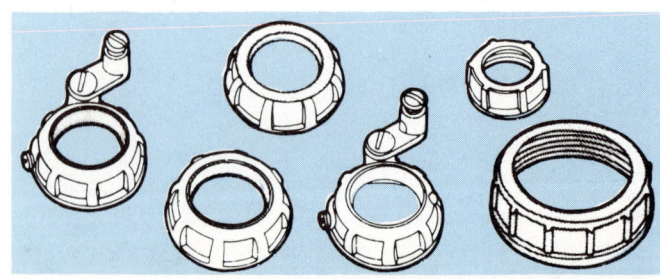

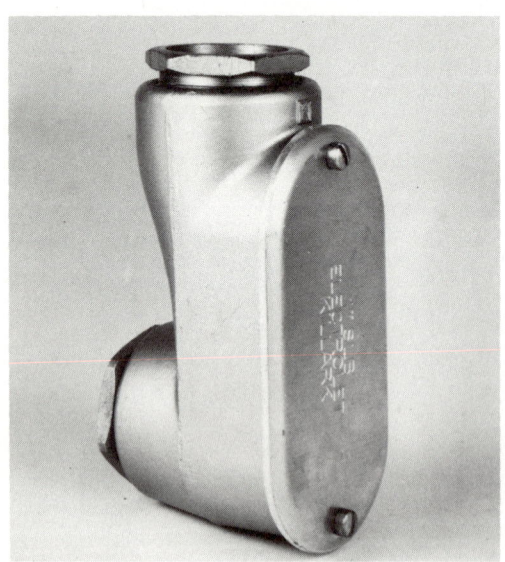

Fig. 7-22. Fittings such as this oval conduit body takes the place of a bend in the conduit. It is preferred when running conduit where space is limited. (Killark Electric Mfg. Co.)

Fig. 7-24. Bushings are used to assure good ground between conduit and boxes. Note that some have lugs for attaching a ground wire. (Electroline Mfg. Co.)

Boxes, Fittings, and Covers

CONDUCTOR ALLOTMENT – METAL BOXES

BOX SIZE	TYPE	WIRE SIZE			
		14	12	10	8
4 x 1 1/4	round or octagon	6	5	5	4
4 x 1 1/2	round or octagon	7	6	6	5
4 x 2 1/8	round or octagon	10	9	8	7
4 x 1 1/4	square	9	8	7	6
4 x 1 1/2	square	10	9	8	7
4 x 2 1/8	square	15	13	12	10
3 x 2 x 1 1/2	switch box	3	3	3	2
3 x 2 x 2	switch box	5	4	4	3
3 x 2 x 2 1/4	switch box	5	4	4	3
3 x 2 x 2 1/2	switch box	6	5	5	4
3 x 2 x 3 1/2	switch box	9	8	7	6

Fig. 7-25. NEC has placed limits on numbers of conductors housed in electrical boxes. Chart summarizes allotment for 11 different sizes.

When mixing different sizes of conductors in the same box, use NEC Table 370-6b. It will specify the required cubic inch allotment for each conductor. An easy method for determining box fill is shown in Fig. 7-26.

BOX COVERS AND ACCESSORIES

Plaster rings are special adaptors which can be used on either ceiling or wall boxes. They may be required because of the small size of some fixture canopies. Fig. 7-27 shows several types of plaster ring for square and round boxes.

All switch boxes and outlet boxes must be covered. No wiring can be left exposed. Because boxes are used for various purposes, there are many different kinds of box coverings. Fig. 7-28 shows some of them.

Easy method of determining box fill

The box fill table below extends the basic data in Table 370-6(b) to speed calculations. Here's how it works:

Example #1 — Conduit boxes: Install 4 #12 current carrying conductors and one device in a box where the conduit and fittings are approved for grounding. You must provide space for 4 conductors and 1 device. Read across the line "No. of Additions" in the box fill table to column 5, then read down to the cubic inches required for #12 conductors. You need a conduit box with at least 11.25 cubic-inch capacity.

Example #2 — Cable boxes: Install two #12-2 non-metallic sheathed cables with ground plus one device in a box with cable clamps. Think of this as 4 #12 current carrying conductors and two ground wires. In this example count:

4 #12 current carrying conductors	4
Ground wires	1
Cable clamps	1
One device	1
Total number of additions	7

Read across the line "No. of Additions" in the box fill table to column 7, then read down to the cubic inches required for #12 conductors. You need a box with cable clamps with at least 15.75 cubic-inch capacity.

When two **different** size current-carrying conductors will be in the same box, calculate the additions for devices, ground wires and clamps for the **largest** conductor, then add the smaller current-carrying conductors.

Example #3: Install 4 #12 and 6 #14 current-carrying conductors plus one device in a box where the conduit and fittings are approved for grounding. Count 4 #12 conductors plus **one** device and you need 11.25 cubic-inches. Now read right to "6" and down to #14 wire and you need an additional 12.00 cubic-inches for a total of 23.25 cubic-inches.

Cubic-inch capacity for standard boxes

Box capacity is given in the tables below. When used with flat covers, this is the total cubic-inch capacity of the box. However, the Code states that the capacity of a box is the "total of the assembled sections." Ganging boxes, adding extension rings and installing **raised** covers are three ways to increase capacity.

BOX FILL TABLE

No. Additions	1	2	3	4	5	6	7	8	9	10	11	12	13	14	15	16	17	18	19	20	21	22	
#14	2.00	4.00	6.00	8.00	10.00	12.00	14.00	16.00	18.00	20.00	22.00	24.00	26.00	28.00	30.00	32.00	34.00	36.00	38.00	40.00	42.00	44.00	#14
#12	2.25	4.50	6.75	9.00	11.25	13.50	15.75	18.00	20.25	22.50	24.75	27.00	29.25	31.50	33.75	36.00	38.25	40.50	42.75	45.00	47.25	49.50	#12
#10	2.50	5.00	7.50	10.00	12.50	15.00	17.50	20.00	22.50	25.00	27.50	30.00	32.50	35.00	37.50	40.00	42.50	45.00	47.50	50.00	52.50	55.00	#10
#8	3.00	6.00	9.00	12.00	15.00	18.00	21.00	24.00	27.00	30.00	33.00	36.00	39.00	42.00	45.00	48.00	51.00	54.00	57.00	60.00	63.00	66.00	#8
#6	5.00	10.00	15.00	20.00	25.00	30.00	35.00	40.00	45.00	50.00	55.00	60.00	65.00	70.00	75.00	80.00	85.00	90.00	95.00	100.00	105.00	110.00	#6

Fig. 7-26. Easy method for finding permissible box fill. (RACO Inc.)

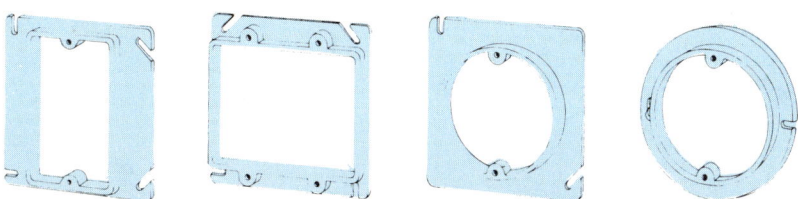

Fig. 7-27. Plaster rings are special fittings designed to adapt box to certain uses. They can also provide room for additional conductors. (Allied Moulded Products, Inc.)

Modern Residential Wiring

Fig. 7-28. Flat and raised switch and outlet box covers. Each is made for a specific device or combination of devices. A — Single tumbler switch. B — Duplex receptacle. C — Single receptacle. D — Duplex and single receptacle. F — Tumbler switch and duplex receptacle. G — Two tumbler switches and duplex receptacle. H — Three tumbler switches and duplex receptacle. I — Raised two tumbler switches. J — Raised two duplex receptacles. K — Raised triplex receptacle. L — Raised two duplex receptacle. M — Raised tumbler switch and triplex receptacle. N — Raised triplex and duplex receptacles. (RACO Inc.)

TEST QUESTIONS — CHAPTER 7

1. Boxes and covers serve to _____ and _____ electrical conductors and electrical devices.
2. Why must all joints, connections, and splices be contained in an enclosure (box)?
3. List the four shapes of boxes.
4. Wall boxes house or support (select correct answer or answers):
   a. Switches and receptacles.
   b. Fixtures.
   c. Splices.
   d. All the above.
5. Explain the need for knockouts and pryouts.
6. A _____ box is sometimes designed to be joined with other boxes. When boxes are joined together they are said to be _____.
7. Which of the following are advantages of PVC and fiberglass boxes: corrosion resistant, lightweight, do not easily break even in extreme cold, rugged, do not require cable clamps.
8. _____ are part of a wiring system which connect cables to boxes.
9. Connectors are available in (straight, 45° angles, 90° angles and 15° angles). Indicate which statements apply.
10. NEC requires a _____ _____ wherever conduit enters a nonthreaded opening on a box.

Chapter 8

INSTALLING BOXES AND CONDUCTORS

Boxes, conduit, cable and conductors are installed in new construction while the walls are still open. This process is called ROUGH-IN WIRING. Except for the faces of the boxes which will hold switches, receptacles and light fixtures, this wiring is eventually concealed behind the drywall or plaster.

PLANNING THE ROUGH-IN

Properly installed wiring will not easily be damaged by other construction work or by later stresses on the building or wiring system. At the same time, the runs should be laid out so that materials are not wasted. The wiring is first laid out with the use of a floor plan. Usually the electrician will receive a set of plans from the contractor. If there is no floor plan, the electrician consults with the owner and makes a sketch showing where various electrical components are to be located. For example, the kitchen plan will show locations of outlets for receptacles, switches, 240 V service for range, and light fixtures. There should be enough detail so there is no room for misunderstanding. See Fig. 8-1.

From this sketch the electrician will make a more complete drawing showing wiring runs. She or he will include information about wire type, gage and assortment of electrical devices. See Fig. 8-2. Before work begins, the electrician will again review the plan with the customer and make any adjustments needed. Chapter 11 has more information on the electrical plan.

LOCATING BOXES

Boxes must be firmly attached to a structural member of the wall, ceiling, or floor. Using the completed plan, lay out the location of the electrical boxes you will need for switches, receptacles, junctions, wire pulling and light fixtures. Use a marking pen or pencil to mark exact location and type of box. Marks are placed on the open frame (studs and joists) of the building.

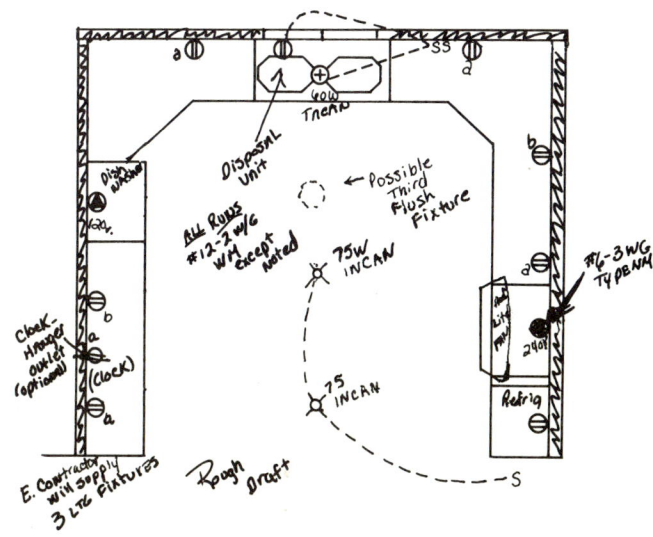

Fig. 8-1. Before beginning a rough-in, an electrician will prepare a rough sketch of a room to indicate where and what type of electrical device is required.

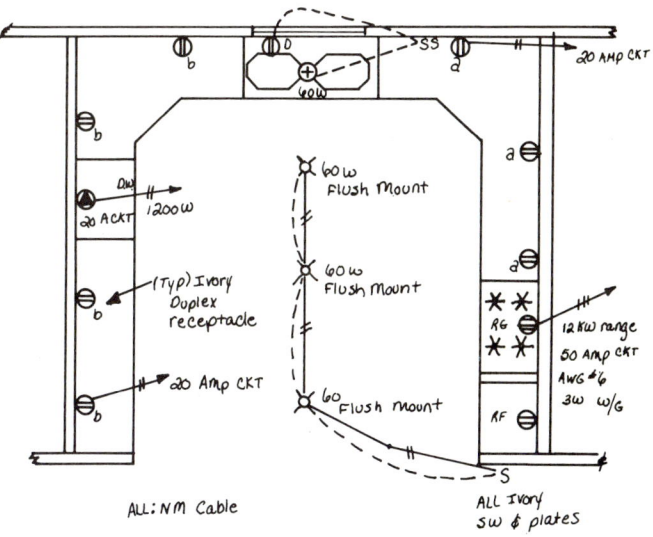

Fig. 8-2. An electrical plan as refined from the information given in Fig. 8-1.

Modern Residential Wiring

Electricians have different ways of marking for identification. A simple method is to use the standard symbols shown in Chapter 11. See Fig. 8-3.

STANDARD BOX HEIGHTS

Switches and other devices intended for standard elbow height are usually marked at 48 in. from the rough floor. A mark at that height on the side of the wall stud will locate the top of the box or its center, as preferred.

Outlet boxes are customarily attached to the stud 12 in. off the floor. You can mark a tape or rule with strips of tape at 12 and 48 in. or mark these points with a felt tip pen. This is a quick way of finding the correct height for both switch and outlet boxes. Some electricians use the handle of their hammer to find height of outlet boxes, Fig. 8-4. You can also mark the heights on a length of 1 x 2.

Most electricians lay out a wiring job room by room and install boxes before any conduit or cable is run. When the boxes are installed, double-check their location with the plan.

It is also a good idea to have the owner approve the box locations at this time. It is easier and less expensive to move them before the rest of the rough-in is done.

ATTACHING BOXES

Boxes have many different arrangements for fastening, Fig. 8-5. Some of these are shown in Chapter 7. Look at Figs. 7-8, 7-10, 7-11, and 7-12.

Boxes must be carefully positioned. Be careful to center ceiling outlets. Brackets, like the one shown

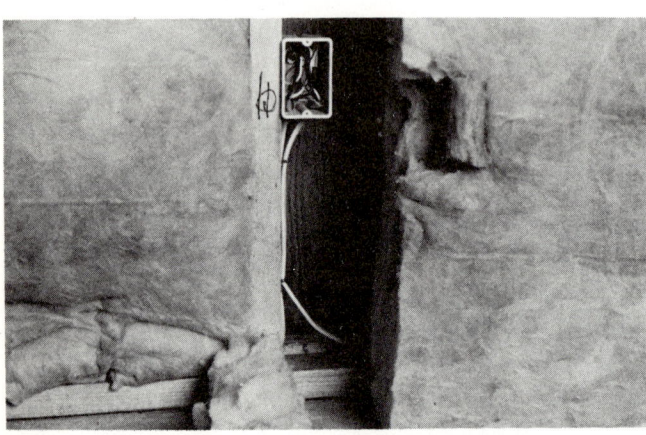

Fig. 8-3. Use a heavy pencil or felt tip marker to mark framing for location and type of electrical box. (Owens-Corning Fiberglas Corp.)

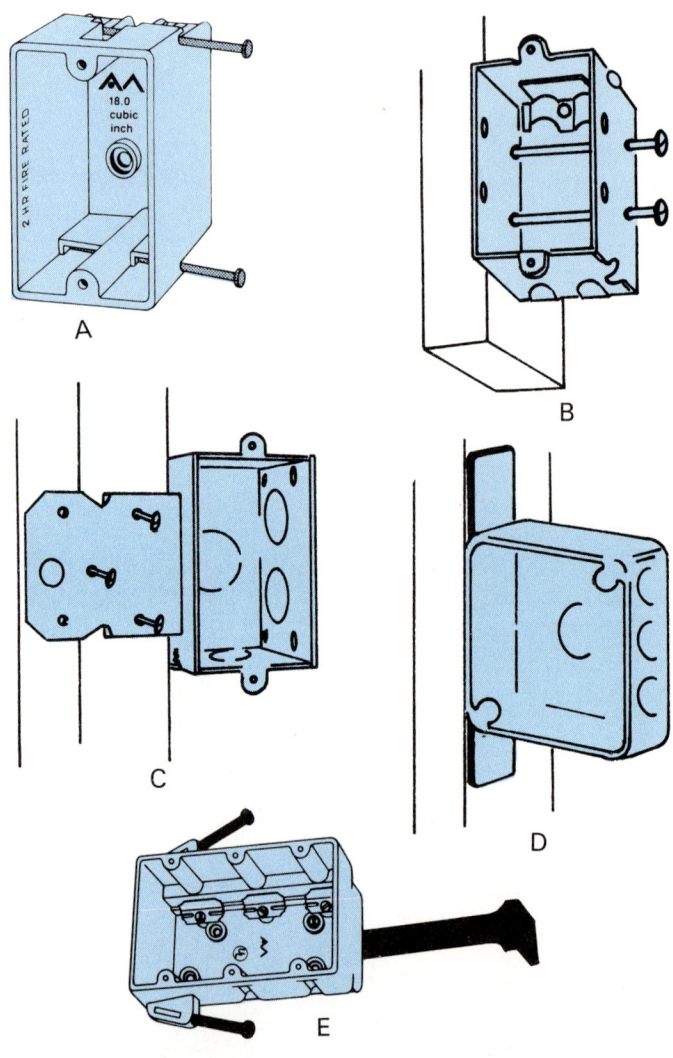

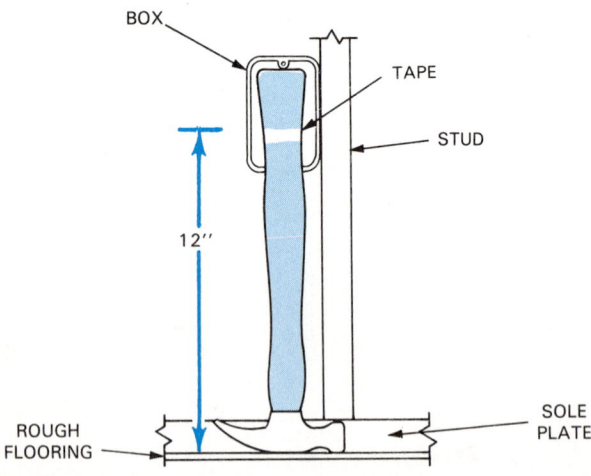

Fig. 8-4. Height of outlet box centerline can be found by length of hammer handle. If handle is more than 12 in. long from head to end, use tape to mark it.

Fig. 8-5. Various fastening methods are used for attaching boxes. A—Box with captive nails in built-in brackets. (Allied Moulded Products Inc.) B—Box with aligned holes for nails. *Code* requires nails to be within 1/4 in. of back or ends of box. C—Bracket designed to mount to face of stud. Note spurs and nail holes for fastening. D—Flat bracket for mounting on side of frame member. E—Captive nails with stabilizing bar.

Installing Boxes and Conductors

in Fig. 8-6, may be necessary. Allowances must be made for thickness of drywall or plaster. See Fig. 8-7.

Be sure to allow clearance when attaching switch boxes near doors. The switch plate must not interfere with door trim. A special bracket may be needed to provide additional clearance.

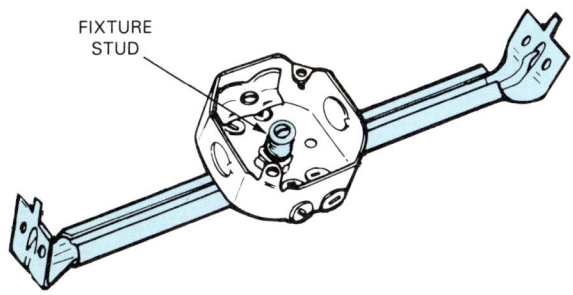

Fig. 8-6. Special brackets center ceiling boxes between joists. Box mounts onto fixture stud which is often part of bracket. (Electroline Mfg. Co.)

Fig. 8-7. Box must be mounted so its face extends beyond stud the thickness of the drywall or plaster surface. (Allied Moulded Products, Inc.)

INSTALLING CONDUIT RUNS

In some areas of the country local codes require that conductors (wiring) be run inside of conduit (rigid, EMT or PVC). This material is described in Chapter 6, Wiring Systems.

Wiring is usually installed circuit by circuit. First the wiring is run from outlet to outlet. Then, the main feed, called the "home run," goes back to the main circuit breaker panel. (The home run is the section from the circuit breaker to the first device in the circuit.)

Before conduit is cut and bent, you will need to plan and measure the best route from box to box and then to the panel.

BORING AND NOTCHING FOR CONDUIT

Wiring must usually be concealed in the building frame behind the surface of the wall and ceiling finish. In some cases, the wiring can run over the tops of ceiling joists when there is to be no flooring above it, Fig. 8-8. In all other situations the frame members must be bored or notched.

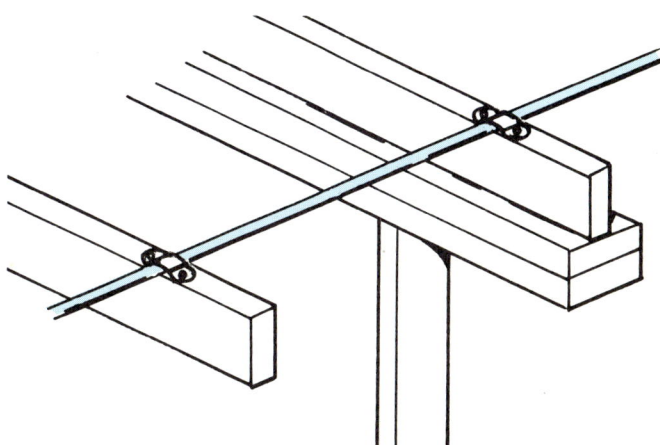

Fig. 8-8. Conduit being run over top of ceiling joists. This is permitted in unfloored attics.

Boring is preferred since it does not weaken the frame as much as notching. However, it is more difficult to install conduit in bored holes. The holes must be drilled large enough so that conduit can be installed at a slight angle to clear the next stud. Conduit may have to be cut in short lengths for easier installation. This requires a greater number of couplings and, thus, more materials and work. EMT can be bowed more than rigid or intermediate conduit. Therefore, it is used almost exclusively in residential wiring, Fig. 8-9.

Notching is allowed but should be used only when necessary since it weakens the structure. Notches should be as narrow as possible and never deeper than 20 percent of the width of the member. See Fig. 8-10. Conduit should be protected at each stud or joist with a metal plate. This prevents accidental piercing of the conduit with a nail, Fig. 8-11.

CUTTING CONDUIT

The hacksaw is normally used to cut all types of conduit. The saw handle should be fitted with a blade having 18 to 24 teeth per inch. The pipe should be held in a portable vise. This will help you avoid broken blades and makes straighter cuts possible. The vise will be necessary also for cutting threads on rigid conduit.

Another tool sometimes used to cut conduit is the plumbers' pipe cutter. Apply a small amount of cut-

Modern Residential Wiring

ting oil where the cut will be made. Slip the cutter mouth over the pipe and adjust the cutter wheel for a shallow cut. Make one revolution at a time and tighten the cutter slightly in between each turn.

Burrs left by either cutting operation must be

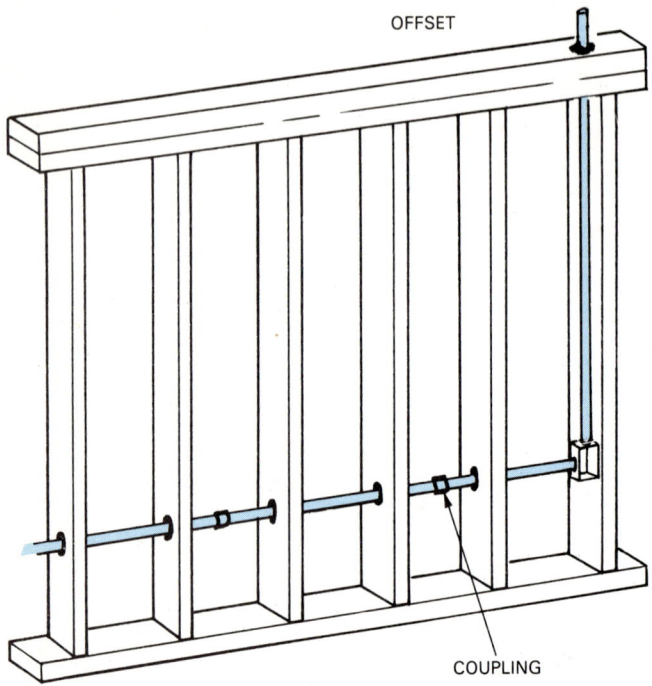

Fig. 8-9. Run of conduit can be installed through bored holes. Make holes larger than the conduit's diameter and use shorter lengths of conduit.

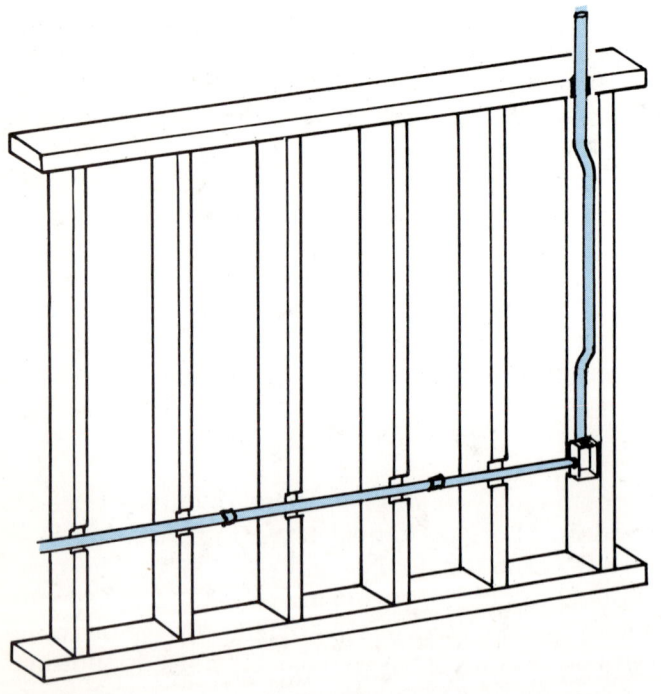

Fig. 8-10. Some electricians prefer to notch conduit into studding. This weakens the wall frame slightly but is permitted.

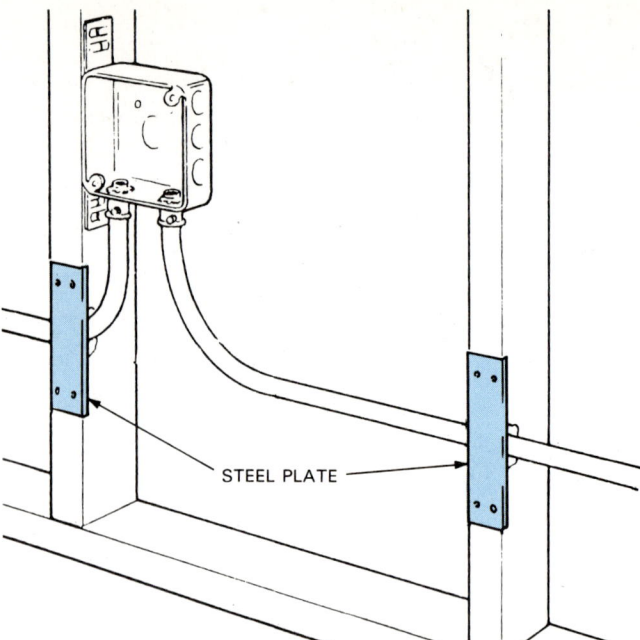

Fig. 8-11. Method of protecting conduit from drywall nails. Plates are let into face so they will be flush with stud surface.

removed so they will not damage or cut the wire insulation when it is pulled through the conduit. The reamer is inserted in a hand brace. Refer again to Fig. 6-3 which shows a suitable reamer.

BENDING METALLIC CONDUIT

Although there are numerous special manufactured fittings available for making turns and bends in conduit, many construction electricians prefer making FIELD BENDS whenever possible. (Field bends are those made on the construction site.) This generally saves time and is far more economical. Bending conduit is an important skill the electrician can easily master with knowledge of basic bending procedures and practice.

TOOLS FOR BENDING

There are two kinds of hand benders used on conduit. One type, called a hickey, Fig. 8-12, is used primarily to bend large sizes of EMT, IMC, or RIGID conduit. The other type, called a bender, Fig. 8-13, is most often used for smaller sizes of conduit. POWER BENDERS or bench-benders are more sophisticated devices. These are used when bending very large sizes of conduit or making many "carbon copy" bends, Fig. 8-14.

RIGHT ANGLE BENDS—SPECIAL CONSIDERATIONS

Rigid conduit cannot be bent to final shape in one complete motion, as can IMC or EMT. It must be

Installing Boxes and Conductors

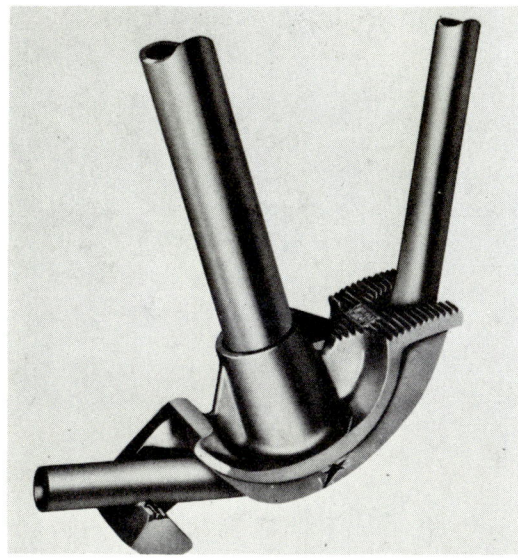

Fig. 8-12. A hickey, or stubber hickey, is employed to bend larger sizes of EMT, Rigid, and IMC. These bending tools require several separate movements per bend. (Appleton Electric Co.)

Fig. 8-14. Field bending of large diameter conduit is accomplished with benders like this. They may be manual or power assisted (hydraulic). These benders are often capable of bending many sizes of EMT, IMC, and Rigid conduit with the use of appropriate accessories and adaptors. (Greenlee Tool Div., EX-CELL-O Corp.)

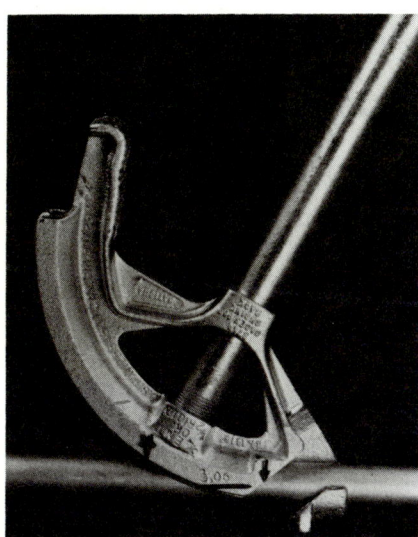

Fig. 8-13. Conduit bender is slid onto the conduit and passes through the hook. This provides a solid positioning from which the bend can begin. (Appleton Electric Co.)

NEC TABLE 346-10

RADIUS OF CONDUIT BENDS (INCHES)		
Size of Conduit (In.)	Conductors without Lead Sheath (In.)	Conductors with Lead Sheath (In.)
1/2	4	6
3/4	5	8
1	6	11
1 1/4	8	14
1 1/2	10	16
2	12	21
2 1/2	15	25
3	18	31
3 1/2	21	36
4	24	40
5	30	50
6	36	61

For SI units: (Radius) one inch = 2.54 millimeters.

Fig. 8-15. NEC specifies the radius of conduit bends. Follow these guidelines or work will fail inspection. (NEC)

bent in stages with a hickey to bring about a smooth, clean bend. Failure to stage the bends in rigid conduit could result in damaged conduit walls. The minimum radius of conduit bends is limited by the *Code*. See Fig. 8-15.

For example, to make a 90° bend in a 1 in. rigid conduit run, the radius must be a minimum of 6 in. See Fig. 8-16. A radius of 12 in. is required for a 2 in. conduit bend, Fig. 8-17. Ninety degree bends are usually called "stubs" or "stub-ups."

These minimum radii, as outlined in Fig. 8-15, apply to all conduit—rigid, IMC, or EMT—and for any degree of bend. See Fig. 8-18.

MAKING THE 90 DEGREE BEND

1. Find the stub height distance. This is the distance the conduit runs in the new direction. Measurement is taken from where the bend is made to the length of the run in the new direc-

tion. Fig. 8-19 helps explain this.
2. Mark the conduit at the stub height distance. For example, if your bend is supposed to leave a run 10 in. high, measure back from the end 10 in. Mark this point with a soft lead pencil, Fig. 8-20.
3. Place the bender on the conduit with the heel directly above the mark as shown in Fig. 8-21.
4. Hold conduit firmly on the floor with your foot just behind the bender. Push down on the bender handle. Bend to a 90° angle. At this

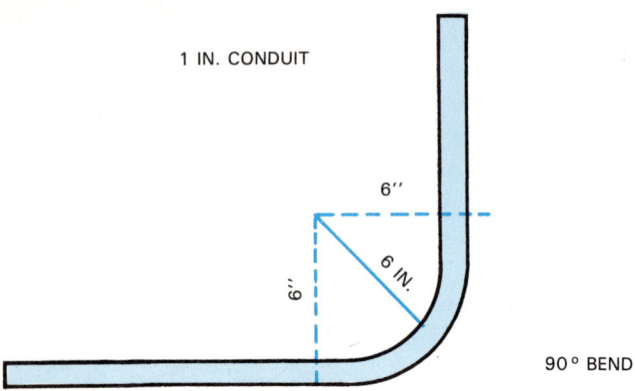

Fig. 8-16. A "stub" or 90° bend. Radius must be no less than 6 in. (15 cm) for 1 in. diameter conduit (see NEC Table 346-10, Fig. 8-15.)

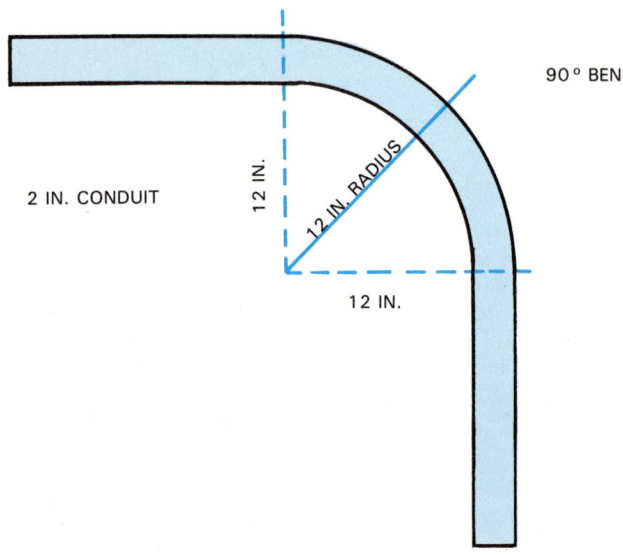

Fig. 8-17. A 2 in. diameter would require a minimum radius of 12 in. (30.5 cm) or twice that of a 1 in. conduit.

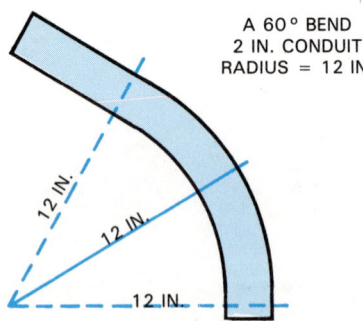

Fig. 8-18. The minimum radii of conduit bends depend solely on the tubing diameter rather than conduit type or degree of bend. a 60° bend of 2 in. conduit is governed by the same radius restriction as a 90° stub.

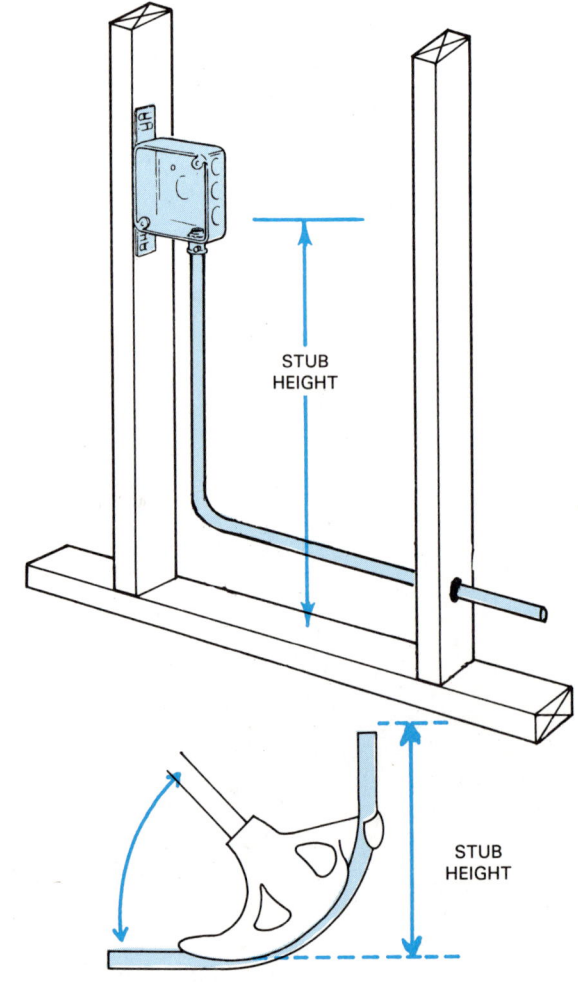

Fig. 8-19. Stub height is length of conduit run measured at right angles from beginning of bend to end of conduit. Top. Dimension line shows stub height. Bottom. Making 90° stub bend with bender. When handle is at 45° to the ground, bend is at 90°.

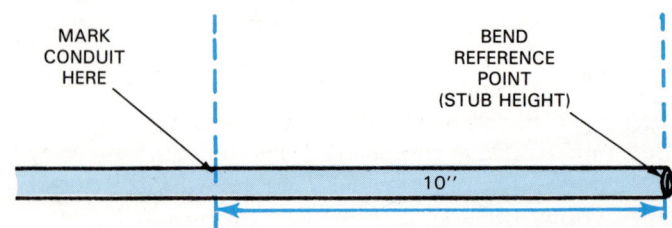

Fig. 8-20. Carefully mark the conduit (soft-lead pencil works well) to indicate stub height.

Installing Boxes and Conductors

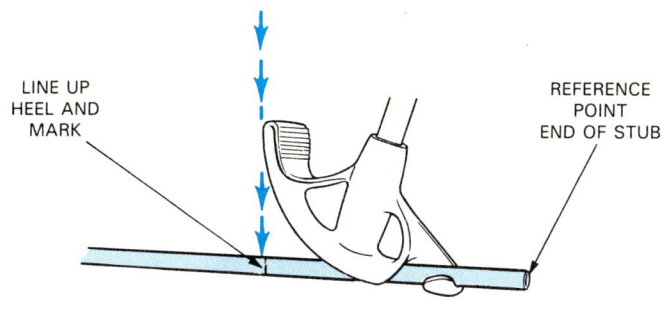

Fig. 8-21. Line up the heel of the bender with the mark on the conduit.

point the handle will be at about a 45° angle from the floor. (This works best on a concrete floor.)

BACK-TO-BACK BENDS

A back-to-back bend is one which produces a bend in back of a 90° bend. That is, another bend is made at a distance away from the previous 90° bend.

The procedure follows:
1. Measure the distance from the back of the prior 90° bend to where the next bend must be made, Fig. 8-22. Mark the conduit.
2. Now, align this conduit mark with the special mark on the bender. This is often called the "star point" or "star," Fig. 8-23.
3. Bend this angle to 90°. Fig. 8-24 shows completed bends.

The star point or star will indicate where the back of any bend will be, regardless of angle. Therefore, the above procedures will work for 30°, 45°, 60°, or any angle of bend.

OFFSET BEND

Another common bend is the offset. It is used when the conduit must go around an obstruction

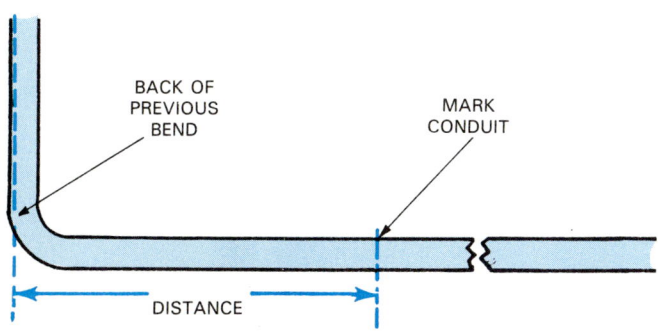

Fig. 8-22. For back-to-back bends, measure the distance from the back of the previous stub to the next bend. Mark this on the conduit.

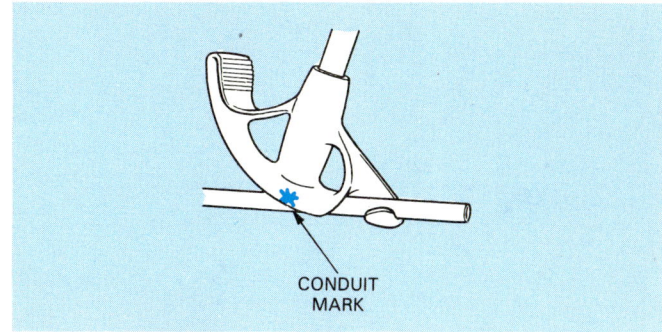

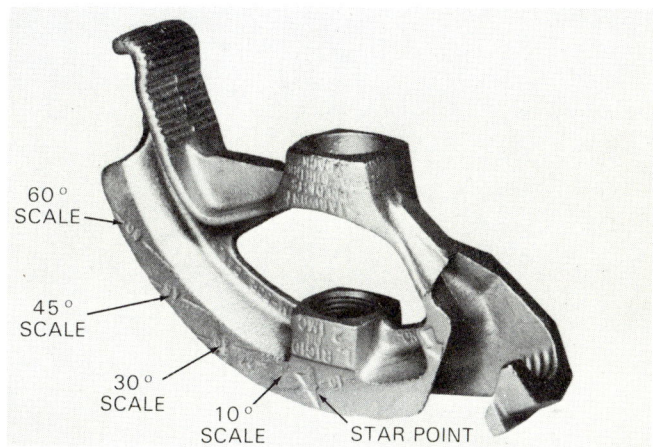

Fig. 8-23. Top. Line up the star on the bender with the measured distance mark. This is the proper position to make the next stub. Bottom. Bender has marks for many different bends. (Appleton Electric Co.)

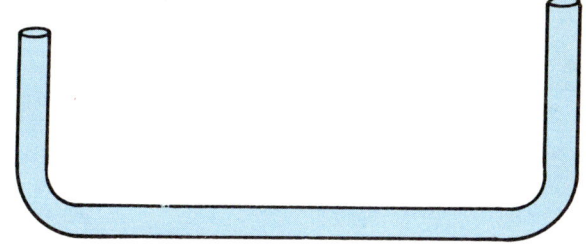

Fig. 8-24. A completed back-to-back bend.

or when the plane of the conduit must change slightly.

Generally, offsets may be done at almost any angle, Fig. 8-25. Shallow bends—those at smaller angles—make wire pulling much simpler. Deep bends, at high angles look better and save space.

Whether you use a shallow bend or a deep one, will most often depend on the depth of the offset. See Fig. 8-26. Moreover, the depth of the offset, the angle of bend, and the distance of the bend are all related.

One other important factor, shrink, must be noted. Shrink is the amount of shortening which takes place in an offset. Shrink need only be considered when working toward the obstruction.

75

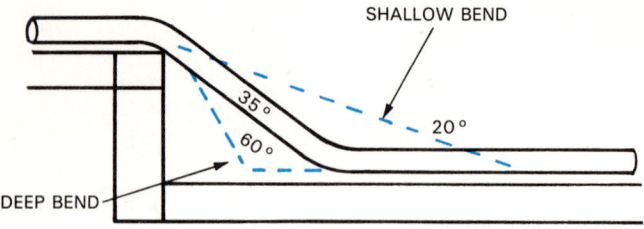

Fig. 8-25. Offsets may be of almost any angle. Large or deep angles make for difficult wire pulling. Small or shallow angles take up more space, but wires pull easier.

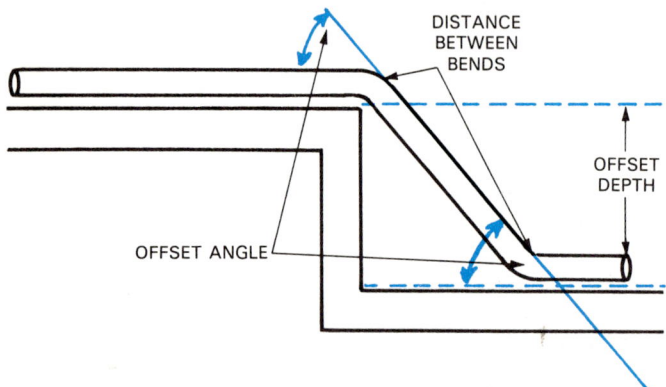

Fig. 8-26. Distance, angle, and offset depth are related.

Attention to four facets, depth, angle, distance, and shrink, if carefully considered, will help you produce a perfect offset. Fig. 8-27 may serve as a guide for the most common offsets.

MAKING THE OFFSET BEND

The following steps are followed when making an offset bend. *Remember, careful measurement is the key to precise bending:*

1. Measure the exact distance from the last conduit coupling to the obstruction. Consider the appropriate angle to use in the offset.
2. To the distance, found in Step 1, add the amount of shrink from the Reference Guide, (Fig. 8-27). Measure off this distance on the conduit and mark it. This is the location of the first bend.
3. From the chart in Fig. 8-27, find the distance between bends. Measure off this distance on the conduit and mark. This is the location of the second bend.
4. Line up the arrow on the bender with the first mark on the conduit. Using the degree scale (refer to Fig. 8-23) on the bender, bend the conduit to the correct angle. *Make offset bends in the air, not on the floor.*
5. Lastly, slide the bender along to the second mark. Line this mark up with the arrow and bend to the chosen angle.

Fig. 8-28 illustrates the five steps in making a 45° offset around an 8 in. high obstruction which is 60 in. from the last coupling.

THE SADDLE BEND

The saddle bend is employed for the same purpose as an offset. It goes around an obstruction. However, unlike the offset bend, the conduit returns to the same level after it passes the obstruction, Fig. 8-29.

The most common saddle consists of one 45° center bend and two 22 1/2° lateral bends, as previously shown.

MAKING THE SADDLE BEND

Executing a perfect saddle bend is easy. The procedure follows:

For example, consider that a 3 in. high obstruction, like the one illustrated previously, is encountered 6 ft. from the last coupling, Fig. 8-30. That is, the center of the obstruction is 6 ft. or 72 in. from the last coupling.

You need to follow two rules for making the saddle bend.

REFERENCE GUIDE FOR COMMON OFFSETS

Offset depth (In.)	Best angle in degrees	Distance between bends (In.)	Amount of shrink (In.)	Shrink per inch of offset depth
2	22 1/2	5 1/4	3/8	3/16
3	30	6	3/4	1/4
4	30	8	1	1/4
5	45	7	1 7/8	3/8
6	45	8 1/2	2 1/4	3/8
7	45	9 3/4	2 5/8	3/8
8	45	11 1/4	3	3/8
9	60	10 7/8	4 1/2	1/2
10	60	12	5	1/2

Fig. 8-27. Chart gives angles and shrink information for various offsets.

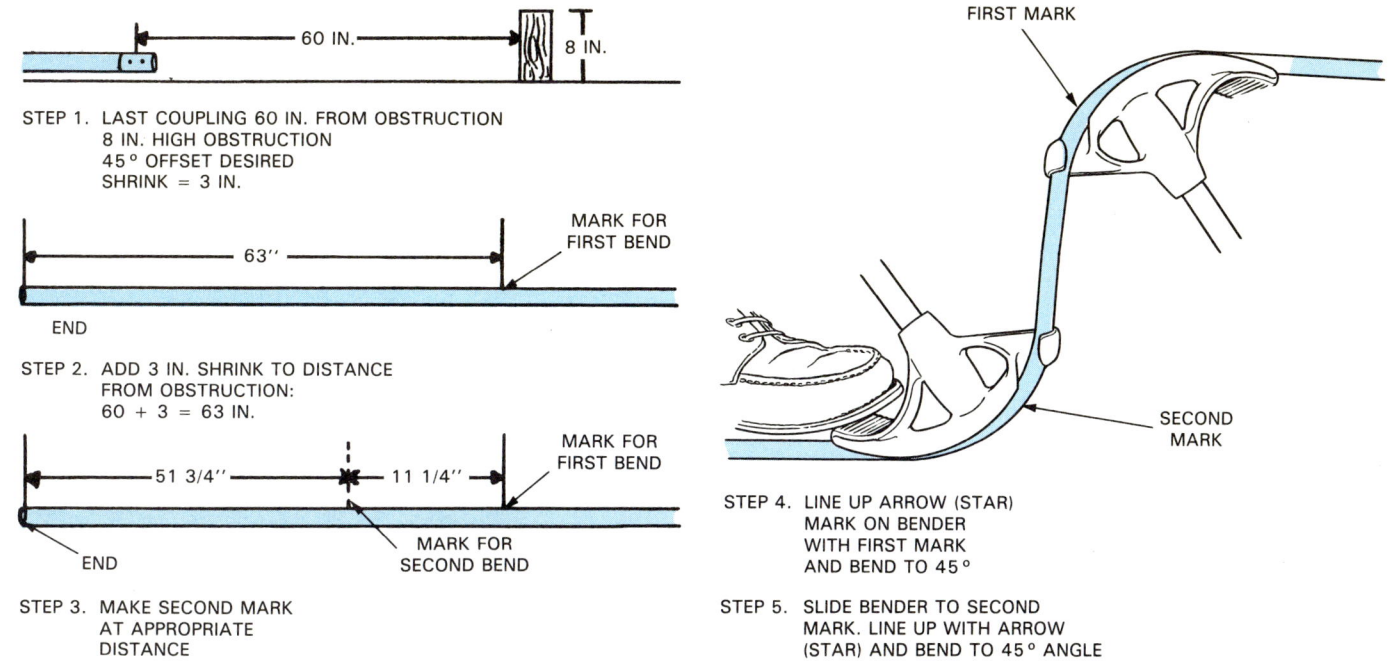

Fig. 8-28. Steps in making an offset bend.

1. For every 1 in. of saddle depth (obstruction height) you must add 3/16 in. and move your center mark *ahead* accordingly.
2. Lateral marks must be made 2 1/2 in. times obstruction height from the center mark.

To make the three bends:

1. The center bend is always made first. Line up the center mark with the RIM NOTCH in the bender rim. Bend conduit 45° as indicated in Fig. 8-31. This bend should be made in the "air" rather than on the floor.
2. Without removing the bender, slide it down one side of the conduit to the lateral mark. Align this mark with the arrow on the bender. Bend to the 22 1/2° angle, Fig. 8-32.
3. Last, slip the bender off the conduit, reverse it

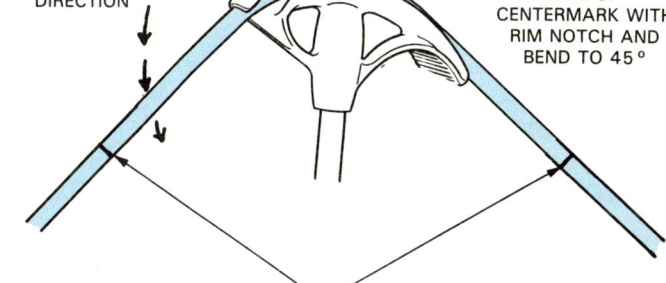

Fig. 8-31. First bend of a saddle is the middle one. This is done in the air, rather than on floor. Align rim notch of bender with center mark to insure accurate bend.

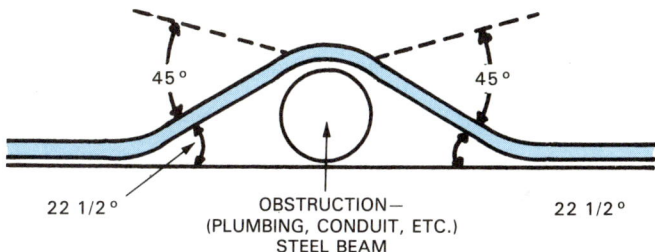

Fig. 8-29. Typical saddle bend. Keep bend angles shallow.

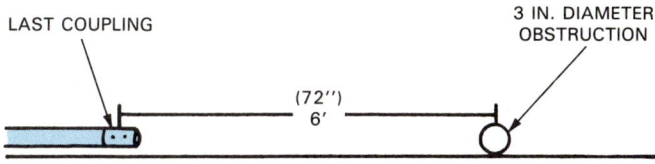

Fig. 8-30. Good measurement results in good bending. The saddle bend begins with measuring the distance from the last coupling to the center of the obstruction.

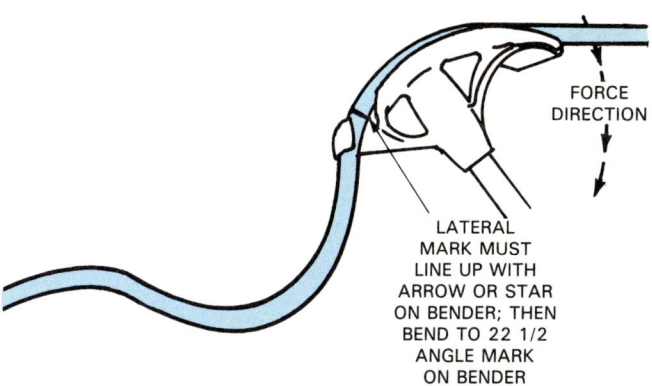

Fig. 8-32. Lateral bends are also done in the air. Lateral marks are matched with arrow on bender prior to bending.

77

and slide it onto the other lateral mark. Line this up with the arrow of the bender and bend once again to a 22 1/2° angle, Fig. 8-33. Saddle bend is now complete.

346-10 and 346-11 of the NEC. One essential point is this: There may not be more than the equivalent of four 90° bends or a total of 360° for each run between outlets of fittings, Fig. 8-35. The reason for this is a very practical one. It would be difficult to pull wiring through conduit with too many bends in it. Possibly the insulation on the wires could become damaged from the force necessary to pull the wires through.

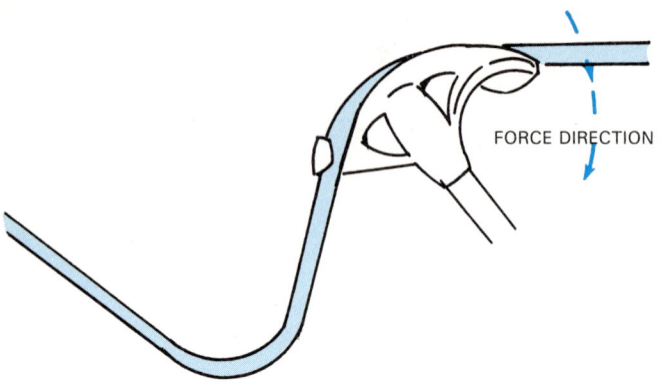

Fig. 8-33. Completion of the saddle bend.

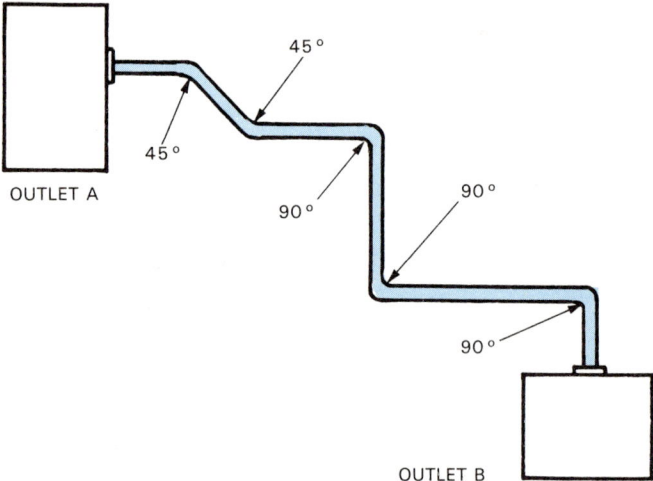

Fig. 8-35. Number of bends between outlets or fittings must not be greater than equivalent of four 90° bends or 360° total.

BENDING NONMETALLIC CONDUIT

Nonmetallic conduit (discussed in Chapter 6) may be bent by heating the conduit, shaping it, and cooling the formed bend. Each of the various types of PVC—Schedule 40, Schedule 80, and Type A, Fig. 6-18, may be heated in an electric heating box like the one illustrated in Fig. 6-19. Once heated, the conduit may be bent by hand (use heat resistant gloves).

For larger sizes, a bending guide is preferred, Fig. 8-34. Generally, the same types of bends, as well as bend specifications (radii, angles, etc.) are used for nonmetallic conduit as for its metal counterparts.

SPECS FOR BENDING IMC and EMT

Rules for bending intermediate metal conduit and electrical metallic tubing can be found in Section

SUPPORTING CONDUIT

The *National Electrical Code* prescribes how conduit shall be supported. Reference should be made to Article 346-12. In general, metal conduit must be supported within 3 ft. of every outlet box and at a minimum of every 10 ft. of run. In larger sizes this last distance can be increased. See Fig. 8-36. For nonmetallic conduit, the distances are less. Refer to Table 347-8 of the NEC.

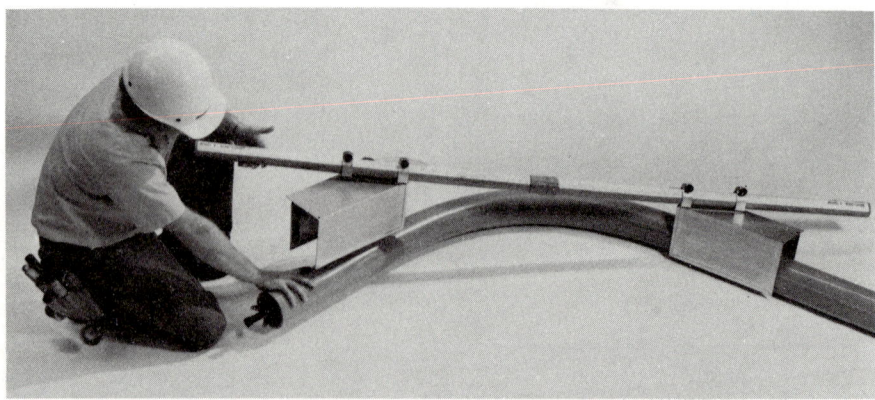

Fig. 8-34. A bending guide is particularly useful in forming bends in the larger diameter rigid nonmetallic conduit sizes. (CARLON, an Indian Head Co.)

Installing Boxes and Conductors

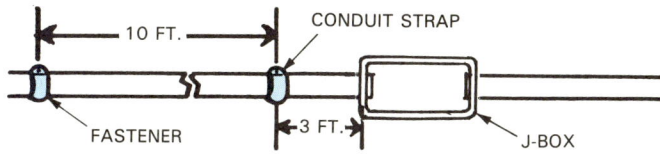

Fig. 8-36. Rigid conduit must be supported every 10 ft. along straight runs and within 3 ft. of each junction box.

INSTALLING FLEXIBLE METAL CONDUIT

Code restricts use of flexible metal conduit. (Greenfield). For example, it cannot be used in wet locations (without lead-covered conductors) or underground and embedded in concrete or aggregate. Diameters under 1/2 in. cannot be used except under certain limited circumstances which are covered in Section 350-3 of the *Code*. When it is used, provision must be made for proper equipment grounding. See Fig. 8-37.

Support must be provided at least every 4 1/2 ft. and within 12 in. of every outlet or fitting. This regulation does not apply when the cable is fished through walls when doing old work. Refer also to Sections 300, 333, and 346 of the *Code* before installing.

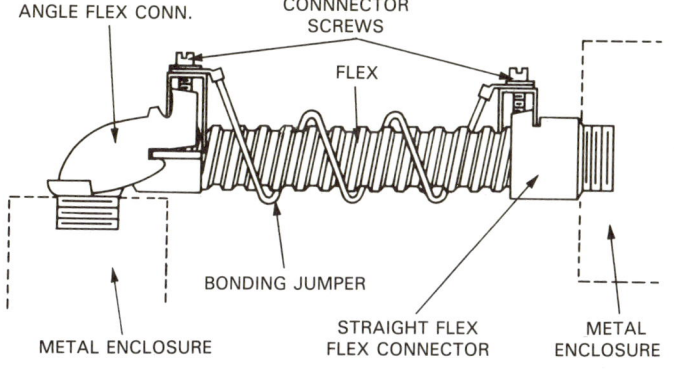

Fig. 8-37. Proper method of creating an equipment ground across a length of flexible metal conduit. The bonding jumper creates a solid electrical path from one metal enclosure to another.

MAKING CONDUIT CONNECTIONS

Code requires that conduit be securely fastened to electrical boxes so that electrical bonding is maintained between the box and the conduit. Couplings should provide the same bonding. A second reason for the security of connections is to insure that no stress is placed on conductors and that conductors do not come in contact with materials that might abrade their insulation or create danger of a short or fire. Figs. 8-38 and 8-39 show various connections and how to make some of them.

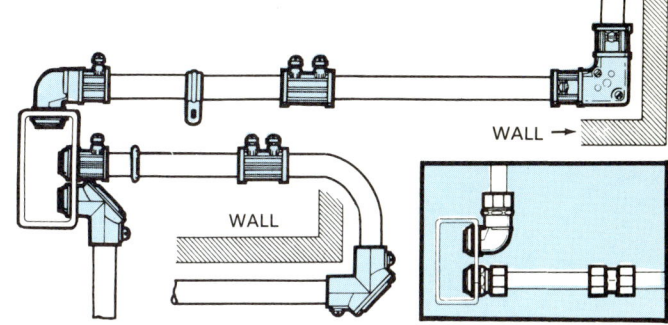

Fig. 8-38. A sample of some EMT fittings. Note angle fittings. (Electroline Mfg. Co.)

TO INSTALL CONNECTORS (To connect conduit to a box):

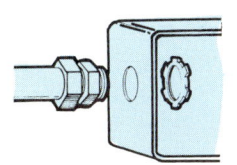

1. Cut conduit carefully with a hacksaw to desired length. Remove any rough edges with a file to avoid damage to wires.
2. Insert conduit into fitting.
3. Tighten setscrew (setscrew type, S 121 series) or torque compression nut until it is tight (compression type, L 21 series).
4. Remove locknut from connector.
5. Push connector through knockout, screw on locknut and tighten by tapping with a screwdriver.
6. Pull wires through conduit.

TO INSTALL COUPLINGS (To connect two pieces of conduit):

1. Cut conduit carefully with a hacksaw to desired length. Remove any rough edges with a file to avoid damage to wires.
2. Insert conduit into fitting.
3. Tighten setscrews (setscrew type, L 122 series) or torque compression nuts until tight (compression types, L 22 series).
4. Pull wires through conduit.

Fig. 8-39. Follow these steps for installing conduit to boxes and fittings.

PULLING WIRE

After boxes are installed and conduit connected to them, the conductors (wire) can be pulled in. On very short runs, small wires can be pushed through from one box to another. However, a fish tape, such as the one shown in Chapter 4, Fig. 4-19, is, in most cases, much faster. The tape is available in various lengths—usually 50, 100, and 200 ft.—of tempered flexible steel in several thicknesses and widths.

The tape may have a hook or a ball on the end, Fig. 8-40. If it has neither, you can fashion a hook with the aid of a pair of pliers. The tempered steel can be brittle, so make the bend carefully or reduce the temper by heating the end with a torch. Let it cool before bending.

USING THE FISH TAPE

Wires are difficult to push through conduit. The fish tape, being somewhat stiff, will snake through

Modern Residential Wiring

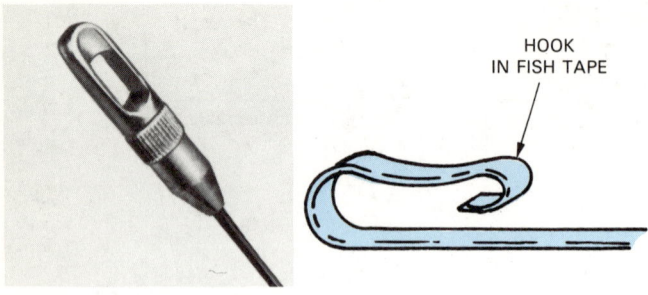

Fig. 8-40. Fish tape may have all or hook end to which wires are attached for pulling. Left. Ball end. (Greenlee Tool Div., EX-CELL-O Corp.) Right. Hook.

bends much easier and can be pushed considerable distances through several bends.

Push the tape through the run and attach the wires being pulled. Fig. 8-41 shows methods of attaching the wire. Be sure to wrap the wires securely so they do not become detached from the stress of pulling. It is better to work from the top down so the weight of the wires works with the pulling rather than against it.

Wires being pulled should be kept straight. Twisted, crossed, or tangled wire will bind at bends, saddles, or offsets. Wire insulation can become abraded or torn. On long runs it may help to coat the wire with wire pulling lubricants such as those containing talc, soapstone, or liquid soap. Special wire pulling lubricants are manufactured for this purpose.

CABLE ROUGH-IN

Chapter 6 describes the various cable systems approved by NEC for electrical wiring. Installation of these materials is somewhat simpler than running conduit and pulling wires. One important difference is that system grounding must be accomplished with a bare grounding wire. This wire must be enclosed within the protective covering along with the ungrounded and grounded conductors (hot and neutral wires).

ARMORED CABLE INSTALLATION

Layout of armored cable (BX) is the same as that described for conduit earlier in this chapter. Cable is considered acceptable by NEC and by most local codes. A few communities limit its use. Fig. 8-42 through Fig. 8-51 show its construction, fittings, and installation steps.

Armored cable is quite flexible and can be pulled through bored holes with greater ease than conduit. The holes should be slightly larger than the cable to avoid strain during rough-in. Fig. 8-43 shows a typical installation which conforms to the *National Electrical Code*.

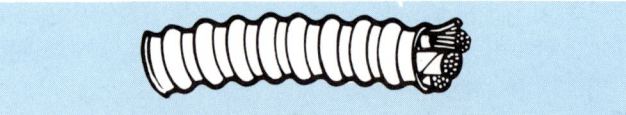

Fig. 8-42. Armored cable consists of a tough, flexible, metal outer shell, insulated conductors, and a bonding strip (bare grounding wire) inside. Follow manufacturer's instructions for installing fittings. (Electroline Mfg. Co.)

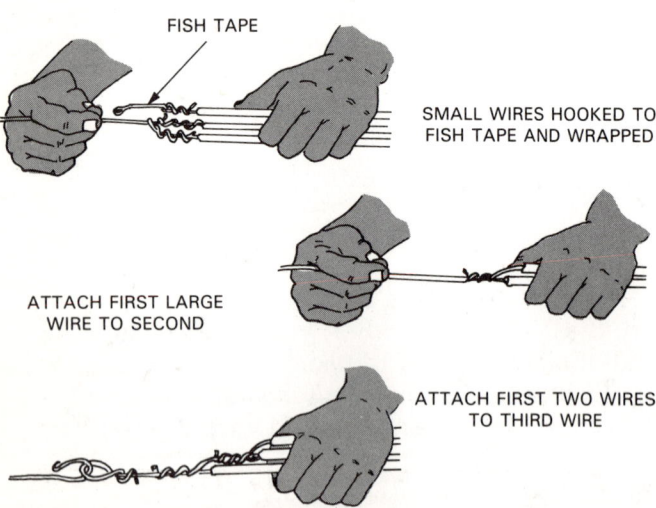

Fig. 8-41. Methods of attaching wire conductors to fish tape. Left Small wire wrap. Right. Stagger larger wires to keep down bulk for easier pulling.

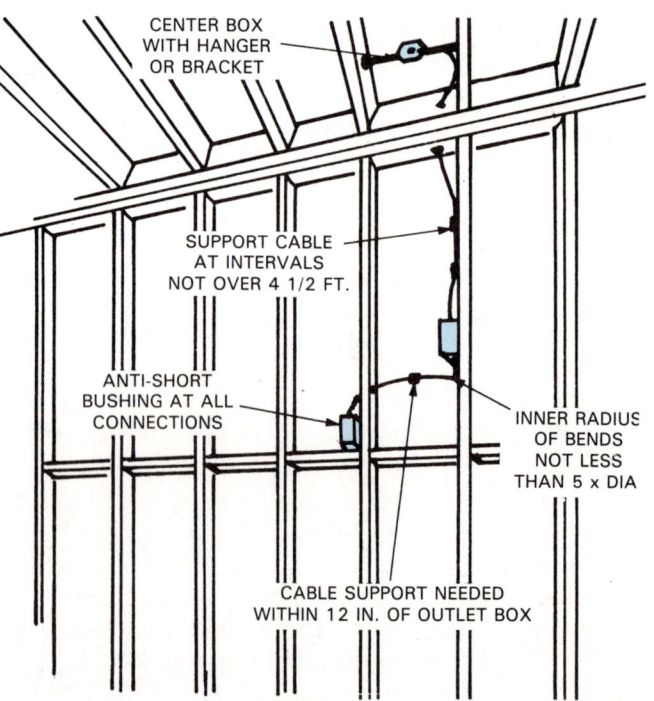

Fig. 8-43. Typical installation using armored cable. Installation conforms to Code.

Installing Boxes and Conductors

WORKING WITH ARMORED CABLE

Armored cable should be measured off and cut before it is pulled through the holes in the framing members. Measure the distance along the route the cable will run and transfer this measurement to the cable. Allow 6 to 8 in. extra for making connections at each end. Cut off the length of cable needed. Use cable cutters or a hacksaw.

STRIPPING CABLE ENDS

There are several methods for cutting through the tough metal armor and stripping the cable end. Perhaps the most common tool for cutting the armor is the hacksaw.

1. Measure back 6 or 8 in. and mark.
2. Saw through a single bead of the flexible metal covering as shown in Fig. 8-44. CAUTION! Do not cut completely through the armor. You are likely to damage the insulation or the conductors and cut the grounding strip.
3. After you have cut partway through the armor, flex the cable back and forth as you would wire. After several bends, the covering will separate and you will be able to slide the section of sheath off the conductors. See Fig. 8-45.
4. Remove the outer wrapping of paper, Fig. 8-46.

Fig. 8-46. Remove outer wrap of paper around conductors. Note grounding wire.

This will expose the insulated conductors and the ground wire.

5. Install an anti-short bushing, Fig. 8-47. This is the red, split plastic sheath between the wires and the rough edge of the armor. This bushing keeps the insulation from rubbing against the sharp edge of the armor. The bushing must be in place to pass any electrical inspection. Usually, a good supply of bushings should be picked up when the cable is purchased.

Many electricians use a special cutter that has been designed to cut the armored cable without leaving rough edges, Fig. 8-48.

To use the cutter:

1. Attach the cutter to the cable and tighten the thumbscrew underneath.

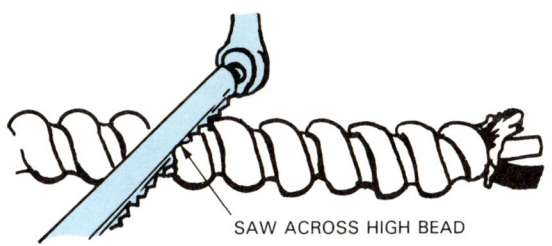

Fig. 8-44. Cutting armored cable with hacksaw. Do not cut so deep that you damage insulated wires.

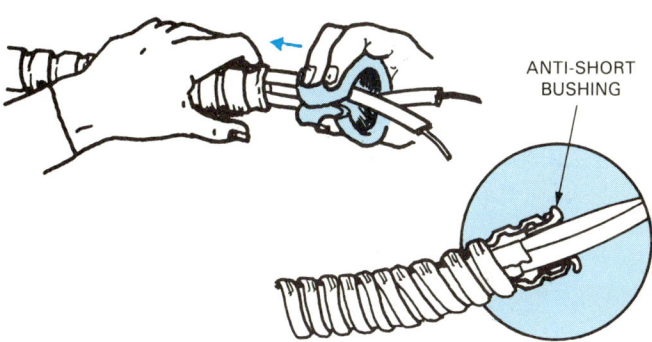

Fig. 8-47. Squeeze anti-short bushing between thumb and forefinger as shown and slide it inside the sheath to protect wires from jagged edges of sheath.

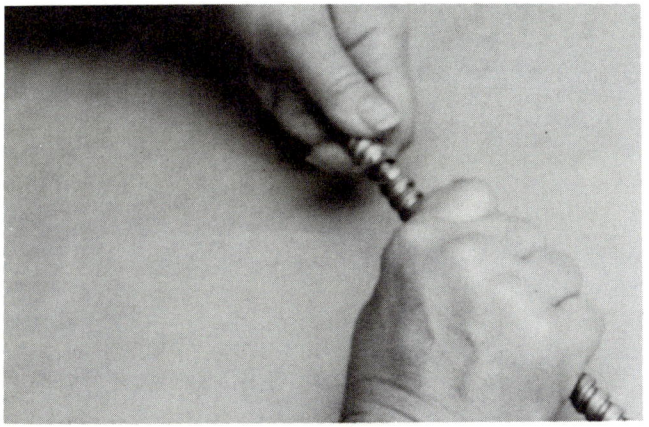

Fig. 8-45. Remove section of armored sheath to expose wires.

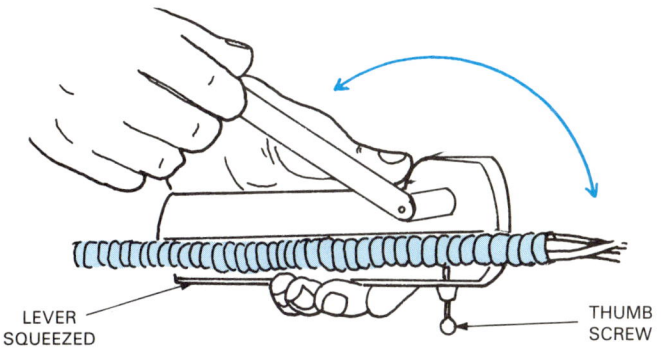

Fig. 8-48. Special cable cutter for armored cable. It must be mounted and clamped to cable.

2. Squeeze the handle on the underside to lower the cutter but do not apply too much pressure if you wish to have a fast, clean cut.
3. Apply additional pressure until force required to handle lessens.
4. Rotate the crank through a short arc forward. Release the crank and relax your grip on the tool.
5. Move the crank slightly and allow it to return to its original position.
6. Remove the tool by loosening the thumbscrew.

A third method of cutting armored cable uses the metal shears. Bend the armor sharply and twist until the armor buckles. Insert the shears through the open loop of the buckled sheath and cut. Trim off any sharp edges.

CONNECTING ARMORED CABLE TO BOXES

When the anti-short bushing is in place, install a connector as shown in Fig. 8-49. Wrap the ground wire around the setscrew to provide a proper ground to the box. Turn in the screw until it holds the cable housing securely, Fig. 8-50. Slip the locknut over the end of the wires. Turn it onto the threaded end of the connector. Tighten it with a screwdriver. The nut should bite into the side of the box for a good ground. See Fig. 8-51.

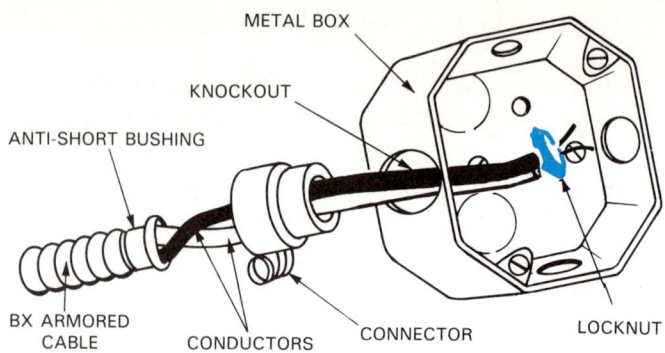

Fig. 8-50. Procedure for installing armored cable to box. Remove knockout from box. Slide connector onto cable. Slide cable through hole in box. Slip locknut over wires and onto connector threads.

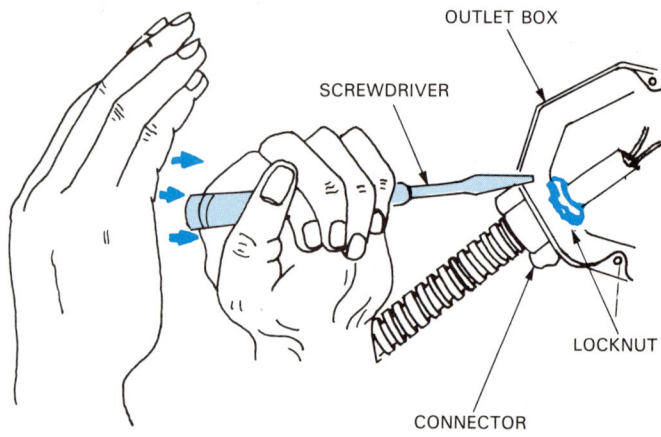

Fig. 8-51. Turn locknut onto threaded end of connector. Use screwdriver and palm of hand to tighten. Nut must "bite" into metal box for good bond.

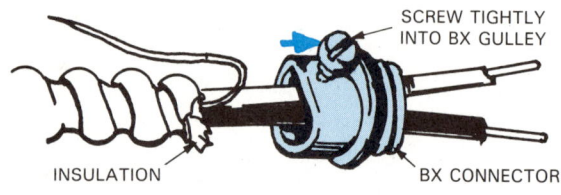

Fig. 8-49. Installing connector to BX. Inside end has peep hole so inspector can see anti-short bushing.

WORKING WITH NONMETALLIC CABLE (NM, NMC)

Nonmetallic cable is installed in walls and ceilings in the same way as armored cable, Fig. 8-52. Like armored cable, a grounding wire is necessary to provide proper bonding through the system. This will be discussed later.

PREPARING CABLE

The outer covering of nonmetallic sheathed cable is usually a plastic or thermoplastic material. It can be stripped off with a knife or special cable ripper as shown in Fig. 8-53. Again, remove about 8 in. or so from the end to allow at least that much to project into the outlet box. Six inches is required. However, it is better to have a little more than not enough. Be careful while removing the sheath not to damage the insulated wires within. Run the wire from outlet box to outlet box as with any other wiring method, but, be sure that you protect the cable from damage by using guard strips (lath) and cable straps wherever the wire crosses exposed studs or joists. In fact, it's best to run the cable along studs and joists. Install cable straps every 4 1/2 ft. for support.

In addition, when using nonmetallic cable, be sure to connect all ground wires to each other, to the box, and to the grounding provision on the device, for a positive and continuous ground, Fig. 8-54. Note the use of the *grounding screw* in the rear of the metal outlet box. Metal boxes have pretapped holes for this screw.

Locknuts, bushings, and connectors for nonmetallic cable are of slightly different design, than those used with conduit or armored cable, but are installed in the same fashion.

Fig. 8-55 shows nonmetallic sheathed cable ground wire attached to the box by means of a *grounding clip.* These clips are simply pushed onto the front edge of the outlet box.

Installing Boxes and Conductors

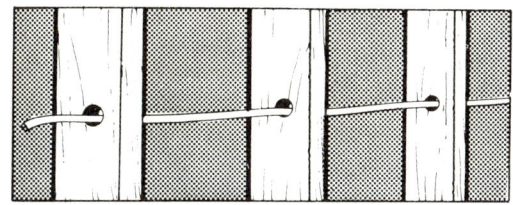

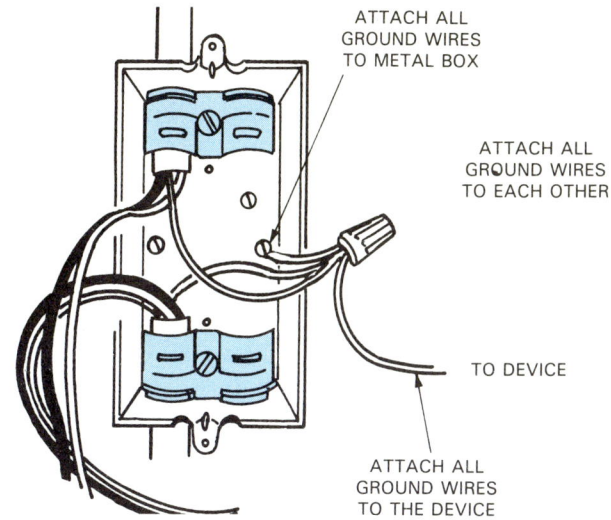

Fig. 8-54. Firmly attach all ground wires together, to the box and to any device installed at the box.

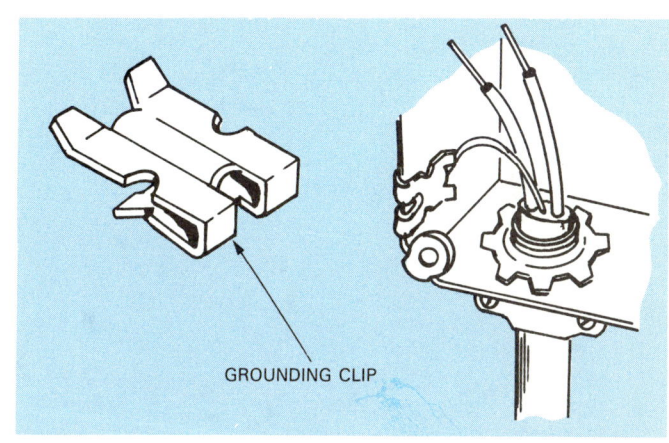

Fig. 8-52. As with conduit or armored cable, nonmetallic cable can be run through holes or notches made in studs and along ceiling or floor joists. Drill holes as straight as possible for easy cable pulling. Top. Holes drilled in studs several feet off floor is conventional method. (Allied Moulded Products Inc.) Bottom. Wiring run through vee notches at sole plate does not interfere with insulation. (Owens-Corning Fiberglas Corp.)

TEST QUESTIONS — CHAPTER 8

1. The process known as roughing-in includes installation of _____, _____, _____, _____, and _____.
2. Before attaching electrical boxes to the building frame, the electrician will (indicate which answers apply):
   a. Call the electrical inspector for permission to go ahead.
   b. Study the electrical plan for location and type of fixtures in each room.
   c. Mark exact location of each box and indicate type of box needed.
   d. Drill holes for running of conduit and/or cable.

Fig. 8-55. Nonmetallic sheathed cable installed and properly grounded in an outlet box. Note the use of a grounding clip. It is particularly useful in "old work" where grounding screws are difficult to install.

   e. Run conduit or cable through the building frame.
3. The proper height for installation of a receptacle (except in a kitchen counter area) is usually:
   a. 8 in.

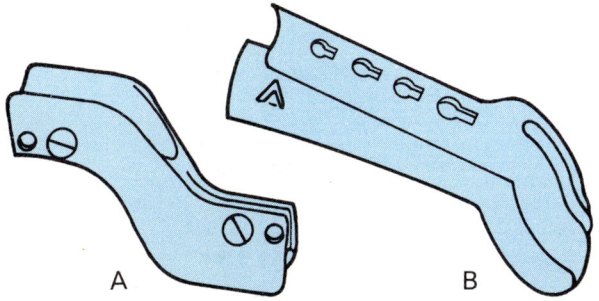
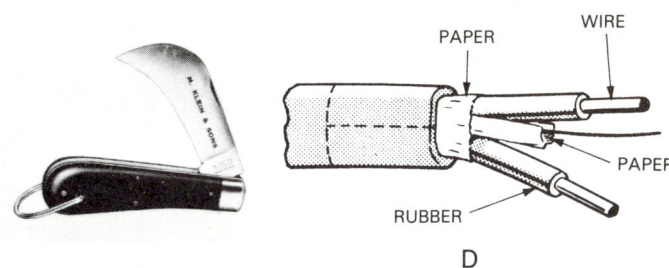

Fig. 8-53. Cable strippers and stripping knife. Cable strippers are handy devices for removing plastic jackets quickly and neatly. A—Type used with UF cable. B—Type stripper used for NM or Romex. C—Stripping knife. D—Method of using knife.

b. 12 in.
   c. 18 in.
   d. 48 in.
4. Wiring is usually installed circuit by circuit. True or False?
5. Describe how conduit is protected where it is notched into framing.
6. A _____ or _____ is used to bend conduit.
7. The radius of a 90° bend in a 1 in. rigid conduit must be no less than:
   a. 3 in.
   b. 5 in.
   c. 6 in.
   d. 8 in.
8. Describe how to make a 90° bend.
9. An offset bend is often referred to as a stub. True or False?
10. _____ is the amount of shortening which takes place in a piece of conduit when an offset bend is made.
11. The type of bend used to bypass an obstruction, such as a 4 in. drain pipe, is known as a (stub, saddle, offset, center) bend.
12. A run of conduit should have no more than the equivalent of four _____ bends.
13. Describe how the fish tape is used to pull wires through conduit.
14. When installing BX (armored cable) an ___-_____ bushing must always be inserted between the armor and the conductors for protection of the insulation on the conductors.
15. Nonmetallic cable must be supported every _____ ft. with a cable staple. (4 ft., 4 1/2 ft., 6 1/2 ft., 10 ft.)

Chapter 9
DEVICE WIRING

Once the electrician has roughed in conductors and boxes and the rough-in has passed inspection, the installation of devices and fixtures begins. This stage is often referred to as FINISHING.

In this stage the final connections are made. Conductors are spliced to one another and connected to switches, receptacles, fixtures, and occasionally, motors. As in other stages, the *Code* has specific requirements for making the system safe.

Article 110 of the *National Electrical Code* sets down various regulations concerning the design and installation of electrical materials and equipment. This set of requirements must be kept in mind at all times by the practicing electrician. Failure to comply with these rules, or any NEC mandates, will most certainly result in a *Code* violation.

One key word used throughout the *Code* is "shall." *It indicates a condition which must be met. It is* mandatory.

EQUIPMENT ACCEPTABILITY

Generally, all electrical equipment and materials must be approved, Fig. 9-1. That is, some inspecting authority must judge the items used in an installation to be suitable. This approval of electrical devices and materials is almost always based on product "listing" or "labeling." This refers to whether or not the item used in the installation has been tested and found acceptable by a recognized testing facility. An unlisted or unlabeled item is rarely used.

In addition, and to reinforce produce approval, the *Code* states that products listed and labeled by such testing agencies as Underwriters Laboratories, Factory Mutual, Canadian Standards Association, and Edison Testing Laboratories, shall be used only for the purpose and conditions as described in the data application directories of those agencies.

WIRING METHODS

The *Code* gives details on various wiring methods in Articles 300 through 384. In these articles it describes, defines, limits, and specifies the many methods of wiring used in the electrical industry today.

MECHANICAL CONSIDERATIONS

The proper mechanical execution of electrical construction is an important subject. *All electrical equipment must be mounted securely.* In addition, the electrical materials and equipment must have a neat and orderly appearance. Conductors, cables, and the like must be carefully routed and sufficiently supported to eliminate "bunching" or "twisting" of the cable and possible conductor damage.

Fig. 9-1. Only with electrical products which carry the label of a testing agency can you be certain of safe, reliable performance. (Underwriters Laboratories, Inc.)

ELECTRICAL CONNECTIONS

Where electrical wires are terminated or spliced, a suitable connection must be made. This requires a clean, secure physical contact between the conductors and/or the device terminals. Connections between unlike metals, such as copper and aluminum, shall not be made unless the electrical devices are suitable for this purpose and are identified to indicate this. *All splices and connections must be covered with an insulation equal to the conductors' original insulation.*

WORKING SPACES

To assure accessibility and to maintain, service or operate electrical equipment, the *Code* insists that "work space shall not be less than 30 in. wide in front of the electric equipment." Further, these spaces must not be cluttered with crates or boxes or in any way used for storage. Storing items in front of electrical equipment would prevent rapid access in an emergency.

IDENTIFICATION FOR SAFETY

Major electric equipment components and the disconnecting means for such devices should be clearly identified. Information as to voltage, amperage, and the particular equipment being controlled should be permanently marked on the disconnect means.

The intent of this requirement is greater safety. Should a circuit or device need to be de-energized in an emergency, the marking would allow for rapid action. See Fig. 9-2.

OSHA standard (Section 1910.303[f]) makes the identification and marking of electrical equipment mandatory for all existing, new, expanded, or modernized systems. The marking must be durable to withstand environmental conditions.

PREPARATION OF CONDUCTORS

Wires are the conductors of electrical energy. They are used to carry this energy from place to place. In this process, they must connect to each other and terminate at various electrical devices. In order for them to move electrons from place to place, conductors are insulated to eliminate as much energy loss as possible. However, where connections are made, this insulation must be removed for proper contact.

STRIPPING CONDUCTORS

Conductors may be an insulated solid wire, Fig. 9-3, or a wire made up of many strands, as in Fig. 9-4. Cable conductors are encased in several layers of insulation in addition to a protective outer covering that may be metal, plastic, or a composition of other materials. There are a variety of methods for removing the insulation from each type so that the conducting wire is bared for a good connection. In removing the insulation you will have to be careful

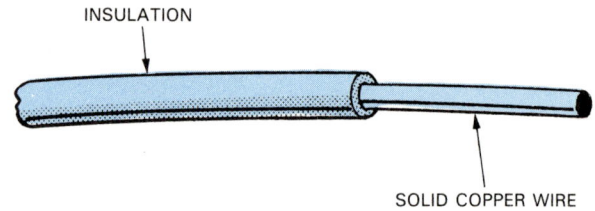

Fig. 9-3. Some conductors (wires are made of a solid length of copper or aluminum.

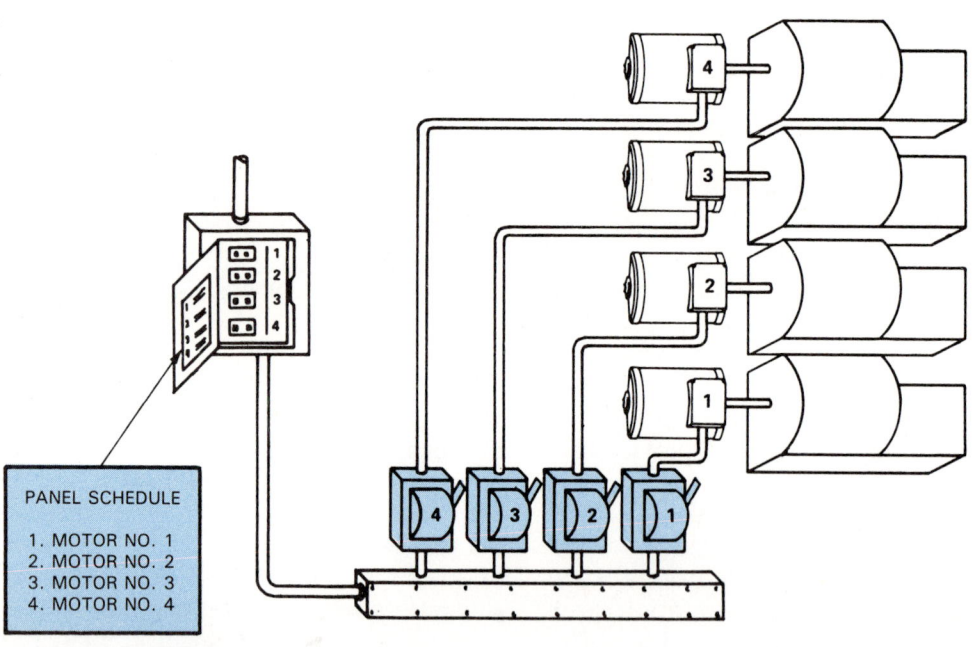

Fig. 9-2. Code requires that disconnect switches for electrical machines and devices should be clearly marked to identify what they control. (OSHA)

Device Wiring

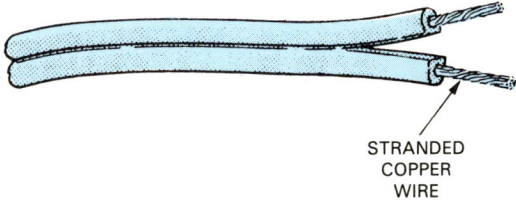

Fig. 9-4. This conductor is made up of many thin strands of wire inside an insulating material.

not to damage either the conductor wire or the remaining insulation.

USING STRIPPING KNIFE

You can remove the insulation from a single wire by carefully stripping it away with a sharp pocketknife or electrician's knife designed for this purpose. Do this carefully so that the blade does not nick or cut away any of the conductor material. Make sloping cuts toward the wire end as shown in Fig. 9-5.

Do not make a circling cut at right angles to the insulation. It is almost impossible to control the depth of cut in this way.

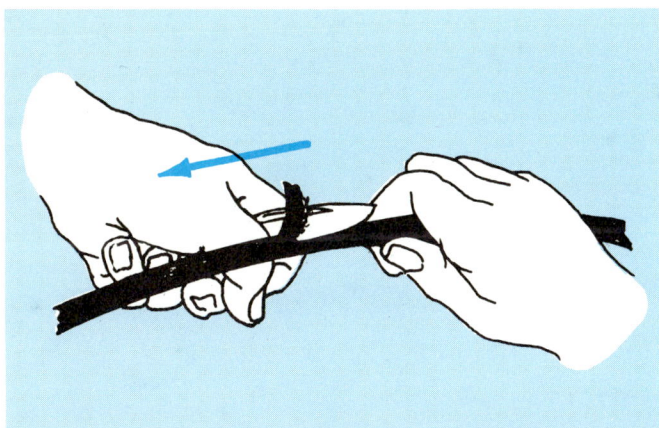

Fig. 9-5. Whittling cuts are used to remove insulation with a knife.

A grooved conductor will break easily if bent at the point of damage. Also, a nick or groove could reduce the current carrying ability of the conductor.

After you have cut away the insulation all around the wire using "whittling" cuts, twist the short piece of insulation. It will separate, as in Fig. 9-6. The stripped end should look like the one in Fig. 9-7.

As a general rule, a little less than an inch of the conductor should be bared. This allows enough bare wire to make the proper connection. Some electrical devices have gages for determining how much insulation to remove. See Fig. 9-8.

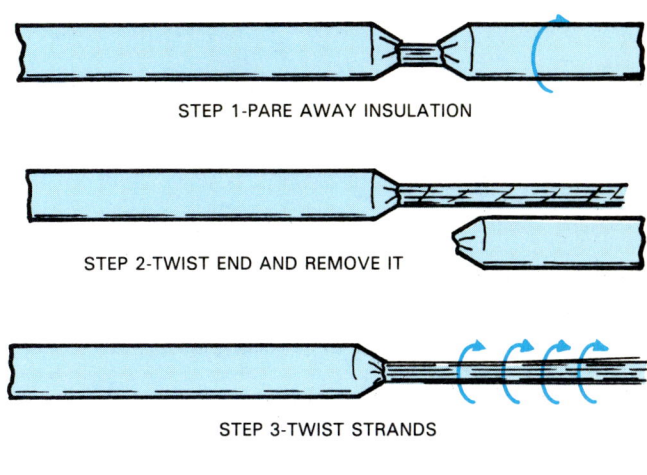

Fig. 9-6. Twist cut insulation to break it away. If conductor is made up of many strands of wire, twist them together as one.

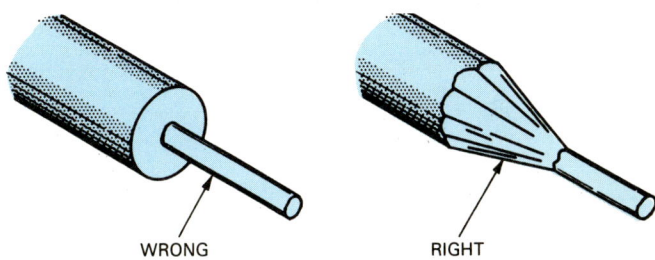

Fig. 9-7. Properly stripped wire shows insulation cut to a taper. About 7/8 in. of conductor should be bared. (U.S. Navy)

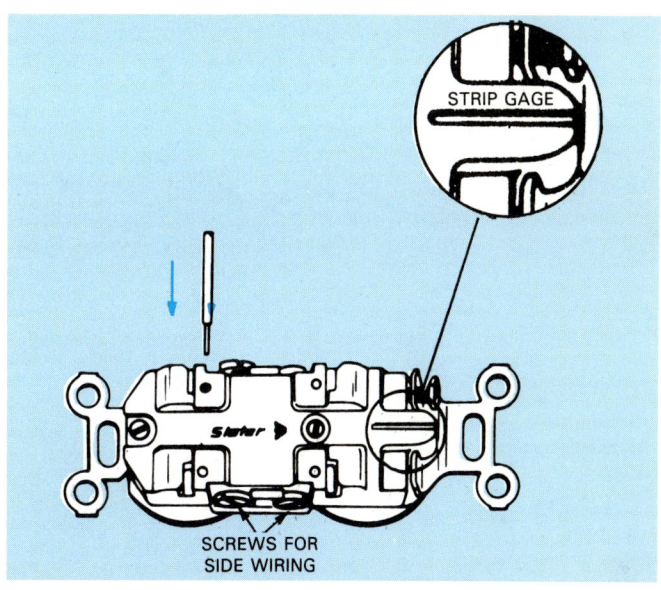

Fig. 9-8. Some electrical devices have a strip gage built into them. Lay conductor against gage to determine how much insulation to cut away. This receptacle also has screws for side wiring. (Slater Electric Inc.)

87

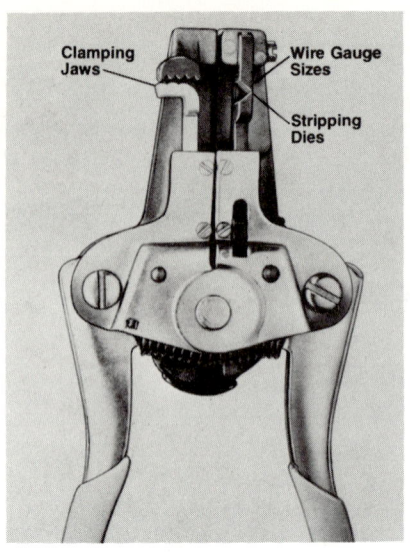 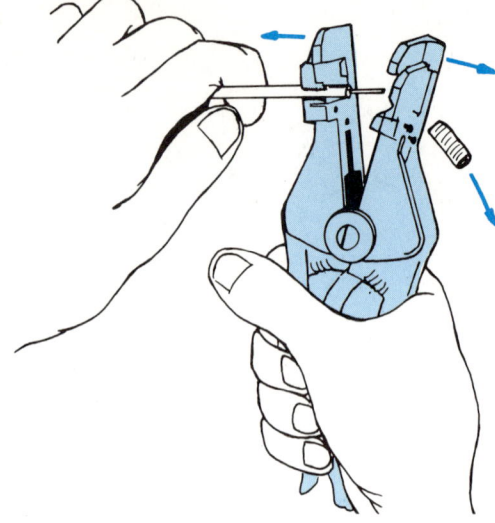

Fig. 9-9. Using a wire stripper speeds up the process. Left. A type of wire stripper. (Vaco Products Co.) Right. Insert wire through stripper jaws and squeeze the handle.

USING WIRE STRIPPER

The preferred way to remove insulation is with a wire stripper like the one pictured in Fig. 9-9. In one model, the end of the conductor is placed in the jaws of the tool. When the handles are closed the jaws will separate and strip away the insulation.

Good wiring technique includes the ability to make approved splices of conductors. The *Code* does not allow splicing or connecting of wires any other place except inside a proper housing. Likewise, connections between a conductor and an electrical device such as a switch or receptacle, must be inside a housing.

ATTACHING CONDUCTORS TO DEVICE TERMINALS

When solid conductors are connected to screw type terminals the ends must be formed in a loop for a proper connection. Use a needle-nose pliers or similar tool to form a curved hook in the conductor as shown in Fig. 9-10. This must be connected clockwise onto the terminal. When the screw is tightened the open end of the loop will be pulled inward, making the connection better. If the screw turns against the open end of the formed hook (counterclockwise) the loop will tend to open. This results in a poor connection which may be unsafe. See Fig. 9-11.

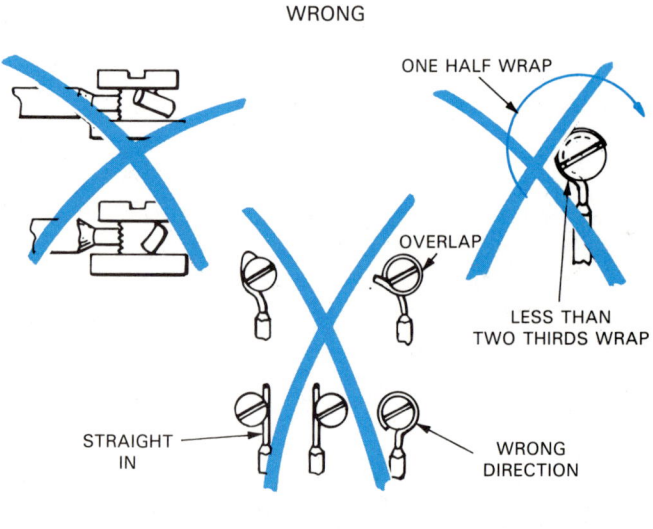

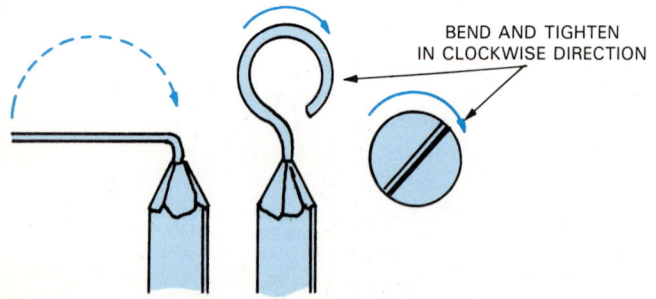

Fig. 9-10. Forming loop in conductor. Bend wire to left just above the insulation. With a longnosed pliers, form a loop to the right.

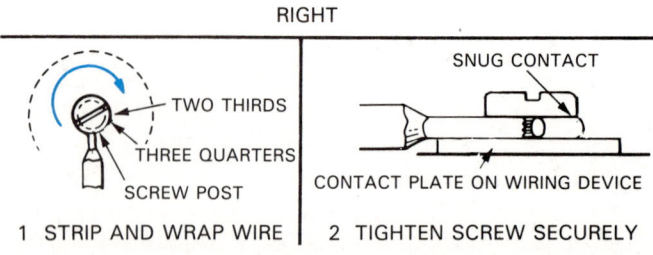

Fig. 9-11. Wrong and right way to attach a conductor to a terminal. Terminal screw must be turned down snugly for a good connection. (GE Wiring Devices Dept.)

Device Wiring

Electrical devices are made to connect with wires in a variety of ways:

1. Terminal screws around which the conductor is wrapped. The screw is tightened to hold the conductor in contact.
2. A variation of the terminal screw which does not require a wrap or loop in the wire.
3. Terminals requiring no screws at all. Contact is maintained, however, with a tension (spring) arrangement.

Several of these devices are illustrated in Fig. 9-12 and Fig. 9-13.

SPLICING CONDUCTORS

Connections (splices) between two wires are made by first twisting them together, Fig. 9-14. These connections must be held together by some method. Just the twisting is not enough. To save

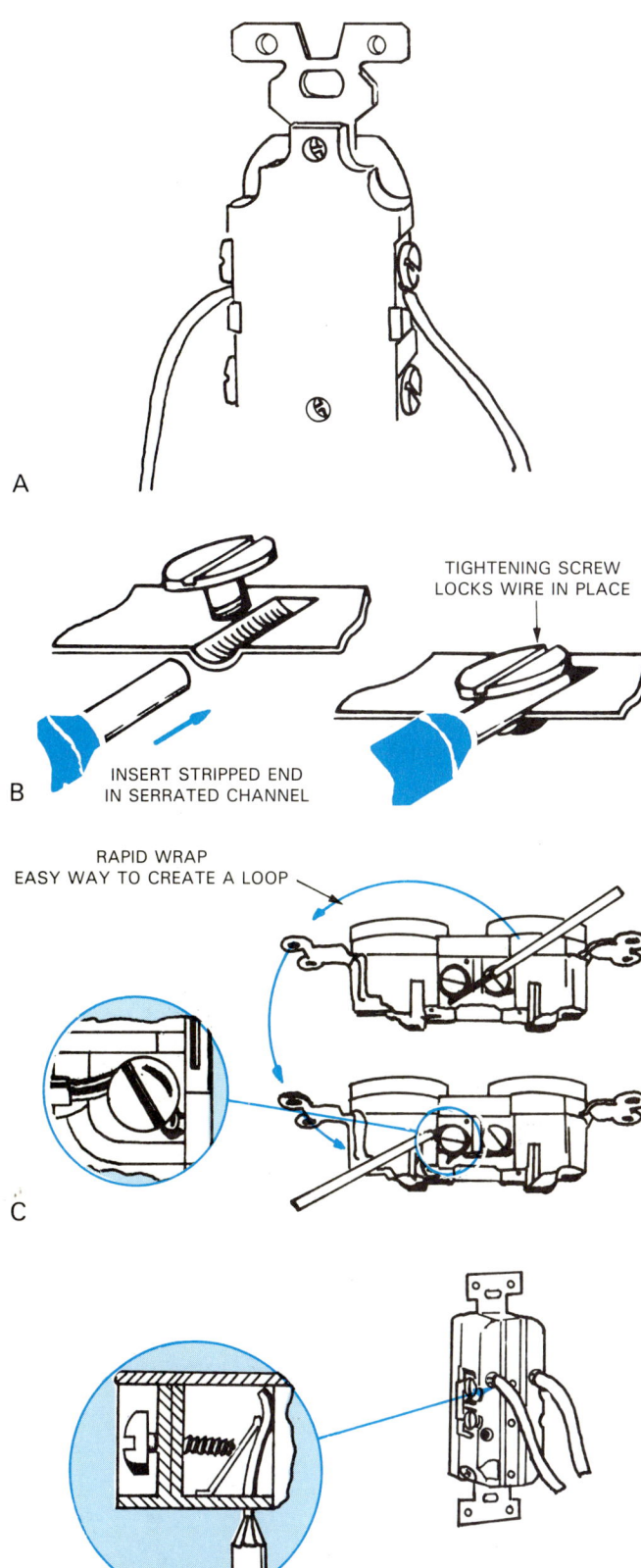

Fig. 9-12. There are several types of screw terminals. A—Receptacle with standard screw type terminals. B—Pressure type. Bare wire is inserted in groove rather than being wrapped around terminal screw. Tightening screw holds the wire. C—Side wired receptacle. conductor is inserted in holes and then snugged around or against screw. (Slater Electric Inc.) D—View of back wired receptacle. Tightening screw wedges wire against spring clip for secure contact.

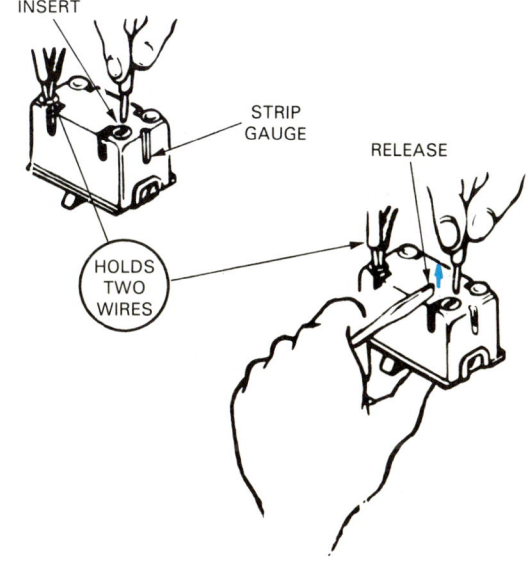

Fig. 9-13. Views of back wired switch using no screws. contact is made by spring tension inside the device. (Slater Electric Inc.)

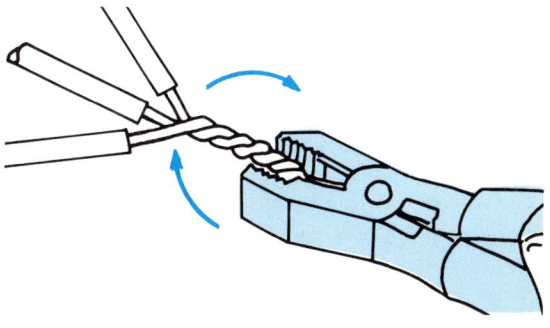

Fig. 9-14. Connections between several wires is made by twisting them together clockwise. After this is done, the splice can be secured and insulated by turning on a wire nut. (GE Wiring Devices Dept.)

time, the splice is usually secured with a wire connector, such as the one in Fig. 9-15. A spring connector is an often-used variation of the wire nut. It works well for connecting wires that will have tension (pull) on them.

WESTERN UNION SPLICE

The Western Union splice was once the most common way to join two small wires. It was especially preferred where wires were placed under heavy strain.

To make the Western Union splice, refer to Fig. 9-16 and follow these steps:
1. Strip away about 8 in. of insulation from each wire. Lay one wire across the other at right angles (more or less).
2. Twist one wire around the other in a tight wrap. In most instances the wire can be wrapped with the fingers and the end bent tightly against the straight wire with a pliers.
3. Wrap the second wire around the first in the same way. The wrapped section of wire should be about 2 in. long. Four or five turns on each wire should be enough.
4. Apply paste flux to twisted wire. Solder the joint carefully as shown in Step 4, Fig. 9-16.
5. When the joint has cooled, tape the entire spliced area. Be careful to overlap the original insulation at either end.

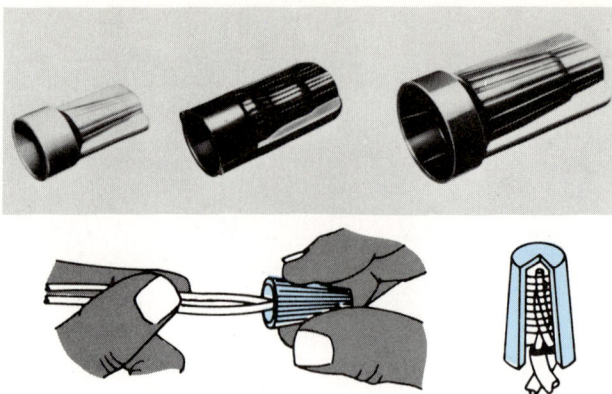

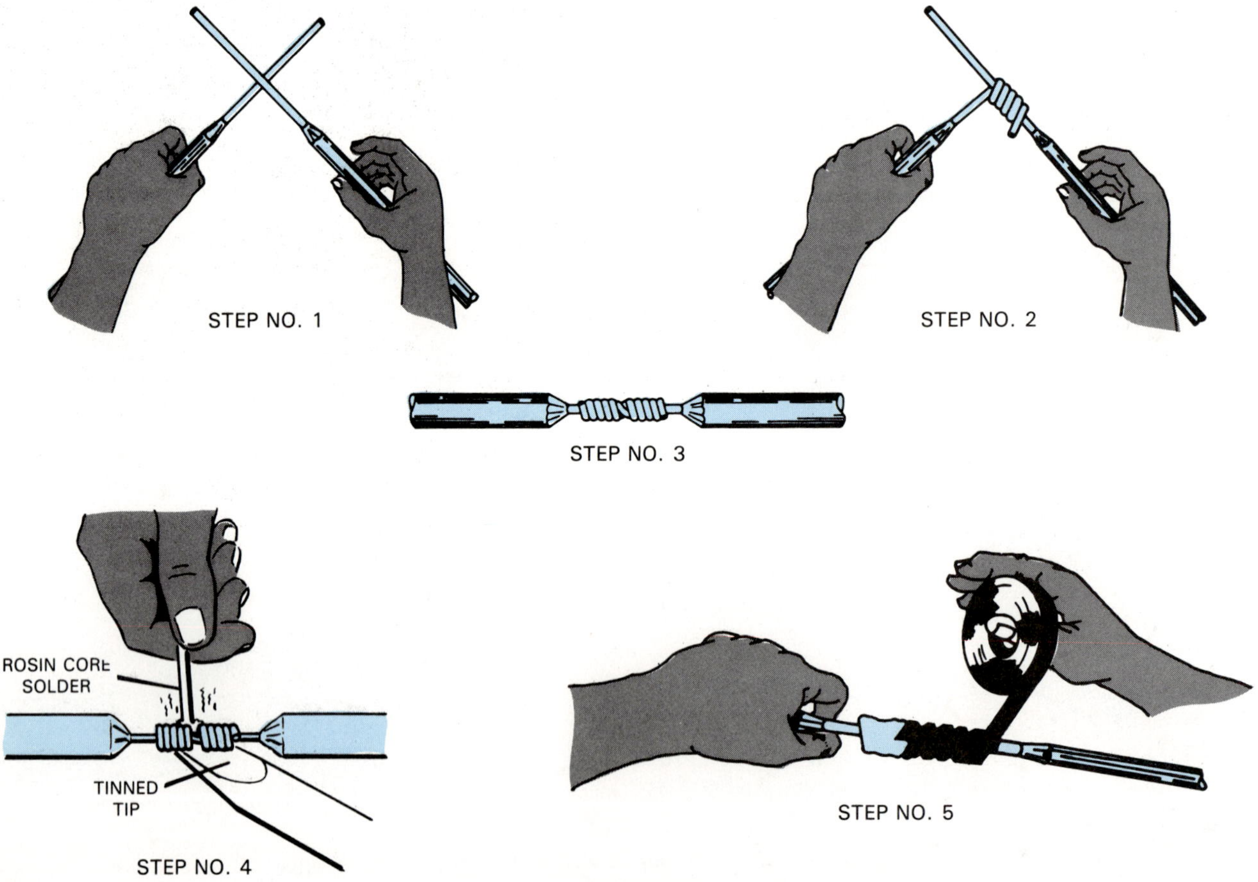

Fig. 9-15. Solderless screw-on connectors are commonly known as "wire nuts." They serve same purpose as a soldered connection and have generally replaced that practice. Wire nuts are made of plastic, Bakelite, or porcelain. Threaded female insert grips wires. Top. Three sizes of plastic wire nuts. (Amerace Corp., Elastimold Div.) Bottom. Method of attaching nut. Cutaway shows threaded, tapered, metal insert. Hood of connector must cover all bare wire. (OSHA)

Fig. 9-16. Western Union splice. NEC code specifies that some splices be made using solder. (U.S. Labor Dept. and U.S. Navy)

Device Wiring

PROPER SOLDERING TECHNIQUE

The key to good soldering is to heat the joint so well, that it melts the solder onto itself. Never drop-melt the solder onto a cold joint. To heat the wire, touch the hot soldering iron, called the "copper", to the joint.

Touch the solder to the joint so that it melts and flows onto it. Allow the joint to cool slowly for a good electrical connection. *Never use acid solder on electrical work. Use rosin-core type only.*

The NEC is very specific on what type of splicing should be done and where such splicings are permitted. It requires the soldering, brazing or welding of certain splices, and insists that all bare wire must be taped with insulation at least equivalent in thickness to that of the conductors' original coverings. See *Code* Section 110-14 (b). In addition, most connectors are marked "CU" for copper wire, or "AL" for aluminum wire or "CU/AL," for either. This coding must be followed strictly for safety purposes.

OTHER METHODS OF JOINING WIRE

Besides the previously discussed methods of joining wire, joints can be secured with other metal connectors such as lugs and split-bolt connectors, Figs. 9-17 and 12-22. Both types are used to join rather heavy wires, such as AWG No. 6 or larger. Metal connectors or split-bolt types must be well wrapped with tape made of rubber, friction, plastic or a combination of these materials.

COMPRESSION CONNECTORS

Compression type connectors and lugs must be attached to conductors with a special tool, Fig. 9-18. In addition, special terminal and splicing systems have been developed by many manufacturers for more efficient and versatile wire connection.

Some advantages of this type of system, shown in the series of illustrations in Fig. 9-19, are:
1. Broader range of wire sizes. A single connector often will eliminate the need for two, three or more standard connecting devices.
2. Less maintenance needed. Connections remain tight.
3. Excellent electrical continuity and conduction.
4. Vibration-resistant connections. Terminations are made with pressure tools and remain secure.
5. Fewer tools needed for installation. A single tool will install all fittings.
6. Constant insulation thickness at splices and terminations.
7. Easy visual inspection of splice or termination.
8. Wire twisting is eliminated.

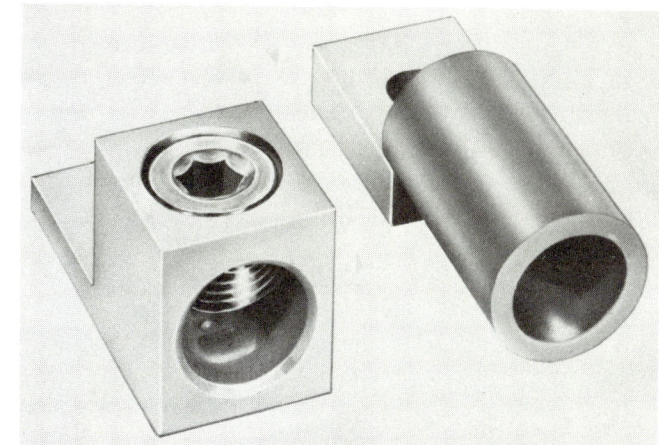

Fig. 9-17. Lugs are designed to join large-diameter conductors. Left. Screw type lug. Right. Compression type lug. (Square D Co.)

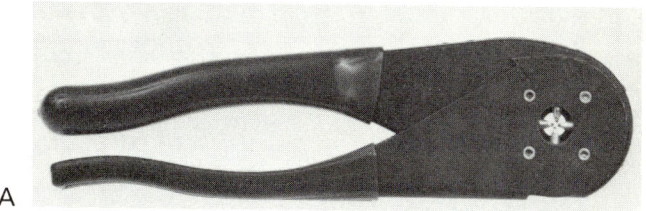

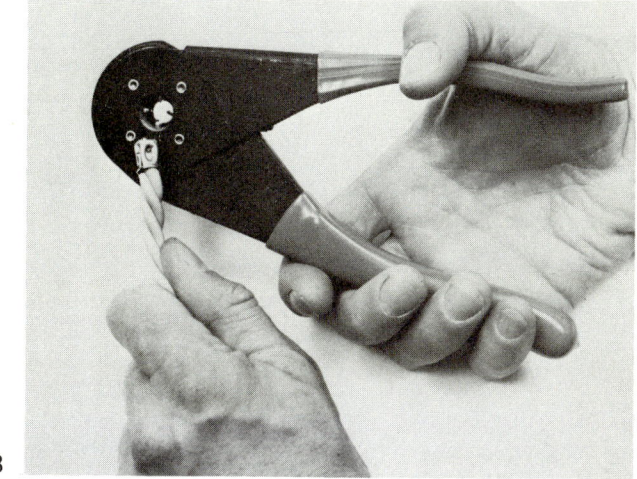

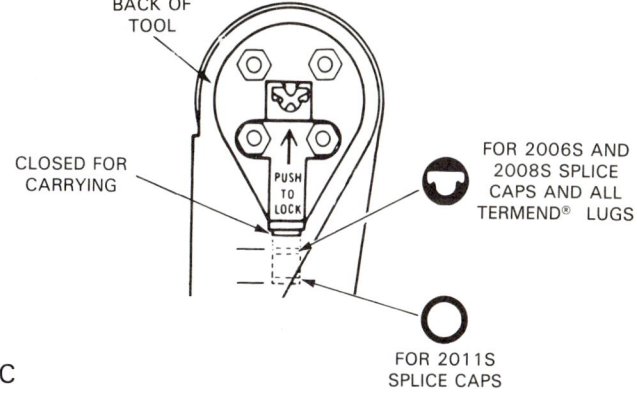

Fig. 9-18. Special tool for attaching compression lugs and connectors. A—In closed position, indenters are visible. B—In open position, indenters retract so splice cap can be inserted. C—Crimp tool is adjustable for different sizes of connectors. Note various positions. (Amerace Corp., Elastimold Div.)

CRIMP TIPS

See chart at far right for wire combinations listed by UL and CSA when using Splice Caps 2006S, 2008S and 2011S applied with C-24 pres-SURE tool.

1 TO SPLICE WIRES

Strip wires approximately 3/4".

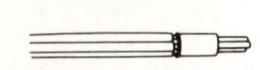

Insert untwisted stripped wires through Splice Cap.

Twist wires (except for 2008S).

Put tool latch in position "A" for 2006S or 2008S, or position B for 2011S (see C-24 tool illustration). Inset in tool. Squeeze to crimp.

Cut wires flush with cap.

Snap on nylon insulator (2007 for 2006S or 2008S, 2014 for 2011S).

2 SPLICING SOLID WIRE TO STRANDED WIRES

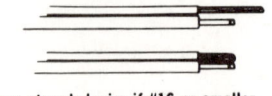

Loop stranded wire if #16 or smaller.

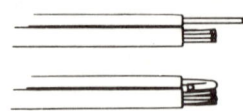

Loop solid wire if smaller than stranded wire.

When joining 2 or more solid wires to a larger stranded wire, twist solid wires together (except when using 2008S).

SPLICE CAPS

Wire Combinations listed by Underwriter's Laboratories, Inc. and certified by Canadian Standards Association. (Partial Listing)

Catalog Numbers	Wire Construction	WIRE SIZE — A.W.G.						
		#18	#16	#14	#12	#10	#8	#6
2006S and 2008S	Stranded and Solid	2 to 10 1 with 1 to 6 #16, 1 to 5 #14, 1 to 3 #12 or 1 #10 1 to 3 with 1 to 5 #16, 1 to 2 #14, 1 #12 or 1 #10	2 to 7 1 with 1 to 8 #18, 1 to 4 #14, 1 to 3 #12 or 1 #10 1 to 3 with 1 to 5 #18, 1 to 4 #14, 1 to 2 #12 or 1 #10	2 to 5 1 with 1 to 7 #18, 1 to 4 #16, 1 to 3 #12 or 1 #10 1 to 3 with 1 to 4 #18, 1 to 2 #16, or 1 #12	2 to 4 1 with 1 to 6 #18, 1 to 4 #16, 1 to 3 #14 or 1 #10 1 or 2 with 1 to 4 #18, 1 to 2 #16 or 1 #10	2 1 with 1 to 5 #18, 1 to 3 #16, 1 to 2 #14 or 1 #12 2 with 1 #18 or 1 #16	—	—
2011S	Stranded	—	—	5 to 11	3 to 7	2 to 5	2 or 3	2 #6
	Solid	—	—	4 to 10	3 to 6	2 to 4	2	
	Stranded and Solid			1 to 3 with 3 to 5 #12 or 3 to 4 #10 5 to 8 with 1 to 3 #12	1 to 3 with 5 to 8 #14 1 to 3 with 3 to 5 #14 1 or 2 with 3 to 4 #10	3 or 4 with 1 to 3 #14 or 1 to 2 #12 1 with 1 #8, 1 #16 or 1 #4	1 with 1 #10 1 #16 or 1 #4	

FOR COPPER WIRE ONLY

3 TO INSULATE SPLICES

Just snap it on! Merely place Nylon Insulator over installed Splice Cap and snap on. (2007 for 2006S & 2008S; 2014 for 2011S)

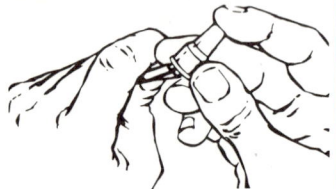

Cut-away below shows metallic retainer which holds insulator securely on installed Splice Cap.

No wrapping! No threading! No vibration worries—JUST SNAP IT ON!

4 TO TERMINATE WIRES WITH TERMEND® LUGS.

Strip wire(s) approximately 5/16".

Slip lug on wire—with tool latch in position "A" (see opposite page).

Insert in tool with flat side of lug in up position so that tongue enters slot in latch.

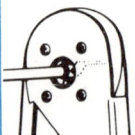

Squeeze tool to crimp.

Termend® lug in proper position for crimping.

5 TO REMOVE SPLICE CAPS or TERMEND® LUGS without damaging wires

By "cutting"

(Most effective when Splice Cap is full of wire.) Snip Splice Cap at both ends of one crimp, counteracting cutting pressure with index finger—then peel off cap.
To remove Termend® lugs, snip off tongue and proceed as for Splice Caps.

By "cold-working"

(Most effective when cap is not full of wire.) Apply pressure alternately at 2 points 90 degrees apart and between crimps—then pull off cap.

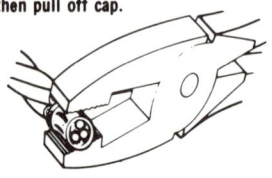

6 TO REMOVE NYLON INSULATORS

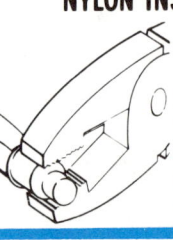

Apply pressure alternately at 2 points 90 degrees apart to "cold work" metallic retainer—or cut up side of insulator body.

7 TO TERMINATE #6 WIRE (Using two Termend® lugs)

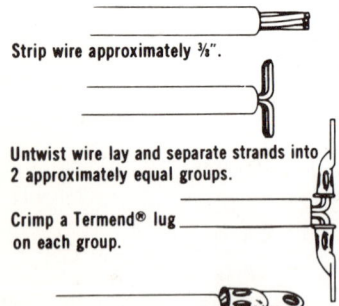

Strip wire approximately ⅜".

Untwist wire lay and separate strands into 2 approximately equal groups.

Crimp a Termend® lug on each group.

Bring flat sides of lugs together.

8 STRAIN RELIEF SPLICE (for service entrance, etc.)

Strip both wires approximately 1½" and untwist wire lay.

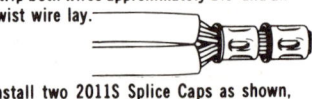

Install two 2011S Splice Caps as shown, leaving about 1/16" between caps.

Cut off the top half of the tip of an insulator and push remainder over cap.

Snap another 2014 insulator over the end cap.

9 REDUCER TAP OFF LARGER WIRE (for parallel street lighting, etc.)

Strip off insulation approximately 2" and squeeze to make wire loop as shown.

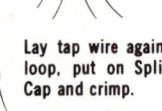

Bend enough strands down so remainder will fit a 2011S Splice Cap.

Lay tap wire against loop, put on Splice Cap and crimp.

Insulate with tape,— or 2014 insulator if wire insulation permits.

10 TEE-TAP (Where slack permits)

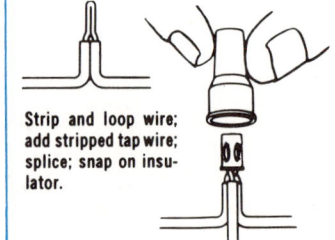

Strip and loop wire; add stripped tap wire; splice; snap on insulator.

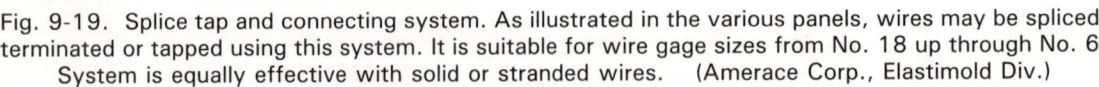

Fig. 9-19. Splice tap and connecting system. As illustrated in the various panels, wires may be spliced, terminated or tapped using this system. It is suitable for wire gage sizes from No. 18 up through No. 6. System is equally effective with solid or stranded wires. (Amerace Corp., Elastimold Div.)

Device Wiring

CRIMP TIPS

11 ADDING WIRE (#12 or Smaller) TO 2006S SPLICE CAP

Loop wire to be added and press into crimped indentation of 2006S.

Force 2011S Splice Cap over joint... Crimp.

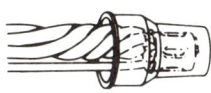

Insulate with 2014.

12 PARALLEL SPLICE

(For attaching one short free length of wire; or for splicing short free length of flexible lead to solid conductor of various shapes as in coil windings, etc.)

Place latch of C-24 tool in position "B" to permit passage of wire thru tool. (On #6 wire remove lay.)

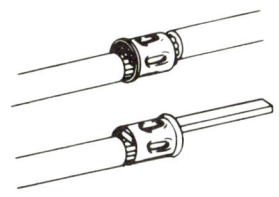

Insert wires into opposite ends of Splice Cap and crimp. Insulate with tape.

13 GROUNDING CONNECTION

Splice wires with 2006S or 2008S Splice Cap, leaving one wire extending thru cap to permite attachment of a Termend® lug.

14 TO TERMINATE TWO OR MORE WIRES

beyond the barrel capacity of a Termend® lug: Splice wires, leaving one or more wires extending thru the cap to permit attachment of lug.

15 TO SPLICE PIGTAIL SOCKETS TO STREAMERS

Strip streamer wires at required intervals. Thread on required number of 2006S or 2008S Splice Caps.

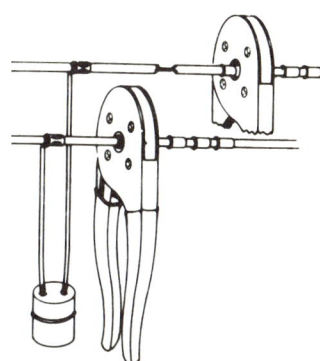

Put tool latch in position "B" and slide tool over streamer. Crimp each cap in succession as shown. Insulate with tape.

16 TO SPLICE TWO #4's (Using two 2011S Splice Caps)

Strip both wires 1½", untwist wire lay and bend approx. ½ of each wire at right angles.

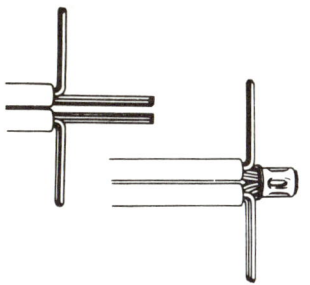

Install Splice Cap as shown and trim off excess wire.

Shape remaining strands around Splice Cap.

Install another Splice Cap.

Cut off the top half of the tip of a 2014 Insulator and remove retainer ring—force over lower cap.

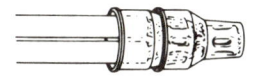

Snap another 2014 insulator over the end cap.

17 TO SPLICE BEYOND RANGE OF 2011S SPLICE CAP (3 or more #6's, 4 or more #8's, etc.)

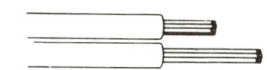

Strip one wire 1½", another wire ½", untwist wire lay.

Install Splice Cap as shown.

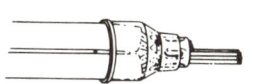

Cut off the top half of a 2014 insulator and push remainder over cap.

(Splicing (3) #6's shown)

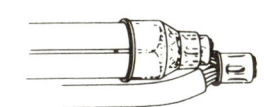

Strip 3rd wire ½" and splice as shown.

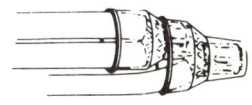

Snap on another 2014 insulator.

18 HERMETICALLY SEALED SPLICE

Fill the insulator 1/3 full of insulating compound and push over installed Splice Cap for perfect hermetic seal.

19 FOR PRE-INSULATED SPLICING

#2002 combines copper Splice Cap and nylon insulator in one unit. Just 1 size splices 2 thru 6 #18 stranded, 3 thru 5 #18 solid, 2 or 3 #16 stranded or solid, 2 #14 stranded and many other combinations. No wire twisting.

Tough see-thru insulator permits visual inspection, easy testing from wire end.

3-indent rolling action crimps evenly without damaging insulation. Use BUCHANAN P-50 pneumatic tool.

Fig. 9-20 illustrates splice caps, insulators, and terminal lugs. All are attached with the crimping tool shown in Fig. 9-18.

WIRING SWITCHES, RECEPTACLES AND LAMPS

In the average building, most electrical boxes will contain either switches, receptacles, lamps, or a combination of these. Therefore, connections must be made between the conductors and these devices.

SWITCH WIRING

Common switches are generally the same in size and appearance. Their function, of course, is to control the flow of electricity to one or more electrical devices. These devices include receptacles, fixtures, lamps, and heaters. Switches are commonly rated at about:
1. 125 volts at 10 amperes.
2. 250 volts at 5 amperes.

Other switches for special purposes are rated at smaller or larger voltages and amperages.

The most common switches are single-pole, three-way, and four-way. See Fig. 9-21. A single-pole switch, as pictured in Fig. 9-22, never has

SPLICE CAPS AND INSULATORS

for Wire Sizes	Splice Caps	Nylon Insulators
(2) #18 thru (4) #12 or (2) #10*	2006S Copper / 2008S† Steel	2007
(3) #12 thru (3) #8 or (2) #6*	2011S Copper	2014

*or equivalent combinations
†For splicing without wire twisting

TERMEND LUGS

Tongue Style	#6 Stud	#8 Stud	#10 Stud	1/4" Stud
Ring	16-8-6	16-8-8	16-8-10	16-8-¼
Spade	16-8-6S	16-8-8S	16-8-10S	—
Locking	16-8-6L	16-8-8L	16-8-10L	—

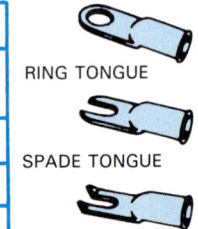

RING TONGUE

SPADE TONGUE

ROCKING TONGUE

Fig. 9-20. Top. Splice caps and nylon insulators create strong splices which resist loosening by vibration, expansion, or contraction. Bottom. Lugs designed for making neat wire endings on larger conductors. (Amerace Corp., Elastimold Div.)

TYPE	SYMBOL	ILLUSTRATION	WIRING DIAGRAM	USE
SINGLE POLE SWITCH	S		Neutral, Hot Leg, Switch, Hot Leg, From Source	Control current going to a load (such as light) when only one location of control if needed
THREE-WAY SWITCH	S₃		Load, Neutral, Hot, Black, Travelers Red/Blue, 3-WAY / 3-WAY, Source	Control current going to load when two control locations are required
FOUR-WAY SWITCH	S₄		Load, Neutral, Hot, Black, Red/Gray, 3-WAY / 4-WAY / 3-WAY, Source	Control current going to load when control is needed from three or more locations. Always used with pair of three-way switches

Fig. 9-21. Comparison of three types of switches used in residential wiring. These types control lights and, sometimes, receptacles used for lamps. (Leviton Mfg. Co. Inc.)

Device Wiring

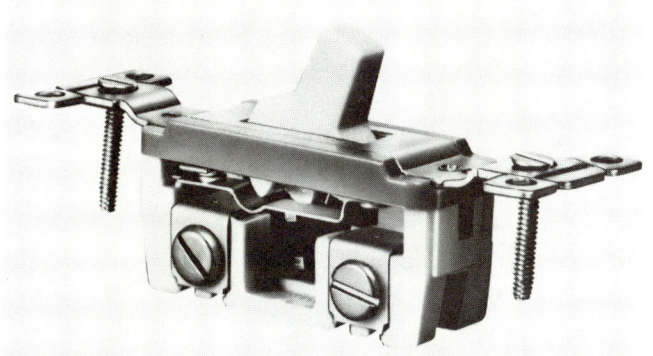

Fig. 9-22. Side view of single pole switch. It never has more than two terminals. (Bryant)

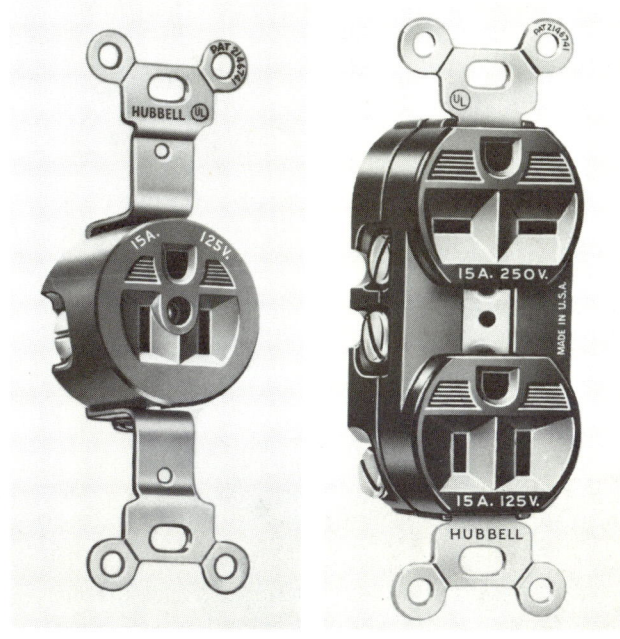

Fig. 9-24. Typical grounding type receptacles. Left. Single receptacle. Right. Duplex type with one receptacle reserved for 250 volts. (Harvey Hubbell Inc.)

more than two terminals. Three and four-way switches have three and four terminals respectively.

The switch function, regardless of type, is to interrupt the hot (black) wire only. It must never be placed on the grounded neutral (white) wire or the ground wire (bare or green). We will explore all types of circuits and how to connect them in Chapter 11.

RECEPTACLE WIRING

Outlet receptacles are used to transfer electrical energy from conductors to such devices as lamps, toasters, radios, television sets, blenders, vacuum cleaners, and numerous other appliances. They are connected with a flexible service cord which has a two or three-pronged (bladed) plug on the end.

MAKEUP OF A RECEPTACLE

A receptacle's current-carrying and grounding parts are arranged in a device that can be connected to a box with two small screws. The parts which conduct current to the appliance cord are enclosed inside nonconducting plastic. Contacts are made of tough alloys which hold their shape and remain springy even after years of service. See Fig. 9-23. When a plug is pushed into the slots, its blades push the metal contacts apart. Tension of the springy metal maintains good electrical connections against the blades. Thus, current may flow through the contacts into the plug easily.

DUPLEX RECEPTACLE

A duplex receptacle, like the one shown in Fig. 9-24, B, will hold two appliance or light plugs. A grounding type receptacle will have five terminals. Two are for the grounded neutral wire and the screws will be light or silver color. Of course, push-in type terminals may have no screw. The terminal will be marked to indicate which is neutral.

Two copper or dark colored terminals are for the "hot" or black wire connections. The fifth terminal is for connecting the bare or insulated green wire that goes to ground. This terminal is electrically connected to the U-shaped grounding slot on the front of the receptacle.

There are other important electrical connections you should know about before wiring the receptacle:
1. Between the two neutral terminals.
2. Between the two hot terminals.
3. Between the grounding slot, green hex-shaped ground screw, and the metal bracket which suspends the receptacle in the box.

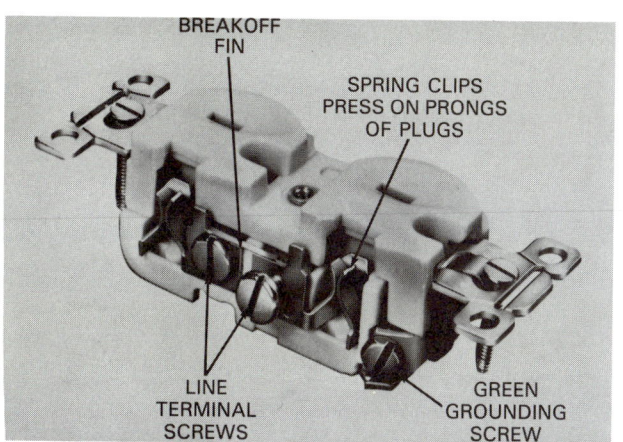

Fig. 9-23. Cutaway of duplex receptacle. Note that contacts are shaped like spring clips. They will keep a light pressure on blades of plugs to assure good electrical contact. Break-off fin ties two like terminals together. (Bryant)

Modern Residential Wiring

The grounding connections are permanent but connections between the other terminals can be interrupted under certain situations. These instances will be covered later.

Receptacles intended for 120 V appliances and other electrical devices have two vertical, parallel slots plus the U-shaped ground openings. Other receptacle outlets have different configurations (shapes). They are designed to accept only locking plugs or other special plugs. See Fig. 9-25.

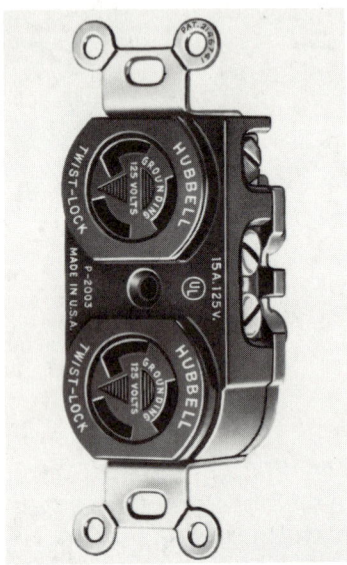

Fig. 9-25. Locking type duplex receptacle. Covers remain closed when outlet is not in use. (Harvey Hubbell Inc.)

MAKING THE RECEPTACLE GROUND

Methods of grounding receptacles vary. It depends on the kind of conductor system used and whether the boxes are metal or plastic. In a metal conduit system, the conduit and metal boxes provide a proper ground all the way back to the service entrance panel. In this case, there is no need for a grounding conductor. When the receptacle is attached to the box, grounding contact is made through the receptacle screws to the box. Also a bonding jumper must be installed between receptical grounding screw and metal box.

Other conductor systems including armored cable (BX), flexible metal conduit (Greenfield), nonmetallic cable, and rigid plastic conduit, require a ground wire. There must be continuity from the grounding screw of the receptacle to the box (if metal) and to the ground wire.

ATTACHING CURRENT-CARRYING CONDUCTORS

The wiring of receptacles is easy and simple. Connect the white neutral wire to one side of the device which has the silver or light-colored screws. Then connect the black or hot wire to the other side. This side will have brass or dark-colored screws. See Fig. 9-26.

When more than one receptacle is to be added onto the circuit from the first one they can be connected as shown in the Fig. 9-27. It is best to make the connections through a pigtail arrangement as shown in view B of Fig. 9-27.

SPLIT—WIRED RECEPTACLES

A split-wired receptacle is one in which one of the outlets in a duplex receptacle is controlled by a switch while the other is always energized. One neutral wire can still serve both outlets, however, two hot wires are required. One will go to a switch; the other to source. The break-off fin must be removed from the hot-side (dark colored) terminals.

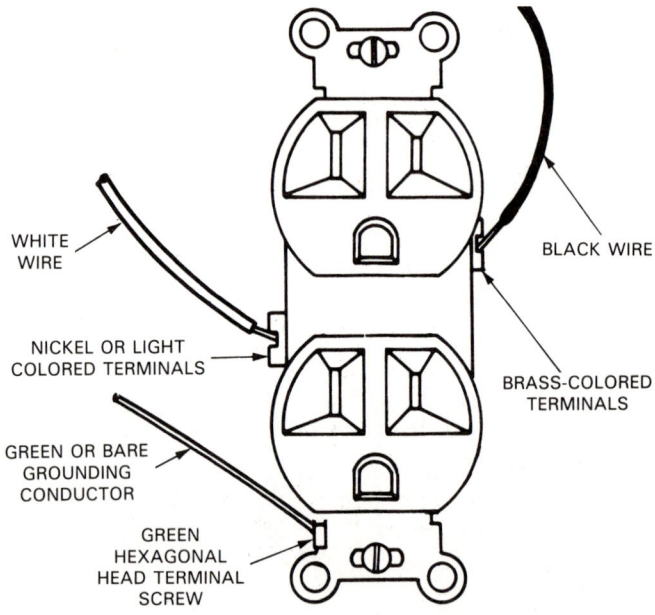

Fig. 9-26. How to connect grounding type receptacle to conductors. Attach bare or green insulated wire to green hex-head screw; black wire to brass terminal screw; white wire to light or silver-colored screw. (OHSA)

Fig. 9-28 shows a break-off fin as well as simple schematics of a split-wired receptacle.

As you know, receptacle outlets are manufactured in various configurations. A few of the more common ones are shown in Fig. 9-29. Other types for locking plug and receptacle configurations are shown in Chapter 25, Technical Information.

FIXTURE WIRING

Fixtures or lamps are wired in the same way as outlets. The neutral or white wire of the source is

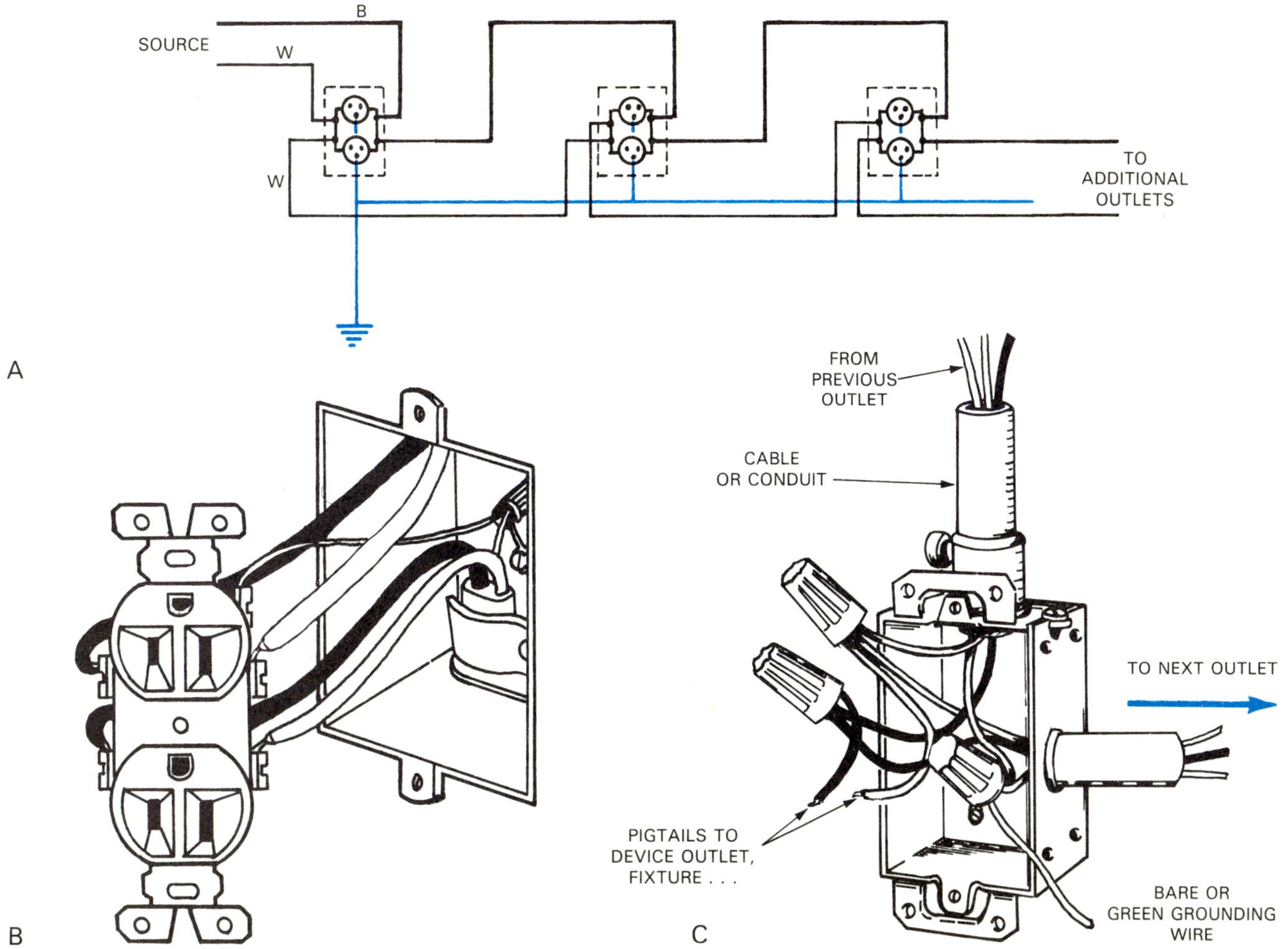

Device Wiring

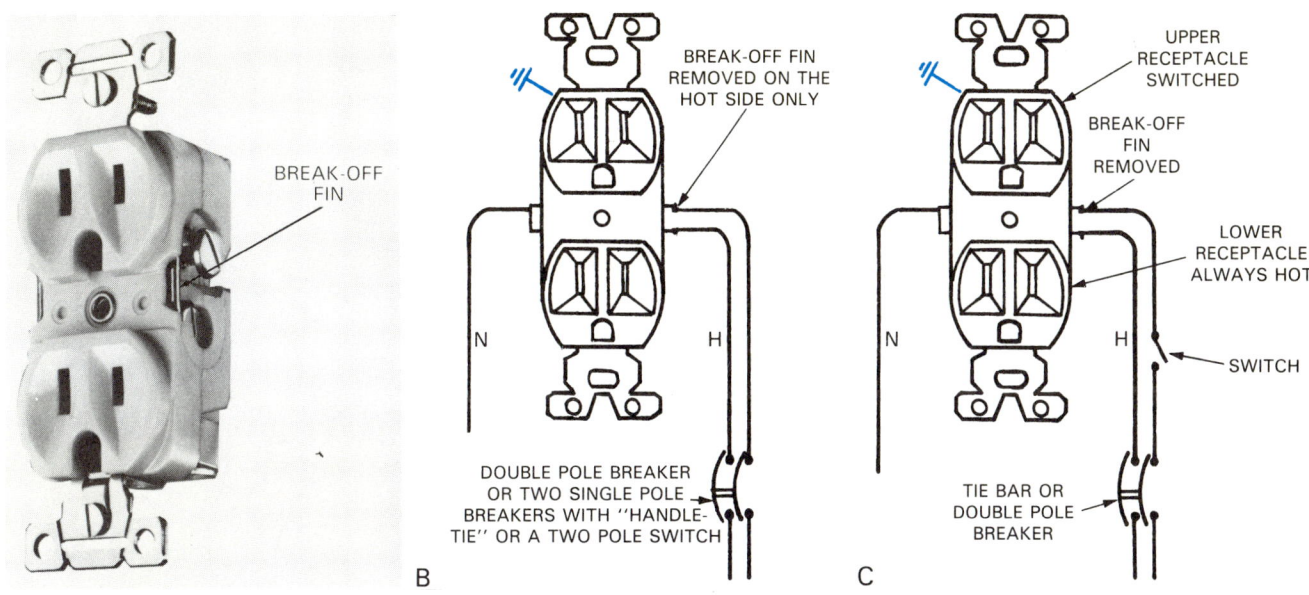

Fig. 9-27. Connecting a number of receptacles in a row on the same circuit. A—Schematic shows wiring arrangement. B—One method of running wires to next receptacle. (GE) C—Pigtail method is preferred to direct connection to the other terminals of the receptacle. Grounding wire is required unless metallic conduit is being used.

Fig. 9-28. Duplex receptacles have break-off fins which can be removed to split-wire the outlet. Then one outlet is controlled by a switch while the other is always hot. A—Duplex receptacle showing break-off fin. B—Hot wires of a split-wired receptacle must be on the same double-pole circuit breaker or two single-pole breakers with a tie bar. C—Diagram of a split-wired outlet with one outlet controlled by a switch. (Bryant)

97

Modern Residential Wiring

		15 AMPERE		20 AMPERE		30 AMPERE		50 AMPERE	
		RECEPTACLE	PLUG	RECEPTACLE	PLUG	RECEPTACLE	PLUG	RECEPTACLE	PLUG
2-POLE 2-WIRE	1 125V	1-15R	1-15P						
	2 250V		2-15P	2-20R	2-20P	2-30R	2-30P		
	3 277V	(RESERVED FOR FUTURE CONFIGURATIONS)							
	4 600V	(RESERVED FOR FUTURE CONFIGURATIONS)							
2-POLE 3-WIRE GROUNDING	5 125V	5-15R	5-15P	5-20R	5-20P	5-30R	5-30P	5-50R	5-50P
	6 250V	6-15R	6-15P	6-20R	6-20P	6-30R	6-30P	6-50R	6-50P
	7 277V AC	7-15R	7-15P	7-20R	7-20P	7-30R	7-30P	7-50R	7-50P
	24 347V AC	24-15R	24-15P	24-20R	24-20P	24-30R	24-30P	24-50R	24-50P
	8 480V AC	(RESERVED FOR FUTURE CONFIGURATIONS)							
	9 600V AC	(RESERVED FOR FUTURE CONFIGURATIONS)							

Fig. 9-29. Receptacles are manufactured for many different purposes. These are a few of the many configurations (shapes to fit special kinds of plugs). (NEMA)

connected to the white or otherwise neutral indicated wire of fixture. The black wire (from the source) is connected to the other terminal of the fixture. Fig. 9-30 shows wiring arrangement for a simple pull chain fixture. This fixture is at the end of a run. A ground wire (bare or green insulated) should be connected between the box and the grounding screw on the fixture. Fig. 9-31 shows the same pull chain fixture further along a circuit run. Note the symbols and schematics for these hookups.

Fig. 9-32 shows a fixture controlled by a switch rather than pull chain. If metal conduit is used, the green grounding conductor is not necessary since the conduit itself becomes the grounding means. Note that the feed source is to the fixture in the pictorial, but to the switch in the schematic.

MOUNTING FIXTURES

It is important to understand how fixtures are safely and securely attached to a wall or ceiling box.

Fixtures may be mounted to boxes in a variety of ways depending on:
1. Type of fixture.
2. Location of fixture.
3. What wall or ceiling is made of (composition).
4. Weight of fixture.

The types of fixtures may range from surface-mount to recessed, chain-supported or not, small-sized to enormous chandeliers. Regardless of type, fixture must be mounted securely to the outlet box.

Whether the fixture will be mounted on the wall or ceiling will affect the manner of supporting it. Usually, wall mounted fixtures are lightweight and can be directly mounted to the box without special provision for a mounting device. Heavier fixtures require substantial box supports and braces.

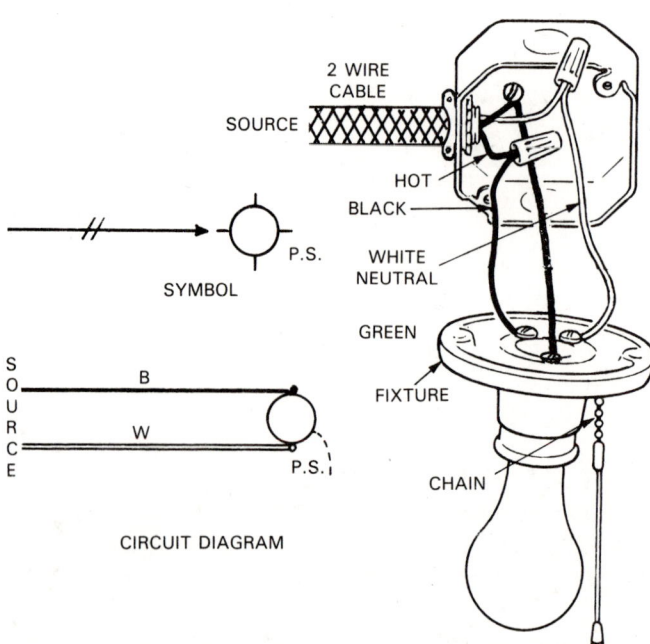

Fig. 9-30. Wiring arrangement for a pull chain light fixture. Compare symbol and wiring diagram with actual fixture. Grounding wire (green) must be used with cable conductors.

98

Device Wiring

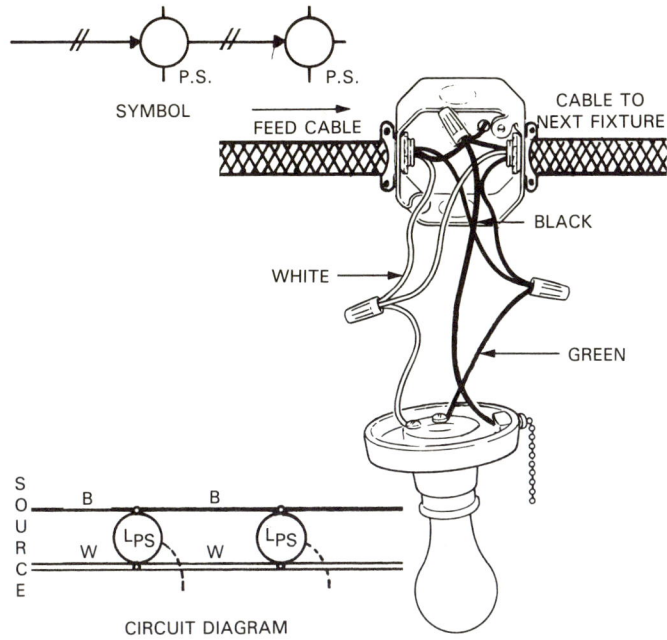

Fig. 9-31. Method of wiring a fixture when conductors must go on to another fixture. Note symbol and diagram.

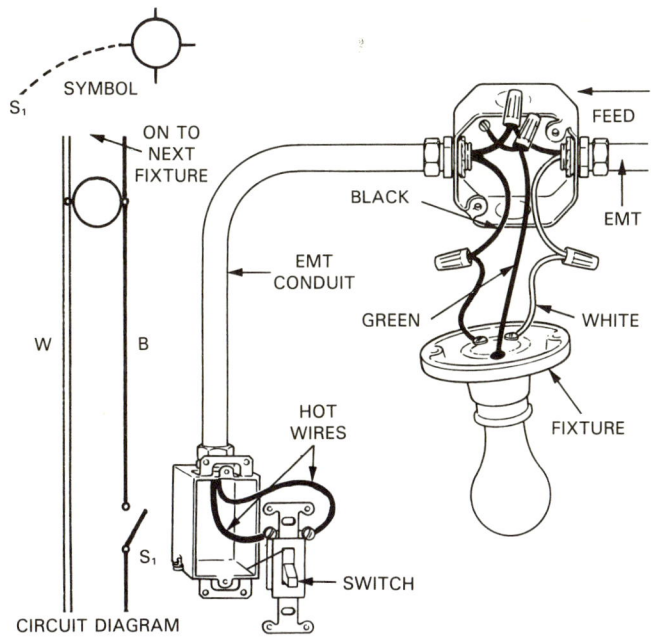

Fig. 9-32. Simple circuit using switch to control a light. Feed is to fixture.

If you are mounting a fixture in plaster and lath, wallboard, sheetrock or any other of the various wall finishings, special considerations must be given to the best procedure to follow. Care must be taken not to damage the fixture or finished surface.

The illustrations in Fig. 9-33 will serve as a guideline for methods and devices used to mount fixtures. Often, special fittings are supplied with the fixture.

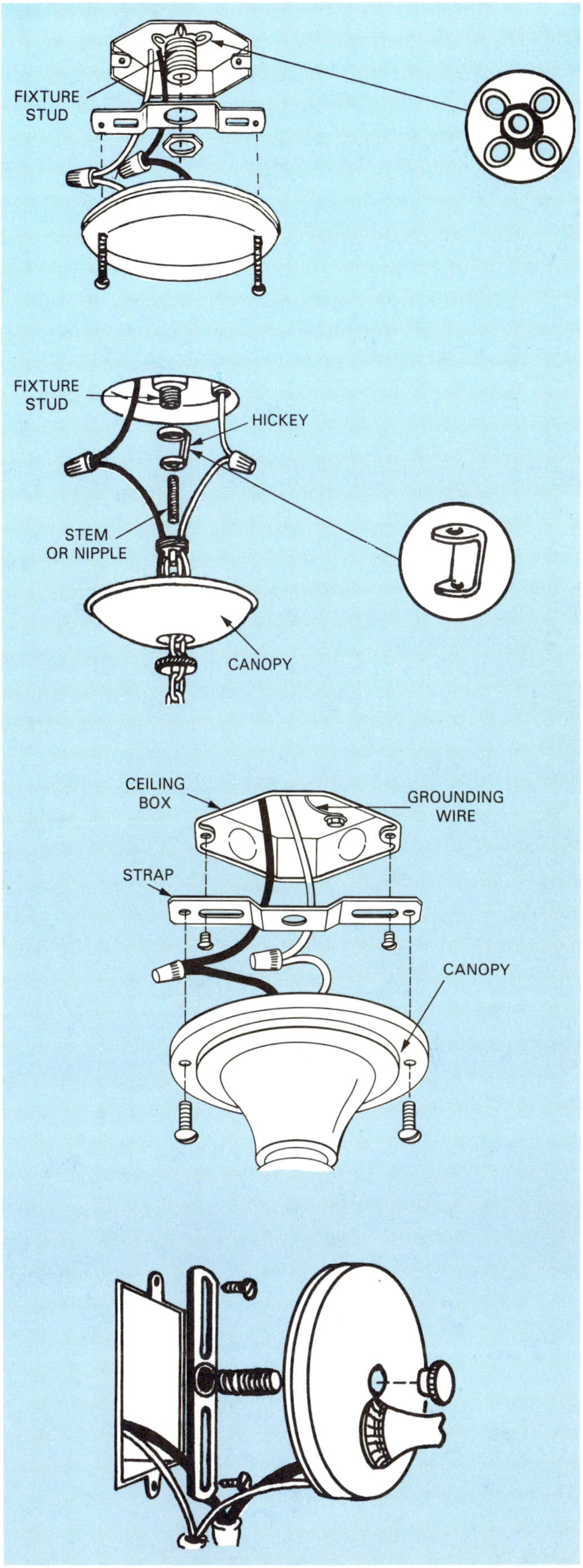

Fig. 9-33. Wiring instructions for wiring and hanging ceiling or wall fixtures.

REVIEW QUESTIONS — CHAPTER 9

1. The word "_____" in the *National Electrical Code* indicates a condition which *must* be met.
2. List two guides for wire size which the electrician uses.
3. *Code* Articles 300 to 384 go into considerable detail on certain aspects of electrical wiring as follows:
   a. Specifies materials to be used under certain conditions.
   b. Deals with grounding and bonding only.
   c. Deals with the methods of wiring to be used by the electrical industry and the electrician.
4. Why should grooving or nicking or a wire be avoided during stripping of insulation?
5. A splice of two conductors is usually secured with _____ _____. These devices are quickly fastened to the splice, require no soldering, and provide insulation over the bared wires.
6. Which of the following is the most commonly used method of joining No. 12 or No. 14 circuit wires?
   a. Lugs.
   b. Split-bolt connectors.
   c. Western Union Splice.
   d. Wire nuts or solderless connectors.
7. When soldering electrical connections, always use _____ core solder.
8. The brass or dark-colored screw on the duplex outlet is for the _____ conductor.
9. The white, neutral, conductor must never be interrupted. True or False?
10. Electrical connections, taps, or splices can be done outside an electrical box or enclosure. True or False?
11. Connectors and connecting points are marked "CU" for _____ and "AL" for _____ wire.
12. Complete the following diagram using a separate sheet of paper:

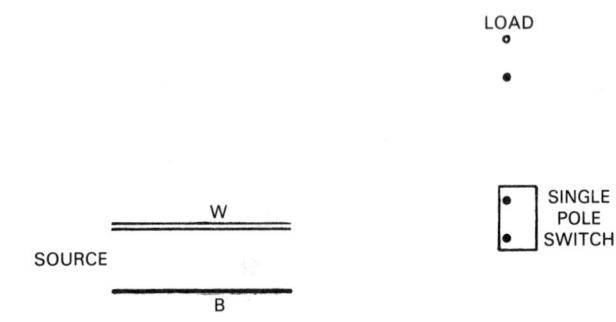

13. Explain the purpose of the green hex screw on a grounded receptacle.
14. How a fixture will be attached to a box depends upon what four things?

Chapter 10

PLANNING BRANCH CIRCUITS

Foresight is the key element in planning and installing a suitable electrical system. In this chapter, we will explore minimum requirements as well as future needs for a well designed wiring system.

The electrician must think of future demands which might be made on the system by the building's occupants. Meeting current demands may be adequate for today but will make the system old before its time.

The planning must follow all the guidelines provided by the *National Electrical Code,* but it should go well beyond the *Code's* minimum requirements. A competent electrical contractor must be prepared to install a system that will be as safe and adequate 20 years from now as it is today.

OBSOLETE WIRING

Obsolete wiring means wiring which becomes inadequate for any reason. It shows up in problems such as the following:
1. Too few wall outlets in rooms.
2. Pull chains on lights where there should be switches.
3. Circuits in which fuses "blow" frequently or where the circuit breakers trip often.
4. Overuse of extension cords.
5. No outdoor, weatherproof, or GFCI-protected receptacles.
6. Service entrances of 60 A or less.
7. Inadequately sized circuit conductors. See Fig. 10-1, for example.

It shows a power panel where wiring insulation is no longer safe and circuitry is no longer adequate.

As you learned in Chapter 1, a circuit is any path along which electrical energy can flow, It is a current-carrying path.

It is possible to place all the electrical loads in a building on a single circuit. That is, every light, every appliance, and every piece of electrical equipment would draw its power from one electrical path.

However, even if this could be done safely, it would not be convenient. Every time the fuse burned out or the circuit breaker tripped, the entire electrical system would shut down. All lights would go out and all appliances would stop working.

To divide the load, the electricity entering a building is distributed into what are called BRANCH CIRCUITS. A branch circuit is a separate electrical path, independent of other electrical paths in the building. It draws its current from the power panel and is protected by its own fuse or circuit breaker. It serves one or more outlets for receptacles, switches, or fixtures. Fig. 10-2 is a schematic of branch circuits coming out of a service panel.

TYPES OF BRANCH CIRCUITS

The types and purposes of branch circuits used—especially in a dwelling—can be divided into several main categories:
1. General purpose (lighting) circuits.
2. Small appliance (kitchen, laundry) circuits.
3. Individual or large appliance circuits. (These are permanent, single-unit circuits that serve water heaters, furnaces, air conditioners, and electric ovens or ranges.)

Fig. 10-1. Obsolete wiring. An inadequate wiring system, such as is indicated by this entrance panel is inefficient, overtaxed, and dangerous. Electrical system components showing deterioration should be replaced. (Scott Harke)

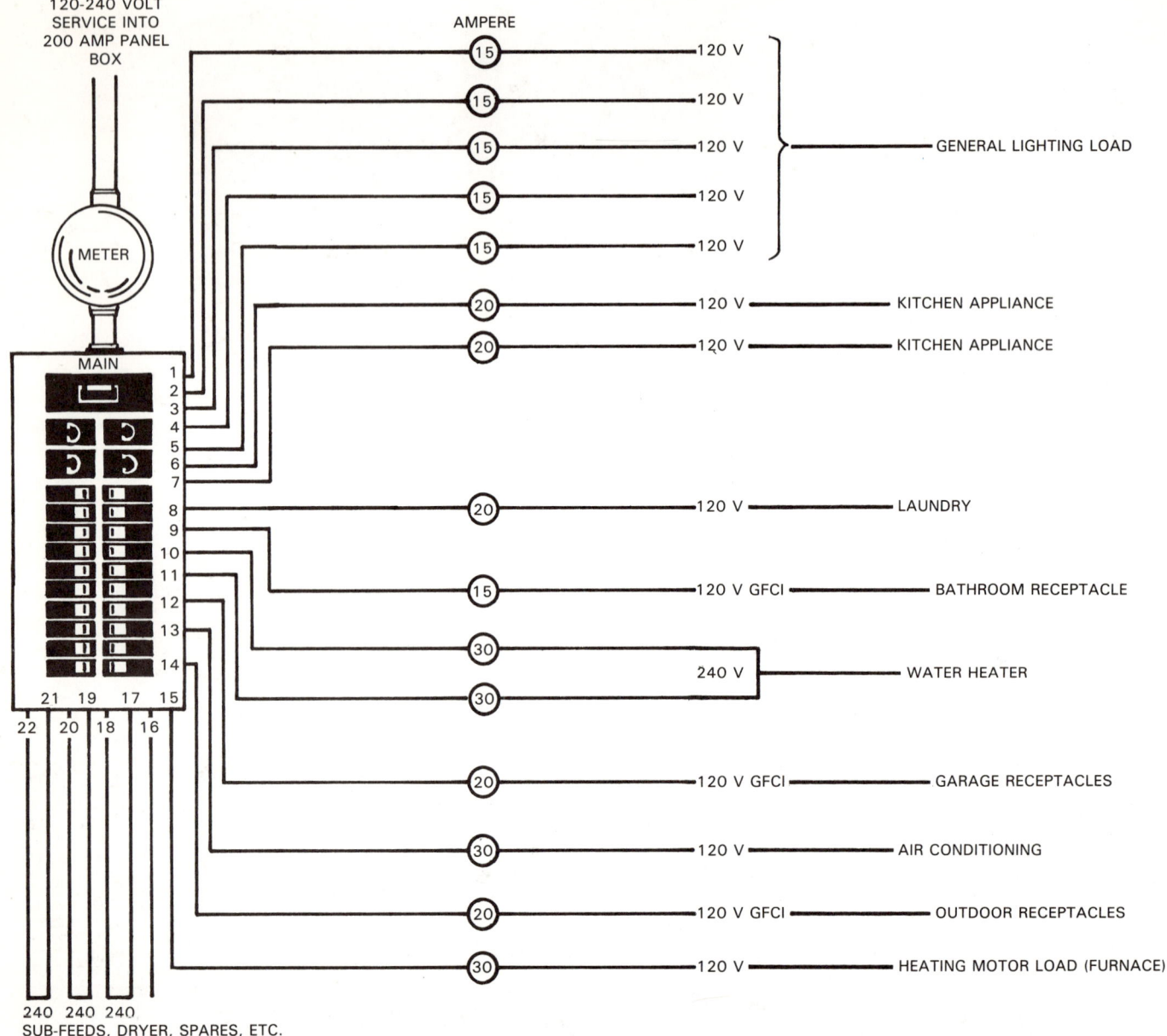

Fig. 10-2. Schematic of a 200 amp service entrance. The 22 branch circuits will meet minimum standards and provide for future expansion. Note that each branch has its own circuit breaker.

We shall investigate each of these types of branch circuits in order to determine the purpose, minimum regulations, and future needs.

First, it is necessary to explore two key factors in planning branch circuits:
1. NEC specifications for loads.
2. Actual load calculations.

MAXIMUM NUMBER OF OUTLETS

The NEC clearly outlines the method of determining the number of general lighting circuits for residences, Fig. 10-3. The *Code* allows 3 VA per square foot of floor area. It does not formally specify the maximum number of outlets permitted for the entire building or even for each circuit. *However, according to Section 220-2, the continuous load on any branch circuit is not to be greater than 80 percent of the circuit rating.* The maximum wattage for each circuit (CKT) is found as follows:

CKT amperage × CKT voltage × .80
= CKT wattage (VA)

A 20 A branch circuit, for example, can have a maximum continuous load of:

20 A × 120 V × .80 = 1920 VA

While this is clear-cut as far as the total wattage per circuit is concerned, some other considerations

102 Continuous = over 3 hrs

Planning Branch Circuits

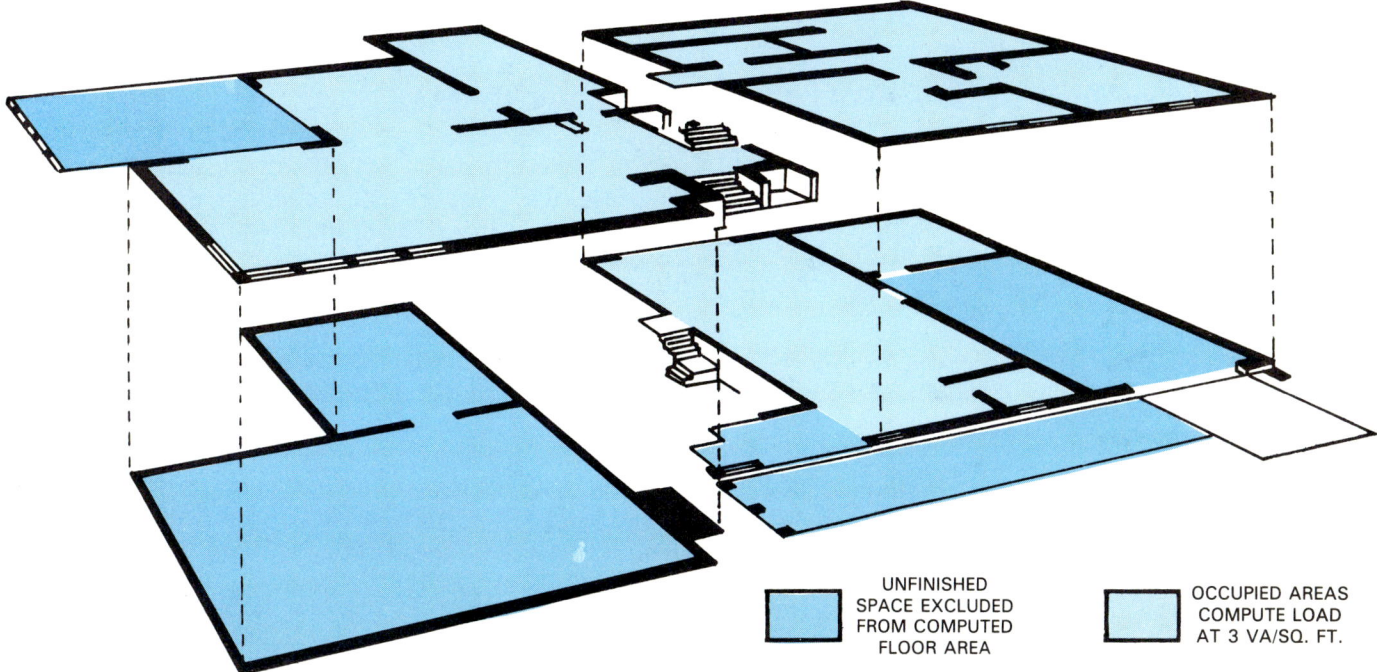

Fig. 10-3. The general lighting load is determined by the unit load per square foot method. For residences this is 3 VA/sq. ft. Occupied areas only are counted.

must be included in order to determine the number of outlets for each circuit.

The load for each lighting outlet can vary considerably. Although some lighting fixtures, like fluorescents, draw a constant load, others to not. Most lighting fixtures can accommodate 25, 40, 60, 75, 100, or 150 W bulbs.

The receptacle outlet load is also variable. Some outlets are rarely used. Others may have no more than a small night-light connected to them. Still others may operate a 100+ W table lamp, floor lamp, or portable heater.

DETERMINING LOADS

These factors are important and lead to the general rule-of-thumb: allow 180 W (volt-amperes) per grouped duplex receptacle outlets. See Fig. 10-4. It assumes maximum bulb rating for each fixture. Total wattage is not to exceed that permitted for the branch circuit amperage rating.

Fig. 10-5 and Fig. 10-6 show a typical entrance and hallway. We shall see if the fixtures and receptacles for these areas can be placed on a single 15 A branch circuit. The following outlets and fixtures are needed:

Outlet	Assumed Load (max.)
1 ceiling fixture (hallway)	150 VA
1 ceiling fixture (entrance)	150 VA
3 receptacles (hallway)	540 VA
1 receptacle (entrance)	180 VA
TOTAL	1020 VA

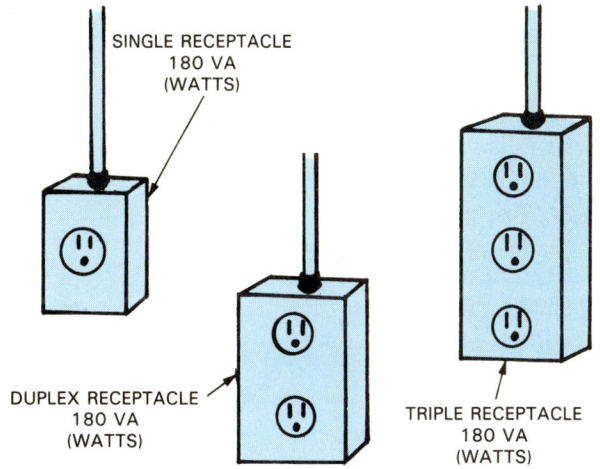

Fig. 10-4. Each single, duplex, or multiple outlet should be considered at not less than 180 W or volt-amperes (VA).

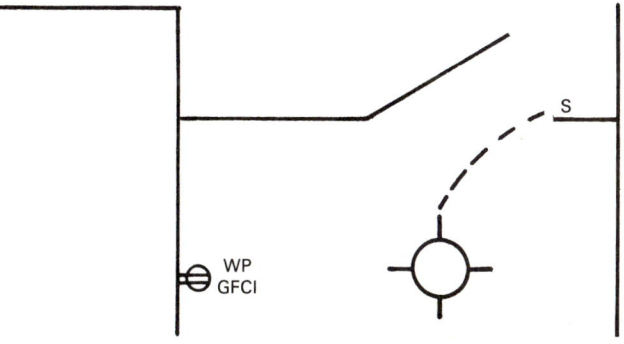

Fig. 10-5. Typical entrance should have switch controlled lighting. A GFCI-protected weatherproof outlet may be desired.

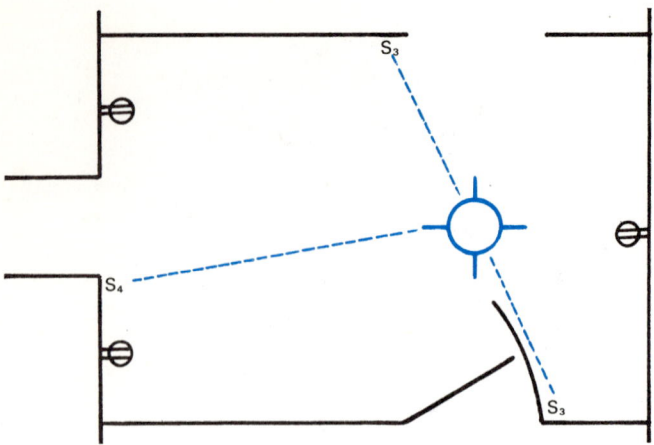

Fig. 10-6. Hallway such as this should be provided with at least one light fixture which can be controlled from every entrance.

This does not exceed the maximum allowable load for a 15 A branch circuit (15 A × 120 V × .80 = 1440 W). The hallway and entrance can be placed on one circuit with a 15 A overcurrent protection device.

BRANCH CIRCUIT DESIGN

Article 210 of the NEC outlines the rules regarding the general provisions and specific requirements for branch circuits. These rules, when followed, will certainly provide the safe wiring of the branch circuits. However, what about more than just adequate wiring?

As outlined earlier, you should plan for more than adequate wiring. With the ever-increasing uses of electrical power, good planning is essential. Remember, the electrician must look well beyond the minimum required.

Besides a well planned service entrance, discussed in Chapter 12, other factors demand careful consideration:
1. Wire capacity.
2. Switches.
3. Outlets (plug-in receptacles).
4. Lighting.
5. Number of circuits.
6. Special outlets.
7. Proper overall design.

WIRE CAPACITY

Although the *Code* specifies No. 14 AWG as minimum wire size, it would indeed be unwise to use anything smaller than No. 12 in homes built today. Lighting and receptacle outlets wired with No. 14 will almost certainly be inadequate in 10 to 20 years. Keep this in mind, not only for lighting and receptacle load considerations, but for *all* feeder calculations. For proper wire ampacities, refer to Fig. 3-11 of this text and NEC Table 310-16.

WALL SWITCHES

Wall switches were once considered convenient luxuries to replace the pull-chain or pushbutton switches on the fixtures themselves. Today, pull-chain or other similarly controlled fixtures are usually inadequate and should be avoided in most locations. All lights should be controlled by a switch or pair of switches located near the entrance(s) to a room.

Wall switches should be placed in locations which are accessible and at a height suitable for all occupants. Generally, wall switches are 48-52 in. above the floor level and on the latch side of a doorway, Fig. 10-7. Article 380 of the *Code* describes additional installation requirements.

PLUG-IN RECEPTACLES

Code requires that outlets be provided in the kitchen, family room, dining room, living room, parlor, library, den, sun room, breakfast room, bedroom, recreation room and other similar rooms in a dwelling. These outlets must be located so that no point along any wall will be more than 6 ft. (183 mm) from a receptacle. This rule applies to wall

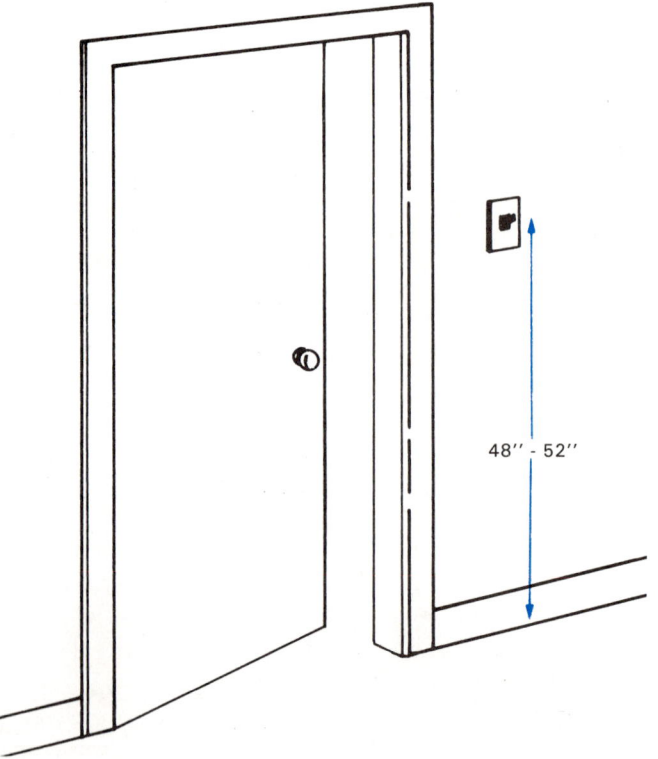

Fig. 10-7. Wall switches must be located at convenient heights. The preferred height, for most installations, is 48 to 52 in. above floor level. Locate the switch on the latch side of the doorway.

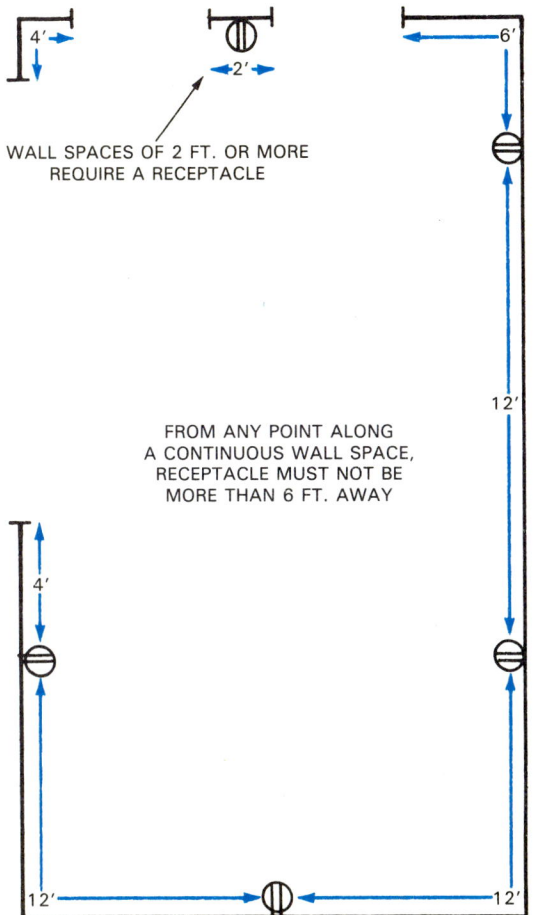

Fig. 10-8. Receptacles must be no further than 12 ft. apart along any continuous wall space.

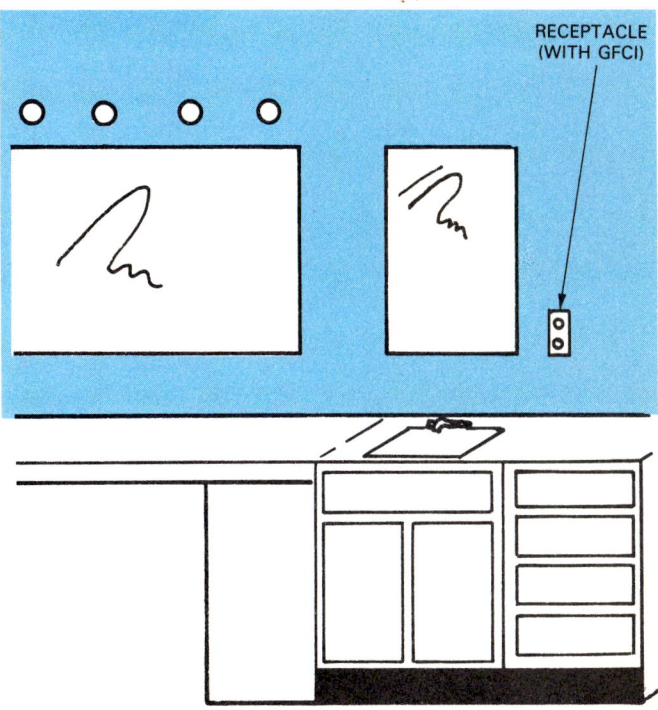

Fig. 10-10. Receptacle near bathroom sink must have GFCI protection.

spaces 2 ft. (61 mm) or more wide. Included are outside wall spaces occupied by sliding panels.

Where fixed room dividers or freestanding counters add to the wall space, the same 6 ft. rule applies. See Fig. 10-8 and refer to Section 210-52 of the *National Electrical Code*.

RECEPTACLE LOCATION

In essence, receptacle outlets are required in the following places as illustrated:
1. Above counter spaces in the kitchen and dining areas, (when wider than 12 in.), Fig. 10-9.
2. Next to bathroom basin. *Code* specifies use of a GFCI (Ground fault circuit interrupter). See Fig. 10-10.
3. Outdoors (with a GFCI), Fig. 10-11.
4. Laundry area, Fig. 10-12.
5. Basement area, Fig. 10-13.

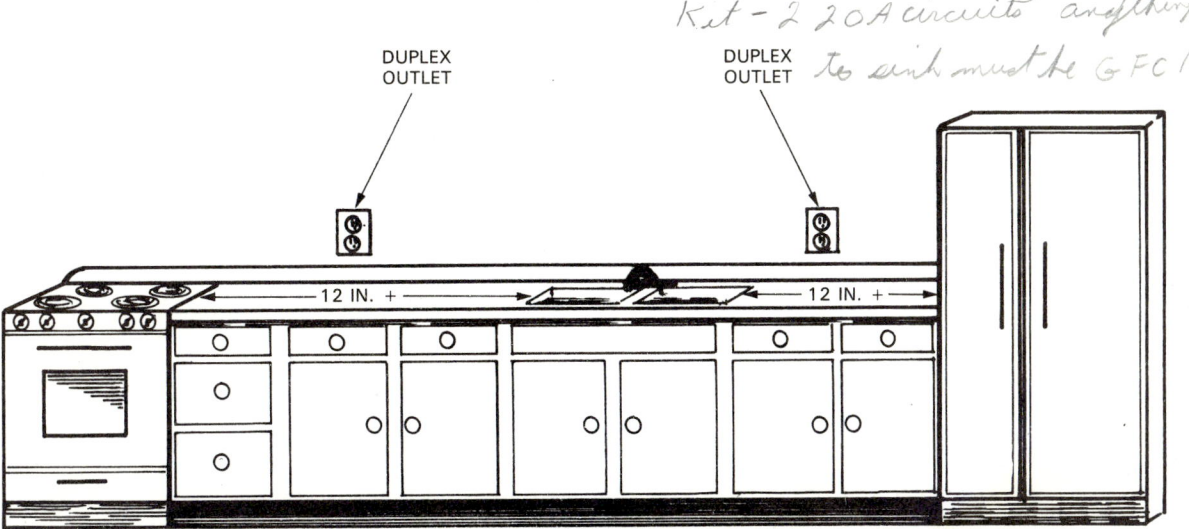

Fig. 10-9. A receptacle is required in every kitchen counter space wider than 12 in.

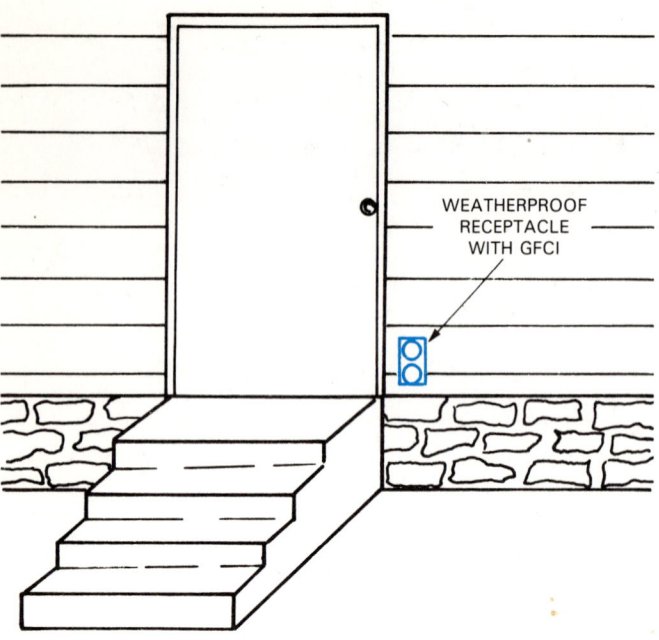

Fig. 10-11. One outdoor receptacle outlet is required. This must be housed in a weatherproof box and must have GFCI protection.

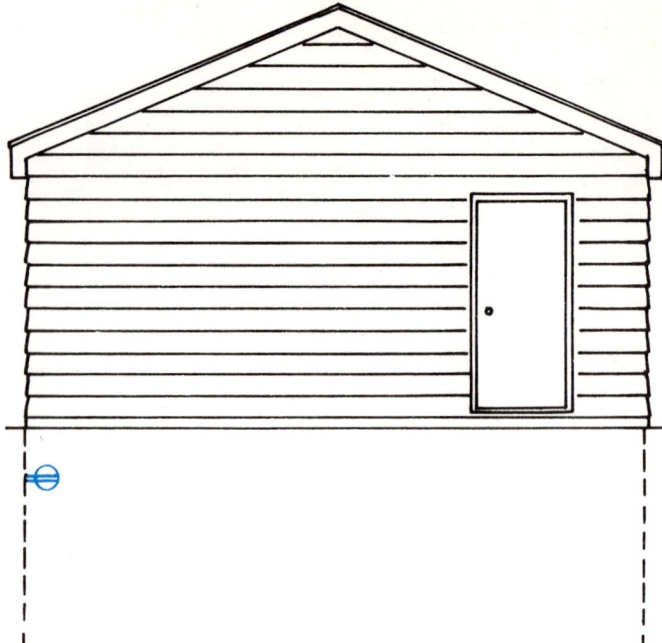

Fig. 10-13. A receptacle outlet should be placed in the basement for general use.

6. Garage (with a GFCI), Fig. 10-14.
7. Any wall space in excess of 2 ft. and along fixed room dividers, Fig. 10-15.

SPLIT-WIRING

Often, receptacles are split-wired so one outlet may be controlled by a switch. The upper half of the receptacle is always hot while the bottom half is switched. This method of wiring receptacles is most frequently used in parlors, dens, and living rooms, as shown in Fig. 10-16.

SPACING

In addition, receptacle outlets should be placed 12 to 15 in. above the floor line and as close to the ends of large wall spaces as is possible. This

Fig. 10-12. Laundry area must have one receptacle outlet.

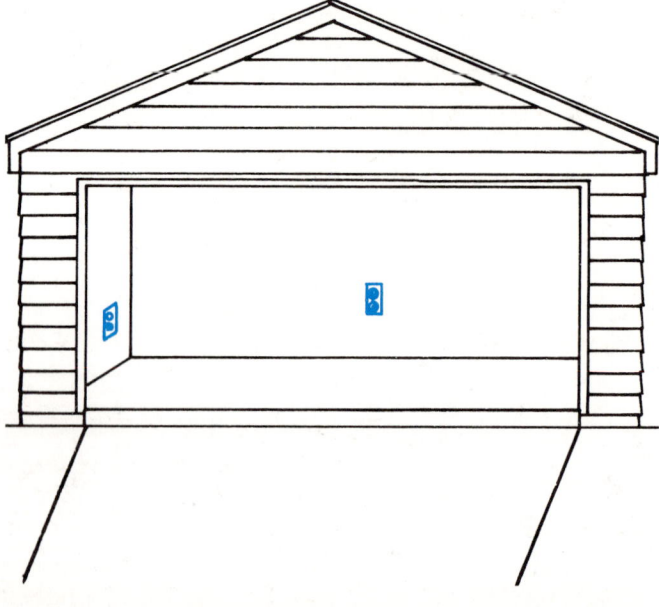

Fig. 10-14. Garage must have at least one GFCI-protected receptacle outlet. Other outlets used exclusively for freezer or garage-door-opener need not be GFCI protected.

Planning Branch Circuits

avoids concealment of the outlet by sofas or other large pieces of furniture. Remember, receptacles must be evenly spaced and so placed that *no space along the adjacent floor line is more than 6 ft. from an outlet.* Once again, refer to Section 210-52 of the *Code* for more details. *Extension cords are poor substitutes for fixed receptacle outlets.*

LIGHTING

Every room as well as the cellar, basement, crawl space, garage, attic, stairway, closet, porches, hallways and the like should have *at least one* lighting outlet. Additional lighting, provided by table or floor lamps, is desirable in places such as the living room, bedrooms, den, library or family room.

The quality of lighting and types of lighting fixtures available are beyond the scope of this book. However, good lighting is essential to personal well-being and should be carefully considered.

Good lighting should provide an adequate amount of light where it is needed. It should not create shadows or glare. In order to understand good lighting the electrician must familiarize himself or herself with the criteria which go into creating good lighting. Terms such as *candlepower, lumens, footcandles,* as well as others, must be understood to get a real insight into the proper quality of lighting for the home.

A detailed guide concerning proper illumination and types of lighting fixtures, or luminaires, as they are more correctly termed, may be found in the publication, "Recommended Practice of Residence Lighting," distributed by the *Illuminating Engineering Society,* (IES).

LIGHTING FIXTURE INSTALLATION RULES

Article 410 of the NEC contains almost 100 Sections dealing with:
1. Fixtures.
2. Pendants.
3. Arc lamps.
4. Discharge lamps.
5. Incandescent filament lamps.
6. Lampholders.
7. The wiring and equipment forming the above lighting and its installation.

NUMBER OF CIRCUITS

The *Code* generally divides the main types of circuits into two broad categories:
1. Those of a special nature serving individual devices like water heaters, space heaters, large appliances, and the like.
2. Those of a general nature which feed outlets for lighting or receptacles.

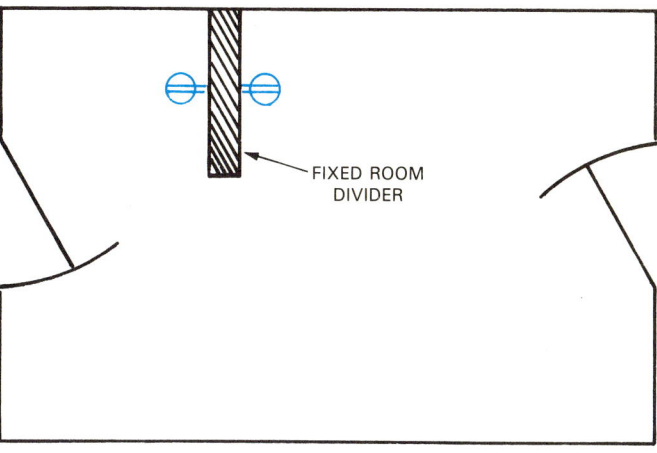

Fig. 10-15. *A fixed room divider shall have an outlet on each side, especially when necessary to maintain the 6 foot rule.

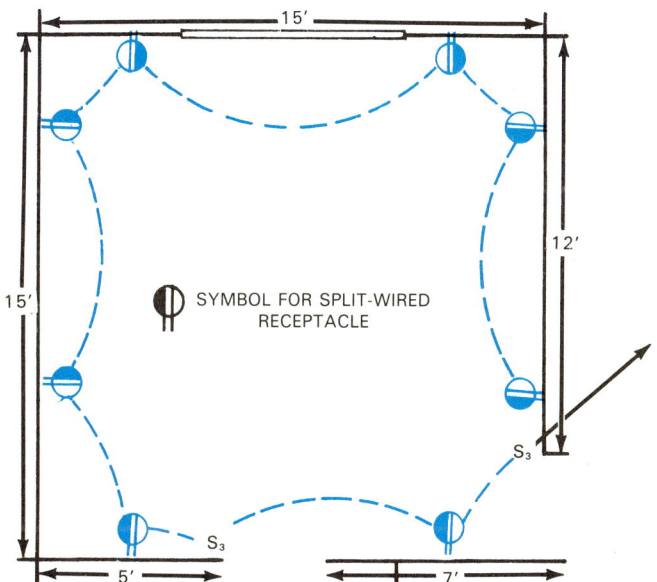

Fig. 10-16. Receptacles may be split-wired. That is, the upper half of the receptacle is always 'hot', while the bottom portion is switch controlled from each entrance. Thus, floor and/or table lamps plugged into the bottom half can be switch activated.

The total number of circuits will be governed by the number of special circuits needed plus the calculated general lighting (and receptacle) requirement. Fig. 10-17 lists the circuits in a typical residence.

APPLIANCE CIRCUITS

Separate circuits should be provided for each of the following:
1. Water heater. Electric water heaters should be connected to a 30 A, 240 V circuit.
2. Clothes washer. This is usually included under the laundry circuit. It requires a 20 A, 120 V circuit.

CIRCUIT DIRECTORY

Circuit 1 Lighting 15 A - 120 V	Circuit 2 Lighting 15 A - 120 V
Circuit 3 Lighting 15 A - 120 V	Circuit 4 GFCI - 20 A - 120 V
Circuit 5 Small Appliance 20 A - 120 V	Circuit 6 Small Appliance 20 A - 120 V
Circuit 7 Laundry 20 A - 120 V	Circuit 8 Spare
Circuit 9 11 Water Heater 30 A - 240 V	Circuit 10 12 Clothes Dryer 30 A - 240 V
Circuit 13 Furnace 20 A - 120 V	Circuit 14 Air Conditioner 20 A - 120 V
Circuit 15 17 Heater 20 A - 240 V	Circuit 16 18 Range 50 A - 240 V
Spares	Spares

Fig. 10-17. Typical single-family residence circuit scheme. Note 240 V circuits set aside for appliances.

Fig. 10-19. Three-pole, three wire, 50 A range receptacle. (Leviton Mfg. Co., Inc.)

3. Clothes dryer. A dryer will most often be rated at around 5 kW. Therefore it should have a separate, 30 A, 240 V circuit.
4. Furnace will require a 20 A, 120 V circuit.
5. Range is generally connected to a heavy-duty 50 A, 240 V power outlet as shown in Figs. 10-18, and 10-19.
6. Heaters. Each should be on a circuit no smaller than 20 A, 240 V. Follow the manufacturer's recommendation.
7. Motors (1/4 hp or more). Motor circuit sizes must be determined on the basis of their nameplate rating. Disconnecting means must be located in sight of the controller, Fig. 10-20. See Article 430 of the NEC regarding motor circuits. Careful study of this Article is essential for proper motor circuit sizing and coordination.
8. Dishwasher. A typical dishwasher is generally placed on the 20 A, 120 V kitchen appliance circuit.
9. Air conditioner. Depending on the size, a circuit of 20 A, 120 V and up is in order. Large air conditioners must be carefully evaluated since they are important criteria in computing the service entrance size.

FIGURING GENERAL LIGHTING CIRCUITS

Table 220-2(b) of the NEC permits a load of 3 VA per square foot for lighting in occupied areas of dwellings. (As you learned earlier in this chapter, you do not count space such as attics, porches, and basements as being occupied.)

Fig. 10-18. A 50 A, heavy duty range receptacle. It is a general purpose, two-pole, three wire type. (Bryant)

Fig. 10-20. Motors should have disconnecting means within sight of person operating motor. (Bryant)

For example, we shall determine the number of general lighting circuits needed for a dwelling with an occupied area of 1500 sq. ft. The ampacity of the circuits may be either 15 or 20 amperes.
1. First, find out how many watts the *Code* requires of the occupied space:
   1500 sq. ft. × 3 VA per sq. ft. = 4500 VA
2. Find the total amperage needed if the supplied voltage is 120 V:
   $$\frac{4500}{120 \text{ V}} = 37.5 \text{ Amperes}$$
3. Now divide by the amperage of the circuits to find the number of circuits needed:
   $$\frac{37.5}{15 \text{ A}} = 2.5 \text{ or } 3, 15 \text{ ampere circuits minimum or:}$$
   $$\frac{37.5}{20 \text{ A}} = 1.875 \text{ or } 2, 20 \text{ ampere circuits minimum}$$

As you can see, you have a choice of either three 15 A circuits or two 20 A lighting circuits. Either would be an adequate minimum for the dwelling. More circuits may be used if desired or if specified by the owner of the dwelling.

FIGURING NUMBER OF SMALL APPLIANCE CIRCUITS

In addition to the lighting load, *Code* requires two or more 20 A, 120 V small appliance circuits. These appliance circuits shall feed receptacles in the kitchen, pantry, dining room, and breakfast room. Also required is at least one 20 A, 120 V outlet in the laundry area.

Refer to Article 220 of the *Code* for specific rules on these circuits. Also, review the tables and examples in Chapter 9, NEC.

BALANCING CIRCUIT LOADS

When planning the hookup of branch circuits to the service panel, it is important to keep a balance in the load between the two hot wires in a three-wire system. Remember that the neutral wire in this system is carrying the unbalanced electrical load from both hot wires.

Fig. 10-21 helps explain what is happening as current flows through the system. The alternating current in the two hot wires does not flow in the same direction at the same time. At the instant it flows one way in the red wire, it is flowing the opposite direction in the black wire. At the same time the flow for each wire will change in the neutral wire. The two currents flowing in opposite directions through the neutral wire tend to cancel each other out.

As long as the current in each direction is nearly equal there is no problem. However, when the current differences are great, there may be trouble. The neutral wire will heat up and the insulation could melt.

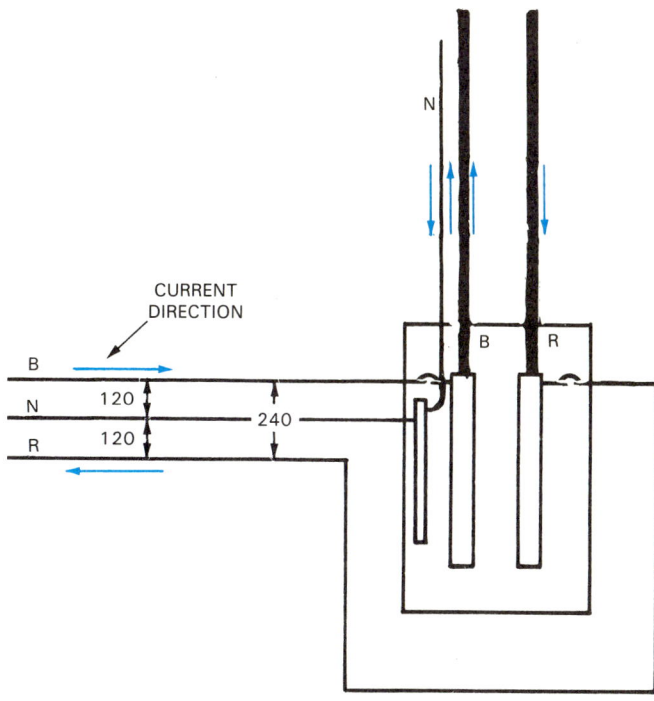

Fig. 10-21. A 120 volt/240 appliance circuit. Current in the two hot wires is always flowing in opposite directions. Thus, opposing currents in the neutral wire cancel each other out.

This situation can arise with appliances that operate on both 120 and 240 volts. These appliances must be evenly balanced between the two hot wires to reduce the likelihood of overloading the netrual conductor. Fig. 10-22 shows a system properly balanced at the entrance panel. If you add up the ampere rating of all the circuits on each side, you can see that the ratings add up to nearly the same total. Also, the same number of large appliance circuits are found on each hot wire.

SPECIAL OUTLETS

As stated, appliances such as ranges, dryers, washers, heaters, and motors require special outlet considerations. Each of these loads are to be considered carefully and then placed on a 20, 30, 40, 50, or 60 A circuit as needed.

OVERALL DESIGN

Proper overall design is nothing more than planning, foresight, and good blueprints. There is no substitute for a well planned installation. Countless hours and dollars are lost by improper planning. A good set of working blueprints or sketches can cut the cost of materials and time substantially. In addition, proper planning eliminates voltage losses and, consequently, inefficient wiring.

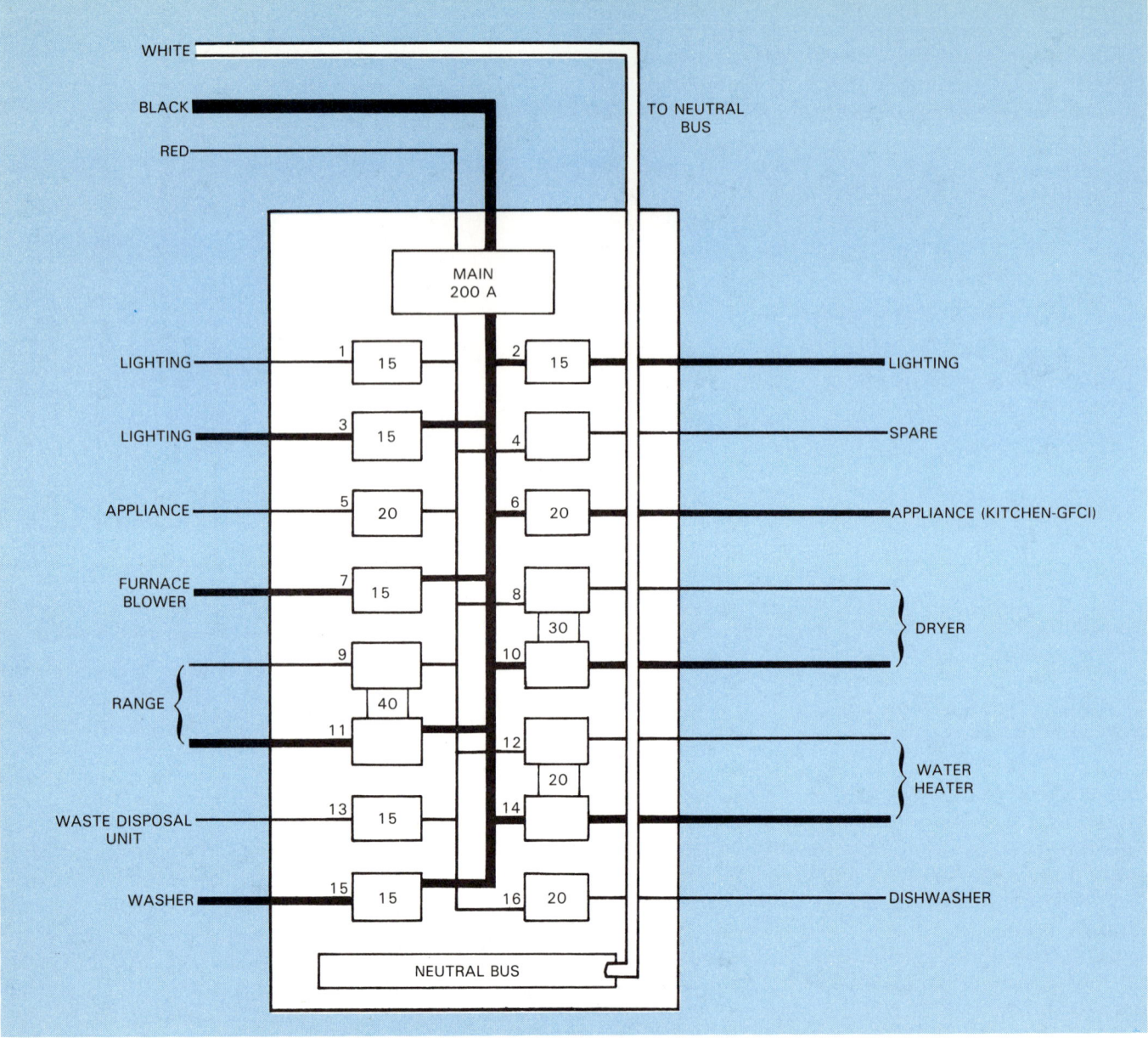

Fig. 10-22. The electrician should be careful to balance off 120 volt circuits on both hot wires of three-wire service. Just add up the total of the rated amperage of each circuit breaker controlling 120 volt circuits.

DETERMINING BRANCH CIRCUIT NEEDS

To determine how many branch circuits will be necessary, it is best to:
1. Carefully list all the places (rooms, hallways, closets, etc.) which will have receptacles, outlets, appliances, and motors.
2. Add up the area of all these rooms.
3. Using this total square footage, figure on one lighting circuit for every 350 sq. ft.
4. Compute the number of general-purpose circuits you will need.
5. Next, add the kitchen-laundry circuits, and the additional individual large-appliance circuits you will need. Altogether, this will represent the number of branch circuits.

Fig. 10-23 shows a floor plan for typical one-family dwelling. Various branch circuit calculations, outlets, and switching arrangements have been included. You can use the plan as a guide in establishing a minimum requirement or base to expand upon and hopefully, in the end, come up with an adequate, modern plan.

BATHROOM CIRCUITS

Because of the temperature and moisture changes which often occur in bathrooms, circuitry for this area warrants special consideration and discussion.
The NEC requires that the bathroom be equipped with recessed ceiling lighting fixtures and a GFCI-protected receptacle outlet.

Planning Branch Circuits

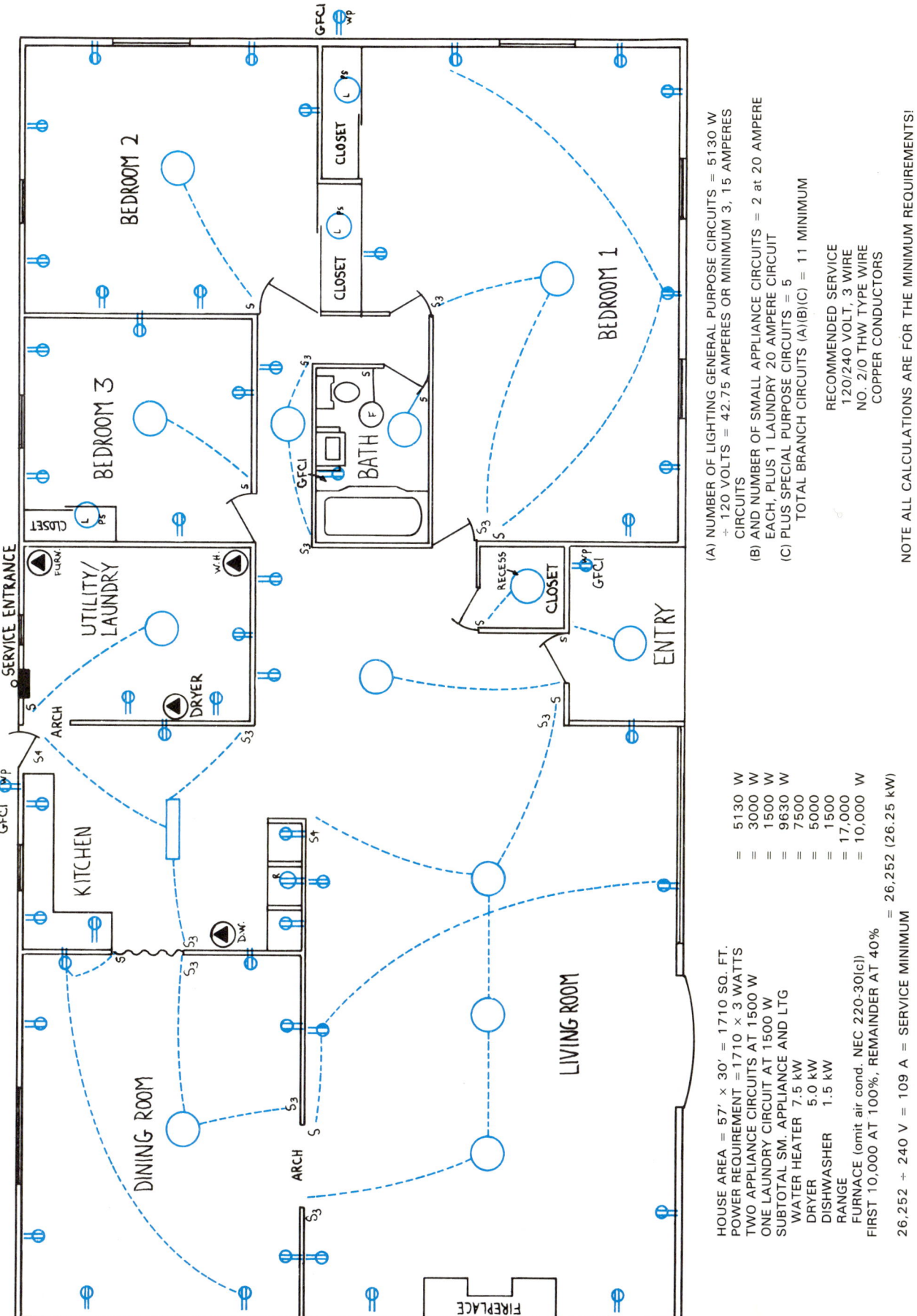

Fig. 10-23. Electrical plan of a single-family dwelling. Plans such as these help the electrician make important calculations concerning size of service, switching, arrangements, and number of general and special branch circuits. A plan such as this should be evaluated before any wiring is done. Considerable time and effort can be saved with careful planning.

(A) NUMBER OF LIGHTING GENERAL PURPOSE CIRCUITS = 5130 W ÷ 120 VOLTS = 42.75 AMPERES OR MINIMUM 3, 15 AMPERES CIRCUITS
(B) AND NUMBER OF SMALL APPLIANCE CIRCUITS = 2 at 20 AMPERE EACH, PLUS 1 LAUNDRY 20 AMPERE CIRCUIT
(C) PLUS SPECIAL PURPOSE CIRCUITS = 5
TOTAL BRANCH CIRCUITS (A)(B)(C) = 11 MINIMUM

RECOMMENDED SERVICE
120/240 VOLT, 3 WIRE
NO. 2/0 THW TYPE WIRE
COPPER CONDUCTORS

NOTE ALL CALCULATIONS ARE FOR THE MINIMUM REQUIREMENTS!

HOUSE AREA = 57' × 30' = 1710 SQ. FT.
POWER REQUIREMENT = 1710 × 3 WATTS = 5130 W
TWO APPLIANCE CIRCUITS AT 1500 W = 3000 W
ONE LAUNDRY CIRCUIT AT 1500 W = 1500 W
SUBTOTAL SM. APPLIANCE AND LTG = 9630 W
WATER HEATER 7.5 kW = 7500
DRYER 5.0 kW = 5000
DISHWASHER 1.5 kW = 1500
RANGE = 17,000
FURNACE (omit air cond. NEC 220-30(c)) = 10,000 W
FIRST 10,000 AT 100%, REMAINDER AT 40% = 26,252 (26.25 kW)

26,252 ÷ 240 V = 109 A = SERVICE MINIMUM

Fig. 10-24 shows a typical floor plan of a bathroom indicating the electrical fixtures. Each vanity is equipped with a soffit-type fluorescent light to illuminate the mirror, and a GFCI receptacle for shaving and hair blow-drying.

In addition, recessed lighting fixtures are installed in certain locations to assure adequate lighting throughout the bathroom. The heater-fan-light device, is another convenient, efficient, and practical item worth consideration. These various bathroom installations are shown in Figs. 10-25 through 10-27.

REVIEW QUESTIONS — CHAPTER 10

1. List five signs of obsolete electrical wiring.
2. A branch circuit is electrically connected to the service panel, has its own overcurrent protection, and serves one or more outlets for _____, switches, or _____.
3. Explain how to determine the number of general lighting circuits.
4. List six individual-circuit appliances.
5. How many 20 A general lighting circuits would be needed to satisfy the "minimum" code re-

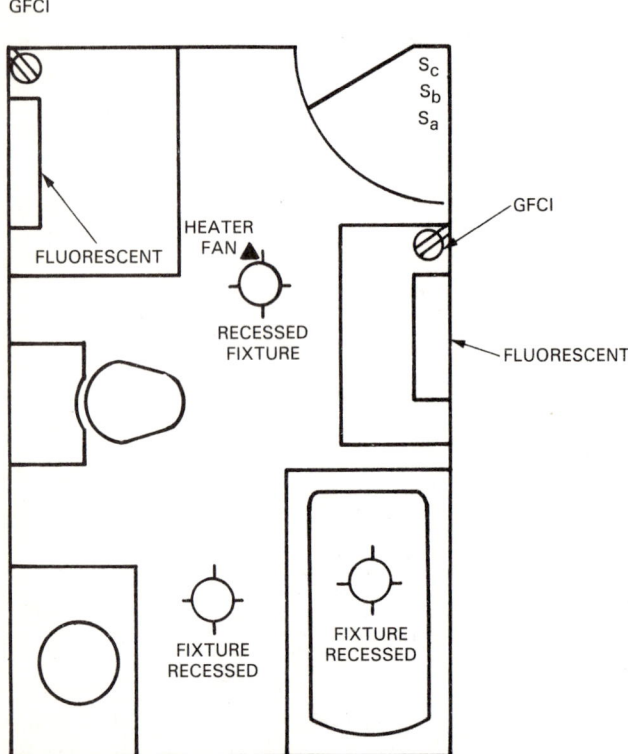

Fig. 10-24. Bathroom floor plan. Note the GFCI location.

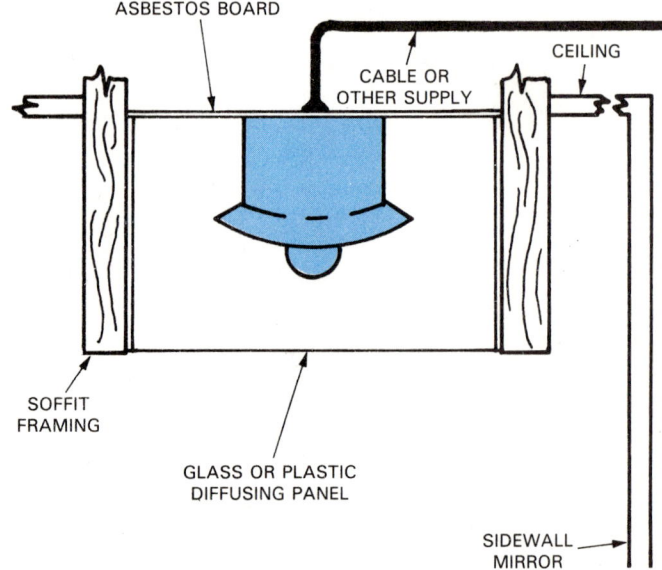

Fig. 10-26. Soffit lighting fixture. A light fixture installed in soffit is commonly used to provide light over the vanity area of a bathroom or powder room.

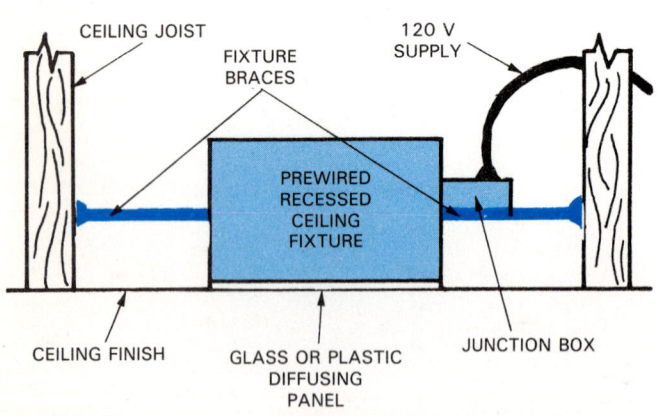

Fig. 10-25. Prewired recessed bathroom ceiling fixture. Special restriction (due to temperature ratings), may be found in Sections 410-64 to 410-72 of the NEC. Recessed fixtures like this are also employed in closets and laundry rooms.

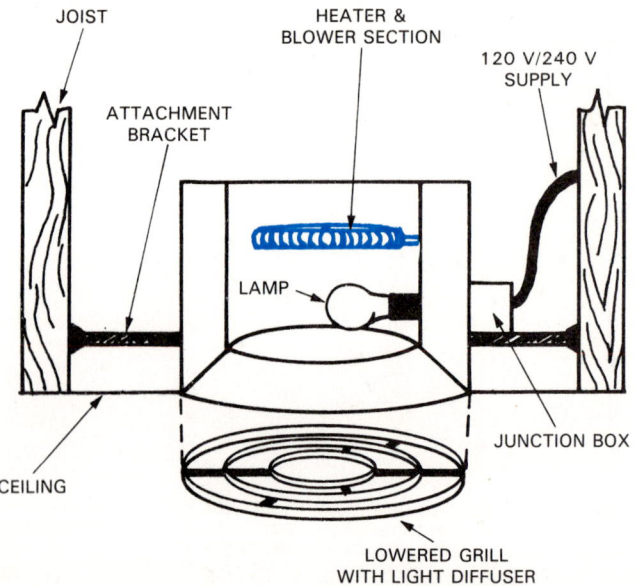

Fig. 10-27. Heater-fan-lighting unit. This device is installed in the same manner as the recessed fixture.

quirements for a 3000 sq. ft. residence? _____

6. The two kitchen and one laundry circuits, should be rated _____ amperes, _____ volts, and wired with AWG No. _____ conductors.
7. Temperature and moisture conditions in bathrooms require special lighting fixtures and receptacle. Name the two requirements.
8. NEC has certain requirements about the distance between receptacles in occupied rooms. Briefly stated, the rule says that no point along a wall shall be more than _____ ft. from a receptacle. The rules applies to every wall space _____ ft. or more wide. It (does, does not) apply to fixed room dividers or freestanding counters.
9. List three locations where GFCIs are required.
10. Counter spaces in kitchens wider than _____ require a receptacle outlet.
11. Receptacles which are half hot and half switched are called split-wired receptacles. True or False?
12. A garage door opener (must have, does not need) a ground fault circuit interrupter.
13. A motor controller (switch, breaker, or fuse panel) should be located _____.

Chapter 11

READING BLUEPRINTS AND WIRING CIRCUITS

If you were to draw a picture of how to wire a ceiling light in a bedroom it might look like the drawing in Fig. 11-1. However, this is hard to draw even if you have the skills. What is more, it would clutter up the blueprint of the house and drawing it would take a long time.

To avoid these disadvantages, electricians use electrical symbols. They are a type of "short hand" consisting of standard shapes and letters. Different symbols stand for the different types of conductors, switches, fixtures, and other devices that are used in wiring a building. Anyone can draw the symbols and they show clearly:
1. Which kind of electrical device or conductor is needed.
2. Where it should go.
3. Which ones are to be connected.

Look at Fig. 11-2. See how simple it was to mark in the switch, wiring, and ceiling light of Fig. 11-1.

The electrician may be supplied an electrical plan by the builder, architect, or owner. See Fig. 11-3. In some cases, the electrician will have to make up his or her own drawing.

If an outlet is controlled by a switch, the electrical blueprint will show a dashed line running from a switch symbol (S) to the outlet symbol. The lines are always curved so that they will not be mistaken for hidden construction details.

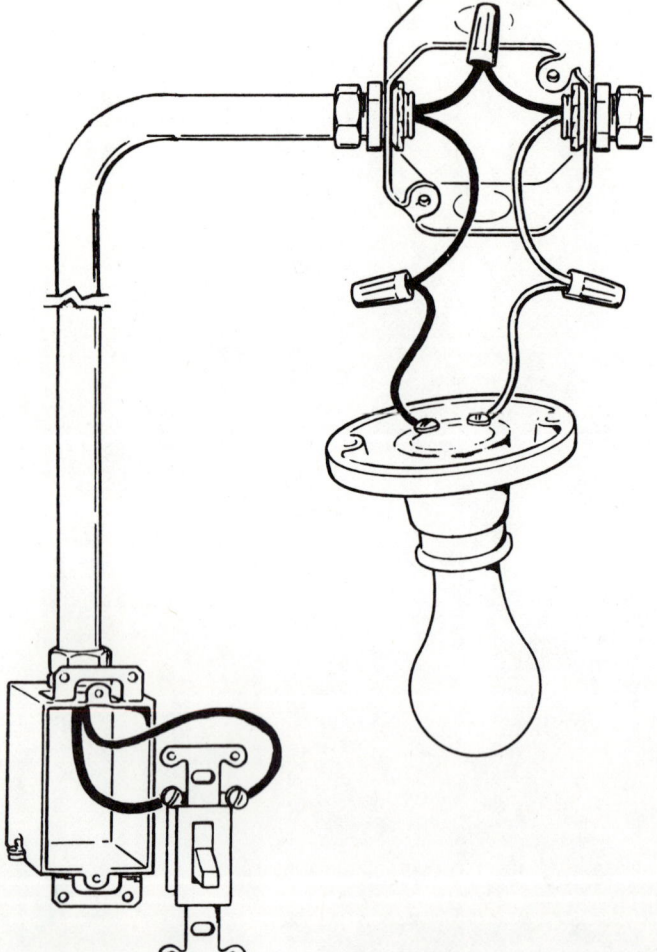

Fig. 11-1. Pictorial view of a switched lighting circuit. Symbols are used in electrical plans to describe each type of circuit.

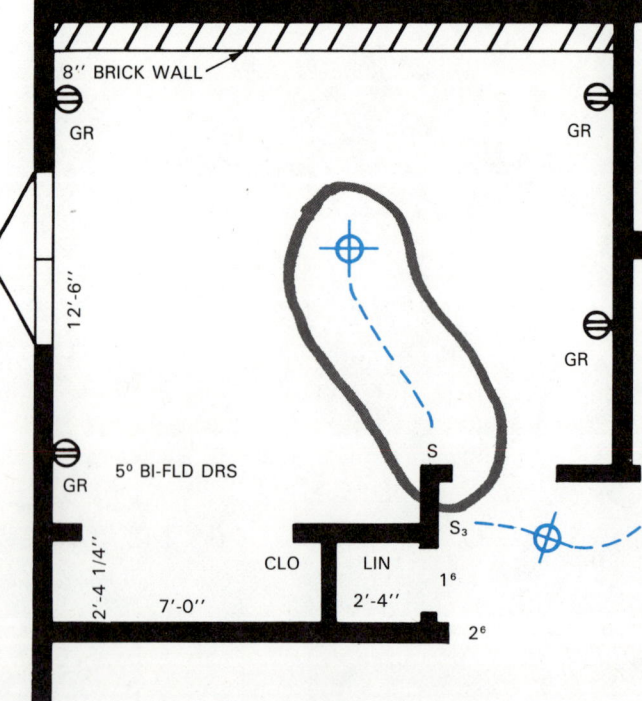

Fig. 11-2. Portion of an electrical plan. Portion of the wiring circled represents the circuit shown in Fig. 11-1.

114

Reading Blueprints and Wiring Circuits

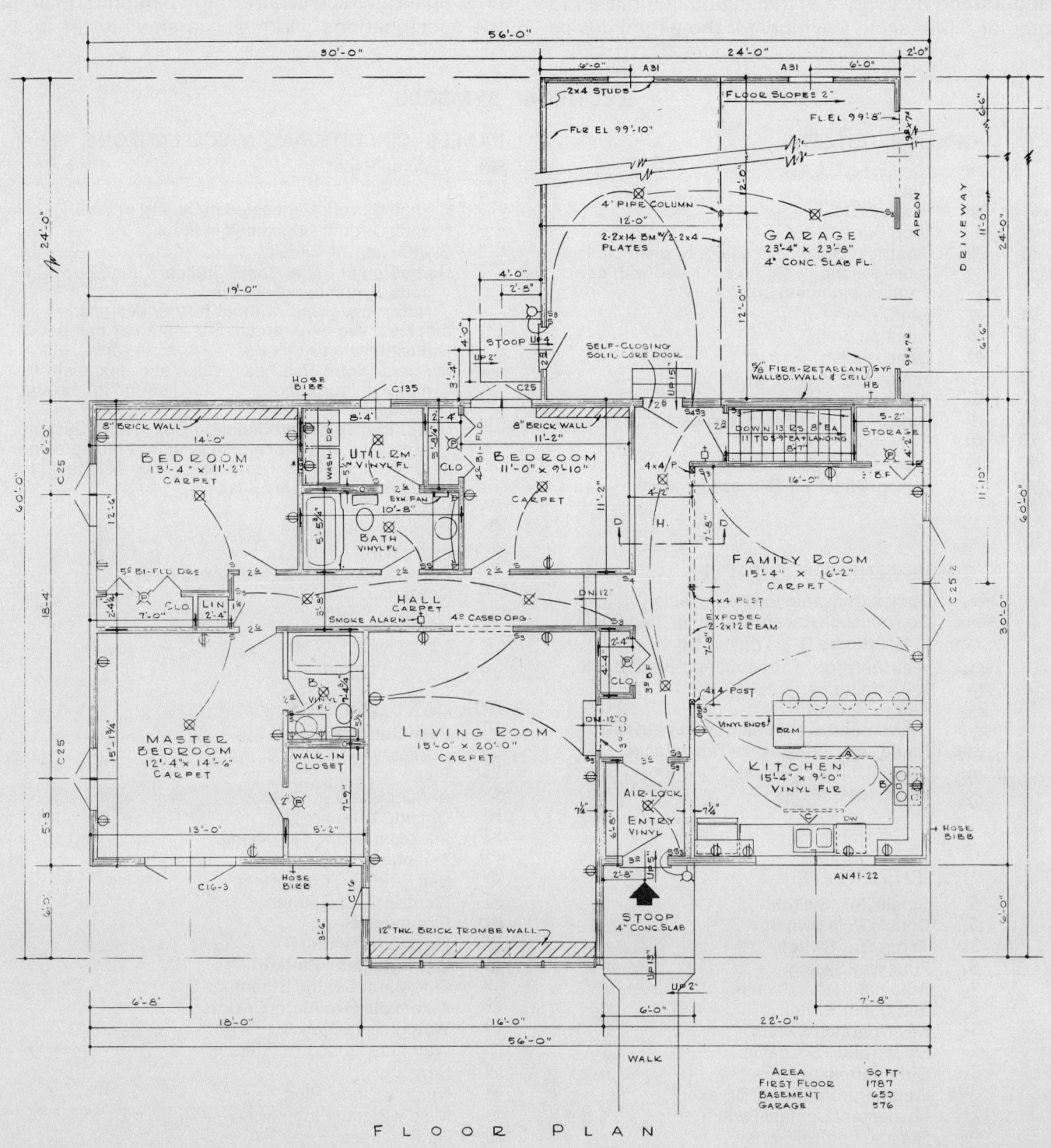

Fig. 11-3. Electrician will work from a plan such as this supplied by the architect or owner. (L.F. Garlinghouse)

Modern Residential Wiring

STANDARDIZED SYMBOLS

Most symbols used on electrical plans are ones that have been approved by the American National Standards Institute (ANSI). These should be understood by every electrician and committed to memory. Mistaking a symbol for the wrong device only makes extra work. Fig. 11-4 shows the commonly used symbols the electrician will need to remember.

Other symbols are sometimes used to designate special outlets. Then an explanation must be made on a legend, which is part of the blueprint, plan, or the specifications. Normally, a small letter (a, b,

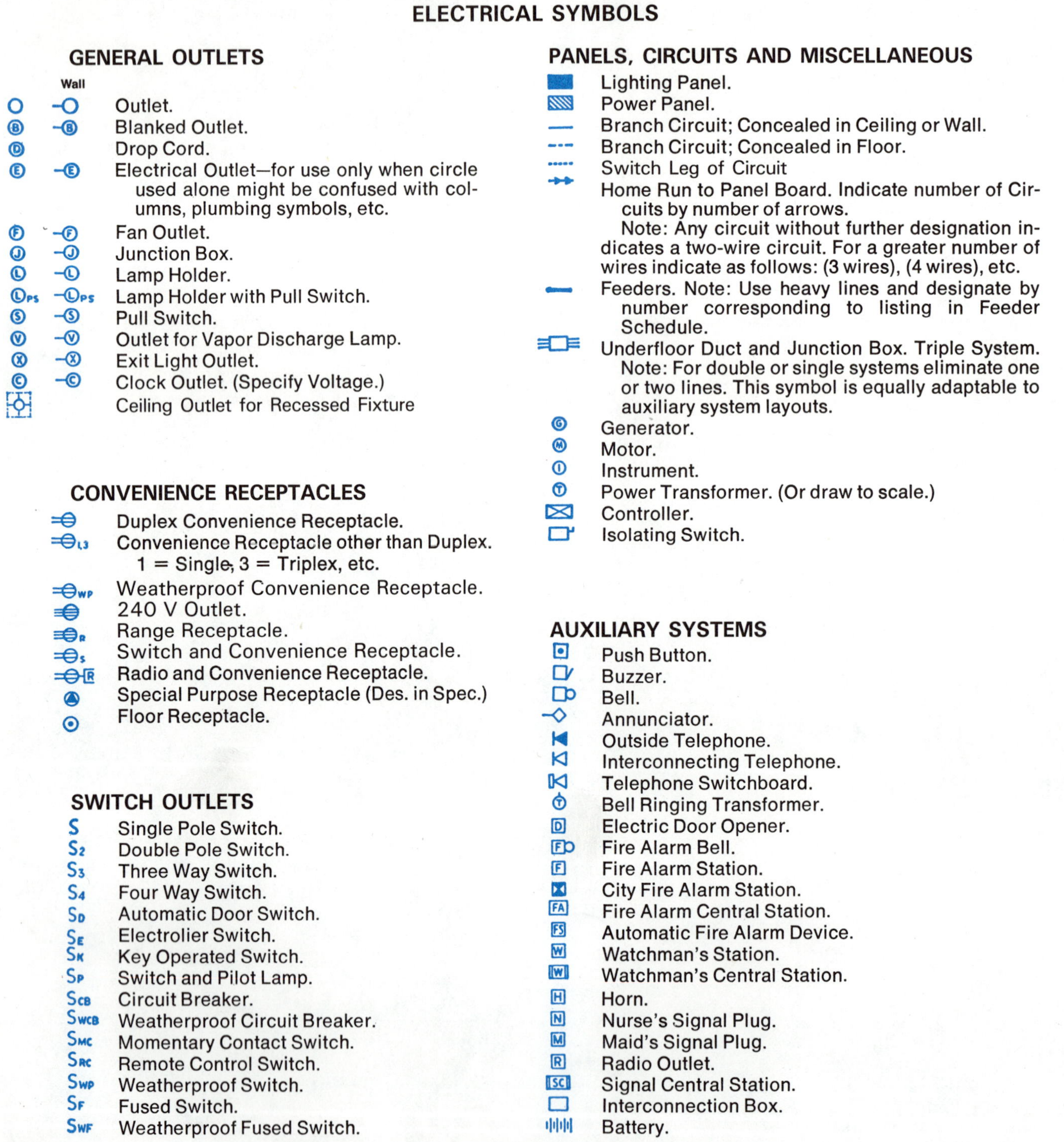

ELECTRICAL SYMBOLS

GENERAL OUTLETS

Ceiling	Wall	
O	-O	Outlet.
Ⓑ	-Ⓑ	Blanked Outlet.
Ⓓ		Drop Cord.
Ⓔ	-Ⓔ	Electrical Outlet—for use only when circle used alone might be confused with columns, plumbing symbols, etc.
Ⓕ	-Ⓕ	Fan Outlet.
Ⓙ	-Ⓙ	Junction Box.
Ⓛ	-Ⓛ	Lamp Holder.
Ⓛ$_{PS}$	-Ⓛ$_{PS}$	Lamp Holder with Pull Switch.
Ⓢ	-Ⓢ	Pull Switch.
Ⓥ	-Ⓥ	Outlet for Vapor Discharge Lamp.
Ⓧ	-Ⓧ	Exit Light Outlet.
Ⓒ	-Ⓒ	Clock Outlet. (Specify Voltage.)
⌗		Ceiling Outlet for Recessed Fixture

CONVENIENCE RECEPTACLES

- ⇌ Duplex Convenience Receptacle.
- ⇌$_{1,3}$ Convenience Receptacle other than Duplex. 1 = Single, 3 = Triplex, etc.
- ⇌$_{WP}$ Weatherproof Convenience Receptacle.
- ⇌ 240 V Outlet.
- ⇌$_R$ Range Receptacle.
- ⇌$_S$ Switch and Convenience Receptacle.
- ⇌$_R$ Radio and Convenience Receptacle.
- ▲ Special Purpose Receptacle (Des. in Spec.)
- ⊙ Floor Receptacle.

SWITCH OUTLETS

- S   Single Pole Switch.
- S$_2$ Double Pole Switch.
- S$_3$ Three Way Switch.
- S$_4$ Four Way Switch.
- S$_D$ Automatic Door Switch.
- S$_E$ Electrolier Switch.
- S$_K$ Key Operated Switch.
- S$_P$ Switch and Pilot Lamp.
- S$_{CB}$ Circuit Breaker.
- S$_{WCB}$ Weatherproof Circuit Breaker.
- S$_{MC}$ Momentary Contact Switch.
- S$_{RC}$ Remote Control Switch.
- S$_{WP}$ Weatherproof Switch.
- S$_F$ Fused Switch.
- S$_{WF}$ Weatherproof Fused Switch.

PANELS, CIRCUITS AND MISCELLANEOUS

- ▬ Lighting Panel.
- ▨ Power Panel.
- —— Branch Circuit; Concealed in Ceiling or Wall.
- ---- Branch Circuit; Concealed in Floor.
- ····· Switch Leg of Circuit
- ↠ Home Run to Panel Board. Indicate number of Circuits by number of arrows.
  Note: Any circuit without further designation indicates a two-wire circuit. For a greater number of wires indicate as follows: (3 wires), (4 wires), etc.
- ▬ Feeders. Note: Use heavy lines and designate by number corresponding to listing in Feeder Schedule.
- ▭ Underfloor Duct and Junction Box. Triple System.
  Note: For double or single systems eliminate one or two lines. This symbol is equally adaptable to auxiliary system layouts.
- Ⓖ Generator.
- Ⓜ Motor.
- Ⓘ Instrument.
- Ⓣ Power Transformer. (Or draw to scale.)
- ⊠ Controller.
- ⌑ Isolating Switch.

AUXILIARY SYSTEMS

- ▫ Push Button.
- ◰ Buzzer.
- ▱ Bell.
- ◇ Annunciator.
- ◁ Outside Telephone.
- ◁ Interconnecting Telephone.
- ◁ Telephone Switchboard.
- ⊙ Bell Ringing Transformer.
- ▫ Electric Door Opener.
- [FB] Fire Alarm Bell.
- [F] Fire Alarm Station.
- [✕] City Fire Alarm Station.
- [FA] Fire Alarm Central Station.
- [FS] Automatic Fire Alarm Device.
- [W] Watchman's Station.
- [W] Watchman's Central Station.
- [H] Horn.
- [N] Nurse's Signal Plug.
- [M] Maid's Signal Plug.
- [R] Radio Outlet.
- [SC] Signal Central Station.
- ▢ Interconnection Box.
- ⊪⊪ Battery.

Fig. 11-4. ANSI electrical symbols. These standard symbols should be studied so they become familiar and instantly recognizable. (ANSI)

Reading Blueprints and Wiring Circuits

SPECIAL RECEPTACLES

 Any Standard Symbol as given above with the addition of a lower case subscript letter may be used to designate some special variation of Standard Equipment of particular interest in a specific set of architectural plans.

When used they must be listed in the Key of Symbols on each drawing and if necessary further described in the specifications.

Fig. 11-5. Special outlet symbols carry a lower case letter alongside. An electrician must know where to look for an explanation. Read all information on the chart. (ANSI)

c, or d) will accompany a special or unusual symbol, Fig. 11-5.

An ELECTRICAL PLAN found in an architect's blueprint will not show the number of conductors in a wire run nor indicate the size of the wire conductor. Sometimes the size of conduit is indicated. It is usually good practice for the electrician to plan these runs on paper or on a floor plan for a single room. Such a plan is called a CABLE LAYOUT. See Fig. 11-6.

Some electricians make several copies of the floor plan and prepare a cable layout of each branch circuit on a separate sheet. For example, three-way switches are always run with three terminal wires while runs to four-way switches require four terminal wires. The neutral may or may not run with these depending on the layout. It is easier to work this out on the plan before you begin the rough-in.

During the actual rough-in, it is common to use the plan as a record of work completed. When a run has been completed a colored line can be penciled through the run on the floor plan. If this practice is followed, no runs will be overlooked. When two or more electricians are working on the same job this kind of record is necessary to keep track of work completed. See Fig. 11-7.

WIRING CIRCUITS

In the rest of this chapter, we shall look at the actual wiring of circuits, by comparing blueprint, schematic, and pictorial drawings. They cover all of the common circuits found in modern electrical construction.

As you study the circuits shown here, note the wiring method used. Trace the different wires in the pictured drawings until you become familiar with

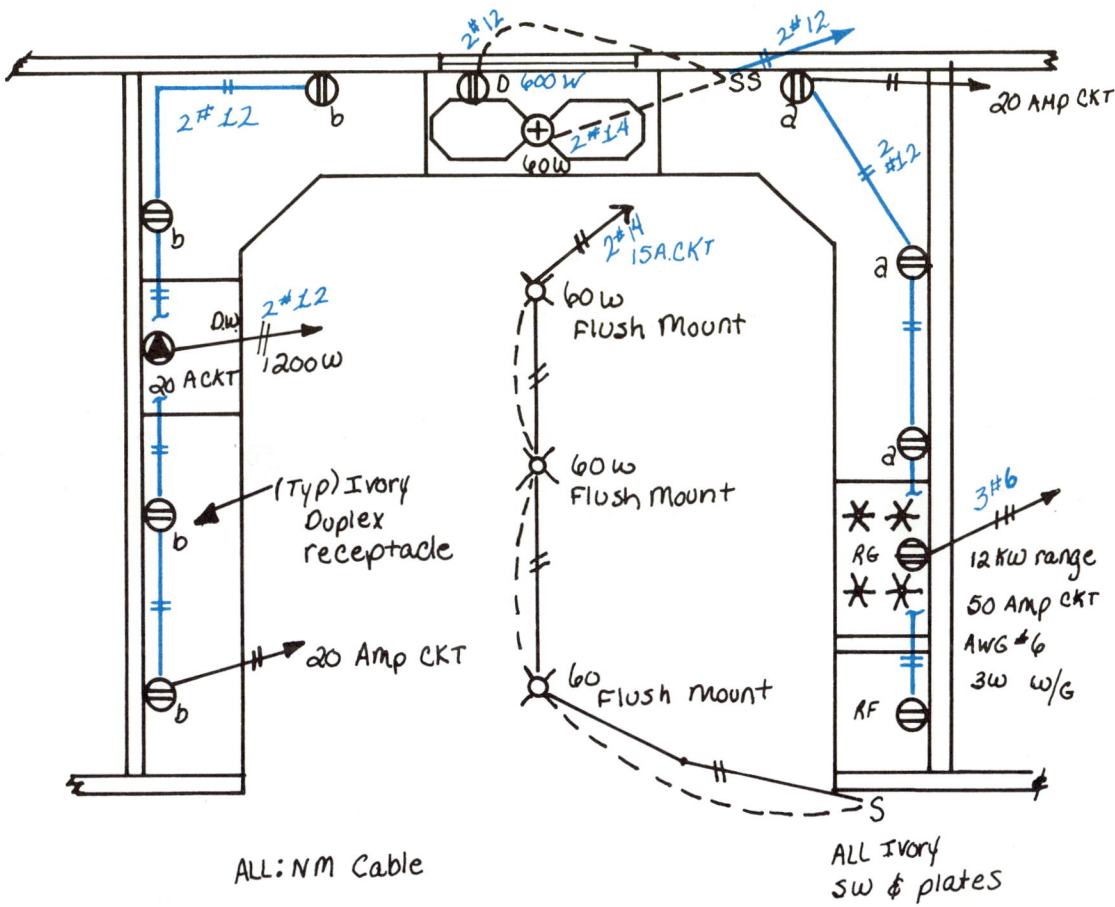

Fig. 11-6. Cable layout is a blueprint of the electrical plan which explains where cable is to run and how many conductors it must have. See Figs. 11-21 and 11-23 for other cable layouts.

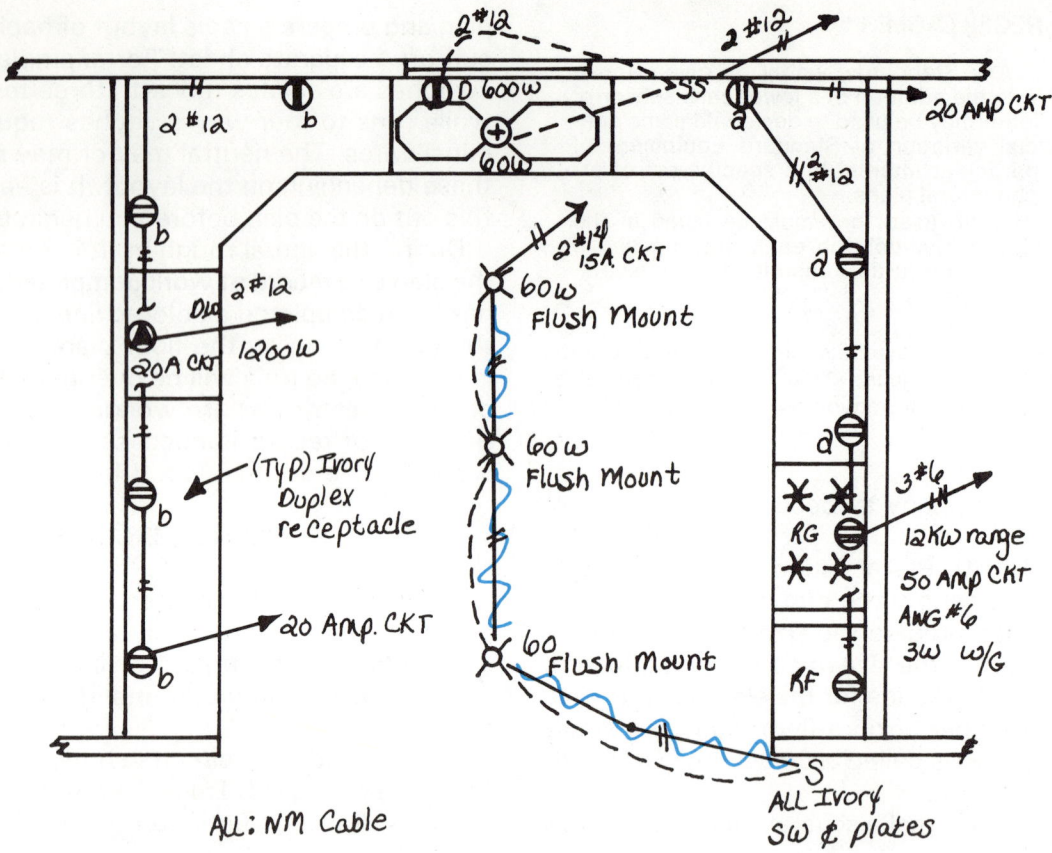

Fig. 11-7. When several electricians are working on the same job, cable plan is used to keep a record of the work done. Colored line is drawn through completed runs.

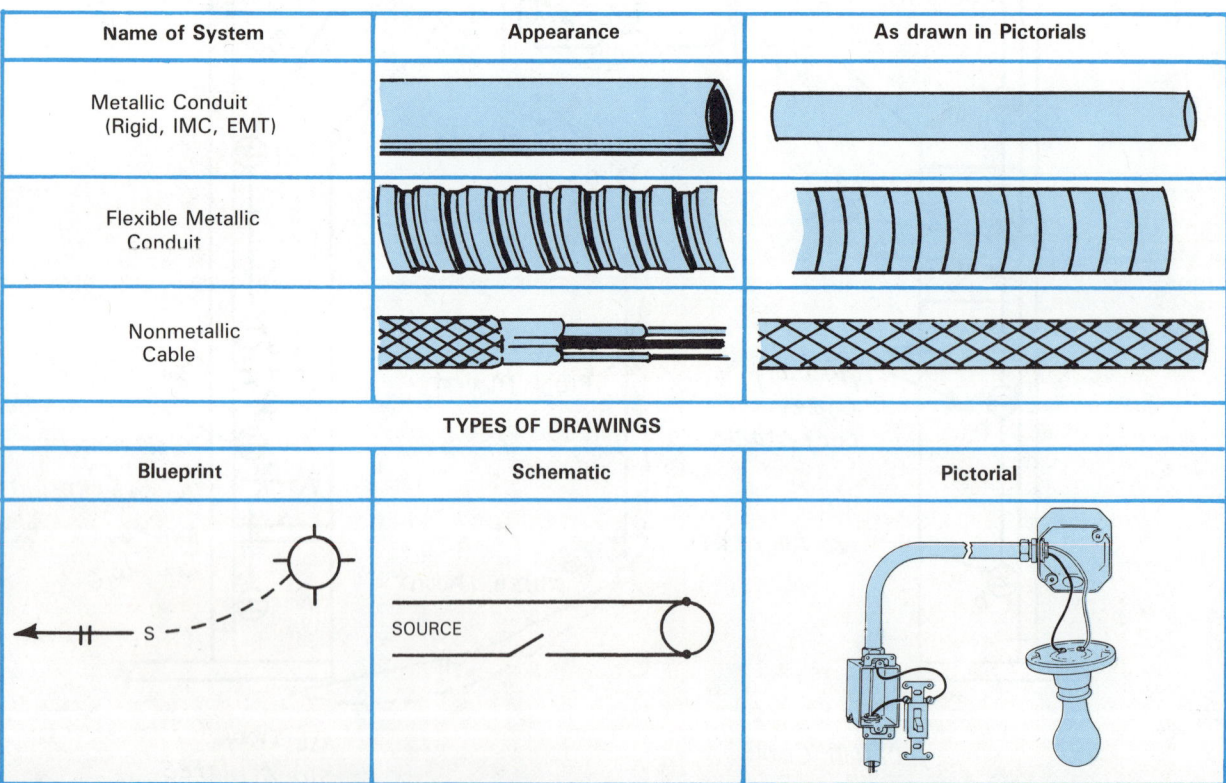

Fig. 11-8. Key to circuit drawings in this chapter. When you can relate the blueprint electrical plan and schematic to the pictorial wiring layout you will be able to wire up any circuit by studying the electrical plan.

the symbols in the blueprint circuit layout and schematic drawings. Compare the circuit in the schematics with the pictorials. Metallic conduit, flexible metallic conduit, and nonmetallic cable will be drawn as shown in Fig. 11-8.

No grounding wire is shown in pictorial drawings for metallic conduit. None is required. The conduit itself serves as the grounding conductor. Blueprints and schematics do not show the grounding conductor regardless of wiring method.

POLARITY IN ELECTRICAL WIRING

A typical 120 V electrical circuit consists of two wires: a "hot" wire and a neutral wire. The hot wire is usually black or red but may be any color except white, green, or green with yellow stripes. The neutral is always white or light gray.

The neutral wire is the grounded conductor. It must never be interrupted by a switch, breaker, or fuse unless the "hot" wire is simultaneously interrupted. This distinction or segregation of the neutral and hot wire is termed POLARITY.

REASONS FOR POLARITY WIRING

There are three primary reasons for polarity wiring:
1. Safety. Since the neutral wire is the grounded conductor, a person making contact with this wire is generally safe from harmful shock.
2. Overload prevention. Since each circuit consists of a hot and a neutral wire, it is unlikely that an overload can occur. See induction next.
3. Induction. Harmful effects of induction, a magnetic action which damages wire insulation through its heating effects, can be prevented by purposely pairing the hot and neutral wire in the same conduit tubing or cable. In simple terms, any magnetic effects are cancelled when the neutral and hot wires are paired. If all hot wires were run in one conduit, and all neutral wires of several circuits were run in another conduit, a dangerous heating condition would occur due to magnetic effects.

JUNCTION BOXES

Fig. 11-9 illustrates a typical junction box with its connections. A J-box, as it is commonly called, is used to route a given circuit in various directions. It meets *Code* requirement that all splices be made in an electrical box. At the same time it saves labor, time, and costs. In wood frame dwellings it once was common to bring the conductors of a circuit to an attic J-box. Then the circuit was split into several subbranches creating a spider-like wiring arrangement, Fig. 11-10.

We will discuss more of this in a later chapter on circuit problems and troubleshooting. For now, it is important to note the color of the conductors and how they are connected in the different circuits.

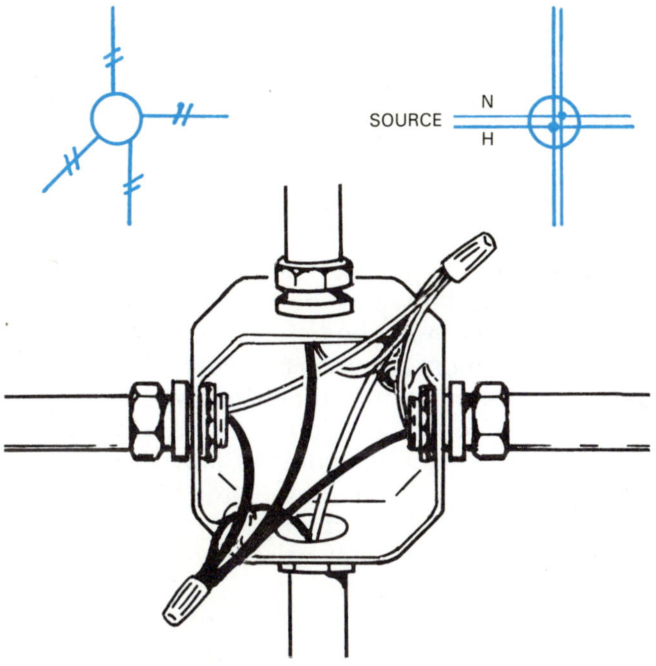

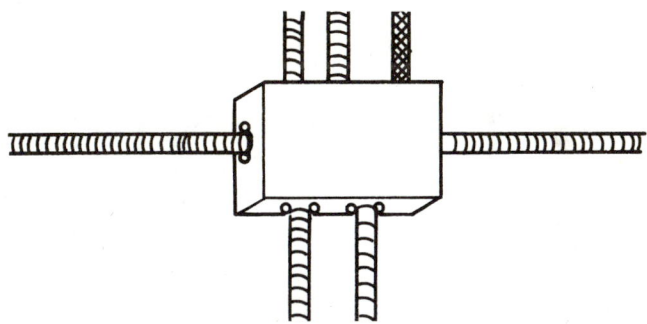

Fig. 11-10. Spider-like arrangement is found frequently at attic junction boxes.

PULL-CHAIN FIXTURE

Although no longer popular, except in closets, attics, and unfinished areas, a simple pull-chain fixture can be wired up at any point along the circuit or at the end of a run as shown in Fig. 11-11. Switched lights have replaced pull-chain types in most areas. In fact, as shown in Fig. 11-12, the same purpose is served in a more convenient form by a single-pole switch.

Fig. 11-9. Junction boxes serve the purpose of dividing a circuit into several directions. Study the cable layout and the schematic and compare them with actual wiring method shown in the pictorial drawing.

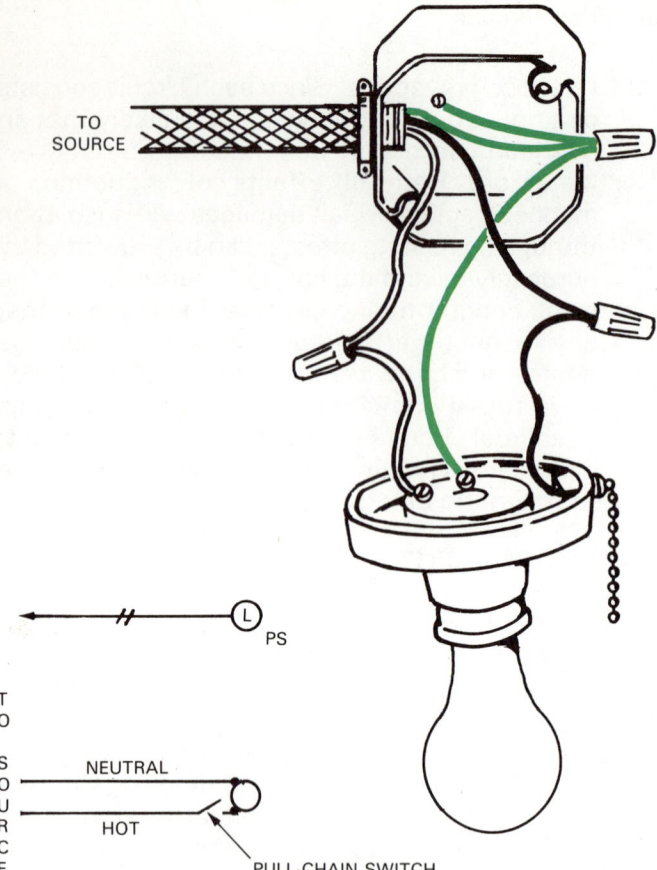

Fig. 11-11. Simple pull-chain fixture at the end of a circuit run. Switch is in the fixture itself.

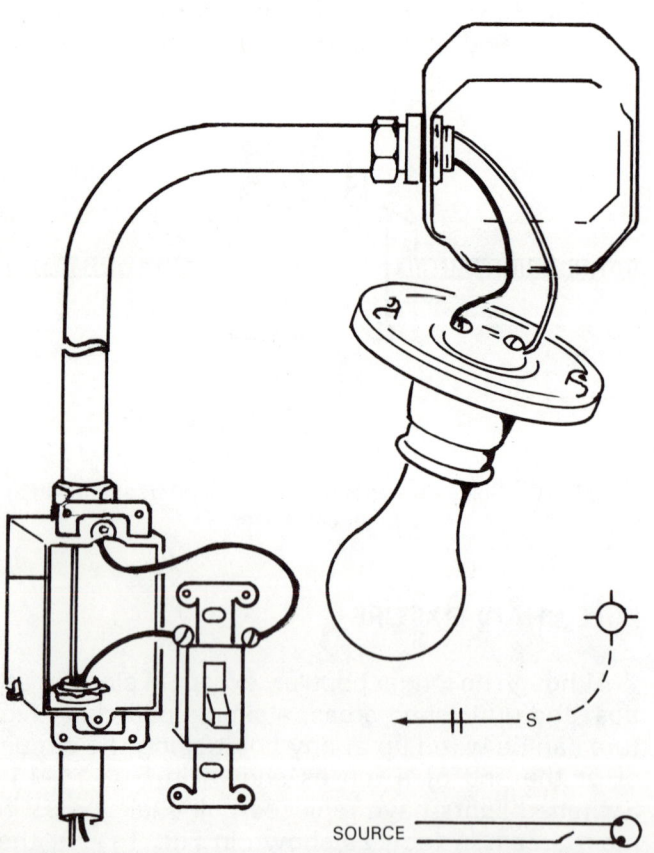

Fig. 11-12. A single light controlled by a wall switch. Feed is to switch.

DUPLEX RECEPTACLES

Duplex receptacles, like that of Fig. 11-13, with provision for grounding can be continuously live or switched. See Figs. 11-14 and 11-15. Note particularly the differences in grounding procedure when using cable, Fig. 11-16, and metal conduit, Fig. 11-17. The ground wire of the nonmetallic

Fig. 11-13. Duplex grounding receptacles are required by Code. (Bryant)

cable is connected to the device box as well as to the ground screw on the receptacle. This provides for a continuous equipment ground system. Metal conduit needs no special ground wire since the casing, itself, provides the grounding path. When conduit is employed, the receptacle ground screw is connected to the box only. This provides continuity of the ground to the receptacle. The connection is made by running a short length of bare or green insulated wire between the receptacle grounding screw and the metal box.

One of the most common circuits has one or two switch-operated light fixtures followed by two or more continuously live outlets. As Fig. 11-16 shows, the wall switch controls only the light fixture. Note the continuity of the ground wire and its connection at each junction point. Should one or more of the devices later be removed, the ground integrity remains intact.

Another often-used circuit controls two or more lights independently from a single location. Fig. 11-17 illustrates this basic wiring technique. Two separate single-pole switches may be used instead, but the one shown is more economical and faster to wire. If lights are to be on separate circuits,

Reading Blueprints and Wiring Circuits

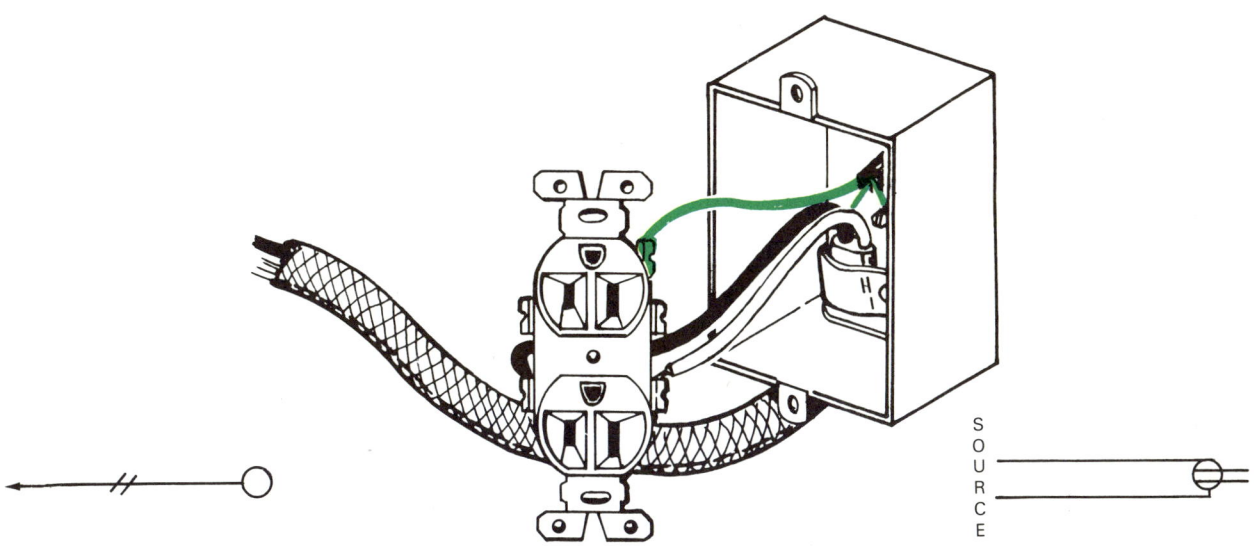

Fig. 11-14. Continuously live duplex, grounding-type, receptacle. Both parts of duplex are served by the circuit wires. (GE Wiring Devices Dept.)

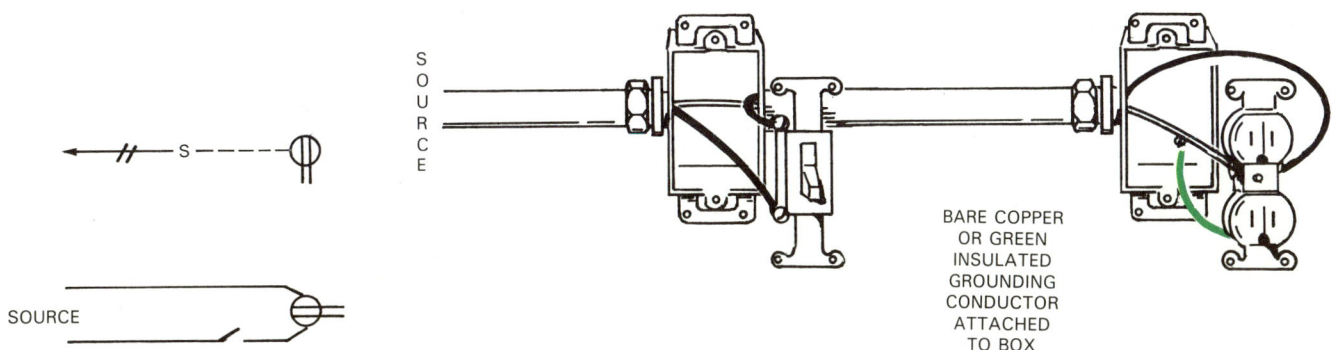

Fig. 11-15. Switched duplex, grounding-type, receptacles are used in rooms where overhead lighting is not desired. Both sides of receptacle are controlled by the switch.

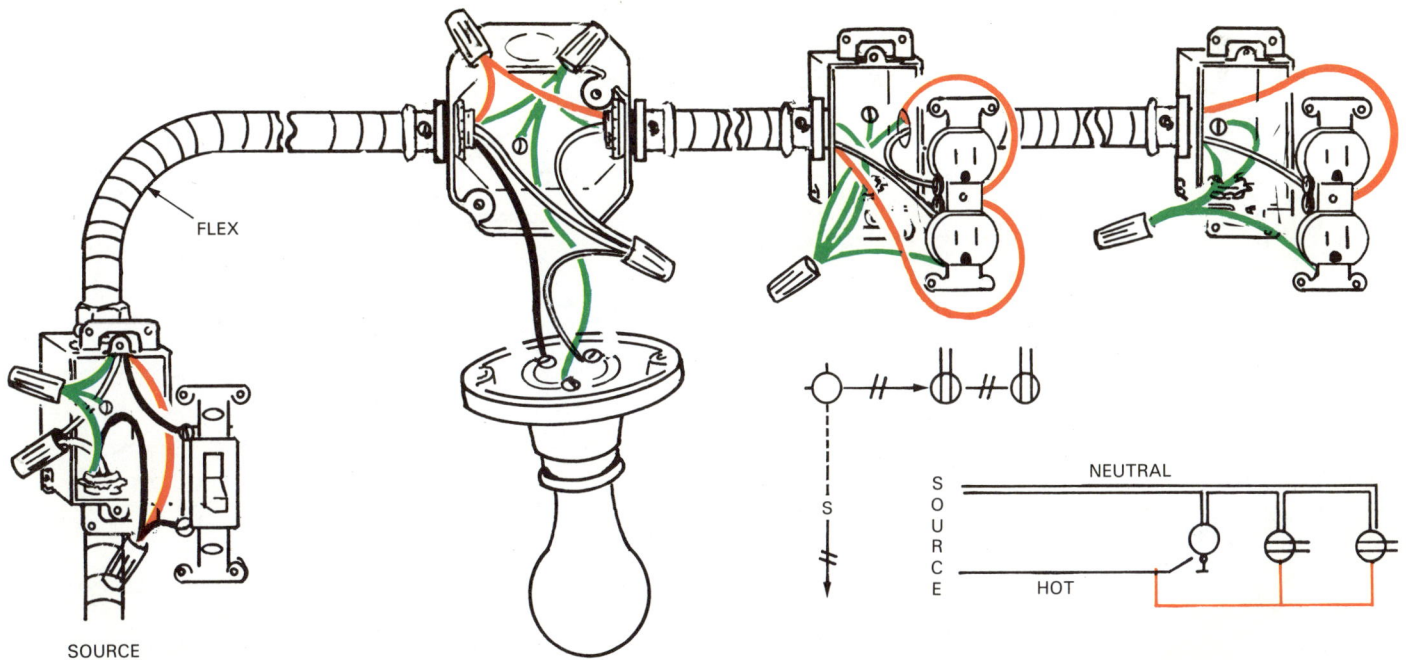

Fig. 11-16. Light controlled by a single pole switch has outlets beyond it which are always live. Source or feed is to switch.

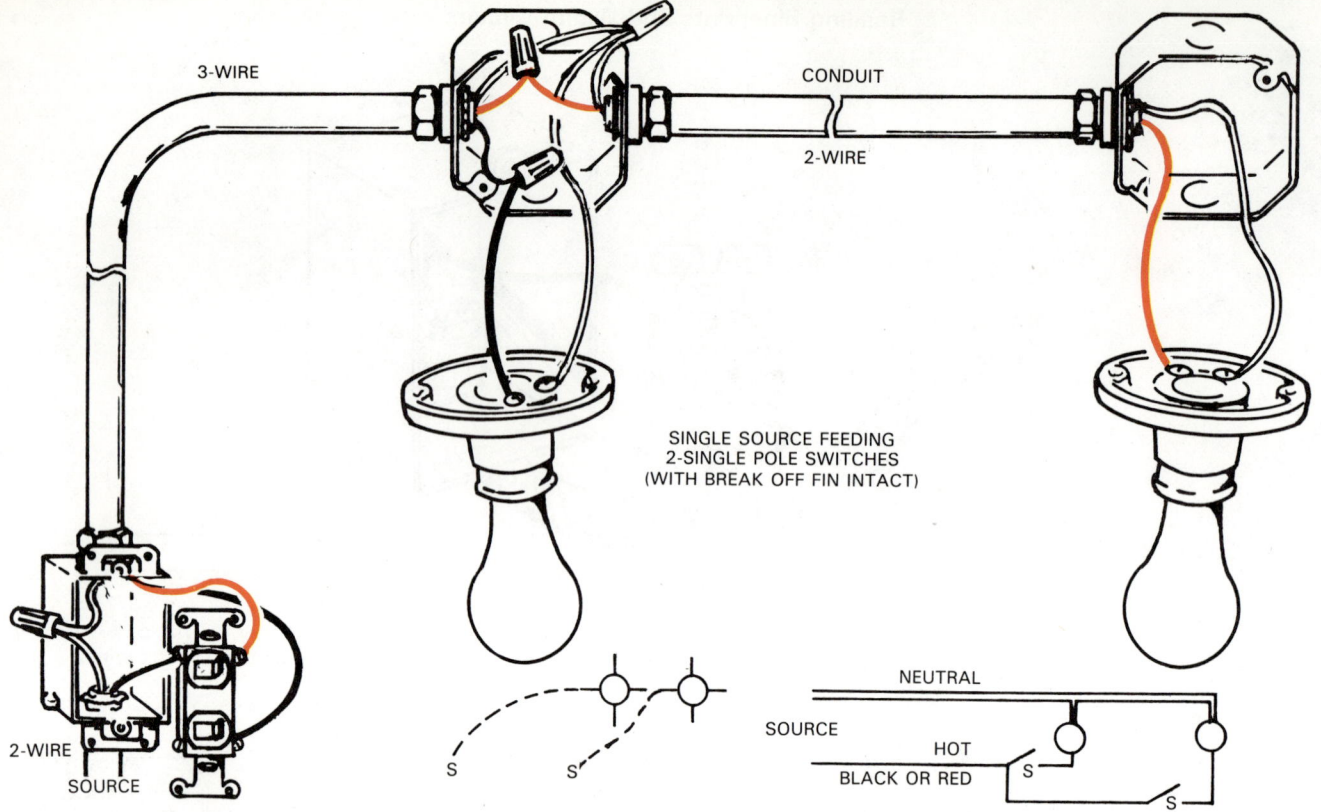

Fig. 11-17. Two single pole switches on the same circuit. Each switch controls a different light.

simply remove the breakoff fin and feed each half independently.

The switch-receptacle combination shown in Fig. 11-18 is used around bathroom vanity locations where space is often limited. *When used in the bathroom, the receptacle circuits must have GFCI protection.*

SPLIT-WIRED RECEPTACLES

Split-wired receptacles, Fig. 11-19, are used mostly in rooms which have no ceiling outlets. Thus, lighting devices, such as lamps, can be switch controlled. Devices such as a radio or TV, can be plugged into the top (always hot) portion.

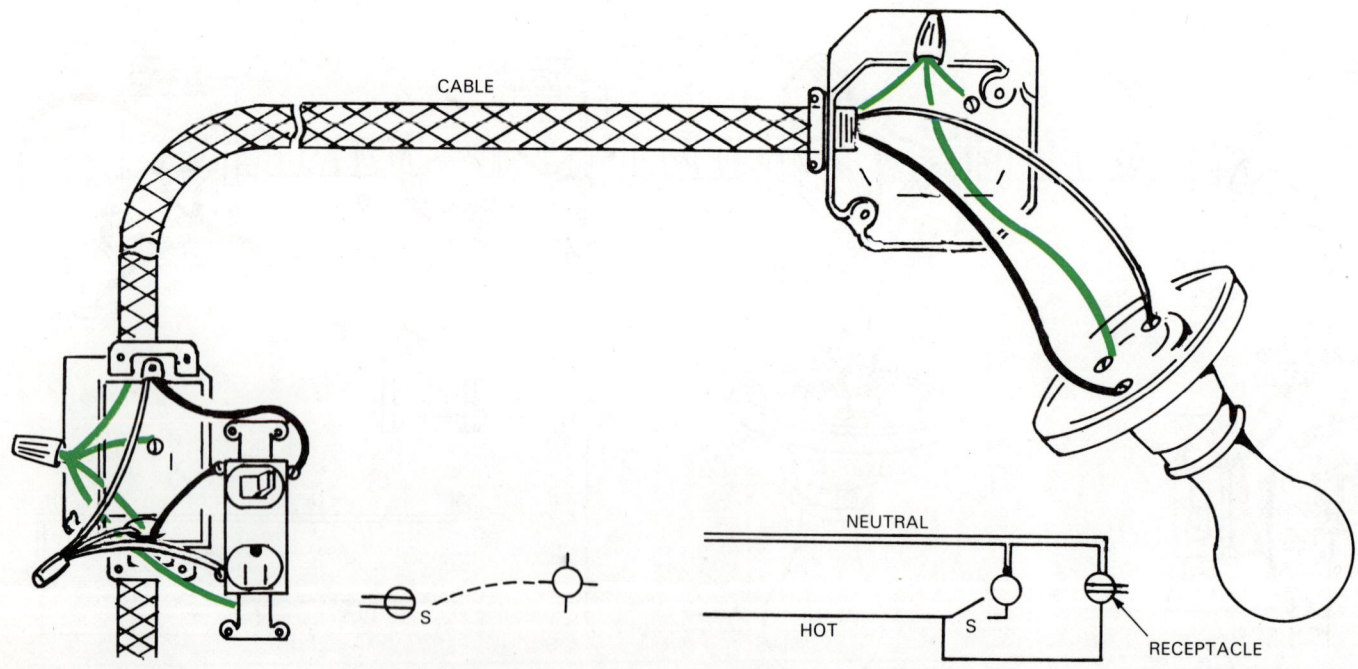

Fig. 11-18. Switch/receptacle combination. Switch controls light while receptacle is always live. A single circuit supplies both the switch and the receptacle outlet.

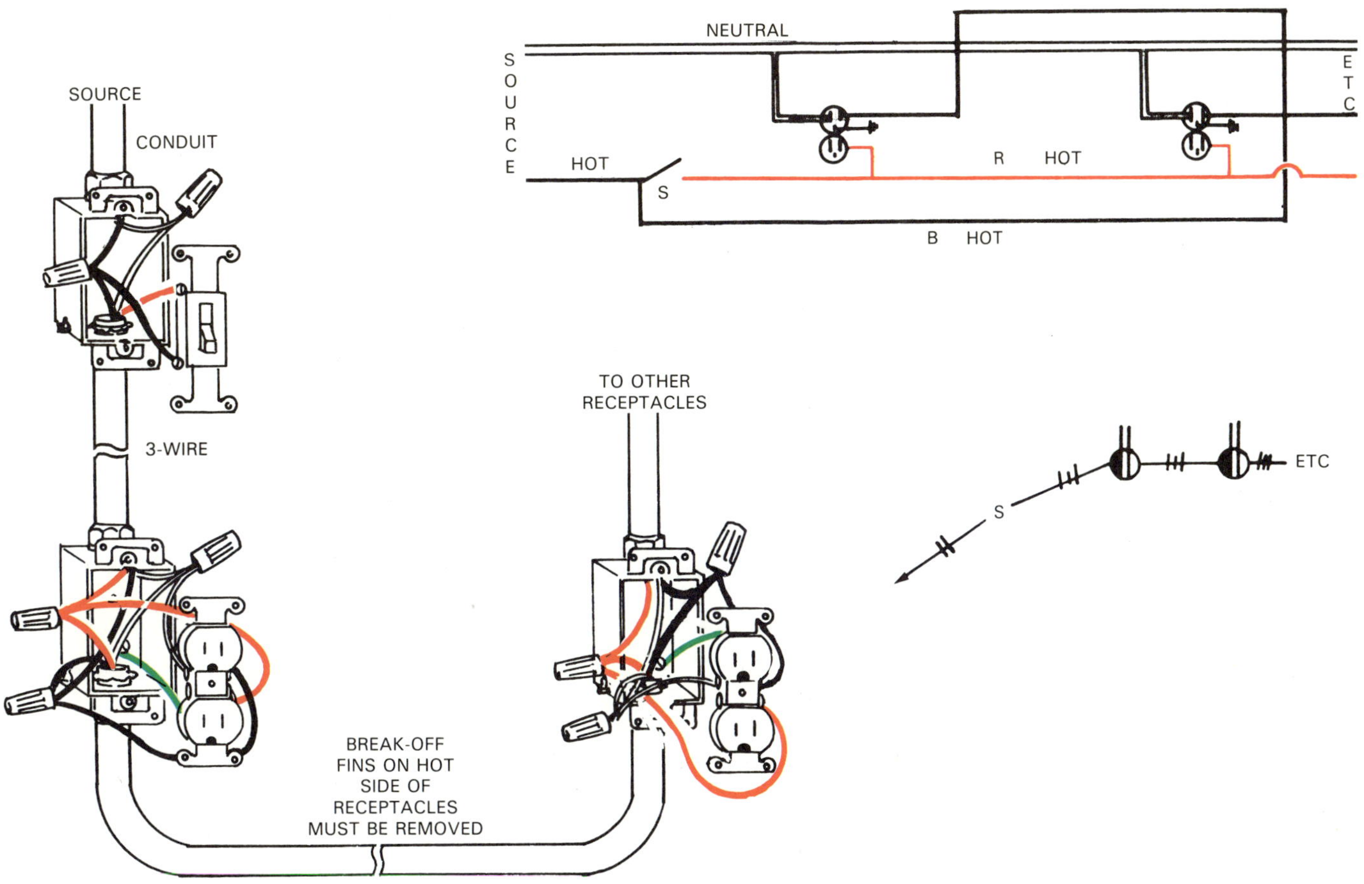

Fig. 11-19. Split-wired receptacles. Top half is always hot; bottom half is controlled by single-pole switch.

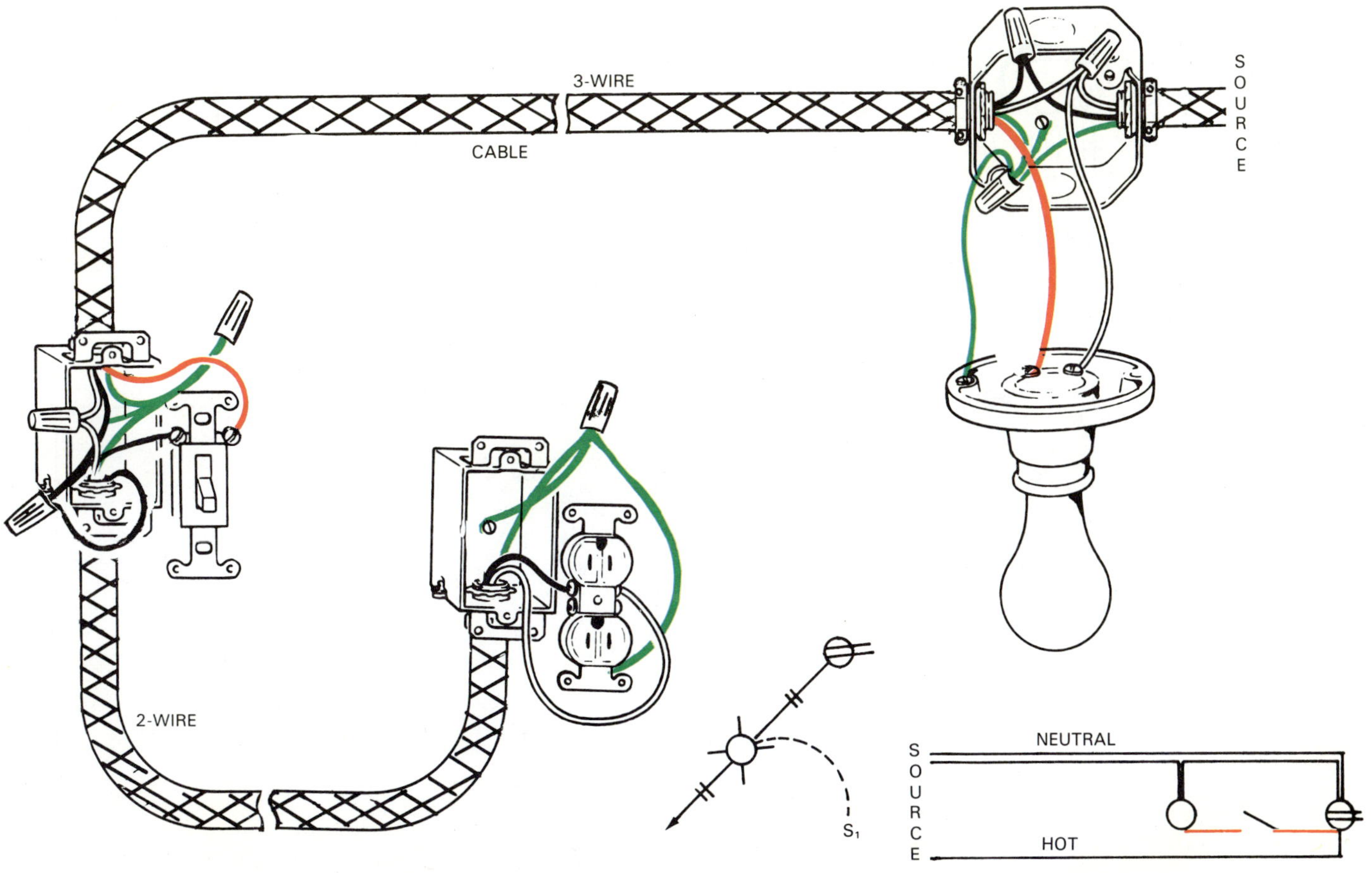

Fig. 11-20. Light controlled by a switch. Receptacle beyond the switch is live. Source is at the light.

123

Modern Residential Wiring

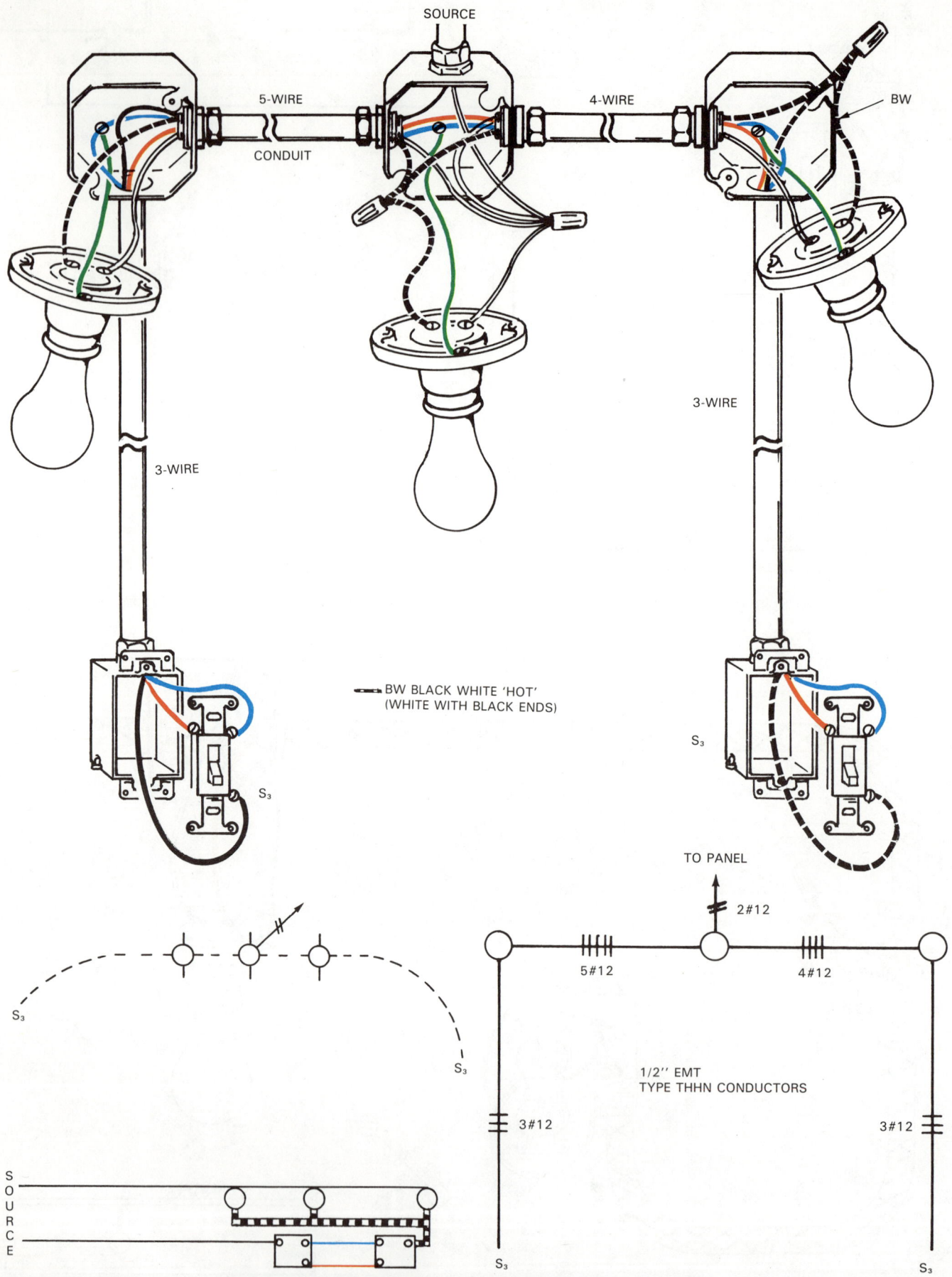

Fig. 11-21. Several light fixtures are controlled simultaneously by a pair of three-way switches. Feed is to the center light fixture. Note cable layout. Refer again to Fig. 11-6.

Reading Blueprints and Wiring Circuits

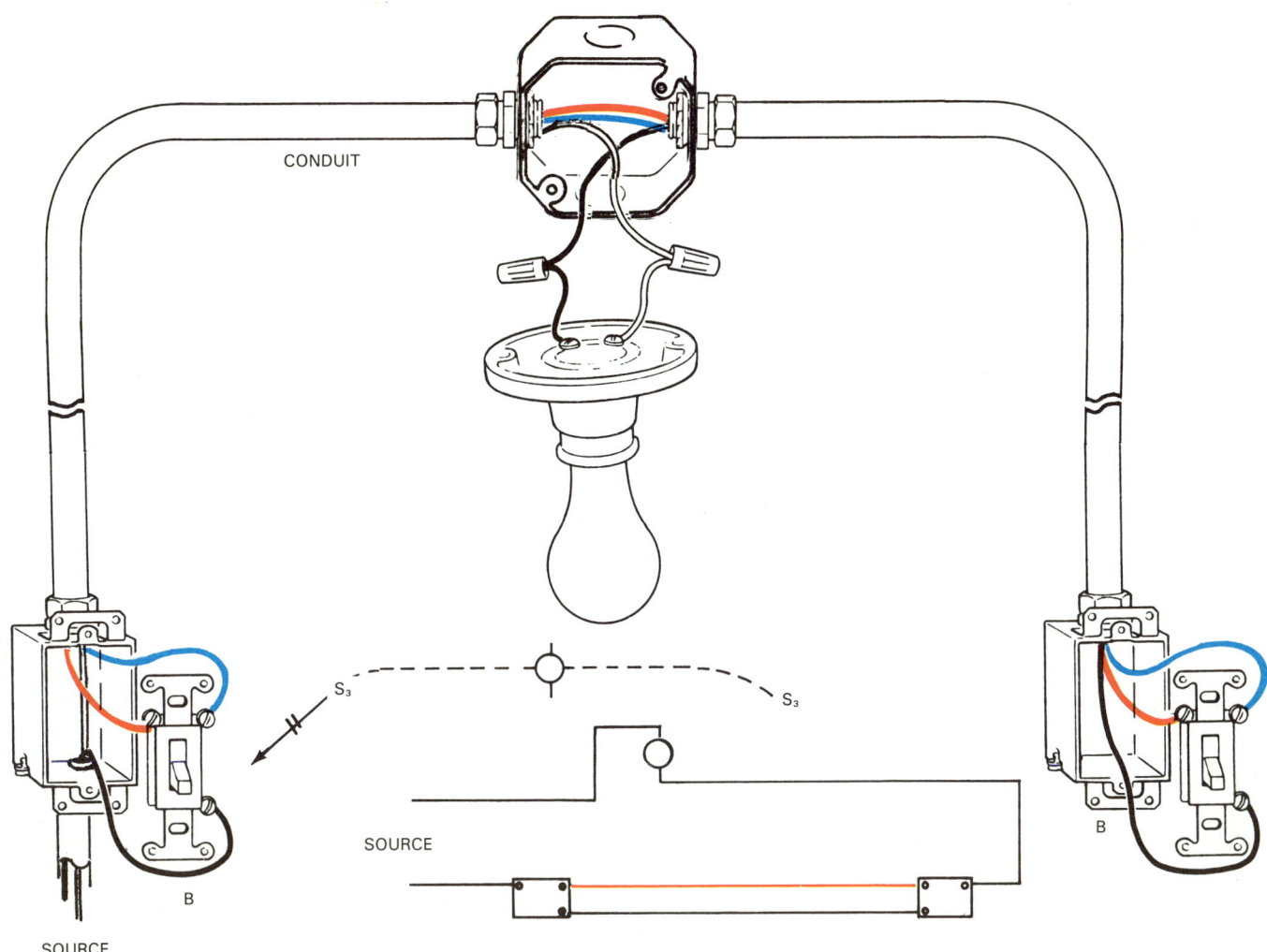

Fig. 11-22. A pair of three-way switches controlling a single light fixture. Source is to one of the three-way switches.

Fig. 11-20 is essentially the same circuit as Fig. 11-16, except that the feed is to the fixture instead of the switch. Note the use of three-wire cable between the lighting outlet and switch.

THREE-WAY SWITCHES

In Fig. 11-21, three-way switches control lights from two distant locations, (for example, in rooms having two entrances). In this case, three lights are controlled by a pair of three-ways. The source, in this instance, is at the center light. Study this illustration carefully. You will undoubtedly encounter this circuit frequently as an electrician.

Fig. 11-22 shows a pair of three-way switches controlling a light. The source is to one of the three-way switches.

FOUR-WAY SWITCHES

Rooms having more than two entrances will, most likely, have switches at each entrance to control the area lighting. Four-way switches, placed between a pair of three-ways, allow for unlimited flexibility in control and location.

Fig. 11-23 shows a light controlled from four separate locations. If more switching locations are needed, more four-ways can be incorporated into this circuit.

Fig. 11-24 illustrates how to get the most use out of a limited space. For example, a narrow entrance hallway, having both a back door and front door, should have a wall switch at each entrance to operate the hallway light. A pair of three-way switches will handle this requirement, but what about the outside light above the front door? The problem is solved by using a single-pole switch to control the outside light. Both switches will be housed in a single two-gang box.

PILOT SWITCHES

Quite often, it is desirable to know whether a light is on or off when we cannot see it from the switch location. An example is a cellar or attic light which is controlled by a switch in the kitchen. For these

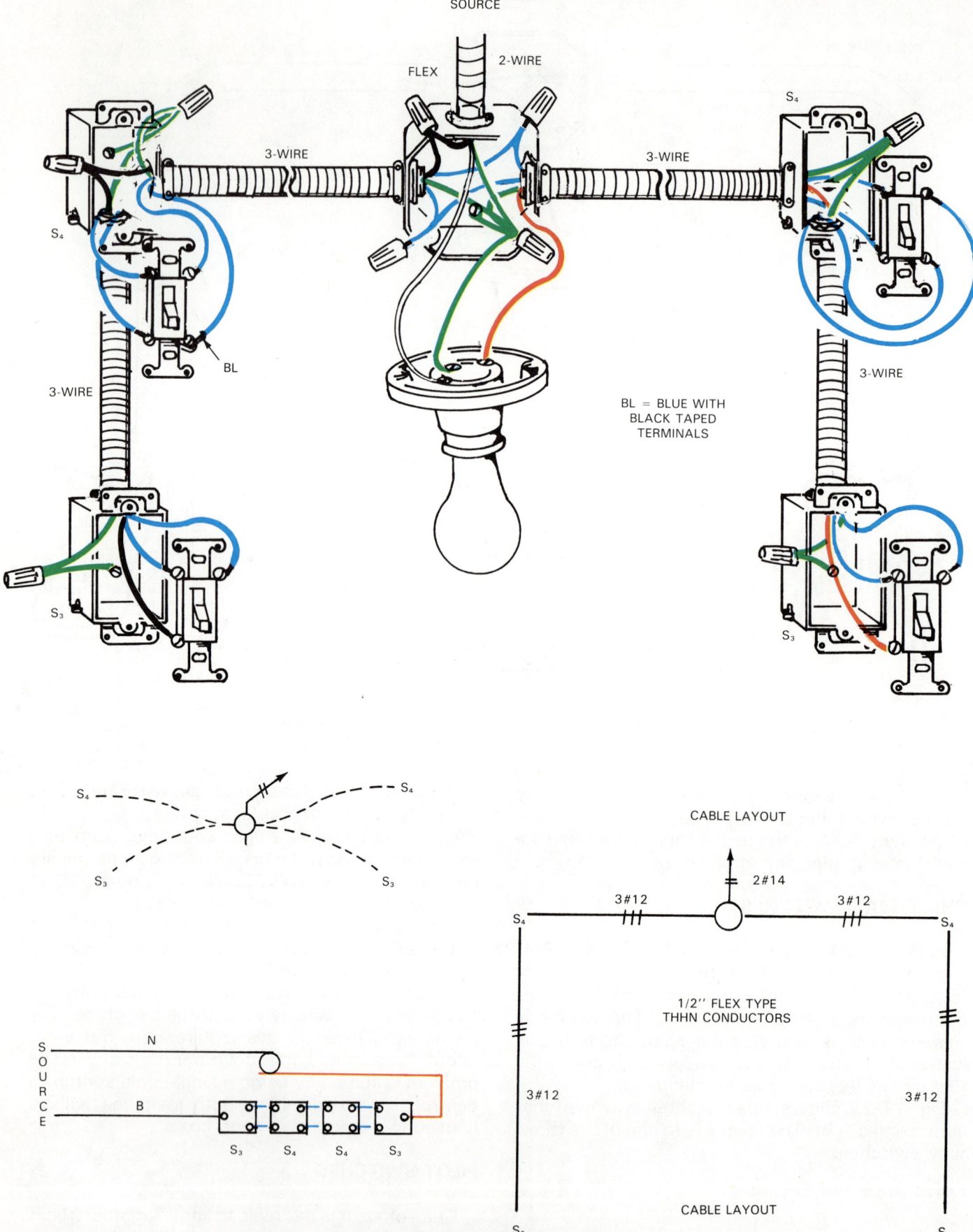

Fig. 11-23. A light fixture controlled from four locations using two three-way and two four-way switches. The feed is at the fixture. Cable layout helps electrician select proper cable for wiring up this circuit.

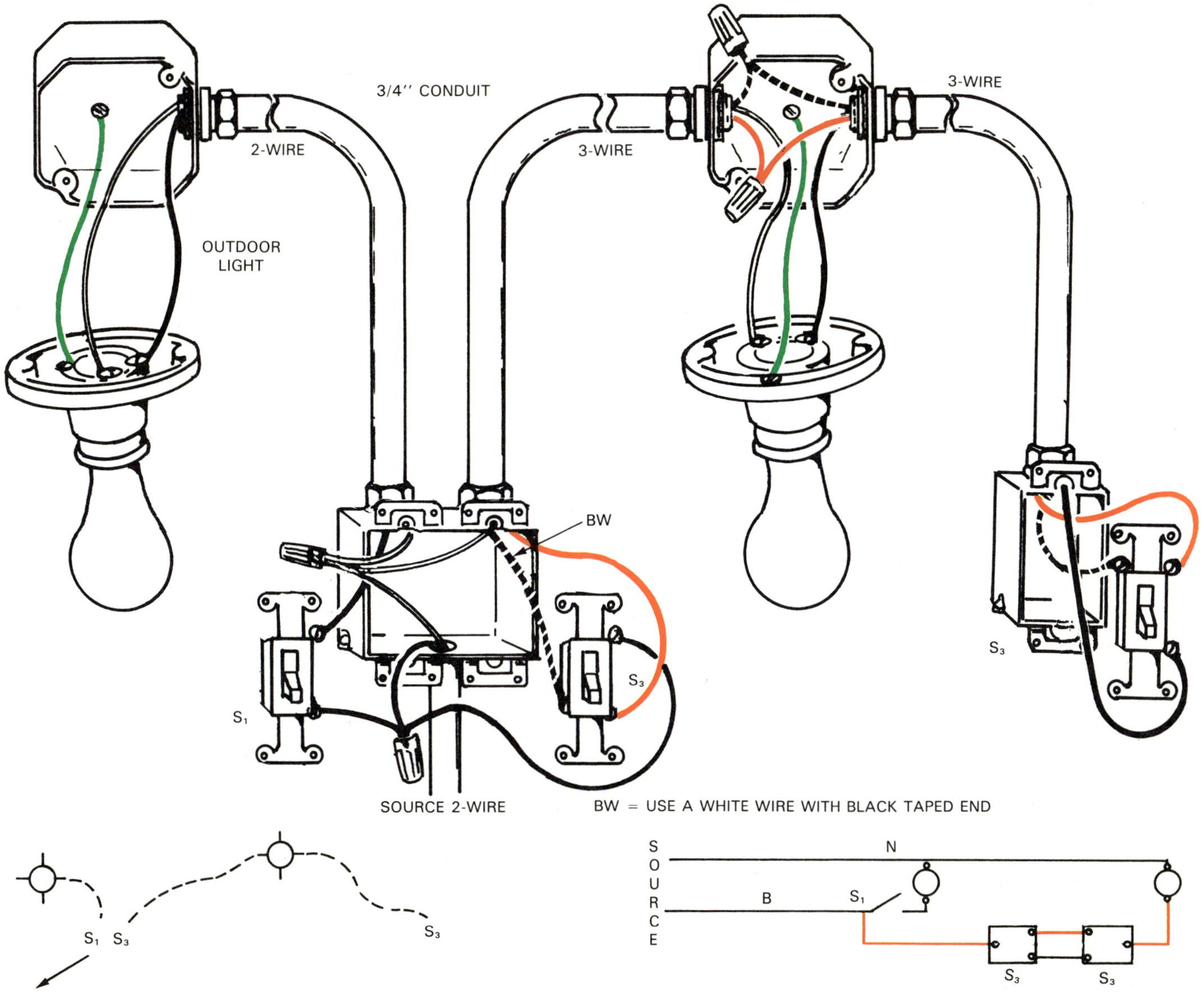

Fig. 11-24. One light controlled by a single-pole switch. One circuit is the source, the single-pole and one three-way switch are housed together.

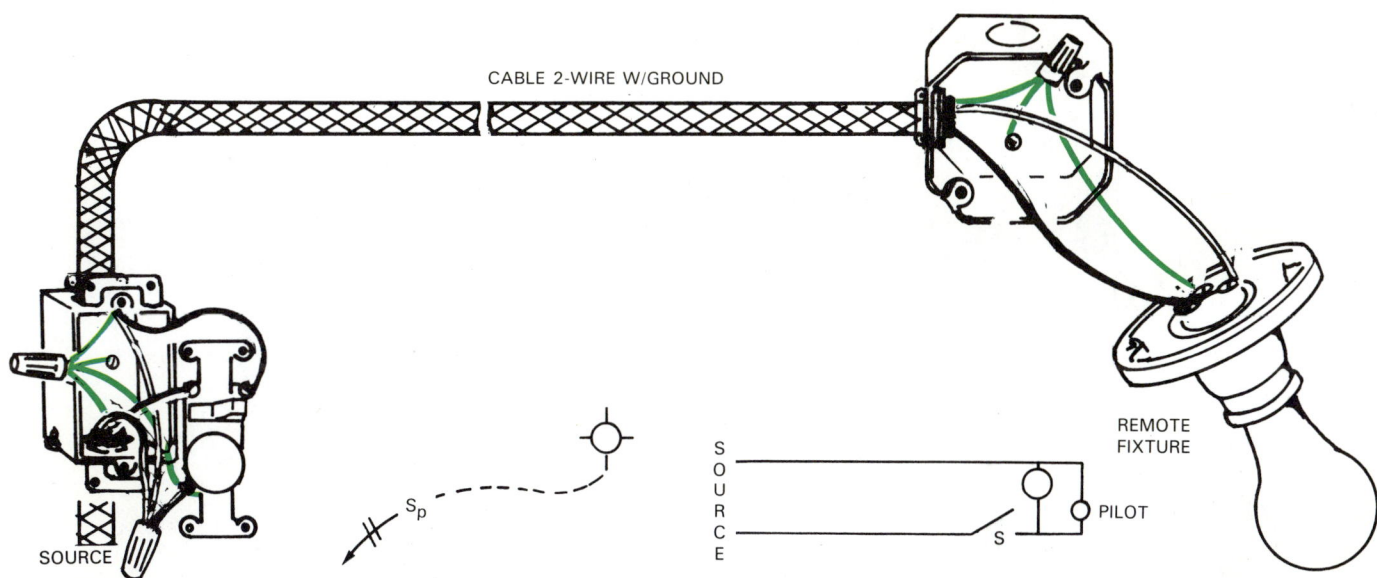

Fig. 11-25. Single-pole switch controlling a remote lighting fixture, which cannot be seen from the switch location. Pilot light is on when fixture is on. Source is at switch. There are different methods of wiring such a switch. Refer to manufacturer's instructions.

127

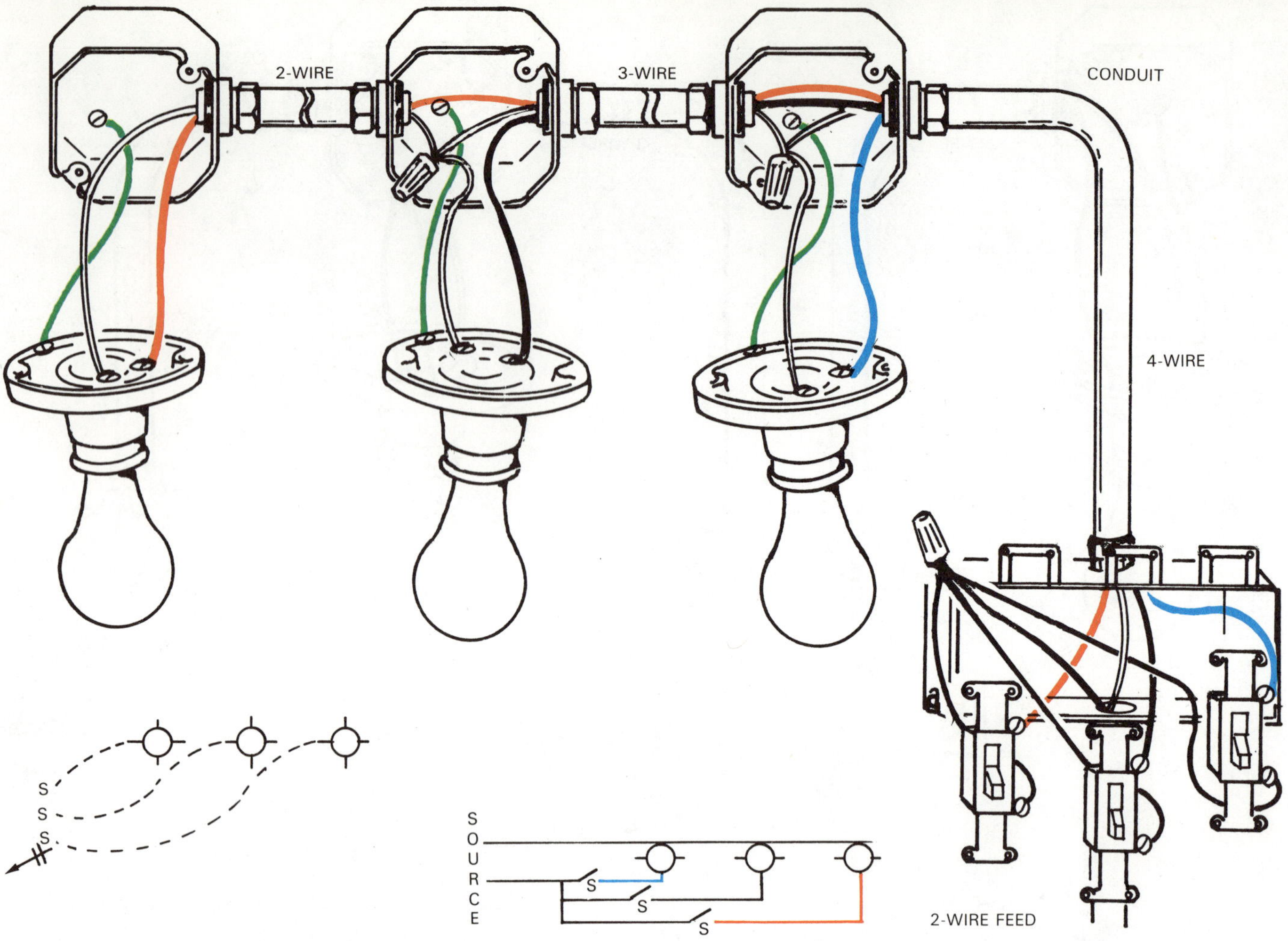

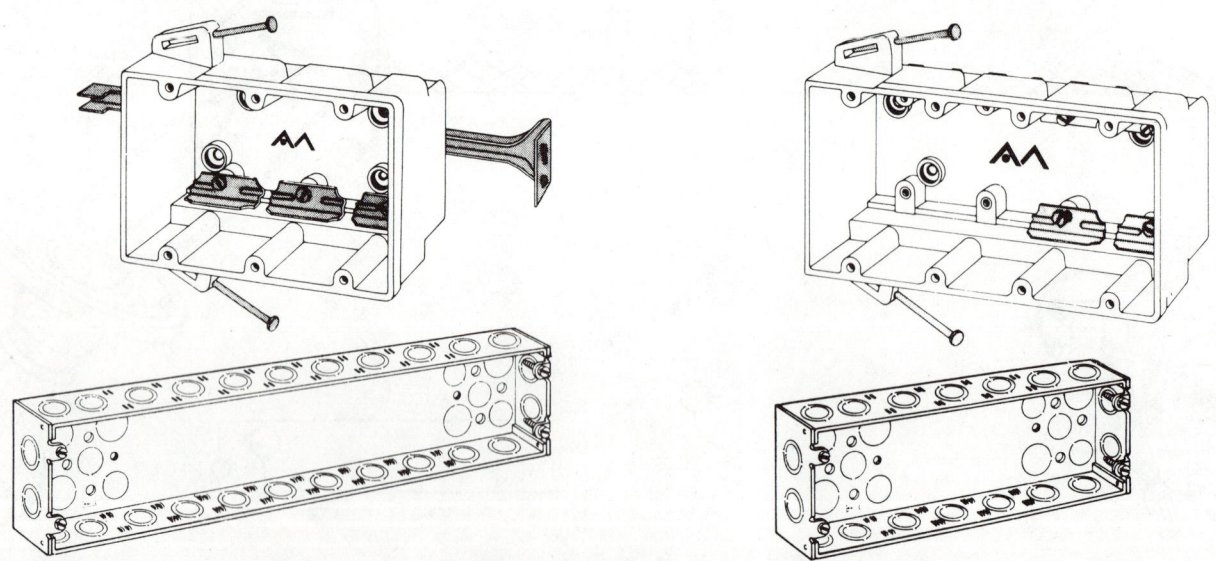

Fig. 11-26. Three switches on one circuit. Each controls a separate lighting fixture. The switches are all at one location in a multi-gang box. The source is at the switch.

situations, a single-pole, pilot switch may be used. This pilot switch is wired so that it glows when the fixture is on, as shown in Fig. 11-25.

Fig. 11-26 shows a single source feeding three switches. Each is separately controlling a light fixture. However, more switches and lights could be done in a similar manner. Four, five, six, and seven-gang switch boxes are not uncommon, Fig. 11-27.

Fig. 11-27. Multi-ganged boxes. Top. Fiberglass boxes. (Allied Moulded Products, Inc.) Bottom. Metal boxes. (RACO Inc.)

GROUND FAULT CIRCUIT INTERRUPTERS

Ground fault circuit interrupters (GFCIs) can be wired into any grounded receptacle outlet and will replace the regular receptacle. The GFCI, Fig. 11-28, will protect an individual circuit or it can be wired up to protect other receptacles that follow the GFCI on the same circuit. To protect these other outlets, the GFCI must be placed in the outlet which is electrically closest to the circuit breaker panel (power source). Fig. 11-29 is a schematic drawing for a GFCI which is protecting three additional receptacles. Study the wiring hookup carefully. Unless properly connected, the GFCI will not provide the desired protection.

To determine if a previously wired receptacle is on the end of a line or if the circuit continues on to other receptacles, remove the plate and examine the wiring hookup. See Fig. 11-30. A receptacle which feeds other receptacles will have additional conductors attached to it which continue on to one or more outlets. Further checking with a neon tester will be necessary to find which outlets are located on the same circuit.

OUTDOOR WIRING PROCEDURES

All homes must have at least one GFCI protected outdoor receptacle. One on each exterior wall would be better and more convenient, but a minimum of one is a necessity and is required by *Code* for new homes.

Regular switches, receptacles, and lamps are not suited for outdoor installation since they would ordinarily be damaged too easily by the elements.

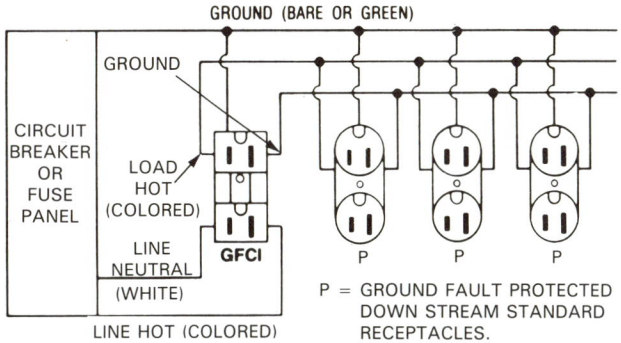

Fig. 11-29. Schematic of a GFCI wired to protect an entire branch circuit. To provide protection for all receptacles, the GFCI must be the first receptacle from the fuse box or the circuit breaker. Any receptacle between the fuse box or the circuit breaker and the GFCI will not be protected. (Arrow Hart Div., Cooper Industries)

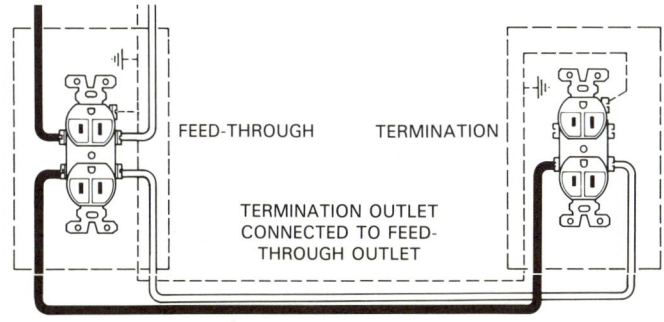

Fig. 11-30. Appearance of a wired-up receptacle will tell you if it is the last (termination) receptacle in the circuit or not. If you find second set of wire conductors, one black and one white, you can be sure that the circuit continues on to at least one more receptacle. Termination receptacle (end of line) has but one set of conductors. (GE Wiring Devices Dept.)

Therefore, the *Code* requires that special housings or weather-proof devices be used.

Standard receptacles and switches may be used outdoors if mounted inside special protective and weatherproof housings, Fig. 11-31. Remember that the receptacle circuits must be GFCI protected.

Outdoor wiring should only be installed using properly protected switches and receptacles as previously mentioned. The preferred type of cable to use is type UF. It is essentially weatherproof. Type XLP and other weatherproof conductors may be buried directly or channeled through conduit.

If outlets are to be located at some distance from the house, type UF cable or conduit can be buried safely around 18-24 inches beneath the surface. For double safety, the UF cable should be encased in rigid metal or plastic conduit particularly where it enters the ground. This protects it against physical damage by bicycles, baseball bats, and other hazards. Fig. 11-32 will serve as a guide to safe outdoor wiring. The exact layout will depend on the purpose of such an installation, but the principles are the same.

Fig. 11-28. GFCI protects individual receptacle or series of receptacles. (Bryant)

Modern Residential Wiring

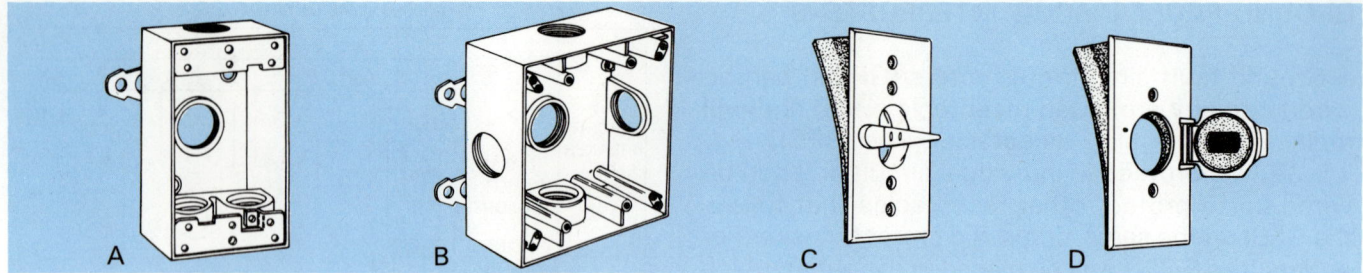

Fig. 11-31. Specially designed weatherproof housings are used to protect devices from the elements. A—Single-gang switch/receptacle box. B—Two-gang box. C—Switch cover. D—Outlet cover. (RACO Inc.)

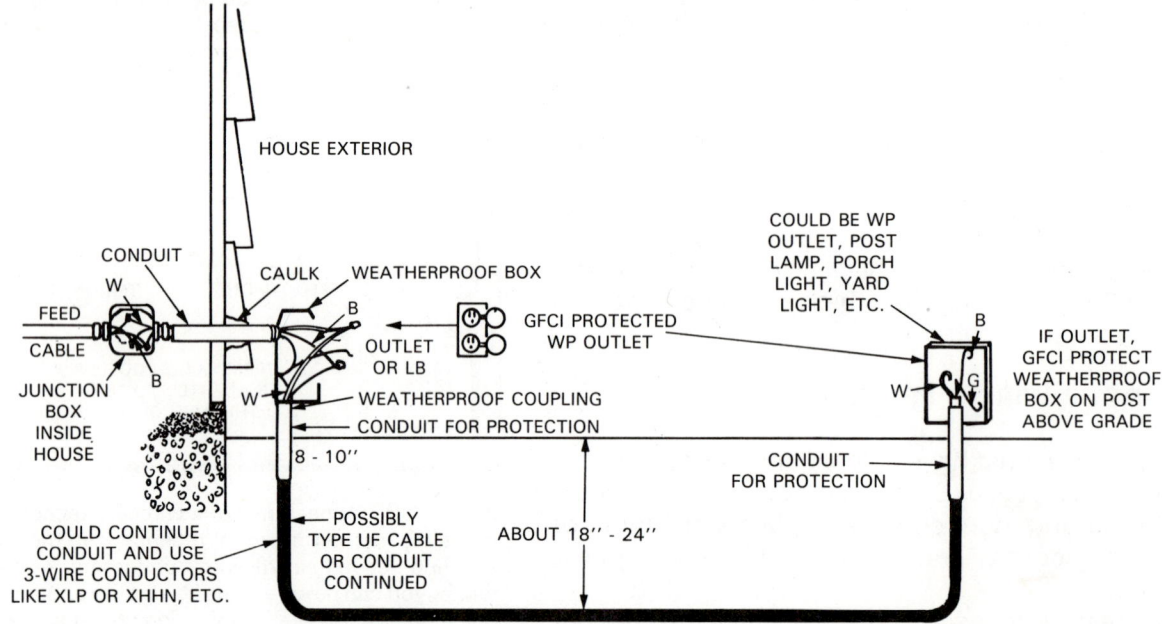

Fig. 11-32. Outdoor wiring diagram. GFCI can be located in weatherproof box at the foundation or at outlet.

REVIEW QUESTIONS — CHAPTER 11

1. Indicate the meaning of each of the following electrical symbols (do not write in the book):

   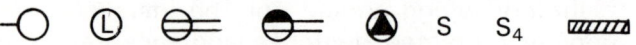

2. If you see a standard symbol used on an electrical plan with a small letter alongside, where would you look for an explanation of its meaning?
3. An electrical plan which shows the size and number of conductors in cable runs is known as a _____ _____.
4. Explain the value of an electrical plan during a rough-in.
5. _____ _____ means that the neutral and hot wires are always totally segregated (kept apart). The neutral wire is grounded and never interrupted by a switch, breaker, or fuse (unless the hot wire is interrupted at the same time).

Complete the following schematic drawings. Make sure that you label the neutral wires ''N'' or ''W'' and the hot wires ''B'' or ''R.''

6. A single-pole switch controlling three light fixtures.
7. A single-pole switch controlling a light fixture with two receptacles always live on the same circuit.
8. A pair of three-way switches controlling a lighting fixture.
9. On a separate sheet draw and complete the following diagram:

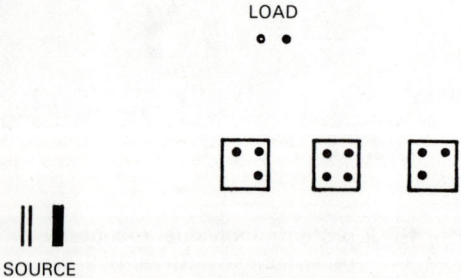

Chapter 12
THE SERVICE ENTRANCE

The service entrance includes all the wires, devices, and fittings that carry electrical power from the power company's transformer to the consumer. All electrical energy supplied to power-consuming devices in the building must first pass through the service entrance equipment.

The service components meter, protect, and distribute the power to the branch circuits.

POLE TRANSFORMER

As you learned in Chapter 1, electrical power is brought from the power plant at high voltage and then, before it can be used in factories, offices and homes, the voltage is reduced through several step-down transformers. The pole transformers shown in Fig. 12-1 reduce the voltage to 120/240 or 120/208 before it is delivered to the customer.

Power is carried from the pole transformer on three wires. Two are hot and one is a grounded neutral wire.

SERVICE ENTRANCE COMPONENTS

The service entrance consists of the following electrical equipment and parts:
1. The power company's wires from the utility pole to the dwelling.

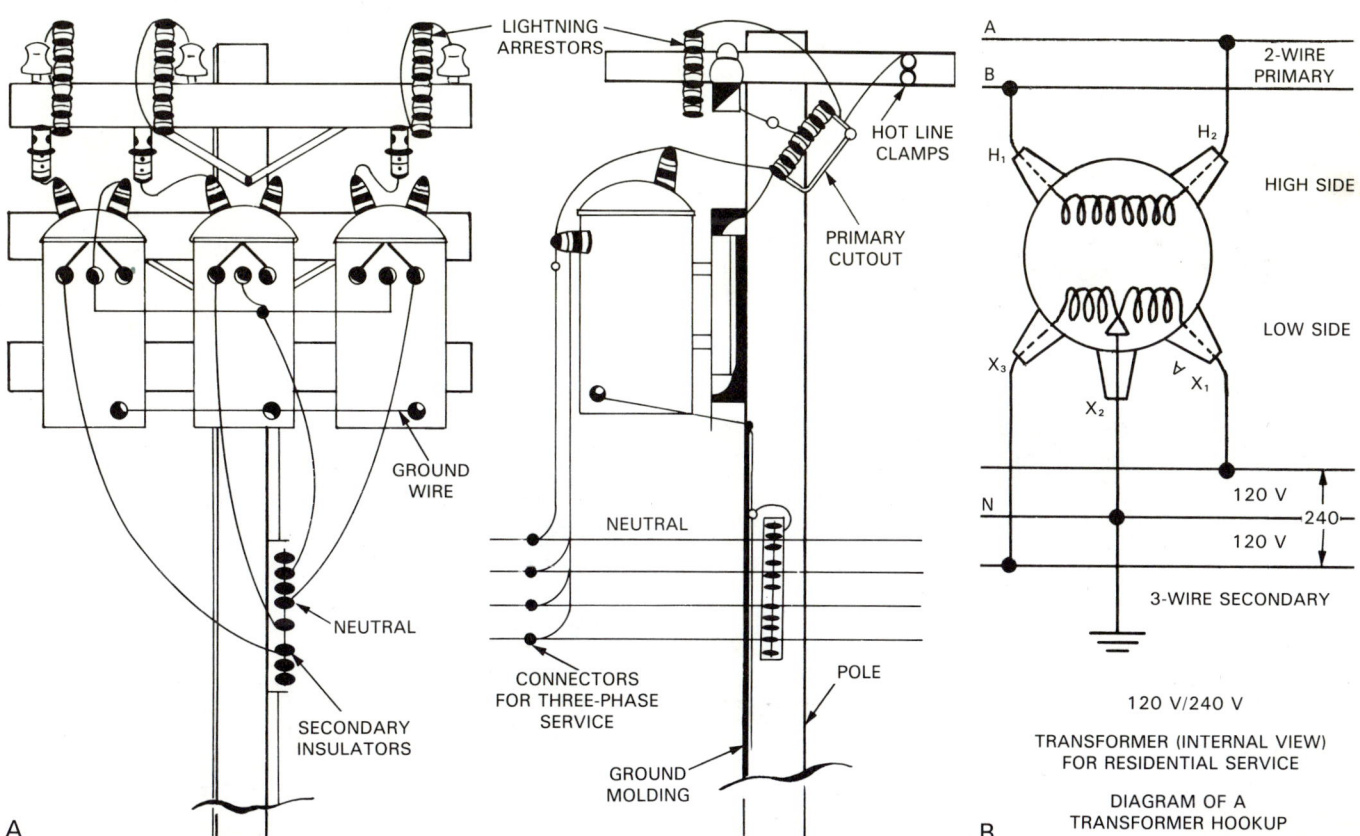

Fig. 12-1. Pole transformers step voltage down so it can be used in buildings. A—Simple sketch of pole transformer hookup. B—Diagram of a transformer hookup. Wires from high side or primary side of coil connect to high voltage conductors from power station. Low side or secondary wires connect to low side of transformer coil.

131

Modern Residential Wiring

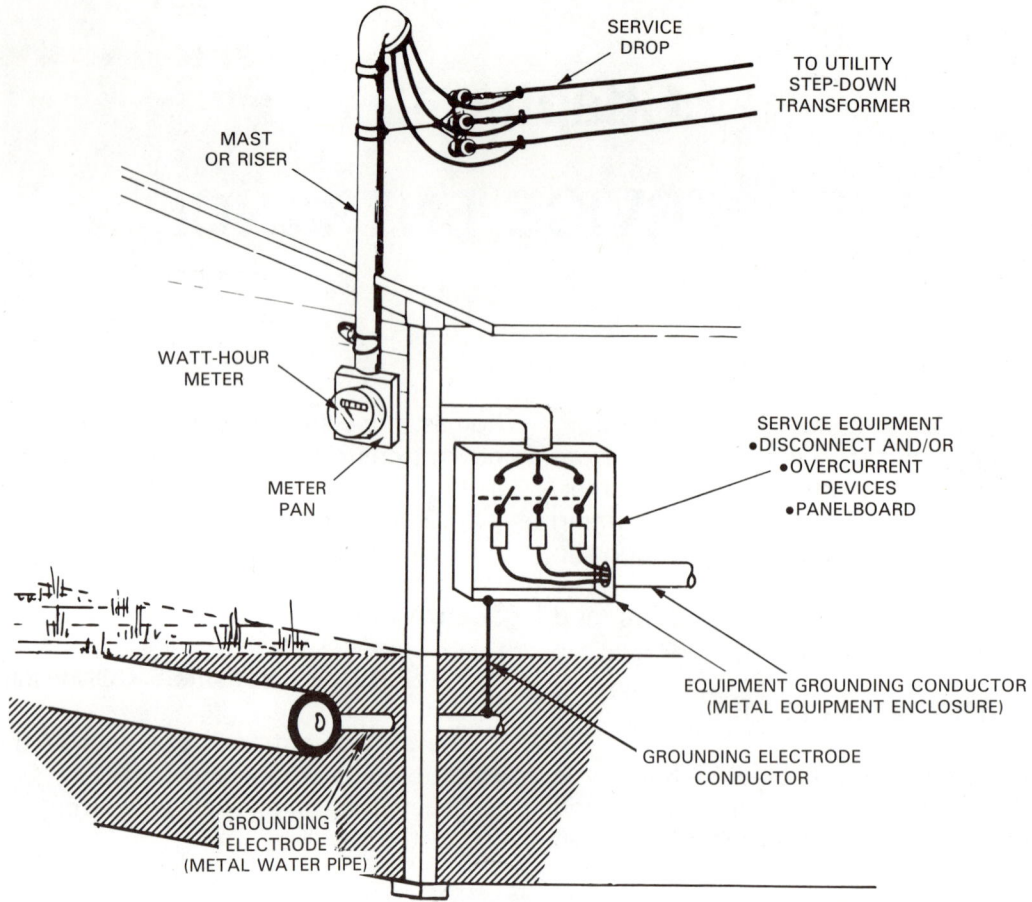

Fig. 12-2. Typical overhead service drop and service entrance. Usually, more than one grounding electrode is provided. Careful study of this illustration will help fix important details in your mind. (OSHA)

2. Service entrance conductors. These are protected within cable or conduit.
3. Meter socket (meter pan, or meter enclosure).
4. Service entrance panel with breakers or fuses and main disconnecting means. The main disconnect may be ahead of the panel and separate from it. Often it is part of the panel itself.
5. A continuous grounding system.
6. Fittings, fasteners, and other hardware necessary to install the service equipment.

SERVICE LOCATION

Service wires brought to the building are run overhead or underground. Those coming in overhead from a utility pole are called the SERVICE DROP. Those routed underground, from either a pole or transformer pad, are called the SERVICE LATERAL. See Figs. 12-2 and 12-3. They show the essential elements of either type of service entrance. In either type, these service wires will be connected to the service entrance wires at the building itself.

Most often, the power company will "spot" or locate the point of attachment of the service en-

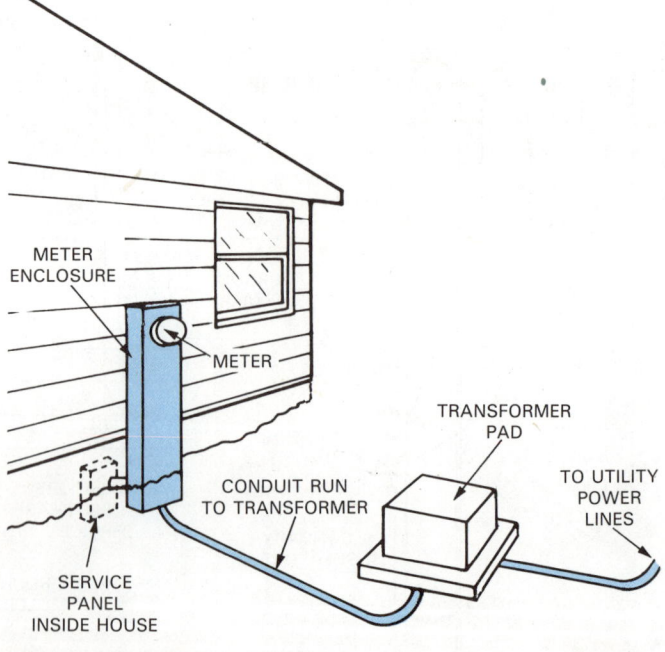

Fig. 12-3. Simplified sketch of an underground installation for a service entrance. The cable from the transformer to the meter enclosure is also known as a service lateral.

132

The Service Entrance

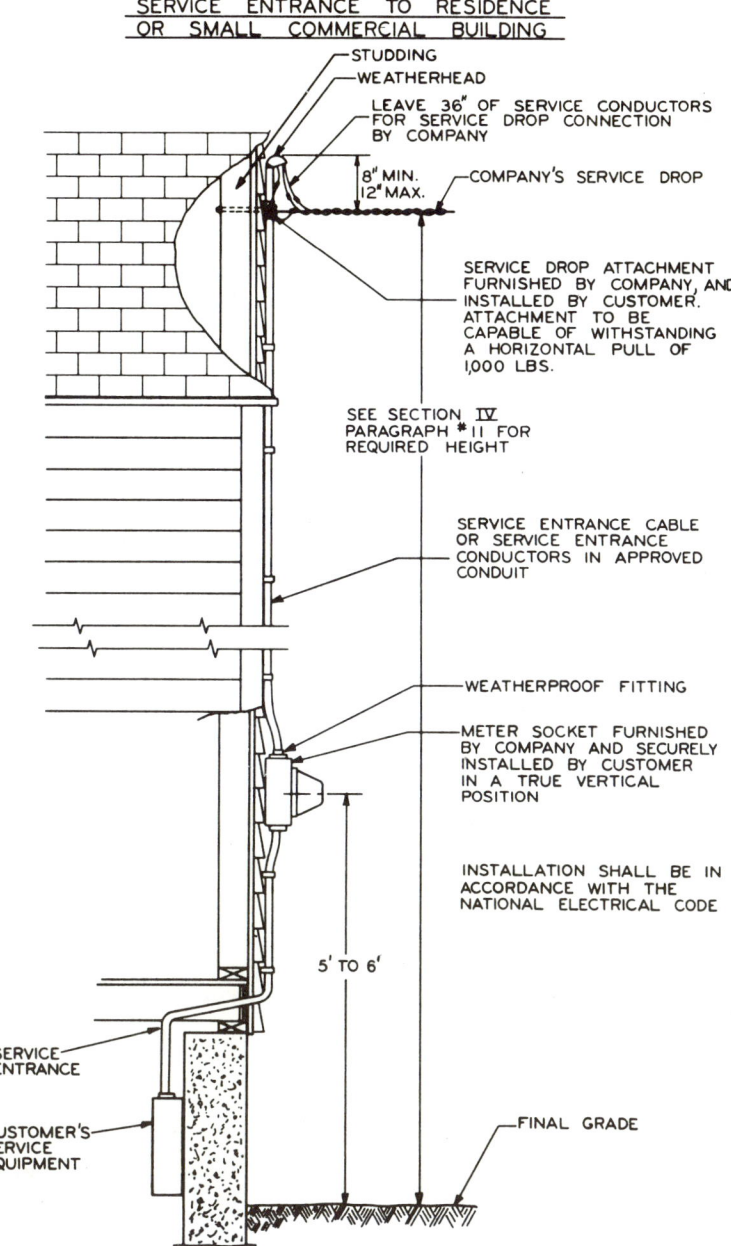

Fig. 12-4. Service entrance specifications like these should be secured from the utility company before the service entrance installation. Refer also to the NEC and consult the local electrical inspector whenever necessary. (New York State Electric and Gas Corp.)

trance for the electrician. In many localities this is required.

Check the service specifications of the power supplier to be sure of any additional requirements before installing the service entrance. Figs. 12-4 and 12-5 are typical utility company service specifications.

WHERE TO LOCATE SERVICE

Should the location of the service entrance be left up to the electrician, the following will serve as a general guide.

1. Service entrance conductors, whether enclosed in cable or conduit, should be installed in as straight a path as possible.
2. Service conductors should be kept as short as practical to avoid voltage losses.
3. Service conductors should enter the building as near the service panel as possible.
4. Service disconnect must be at or very near the point of entry.
5. The service panel should be located in a central accessible area, near the major electrical equipment of the building.
6. The service equipment should be protected from

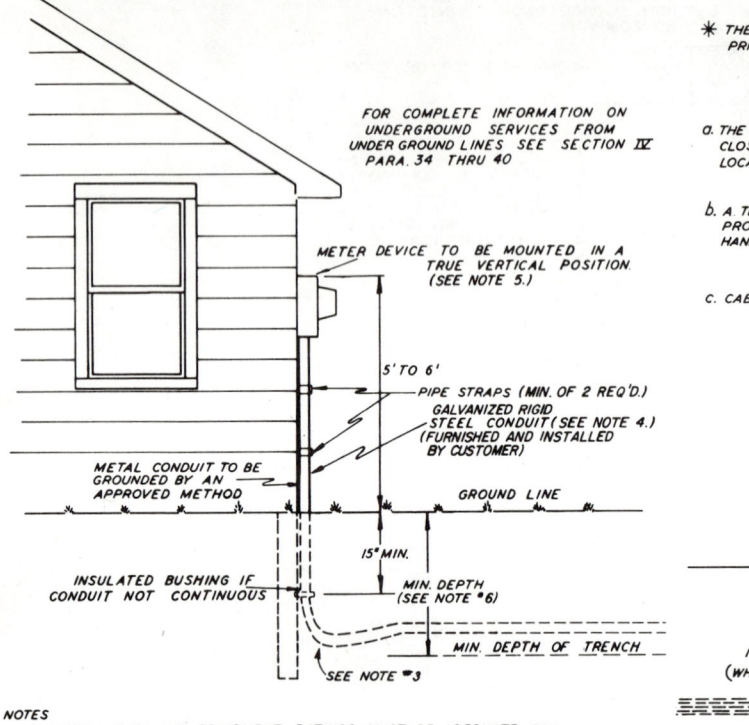
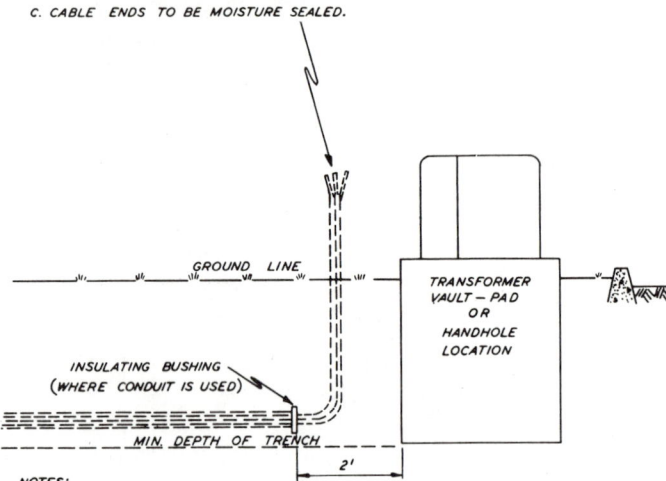

Fig. 12-5. A typical specification for installation of underground service. (New York State Electric and Gas Corp.)

physcial damage or exposure to water, dust, etc.
7. Service equipment should not be placed in bathrooms, storerooms, closets or damp cellars.
8. The meter enclosure should be mounted approximately 65 in. above grade. between 5–6'

POWER COMPANY WIRES

The power company wires are provided, usually without cost, from the utility pole or pad mounted transformer to the customer's service entrance conductors. See Figs. 12-6 and 12-7. Connection is completed by the utility upon inspection and approval of the installation.

SERVICE LOAD

The size (capacity) of a service entrance depends on the power requirement or load of the various circuits of the dwelling. To size the service correctly, you must know what this load will be.

For new single family dwellings, the *National Electrical Code* requires at least a 100 A service. For

Fig. 12-6. Service connection is made by the power company. Here line crew is connecting service drop conductors to pole transformer.

The Service Entrance

Fig. 12-7. Service lateral conductors are carefully buried in a trench. (New York State Electric and Gas Corp.)

homes with electrical heating systems or where future expansion is likely, 150 or 200 A service is best.

Details and data regarding service entrance load requirements can be found in *Code* Article 230. Commercial structures usually require considerably higher service ratings. When the electrical demands are known, Fig. 12-8 and Fig. 12-9 show the steps for figuring total loads and feeder size.

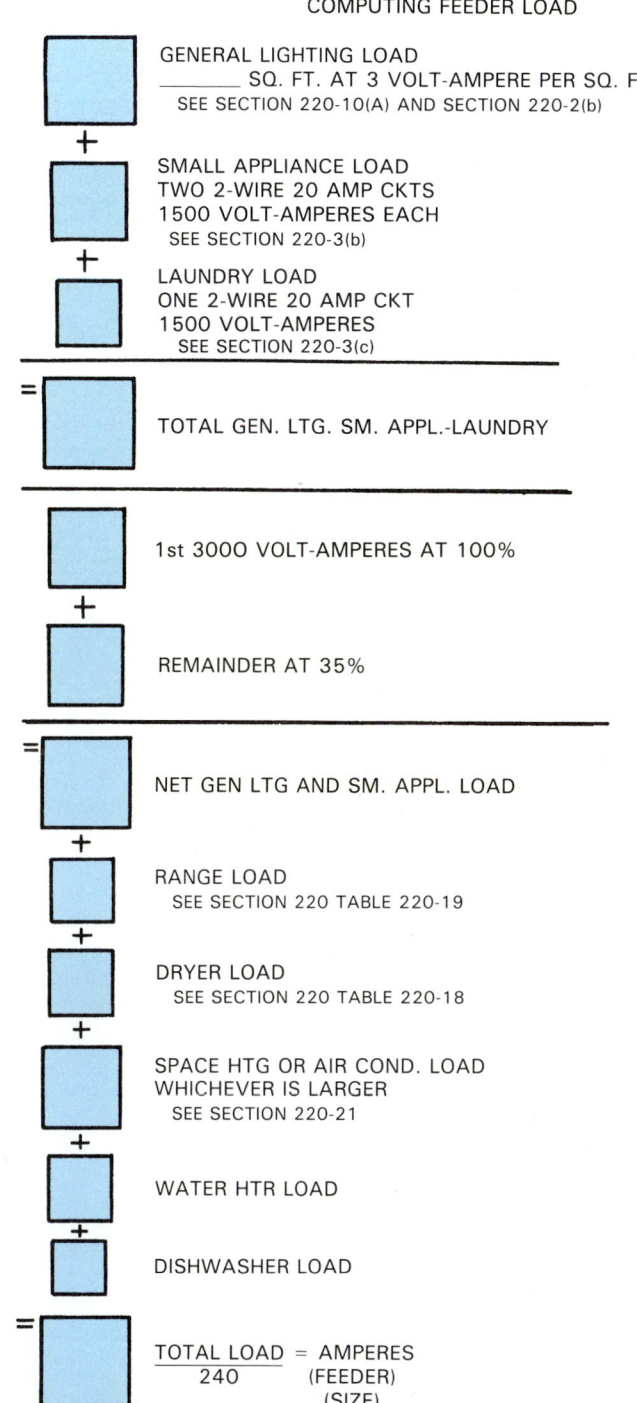

COMPUTING FEEDER NEUTRAL
SEE SECTION 220-22

- LIGHTING/SM. APPL./LAUNDRY NET LOAD
- RANGE LOAD AT 70%
- DRYER LOAD AT 70%
- OVENS & COOKING UNITS SEE TABLE 220-19
- DRYER AT 70%
- DISHWASHER 100%
- TOTAL / 240 V = AMPERES NEUTRAL FEEDER SIZE

Fig. 12-8. Figuring load on the neutral and size of the neutral conductor.

Fig. 12-9. Use these steps for figuring load and feeder size.

Modern Residential Wiring

SERVICE ENTRANCE CONDUCTORS

Service entrance conductors may be types RHH, THW, THWW, XHHW, RUX, XLP, or RHW. They can be run through conduit or enclosed in a cable assembly called service entrance cable, Fig. 12-10. Many service drop installations are made with triplex cable, Fig. 12-11.

When in conduit, all the conductors are insulated. The neutral is white, yellow, or gray. The hot wires are black and/or red. Service entrance cable is similar, but the neutral is uninsulated or bare.

Whether conduit or cable is used will depend upon customer preference, local codes, building structure, environmment, utility specifications, and the electrician's decision. In any case, the conduit or cable must be securely fastened to the building with the proper clamps or supports as shown later in this chapter.

SIZING SERVICE CONDUCTORS

In addition to proper types of service conductors, the right conductor size is essential. As was mentioned earlier, Table 310-16 of the NEC establishes the allowable ampacities of insulated conductors. Note 3 of the table instructs us on proper types and sizes of conductors for three-wire, single-phase dwelling entrance services. See Fig. 12-12.

CONDUCTOR TYPES AND SIZES RH - RHH RHW THW THWN THHN XHHW		
Copper	Aluminum and Copper - Clad Al	Service Rating in Amperes
AWG	AWG	
4	2	100
3	1	110
2	1/0	125
1	2/0	150
1/0	3/0	175
2/0	4/0	200

Fig. 12-12. Service entrance cable sizes are specified by the *National Electrical Code.*

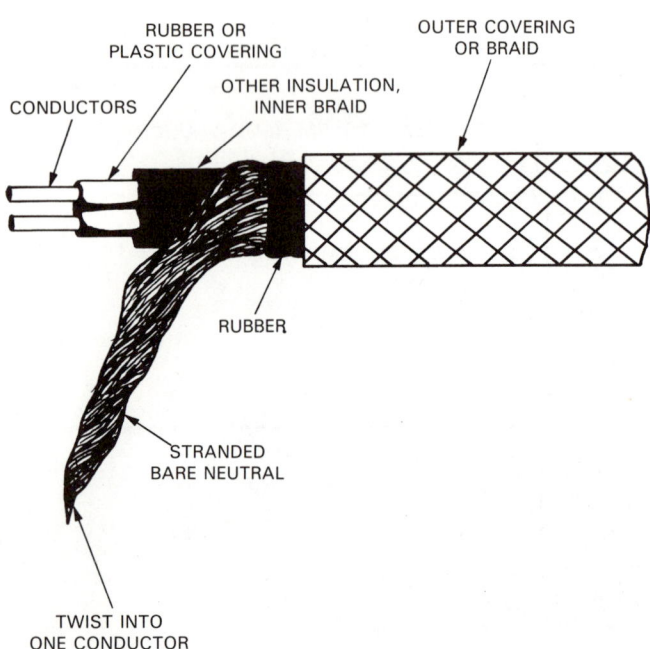

Fig. 12-10. Service entrance cable (SEC). Bare, stranded neutral wire must be twisted into a single strand before being connected.

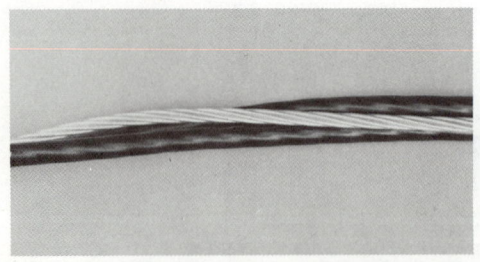

Fig. 12-11. Most service drops in new homes are made with triplex cable. It consists of a bare neutral wire around which two insulated cables are loosely wrapped. The neutral wire supports the insulated wires.

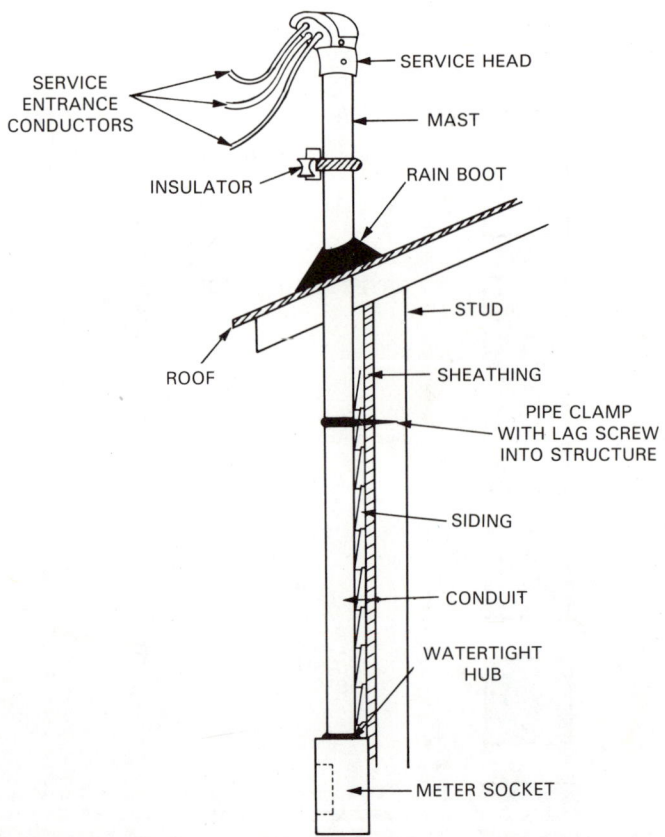

Fig. 12-13. Service entrance installed in conduit. Housing consists of the conduit which is connected to the service head on one end and the meter enclosure at the other. Mast must be securely fastened to building stud with clamps.

The Service Entrance

SERVICE DROP MAST AND INSULATOR

Sometimes the service drop mast (or riser) and insulators are supplied and installed by the power company. In most situations, the electrician makes the installation.

Insulators may be attached to the mast or to the building. See Fig. 12-13 and Fig. 12-14.

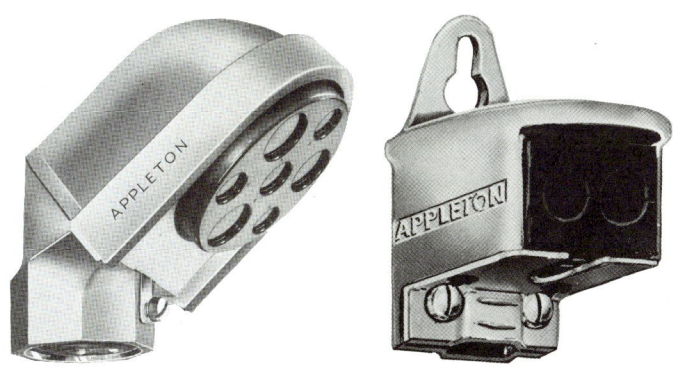

Fig. 12-15. Service heads are fittings designed for keeping water from entering the service entrance cable or the conduit leading to the meter enclosure. Left. This type is installed on conduit. Right. Head designed for use with nonmetallic service entrance cable. (Appleton Electric Co.)

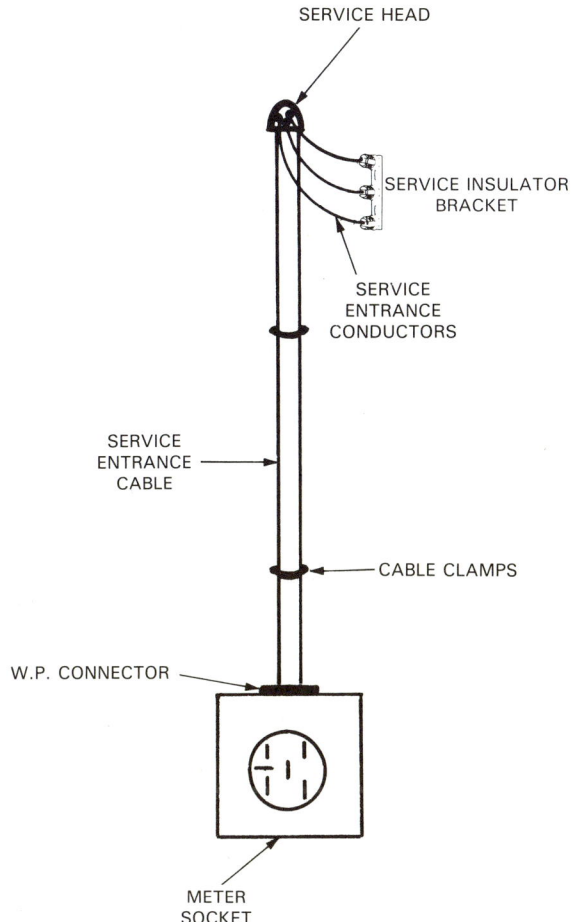

Fig. 12-14. Front view of a service entrance cable assembly and lag screws. Insulator rack is used here to secure the service drop to the side of the building. See Fig. 12-8 for details of the cable.

SERVICE HEAD

A service head like those shown in Fig. 12-15, is a fitting installed at the top of the conduit or service entrance cable to prevent water from entering and shorting out the conductors. The service conductors should extend through the head for approximately 3 ft. to provide a suitable DRIP LOOP. A drip loop is merely a formed curvature of the service conductor to prevent water from entering the service head around the cable or conduit. Refer to Fig. 12-16. Service heads are designed to be used with either conduit or cable. See Fig. 12-17.

Fig. 12-16. Drip loop on an actual installation. Cables drop a foot below the head which is angled to keep out moisture (inset).

Modern Residential Wiring

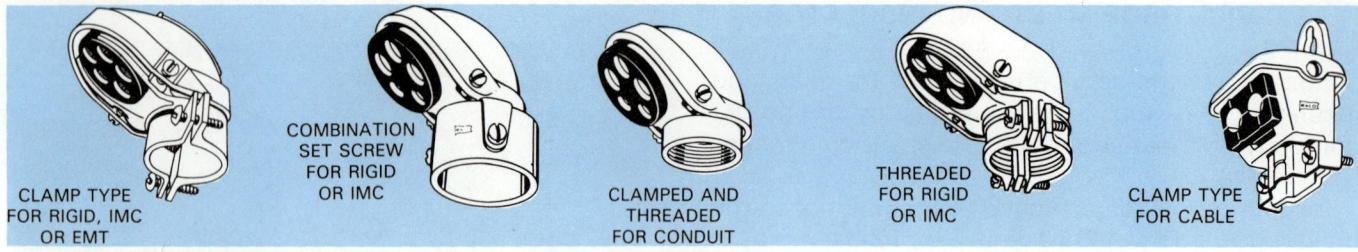

Fig. 12-17. Service heads are designed to work with any style of service entrance conductors Some are fitted with clamps to work on cable. Others are threaded or have a clamp mechanism for use on conduit. (RACO Inc.)

METER ENCLOSURE

This device houses the meter, Fig. 12-18, which is used to monitor the power consumed. Essentially this is a kilowatt-hour meter. Both the meter enclosure and meter are provided by the utility, often without charge.

Service conductors (cables or wires) enter this enclosure and are secured to the line-side lugs. Note that the neutral conductor is always connected to the center contact and continues essentially uninterrupted through the meter socket on its journey to the service main disconnect or panel. The meter itself will be installed by the utility when the service drop or lateral is completed.

WEATHERPROOF CONNECTORS

Fittings, called weatherproof connectors, are used to connect service entrance cable to the meter pan. One is shown in Fig. 12-19. The purpose of this fitting is to exclude moisture or water from the meter enclosure.

INSULATORS

An insulator, Fig. 12-20, is a porcelain or other nonconductive device which can be clamped to the conduit or attached to the exterior of the building. It is used as a means of attaching the service drop wires to the service entrance conductors. At the

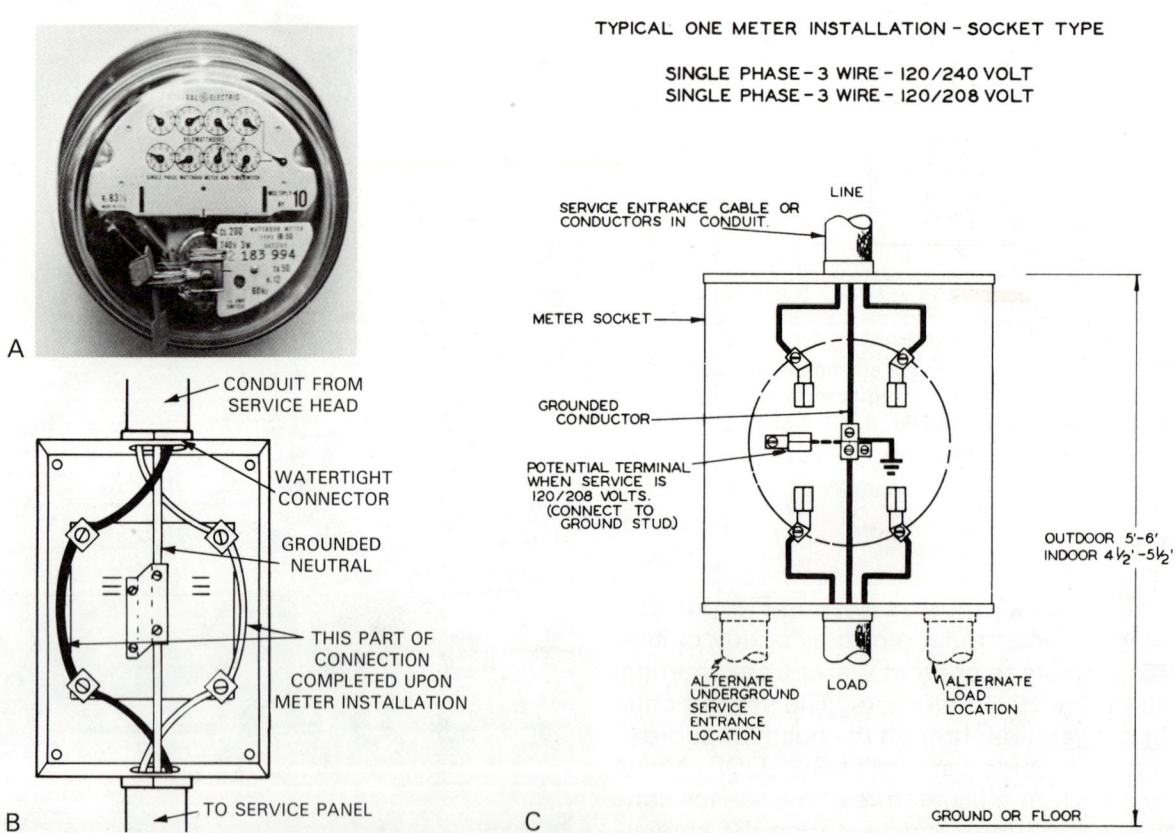

Fig. 12-18. The watt-hour meter keeps records of the power consumed. A—View of an actual meter. (Scott Harke) B—Meter socket. Meter completes the "hot" conductor circuits. Neutral is uninterrupted. C—Specifications for installation of a single phase three-wire meter. (New York State Electric and Gas Corp.)

The Service Entrance

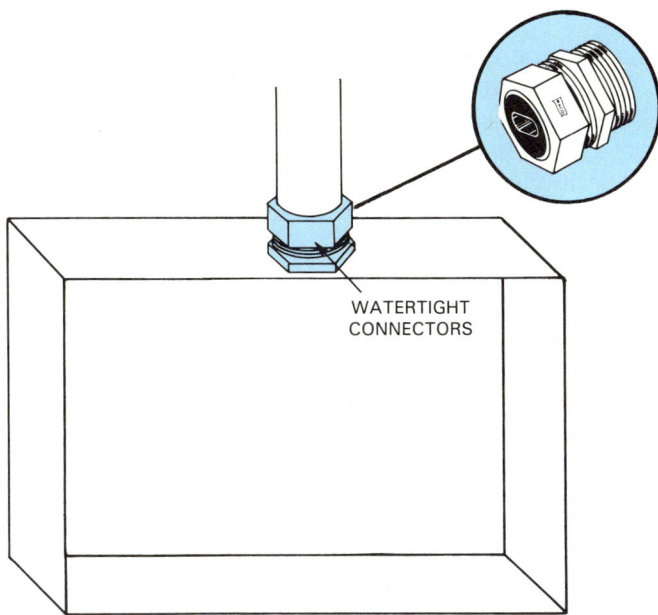

Fig. 12-19. Weatherproof or watertight connector is used to prevent water from running down cable or conduit into the meter enclosure. This type of fitting must always be used when service entrance cable enters meter enclosure from the top or side. (RACO Inc.)

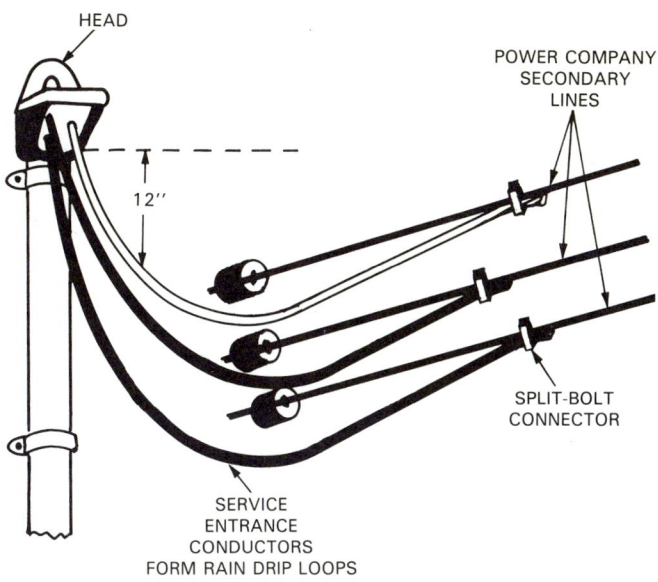

Fig. 12-21. Insulators should be placed well below the head. Cables are spliced with a split-bolt connector like the one shown in Fig. 12-22.

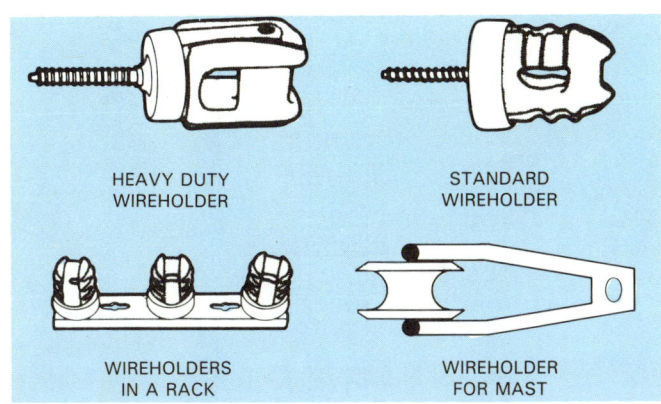

Fig. 12-20. Service drop wires are often attached to the building. Insulators or wireholders such as these are fastened into studding or attached to the mast. (Electroline Mfg. Co.)

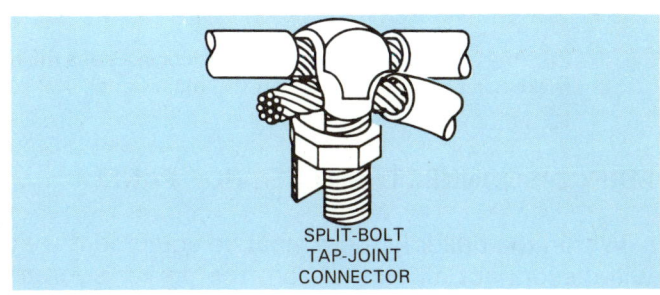

Fig. 12-22. Close-up of a split-bolt tap-joint connector. It assures good electrical connection between large conductors. (OSHA)

same time it eliminates strain between those wires. Insulators are placed approximately 12 in. below the service head, Fig. 12-21.

TAP JOINT CONNECTOR

A split-bolt connector is normally used for large wires, Fig. 12-22. The wire ends are stripped and overlapped in the connector. Then the nut of the split bolt is tightened. This type of connector is useful in making splices between the conductors of the service drop and those at the entrance head. Usually such splices are then wrapped in layers of electrical tape to the same thickness of the original insulation.

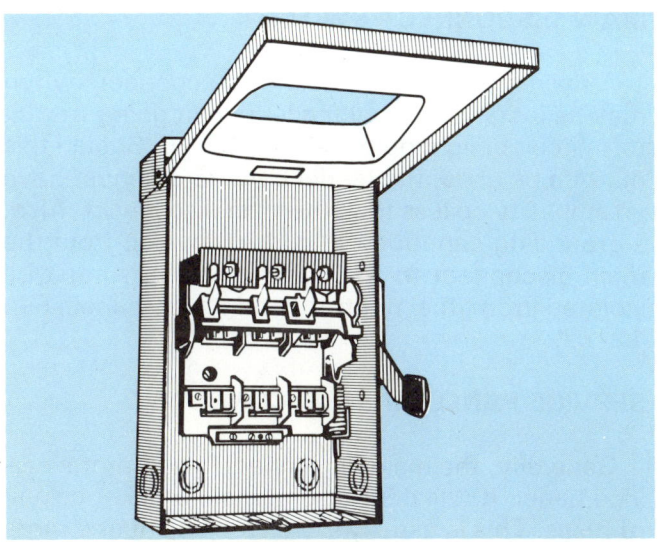

Fig. 12-23. Within every structure, there must be a means of disconnecting all conductors from the service conductors. These disconnects must be clearly marked and easily accessible. (Wadsworth Electric Mfg. Co., Inc.)

neutral buss serves as disconnect for neutral conductor in single family dwelling

139

Modern Residential Wiring

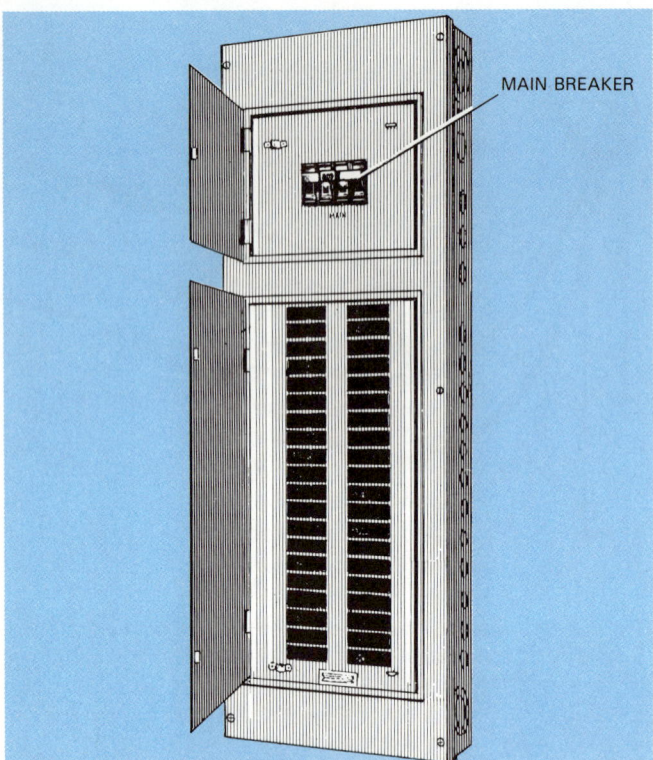

Fig. 12-24. Very often, the main disconnect consists of a main breaker or pair of fuses within the main panel.

MAIN DISCONNECT AND SERVICE PANEL

Within the building there must be equipment that will disconnect all wiring from the power source. This can be arranged by using a single main disconnect switch, Fig. 12-23, or a main circuit breaker that is part of the service panel, Fig. 12-24. Note the wiring diagram in Fig. 12-25.

MAIN DISCONNECT SWITCH

A main disconnect or service disconnect switch may be installed with a feeder circuit going from it to a circuit or lighting panel, Fig. 12-26. Should this method be used, the feeder conductors must have an ampacity no less than the main disconnect. Also, a grounding conductor must be provided from the main disconnect to the lighting panel. It must be isolated from the neutral within that panel, Fig. 12-27.

SERVICE PANEL MAIN DISCONNECT

Generally, the main disconnect is part of the service panel. It will consist of a main breaker or pair of fuses. This is usually more convenient and more economical. The service conductors enter the service panel directly from the meter enclosure.

Whichever method is used, the disconnect must be located in an accessible place. Further, it must be located as close as possible to the point where the service conductors enter the structure.

Modern entrance panels house circuit breakers or fuses controlling individual circuits, Fig. 12-28. Four large terminals are provided for connecting the three incoming service conductors and the important

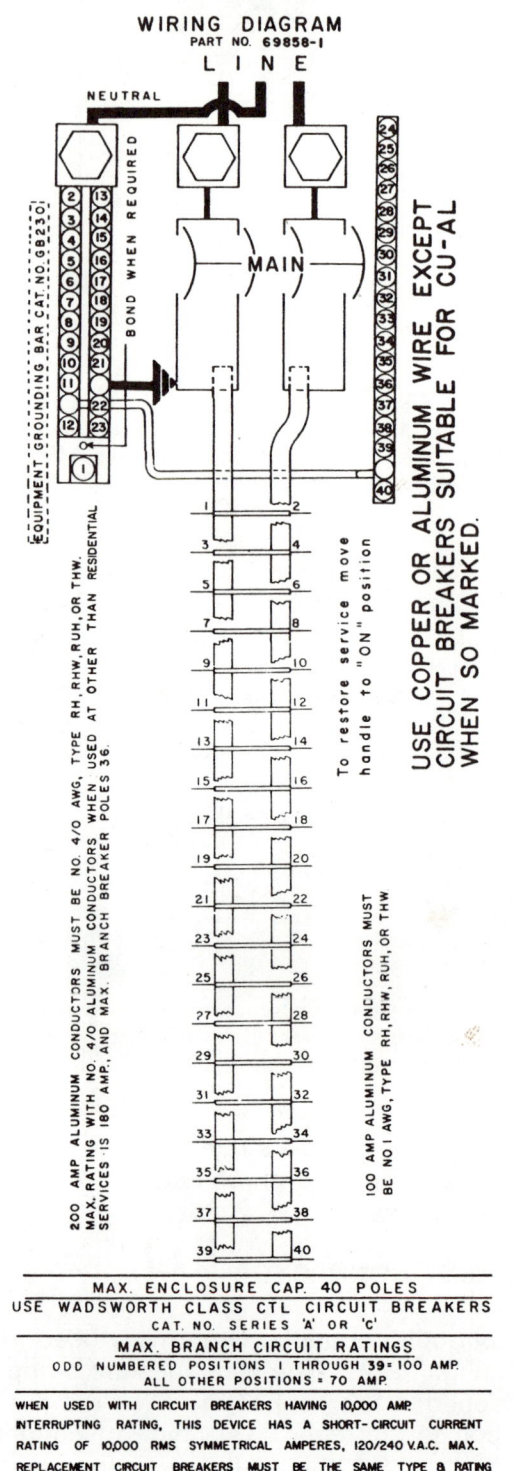

Fig. 12-25. As shown in this wiring diagram, the main breaker is located ahead of the circuits. All wiring can thus be disconnected from the service conductors by opening the main. (Wadsworth Electric Mfg. Co., Inc.)

The Service Entrance

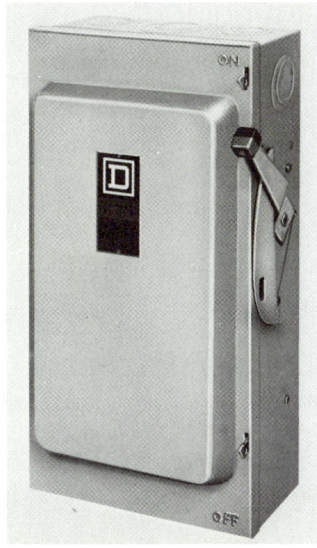

Fig. 12-26. A main disconnect may be placed ahead of a lighting panel. In that case, the supply conductors (feeders) must have an ampacity (rating) no less than that of the main breaker or service switch. (Square D Co.)

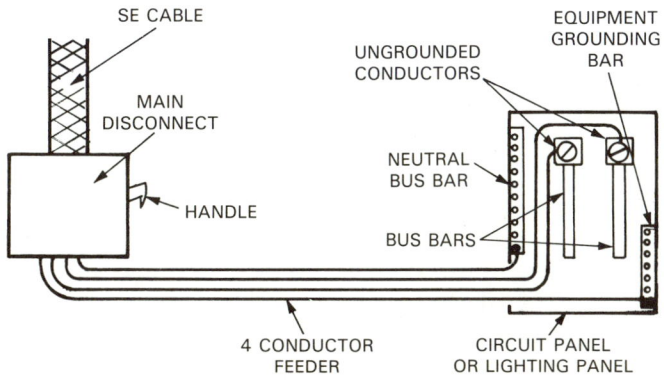

Fig. 12-27. Four conductors must be provided between a main disconnect and the lighting panel. The fourth, called a grounding conductor, must be isolated from the neutral or grounded conductor.

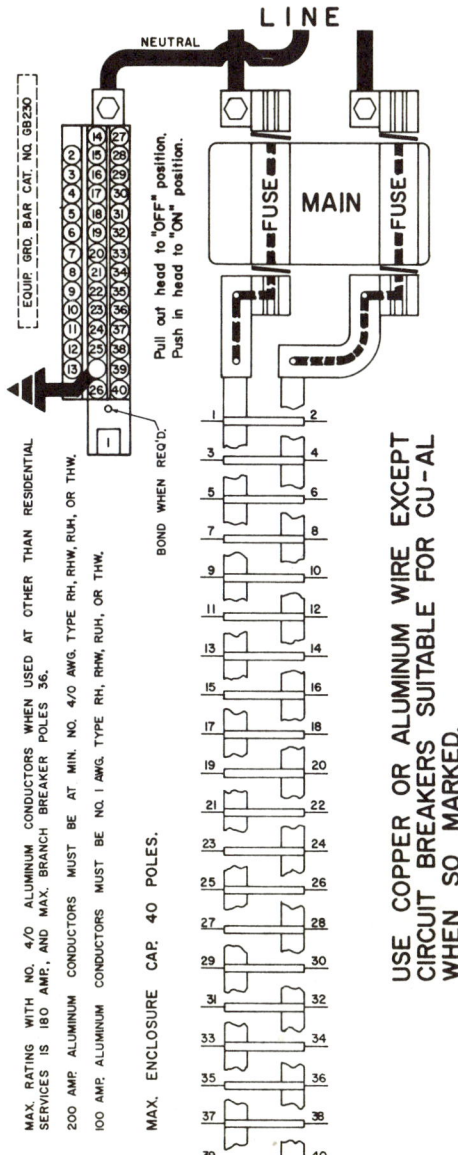

Fig. 12-28. Schematic of panel using pair of fuses for main disconnect. (Wadsworth Electric Mfg. Co., Inc.)

grounding conductor. Two of the large terminals are located on the main fuses or breaker for connecting each of the hot wires. The remaining two are on the neutral bus bar for attaching the neutral service conductor and the grounding conductor.

Tight connections, in all cases, are necessary. This is very important for safety.

SERVICE GROUNDING

Grounding the service entrance is vital to the safety of the entire electrical system. Careful attention should be given to this section to insure a thorough understanding of the procedures involved. In addition, you should study all sections of Article 250 of the *Code* before undertaking any electrical installation.

Grounding the service entrance involves connecting the neutral or white service conductor with the earth. This is done by means of the GROUNDING CONDUCTOR. This conductor connects the neutral lug of the meter housing or the neutral bus bar of the service switch (service panel) to the metallic water supply pipe. See Fig. 12-29.

In addition to the water supply pipe, or, if one is not available, other grounding electrode systems shall be used:

1. Metal frames of buildings.
2. Metal electrodes encased in 2 in. of concrete at the bottom of a concrete foundation.
3. 20 ft. of No. 2 AWG bare copper wire buried not less than 2 1/2 ft. below the surface.
4. Uninterrupted and electrically continuous gas piping.

Modern Residential Wiring

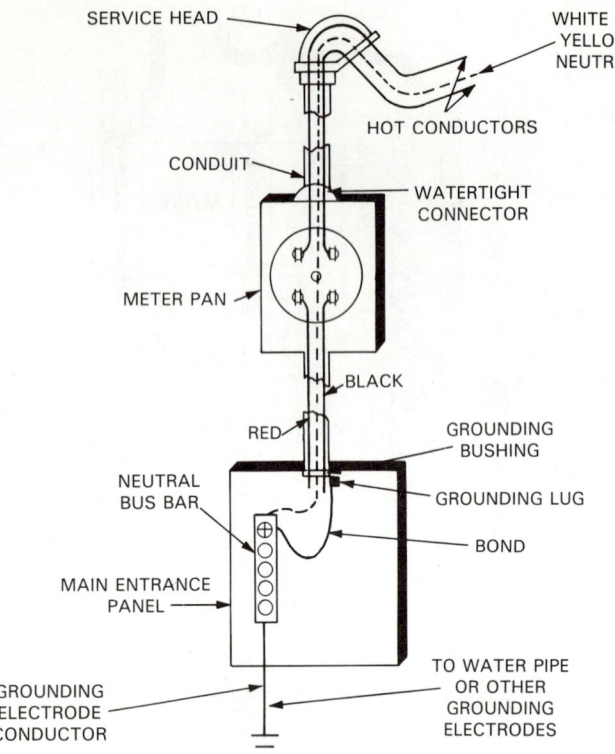

Fig. 12-29. Service entrance layout. It must have continuity of the grounded neutral and grounding electrode through grounding electrode conductor. Metal conduit must also be bonded to grounding means.

5. Underground metal tanks.
6. Rod or pipe electrodes which are at least 8 ft. long, fully driven into the ground or buried not less than 4 ft. below the surface.
7. Metal plate of no less than 2 sq. ft. surface area.

Where two or more grounding electrodes are used, they should be separated by a distance of 6 ft. and bonded (linked with a copper conductor). The less resistance the better. In fact, the resistance must not exceed 25 ohms.

GROUNDING ELECTRODE CONDUCTOR FOR AC SYSTEMS			
Size of Largest Service-Entrance Conductor or Equivalent Area for Parallel Conductors		Size of Grounding Electrode Conductor	
Copper	Aluminum	Copper	*Al or Copper-Clad Aluminum
2 or smaller	0 or smaller	8	6
1 or 0	2/0 or 3/0	6	4
2/0 or 3/0	4/0 or 250 MCM	4	2

*See installation restrictions Section 250-92 (a) NEC

Fig. 12-30. Use NEC chart to determine size of grounding electrode. (National Fire Protection Assoc.)

GROUNDING CONDUCTOR

As already stated, the grounding electrode conductor should provide little resistance. For that reason, the conductor should have the following characteristics:
1. Copper or aluminum or copper-clad aluminum.
2. Stranded.
3. Bare.
4. Uninterrupted.

SIZING GROUNDING ELECTRODE CONDUCTOR

Table 250-94 of the NEC indicates the proper sizes for the grounding electrode conductor. This is based on the service entrance conductor size. See Fig. 12-30.

For example, the installation of a 200 ampere service entrance using No. 3/0 AWG copper service wires, would require a No. 4 AWG copper grounding electrode conductor.

SERVICE CLEARANCES

Where overhead services are installed, certain regulations concerning clearances must be strictly observed. The NEC indicates that clearances for overhead, service drop conductors shall be as follows:
1. Above roofs — see Fig. 12-31.
   a. 8 ft. as measured vertically from all points.
   b. 3 ft. if the slope is not less than 4 in. in 12 in.
   c. 18 in. over an overhanging portion of a roof. This clearance is limited to no more than 4 ft. of the overhang and only if they terminate at a raceway which goes through the roof or other support.
2. Above final grade — see Fig. 12-32.
   a. 10 ft. above sidewalks.
   b. 12 ft. above residential driveways.
   c. 18 ft. above public alleys or thoroughfares subject to truck traffic. This can be reduced to 15 ft. if there is no truck traffic.
3. At building openings — see Fig. 12-33.
   a. Not less than 10 ft. above a finished surface, platform, or other accessible surface.
   b. Not less than 3 ft. from a window in any direction (except above the window).

Additional, and more detailed discussion of the regulations on clearances is found in Section 230-24 of the NEC.

SERVICE COMPLETION

When the service entrance installation or rough-in is completed, it will be inspected by the electrical inspector. After this, the power company will install the service drop or lateral by connecting the

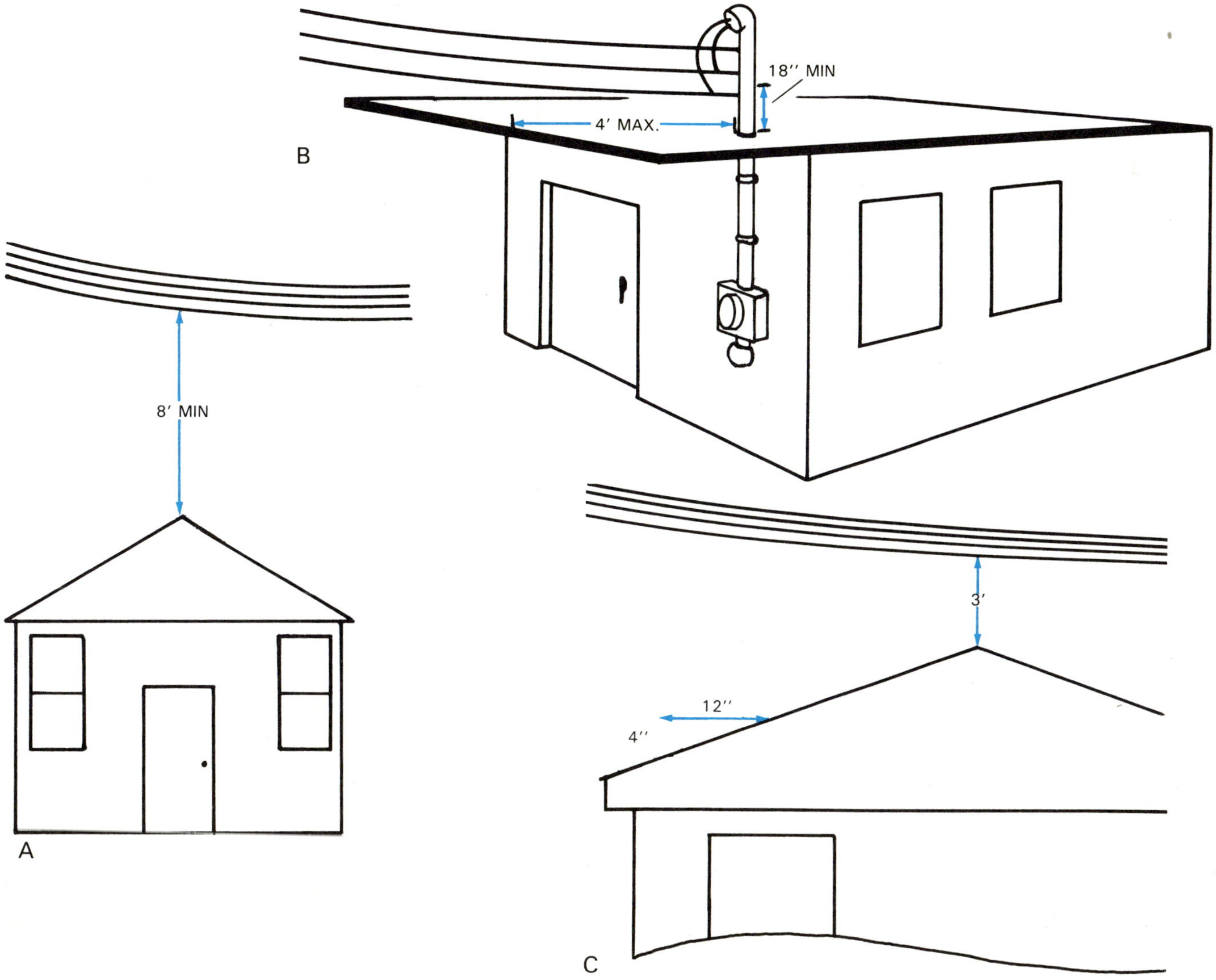

The Service Entrance

Fig. 12-31. Above-roof clearance for overhead conductors. (OSHA)

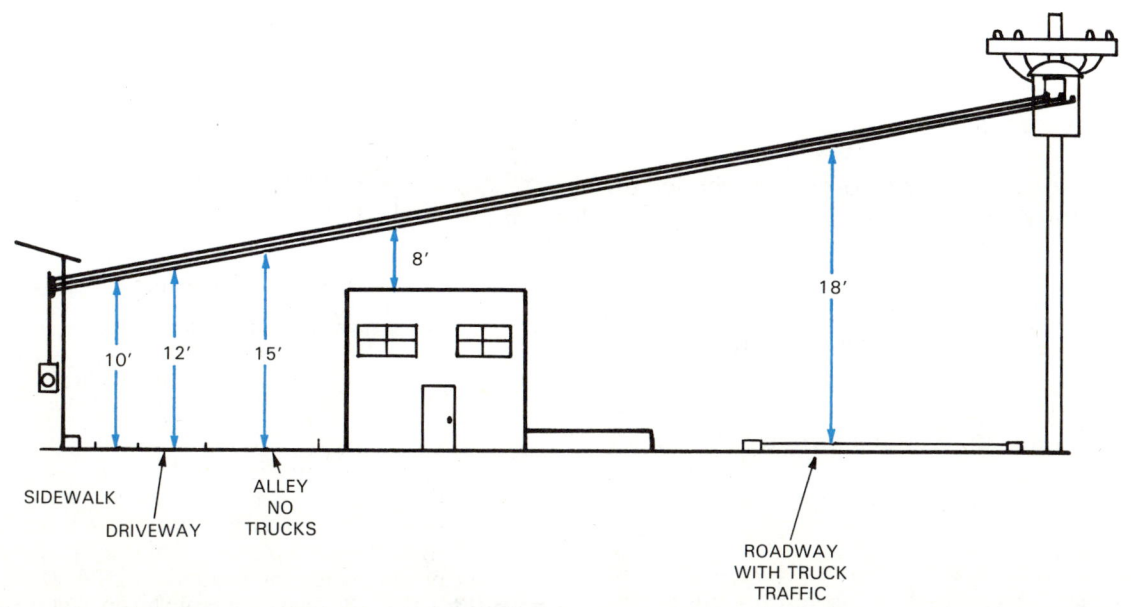

Fig. 12-32. Above-grade minimum clearances.

143

Modern Residential Wiring

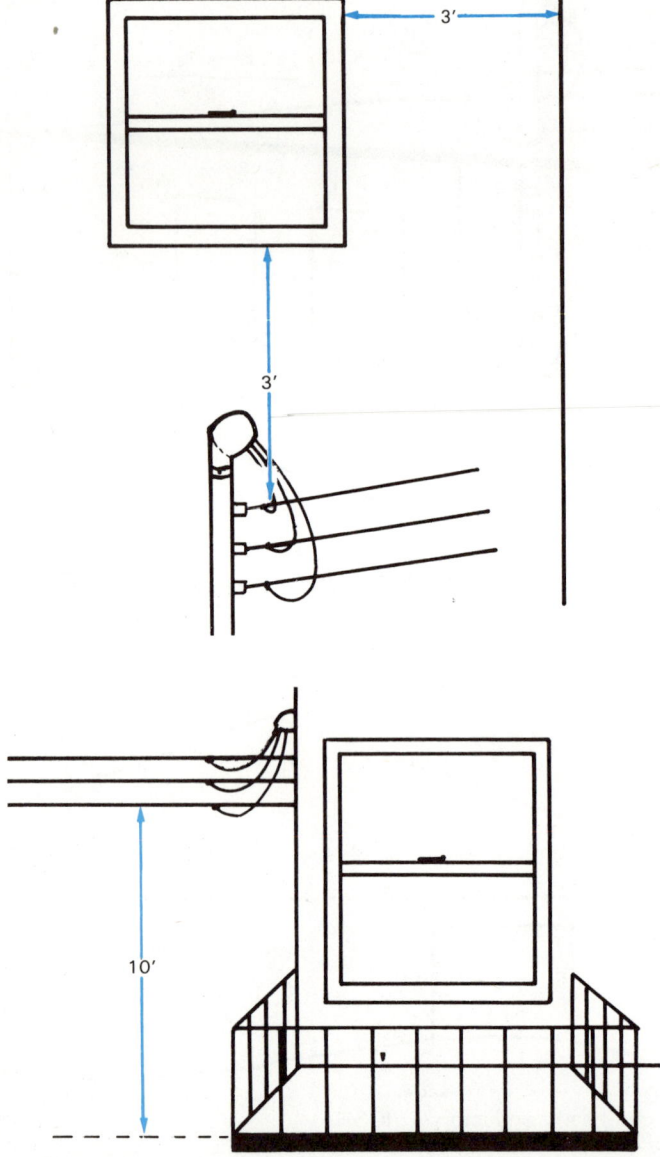

Fig. 12-33. Clearances around building platforms and openings.

CIRCUIT BREAKER

A circuit breaker is defined under NEC Article 100 as "a device designed to open and close a circuit by nonautomatic means and to open the circuit automatically on a predetermined overcurrent without injury to itself when properly applied within its rating."

Circuit breakers are illustrated in Fig. 12-35. A single-pole breaker serves a 120-volt circuit. A double-pole breaker serves a 240-volt circuit.

Typically these breakers are plugged or bolted into the service panel and connected to the "hot" wire(s) of the particular circuit they are to protect. The neutral conductor of each circuit is not connected to the breaker, but fastens directly to the neutral, grounded bus bar. The amperage rating of

wires from their transformer to the service entrance conductors.

Fig. 12-34 shows a typical completed service entrance. Note the drip loops discussed earlier, as well as the positioning of other components in the finished service system.

SERVICE SUPPLIES AND FITTINGS

The major fittings and supplies necessary to install a service entrance are shown in Fig. 12-34.

OVERCURRENT PROTECTION

Both circuit breakers and fuses are devices placed in a circuit. They protect the circuit wires and the load which they supply from an overcurrent.

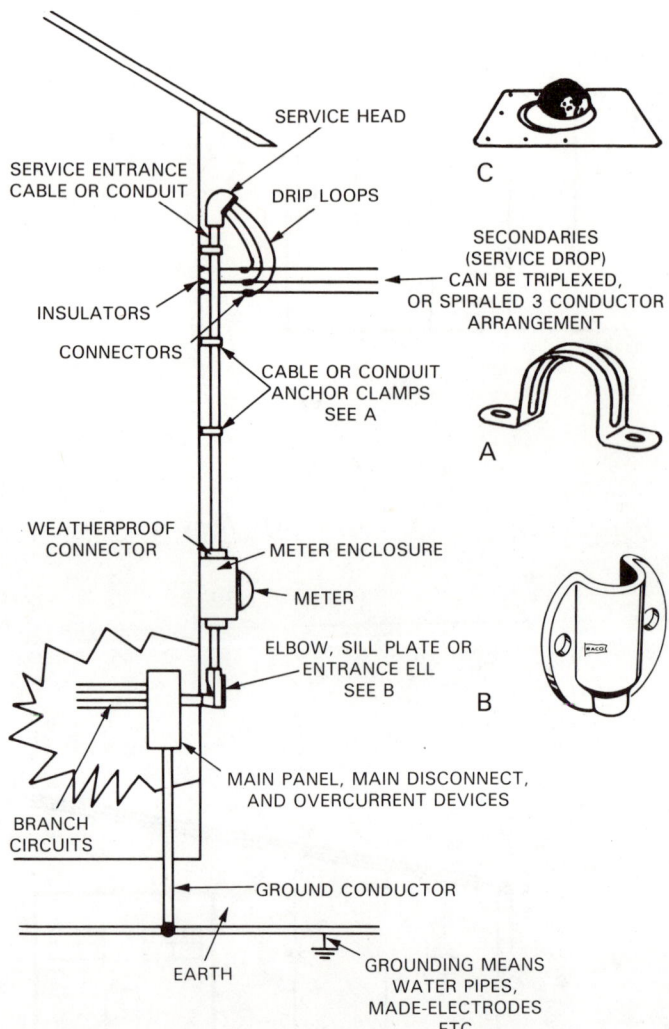

Fig. 12-34. Completed service entrance and various fittings. Note relative position of components and special fittings. A—Cable, conduit, or anchor clamps fasten cable and conduit to structure. B—Elbow, sill plate or entrance ell protect conductors at point where they enter structure. (RACO Inc. and Electroline Mfg. Co.) C—Rain boot (not shown in drawing) is used to protect from leaks when service mast goes through roof.

Modern Residential Wiring

PHASE CONCEPT

Phase refers to the angle between various generated ac currents. The term three-phase indicates three separately derived ac currents which are "out-of-step" from each other by 120 electrical degrees.

Two phase current occurs when two alternating currents are generated 90 degrees apart. The typical single-phase current exists when one current is produced at regular intervals. Fig. 12-37 illustrates this important variance in electrical power generation.

Single-phase, two-phase and three-phase are abbreviated as 1φ, 2φ, and 3φ.

Single-phase and three-phase current are used commonly in electrical generation. Two-phase has little practical application in industry today.

MULTIPHASE SYSTEMS

The supply or source for an entrance service begins at the TRANSFORMER. A transformer is a device which is capable of changing voltage and amperage. The utility company uses transformers to distribute and change electrical power throughout their service area.

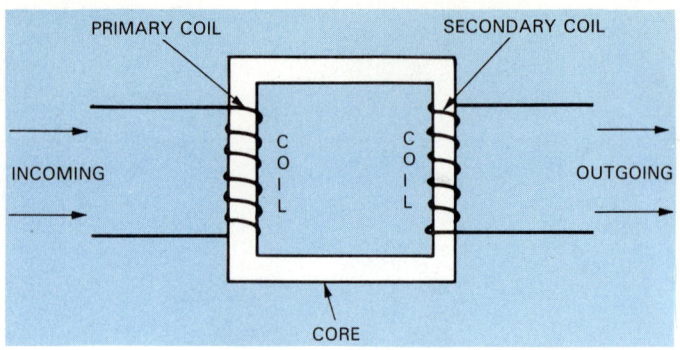

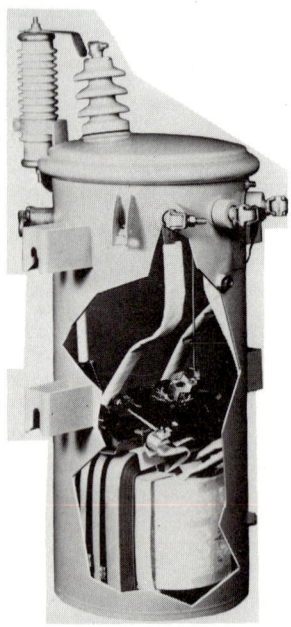

Fig. 12-38. Transformers have a simple makeup. Basically they consist of a metal core with windings or turns of wire around opposite sides.

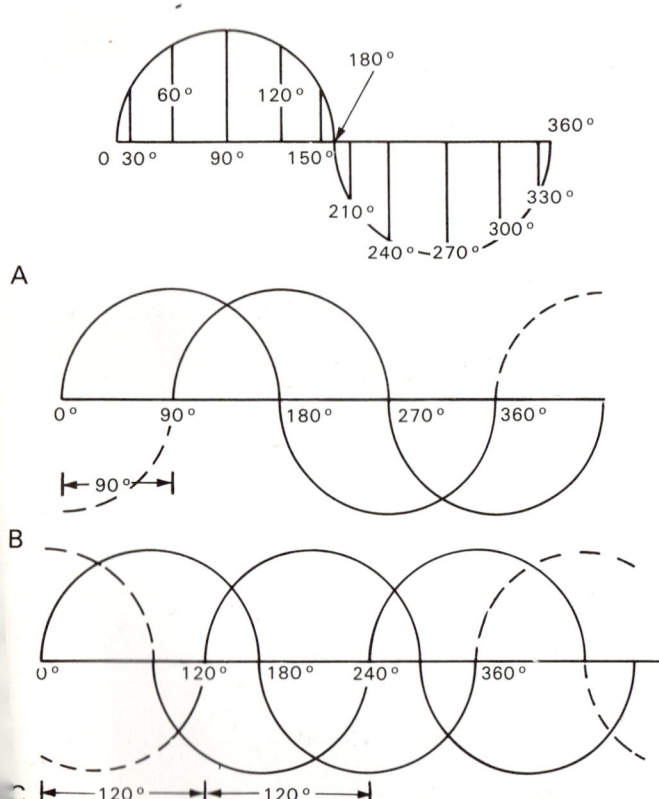

Fig. 12-37. The meaning of phase. Electrical energy may be expressed graphically by using sine-wave forms. A—Single-phase electrical energy generation. B—Two-phase current is accomplished by simultaneously generating two single-phase currents 90 degrees apart. C—By generating three, single-phase currents 120 degrees apart, three-phase current is created.

TRANSFORMER OPERATION

While the electrician who works on residential systems will not be working on transformers, it is helpful for you to understand how they are made and how they work. Because of the voltages involved, you should be careful working around them. These units are installed and serviced by trained lineworkers employed by the power company.

A transformer has a relatively simple structure, Fig. 12-38. Basically it is two COILS or WIRE WINDINGS side by side, with a common metal CORE. These windings are called the PRIMARY and SECONDARY. They are merely turns of wire.

When an alternating current is sent through one of the coils, the primary, it magnetizes the metal core, Fig. 12-39, causing regular surges of magnetic flux. These surges are constantly reversing directions. This magnetic flux also moves through the other coil which is called the SECONDARY. A current is created there too. Since the same magnetic flux exists throughout the core, the same

The Service Entrance

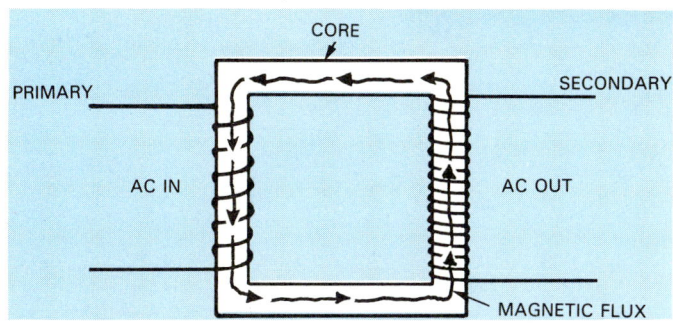

Fig. 12-39. Application of alternating current causes a magnetic flux within the transformer core. This magnetic flux creates a current in the secondary side (outgoing) of the transformer.

voltage per wire turn is created. Therefore, any voltage difference is the direct result of differences in the number of turns in the primary and secondary windings.

TRANSFORMER RULE

Mathematically stated: the total voltage in the primary coil is to the total voltage in the secondary coil as the number of turns in the primary is to the number of turns in the secondary. This can be expressed as:

$$\frac{\text{Voltage of Primary}}{\text{Voltage of Secondary}} = \frac{\text{turns of primary}}{\text{turns of secondary}}$$

Power into and out of a transformer is (ideally) the same. Since voltage changes within the transformer, so must amperage. Remember, power equals volts × amperes. For this reason, the amperage must change proportionately, but inversely as does the voltage. This can be stated mathematically as:

$$\frac{\text{Primary Voltage}}{\text{Secondary amperage}} = \frac{\text{secondary voltage}}{\text{primary amperage}}$$

For example, suppose the primary voltage entering a transformer is 12 volts and the primary amperage is 20 amperes. What will the secondary amperage be if the secondary voltage is 120 volts?

Using the equation just given:

$$\frac{12 \text{ Volts}}{(X) \text{ Secondary Amperage}} = \frac{120 \text{ Volts}}{20 \text{ amperes}}$$

by cross-multiplying:

$$\frac{12 \text{ V}}{X} = \frac{120 \text{ V}}{20 \text{ A}}$$

$$120 X = 240$$
$$X = 2 \text{ Amperes}$$

Since transformers change voltage, they can either increase or decrease the primary voltage. Those which increase the primary voltage are called STEP-UP transformers. Those which decrease the primary voltage are called STEP-DOWN transformers, Fig. 12-40. Directions for selecting the proper transformer to match the electrical load are found at the end of this chapter.

SERVICE DESIGNATION

The power delivered by the power company is actually routed through a chain of transformers, Fig. 12-41, up to the last pole transformer near the building it supplies, Fig. 12-42. These transformers are designed to supply the following service ratings:
1. Two-wire, single-phase 120 volts. (Such installations are no longer made.)
2. Three-wire, single-phase 120/240 volts.
3. Four-wire, three-phase 120/208 volts.
4. Four-wire, three-phase 120/240 volts.

Three-wire, single-phase 120/240 volt systems are, perhaps, the most common and most familiar. We deal with this type of service throughout this text. It provides both 120 volts for general lighting and receptacle loads as well as 240 volts for large motors, welders, heavy duty appliances, and other electrical equipment requiring this voltage.

COMMERCIAL SERVICE ENTRANCE REQUIREMENTS

Commercial structures, such as apartment buildings, factories, industrial complexes, hotels,

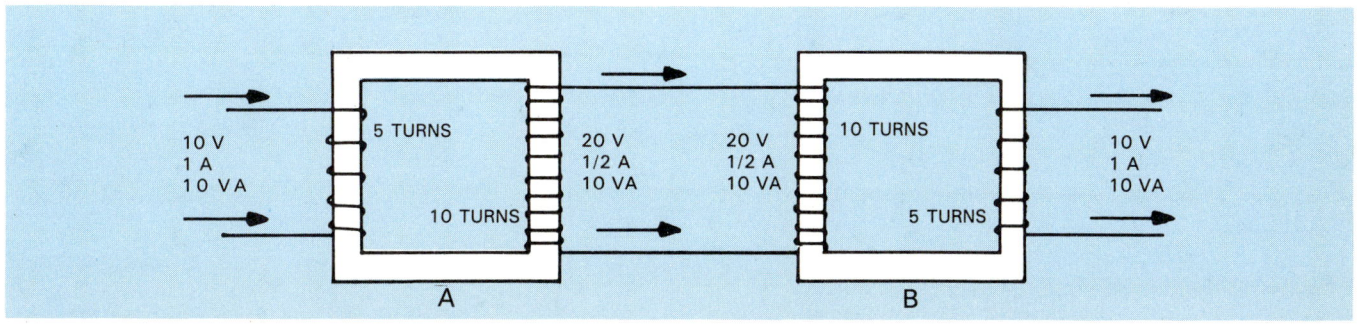

Fig. 12-40. Transformers may increase the voltage or they may reduce it. A—Those designed to increase voltage are called step-up transformers. B—Step-down transformers reduce voltage.

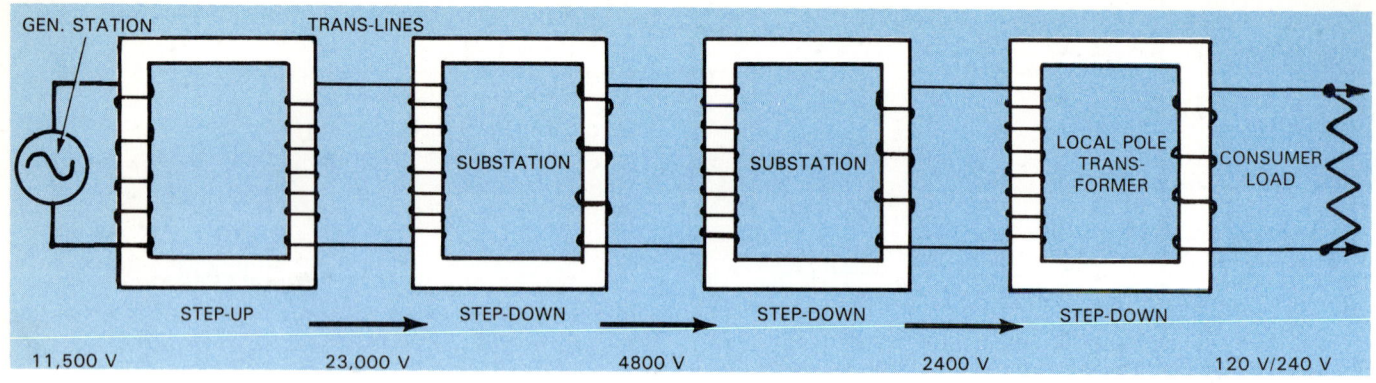

Fig. 12-41. Power companys rely on transformers to efficiently transmit electricity over long distances. Actually, a chain of transformers links the generating plant with the consumer.

motels, garages, and the like, require more circuits and higher levels of general lighting. In addition, there are larger motor loads and more heavy-duty electrical equipment of all types. All of these factors contribute to a need for many more circuits and much larger circuit wires when using the conventional three-wire, single-phase service entrance. So, out of necessity, other electrical services, which satisfy these needs were developed. These are the four-wire, three-phase systems.

FOUR-WIRE THREE-PHASE SYSTEMS

Fig. 12-43 illustrates a wye-connected four-wire, three-phase 120/208 volt arrangement. Such a system can supply both single-phase 120 volt circuits and three-phase 208 volt circuits.

Study the main panel wiring diagram shown in Fig. 12-44. Note particularly the different circuit arrangements that are possible. In this type of panel, there are three hot bus bars. Each has 120 volts to ground when connected to the neutral. For 120-volt circuits, the circuit breaker connections are the same as in three-wire single-phase services. The black (hot) circuit wire goes to the circuit breaker terminal and the white neutral circuit wire is connected to the neutral bus bar. As with single-phase systems, the 120 volt circuits should be balanced or equally arranged among the three hot buses.

Two-pole circuit breakers are also installed in the same manner as with three-wire, single-phase service panels. That is, the double-pole breaker is attached to any two of the three hot bus bars. Again, carefully balance these between all the buses. These comprise the two-wire, single-phase 208 volt circuits.

MAKING SINGLE-PHASE CONNECTIONS

A more versatile three-wire, 120/208 volt, single-phase circuit can be formed by connecting two circuit wires to a double pole breaker as previously described and adding a grounded conductor to the circuit. A circuit such as this can be used for 120 volt receptacles and lighting outlets. It can also supply 120/240 voltage for electric ranges, counter cook tops and other devices designed for that purpose.

Three-phase, three-wire 208-volt circuits are derived by connecting the circuit wires to a three-pole circuit breaker. These circuits are specifically designed to operate special three-phase devices such as heaters and various heavy-duty motor-driven equipment.

Last, but not least, a four-wire three-phase 120/208 volt circuit is connected in the same manner as the three-wire 208 volt circuit with the addition of a grounded wire connected to the neutral bus. This type of circuit is extremely useful when

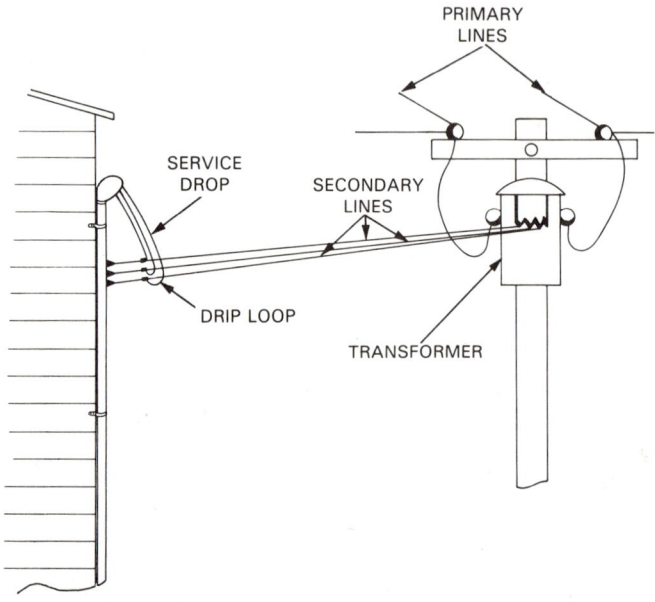

Fig. 12-42. The final step-down occurs at the local pole transformer near the structure it serves. Overhead conductor provides the last link to the customer's service wires.

The Service Entrance

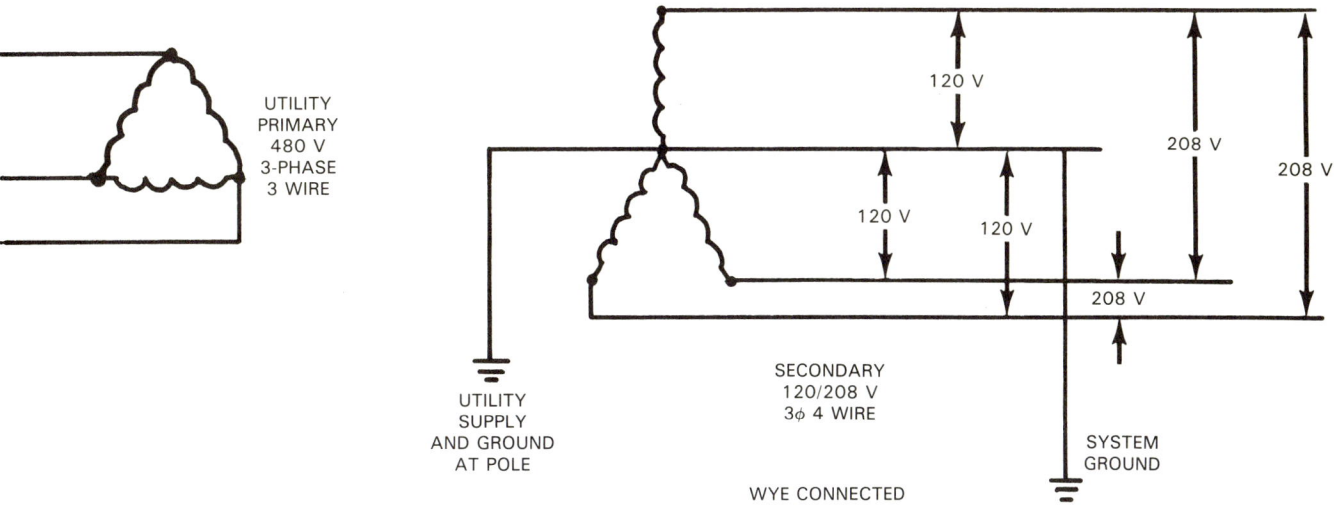

Fig. 12-43. A more versatile alternative to single-phase wiring is three-phase. This wye-connected, three-phase, four-wire arrangement provides 120 V/208 V circuits.

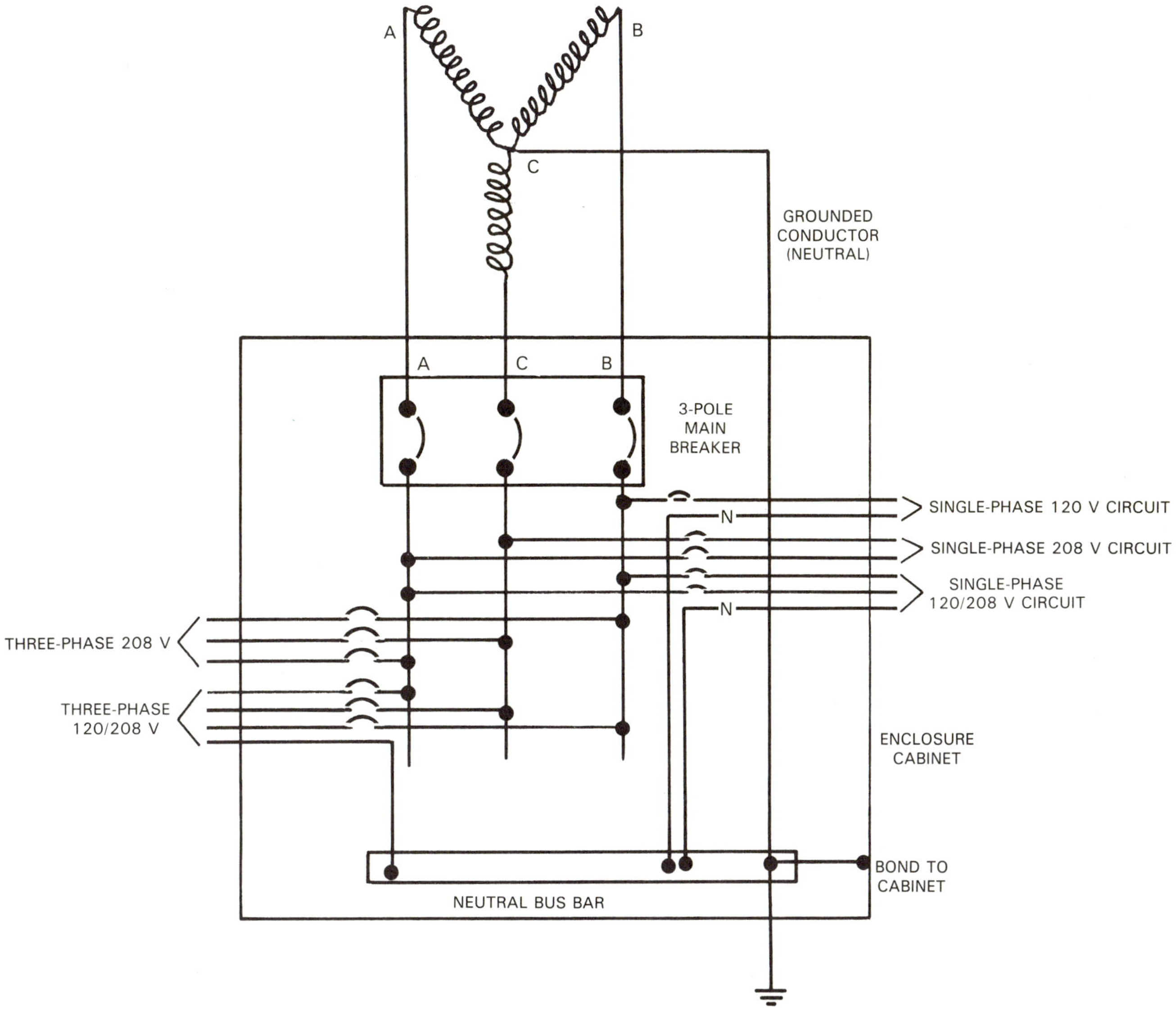

Fig. 12-44. Main breaker and panel circuit connections for a wye-connected three-phase, four-wire service. Hot bus bars are marked A, B, and C.

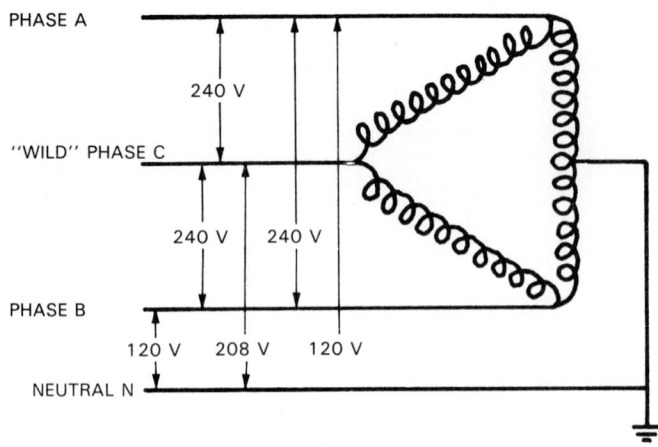

Fig. 12-45. Popular delta-connected three-phase, four-wire service supplies 120 V single-phase, 240 V single-phase, and 240 V three-phase circuits. The "wild" phase C must always be identified at all terminations and accessible points. Usually it is indicated by the color orange.

extending an electrical system to other circuit panels.

DELTA FOUR-WIRE SYSTEMS

Another service alternative is the delta-connected, four-wire, three-phase 120/240 volt system. A schematic for this type is illustrated in Fig. 12-45. It can provide 120 volt single-phase, 240 volt single-phase and 240 volt three-phase power.

The panel connections, as shown in Fig. 12-46, are made as follows:

1. For 120 volt single-phase circuits, connect one circuit wire (white) to the neutral and the other (black) to an ordinary single-pole breaker attached to phase A or B only. Note: the breaker must not be attached to phase C for this type of circuit since this would provide 208 volts,

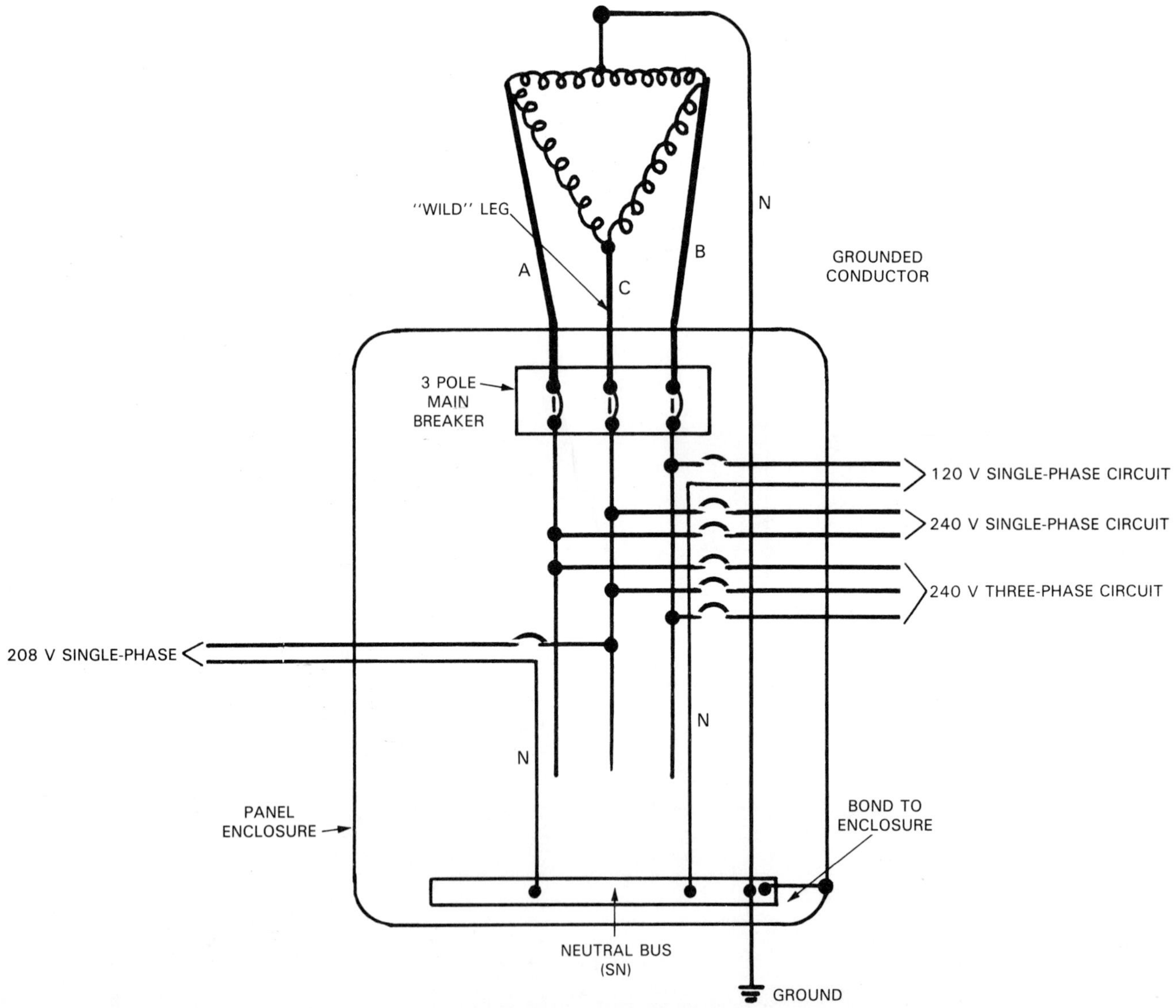

Fig. 12-46. Delta three-phase, four-wire panel connections. Note the variety of circuit possibilities.

as shown in Figs. 12-43 and 12-44. Equipment designed for 120 volt operation would be ruined.

The phase C wire of this system is often called the "wild leg" or "high leg" and must be identrified as such at all accessible points with an orange-colored indicator. See NEC, Section 215-8. At the panel, wire C is connected to the center bus bar.

2. To derive power at 240 volt single-phase, you simply connect both circuit wires to a double-pole circuit breaker attached to any two phases. Do not include a neutral grounded wire in this circuit.
3. For 240 volt, three-phase circuits, run three circuit wires to a three-pole breaker attached to all three phases. Again, ignore the neutral for this type of circuit.

Other variations of multiphase service systems are possible and, in fact, available in many locations:

1. Four-wire, wye-connected, three-phase, 277/480 volt. This is very similar to the 120/208 volt system already discussed. The major difference is that the transformer secondary supplying such a system has a higher voltage. Since most major appliances, lighting fixtures* and heating units do not operate at more than 240 volts, this system has limited use.
2. Three-wire, delta-connected, three-phase, 240 volt. This system has strong limitations in actual use. It may supply 240 volt three-phase or 240 volt single-phase circuits only.

*Discharge lighting, like fluorescents, are rated at 277 volts and would require this voltage source.

REVIEW QUESTIONS — CHAPTER 12

1. List five major components of a service entrance.
2. The minimum service entrance rating for new builidngs is _____ amperes.
3. The proper size service for 200 ampere service is No. _____ AWG copper wire.
4. Explain what is meant by:
   a. Service drop.
   b. Service lateral.
5. The neutral bus bar at the service entrance panel provides an interconnection means for the _____ conductor(s) and _____ conductor.
6. The minimum clearance of service wires above a sidewalk is _____ ft.
7. _____ are devices used to change voltage.
8. Name two types of transformers.
9. Single-phase, three-wire, 120/240 volt systems are commonly used to provide power to _____, while three-phase, four wire 120/208 volt systems are found supplying _____.
10. The minimum size grounding electrode conductor is No. _____ AWG.
11. Describe, in your own words, the purpose and location of the following service entrance components:
    a. Meter enclosure.
    b. Service head.
    c. Main disconnect.
12. Based on the transformer illustrated below, indicate the voltage, amperage and power of the secondary.

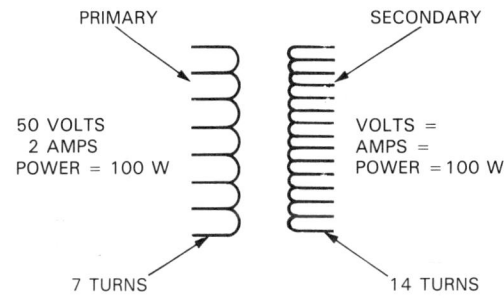

13. What is the "wild leg" and how is it connected into the panel?
14. How many "hot" bus bars are there in three-phase lighting panels?

Open type package substation. Units like this reduce voltage of electrical power for distribution to still other step-down transformers. (McGraw-Edison Co., Power Systems Div.)

Chapter 13
APPLIANCE WIRING AND SPECIAL OUTLETS

Wiring for large appliances such as dishwashers, clothes washers, clothes dryers, garbage disposal units, water heaters, etc, requires special consideration. In fact, these units need separate individual circuits. In this chapter we will investigate these special circuits and outlets.

GENERAL CONSIDERATIONS

The separate appliance circuit may be designed and connected to supply both 120 and 240 volts for devices such as electric ranges and clothes dryers. These appliances may have differing voltage needs at certain times during their operation. For example, clothes dryer circuits are varied and may be split, so that high heat is supplied at 240 volts while low heat is supplied at 120 volts. Fig. 13-1 is a schematic for such a circuit.

Quite often, the appliance is connected directly to the power panel. It is protected by the branch circuit breaker which also may serve as a disconnecting means.

Other hookups are made with a pigtail plug and a receptacle. This is a common practice with clothes dryers and electric ranges, Fig. 13-2.

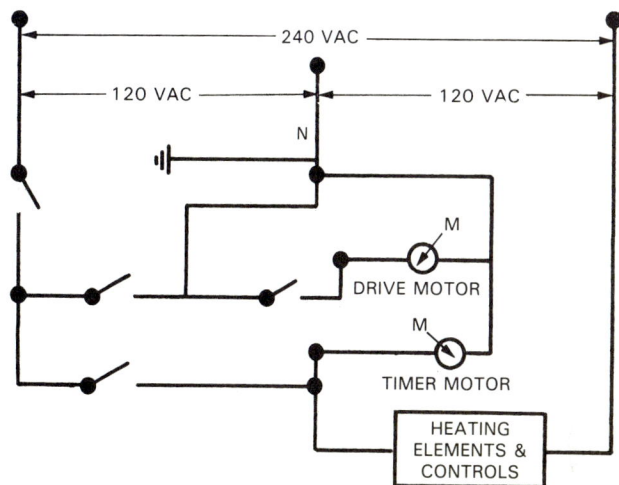

Fig. 13-1. Clothes dryers are wired for 120 and 240 V. Motors and lights operate on 120 V, while heating elements operate at 240 V.

As shown in Fig. 13-2, the pigtail-receptable method involves running the conductors from the service panel to the receptacle, Fig. 13-3. Then, through the use of the pigtail plug, the circuit is com-

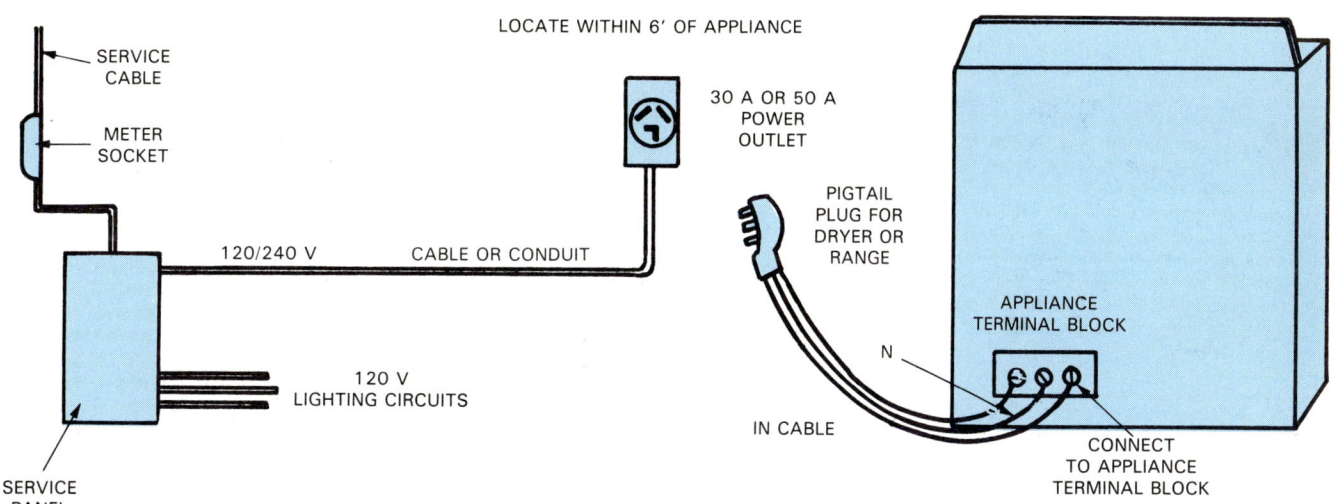

Fig. 13-2. Pigtail receptacle hookups commonly used on dryers and ranges. Most ranges and dryers have a terminal block, the place where the pigtail plug or circuit conductors are connected.

Fig. 13-3. A typical 30 A receptacle outlet.

pleted to the appliance.

Still another method used is the system illustrated by Fig. 13-4. Here, providing the appliances are close to one another, an appliance power panel is installed. The various appliance circuits are run from the panel. This works well in small kitchens with built-in units, more or less tandem to each other, or in utility rooms.

In general, the hookup for a range is a three or four-wire pigtail which plugs into a 50 A heavy-duty receptacle. The fourth terminal is for the equipment ground.

Receptacles can be either surface mounted or recessed for flush mounting. The latter is usually preferred for new work. A 4 in. square box is needed for flush mounting.

In general, appliances should be placed on a separate circuit if, as in the case of washers, dryers, ranges, garbage disposals, dishwashers, pumps, motors, and the like, they are rated at:
1. 120 volt, 12 amperes plug.
2. 1/8 plus hp.
3. 240 volts or more. See Fig. 13-5.

APPLIANCE CLASSIFICATION

At one time, rules regarding appliances provisions were based on their classification of the various types of appliances:
1. Stationary.
2. Fixed.
3. Portable.

Stationary appliances were those not permanently attached in a specific place, but yet were not ordinarily moved. These included clothes dryers, washing machines, and ranges. Fixed appliances were those like water heaters, motor-operated oil burners, and built-in electric heaters which could be permanently installed. Portables were the commonly moved type of appliances: blenders, mixers, toasters, and radios.

However, despite classifications given by these older rules, many appliances fell ''between'' these designations and had to be connected as if they were one type or another. When this occurred, the electrician had to use his/her best judgment based on the amperage, voltage, horsepower, and wattage rating of the appliance. The 1981 NEC dropped those definitions, but they may still serve as a guide.

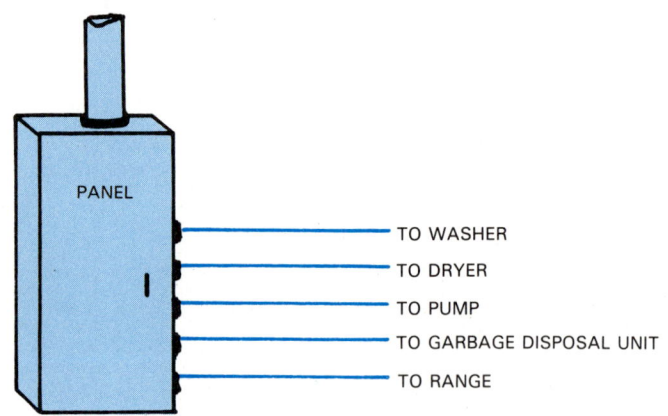

Fig. 13-5. Large appliances require separate circuits.

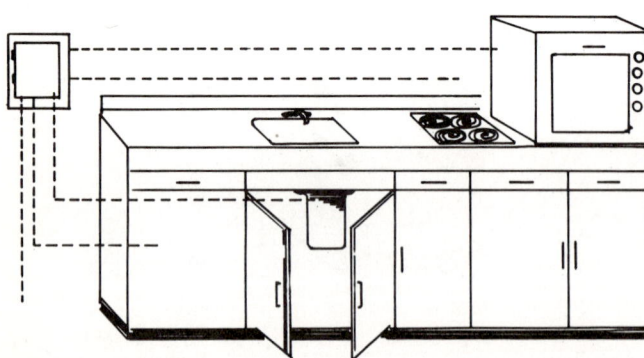

Fig. 13-4. Appliance power distribution method is particularly useful when appliances are close to each other.

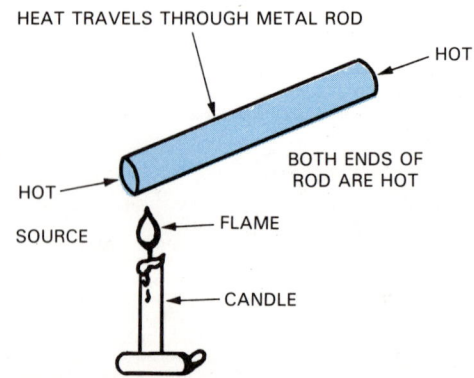

Fig. 13-6. Conduction of heat. This is how heat travels through solids.

HEATERS

A frequent installation or repair performed by electricians involves electric heaters. Heaters are devices containing resistors through which an electric current passes to produce heat. This heat is subsequently given off to the surrounding area by a process called either conduction, convection, or radiation. These are the three methods by which heat travels.

Conduction is the means or method by which heat passes through a substance by the contact of its molecules. That is, the heat energy is tranferred from point to point within the substance by molecules touching each other. Due to the nature of this type of heat transfer, it takes place almost exclusively in solid substances.

Convection is the process by which heat is transferred through fluids, liquids, and gases, by the movement of masses having unequal temperatures. Warm fluid masses rise because they are less dense than the surrounding fluid. Cool fluids sink because they have greater density than their surroundings.

Radiation is the primary method by which heat is transferred through energy waves. The waves come out from the source in straight lines, and move in all directions from the source. It is the method by which the sun's energy reaches earth. Models of these three heat-transferring methods are seen in Figs. 13-6 through 13-8.

HEATER INSTALLATION

Typical home heaters are usually built into floors, ceilings, and walls or, in some cases, they are surface mounted. These heaters are almost always automatically controlled by thermostats which are located close by the heating unit or built into it. Some are provided with circulating fans to help force the warm air away from the unit while drawing in cooler air to be warmed. Many heating units are designed to be completely portable so they can be easily relocated.

TYPES OF HEATERS

Home heating units are available in a great variety of shapes, sizes, and power ratings. Units may

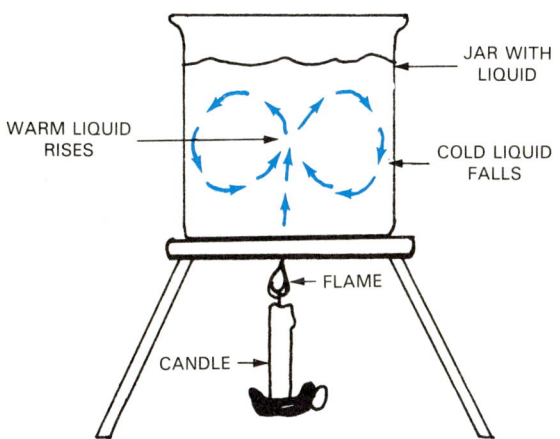

Fig. 13-7. Convection of heat. Heated fluids, being lighter, rise. Cooler fluids drop. Both liquids and air circulate this way.

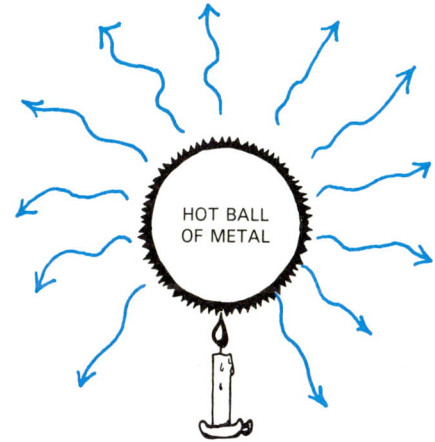

Fig. 13-8. Radiation of heat. This is how heat travels through air and space.

Fig. 13-9. Baseboard heater unit. (Square D Co.)

be square, rectangular, or circular. The heater's size has traditionally been given in watts or kilowatts of power used. Latest NEC publications prefer giving size in volt amperes. (It is the same as the watt since volts times amperes equals watts.)

The units may range in size from 750 W (750 VA) to 7500 W or more. Floor heaters and baseboard units, Fig. 13-9, are varied. The size and type chosen depends mostly on room size and comfort needs. Fig. 13-10, illustrates additional types of heater units.

Wall units

Wall units are common, Fig. 13-11. They can have ratings up to 6 kW (6 kVA) or more. They may be flush mounted, recessed, or surface installed.

Fig. 13-11. Wall mounted heating units. (Markel Nutone Div., Scovill)

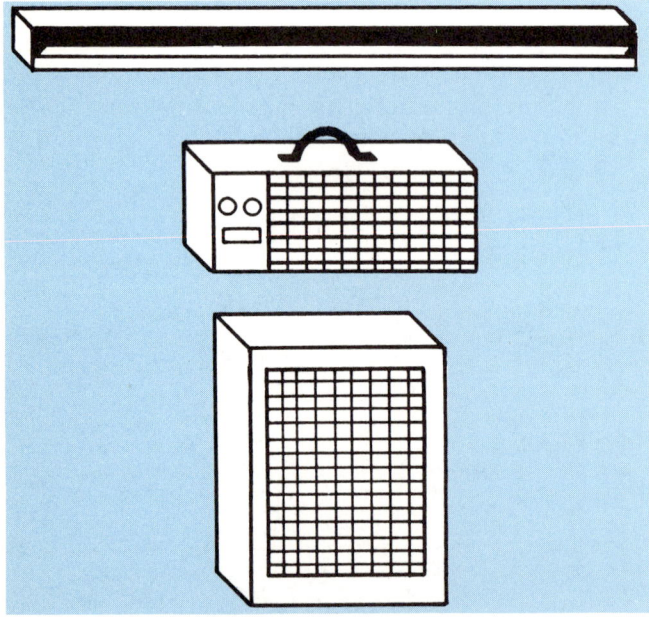

Fig. 13-10. Electric heating units are available in a variety of sizes, shapes, styles, and power ratings.

Various styles of wall units are available. They can operate on the radiation principle (called radiant heaters) or they can be fan assisted.

Control may be a wall-mounted or integral thermostat.

Floor units

Floor heating units are primarily of two types:
1. Floor heating cable.
2. Floor duct or drop-in units, Fig. 13-12.
    These units have general ratings of 120 to 240 volts.

The floor heating cables require special installation procedures. Consult the *Code* as well as the

Fig. 13-12. Floor heating unit.

local inspector.

Floor duct units are simple to install. The procedure is carefully explained and illustrated by the manufacturer. Heat is moved by natural convection or a fan. Thermostats are generally wall mounted.

Appliance Wiring and Special Outlets

Baseboard units

Perhaps the most common residential and small commercial type heaters are the baseboard units. See Fig. 13-9.

Again, these operate at 120, 240, or 208 volts, and have a load rating ranging from 35 (35 VA) to about 4.5 kW (4.5 kVA), depending on size and construction. Thermostat control is possible from either a wall-mount or integral positions (on the appliance).

Other units

Although only a few specific types of heater units are discussed here, there are many other types available. The choice depends on many factors:
1. Size of area to be heated.
2. Heat loss factors.
3. Insulation present.
4. Climatic conditions.
5. Desired temperature of area to be heated.
6. Style preference.

Other types of electric heaters marketed today are:
1. Heat pumps.
2. Electric furnaces.
3. Electric fireplaces.
4. Ceiling heaters.
5. Outdoor roof-mount central heating.

HEATER CIRCUITS

Because of the variations in types, makes, and models of heaters, it would be impossible to describe all the heater circuits possible. However, the common types of heaters available for home or office installation are similar enough so that they can be shown and described here.

Fig. 13-13 shows the circuitry found most commonly in baseboard units. Generally, these may be wired for 120, 208, or 240 volts. Circuits must be properly grounded at all times.

Fig. 13-14 shows a typical wall or ceiling heater circuit. Note the similarity in circuitry of the two units. The relay, in each case, is used because of the time-delay needed to compensate for the slow moving contact points within the thermostat. Most units have built-in relay switches so there is no need for auxiliary wiring. Fig. 13-15 shows a wiring schematic for a heater with fan-assisted circulation.

WATER HEATERS

A single-family home or small commercial enterprise will usually have a 30, 40, or 50 gallon hot water heater. Electric water heaters require a separate 240 volt circuit.

In most instances, the water heater is connected to a two-pole 20 A or 30 A breaker at the main panel or a nearby subpanel. 20 A should be used

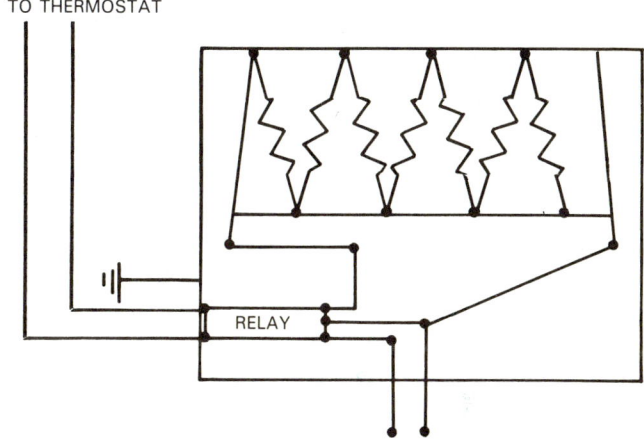

Fig. 13-14. Wall or ceiling heater circuit. Note circuitry for relay.

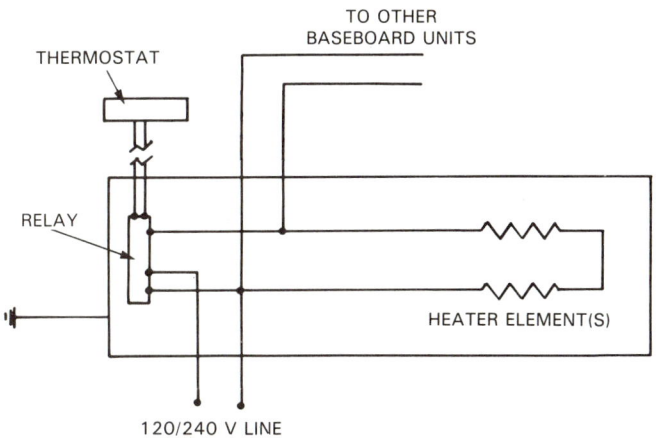

Fig. 13-13. Baseboard heater circuit. Baseboards are perhaps the most common heaters used in both residential and light commercial application.

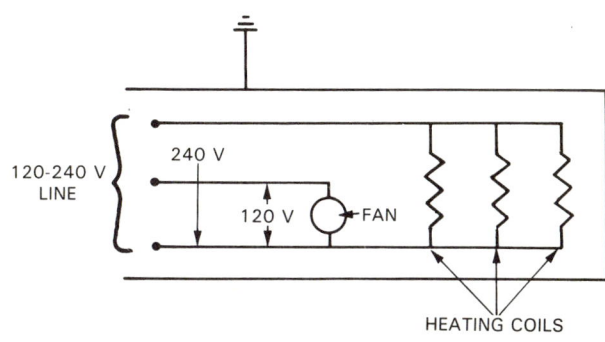

Fig. 13-15. Typical circuit for a heater with fan-assisted circulation.

for heaters up to 4 kW (4 kVA), while 30 A is appropriate for those up to about 6 kW (6 kVA). These would be wired with No. 12 or No. 10 AWG conductors respectively, Fig. 13-16. This manner of connecting the water heater will provide 24 hour heating.

Another alternative is shown in Fig. 13-17. It is called off-peak water heating. It allows water to be heated between the hours of 10 p.m. and 7 a.m. approximately. In addition, with minor modification, a manual override can be installed for those times when more than the usual amount of hot water is needed.

Water heaters, like most electrical equipment, must be properly grounded. This is accomplished simply by bonding the water heater unit casing to the equipment grounding conductor of the cable or to the conduit, if supplied in that manner.

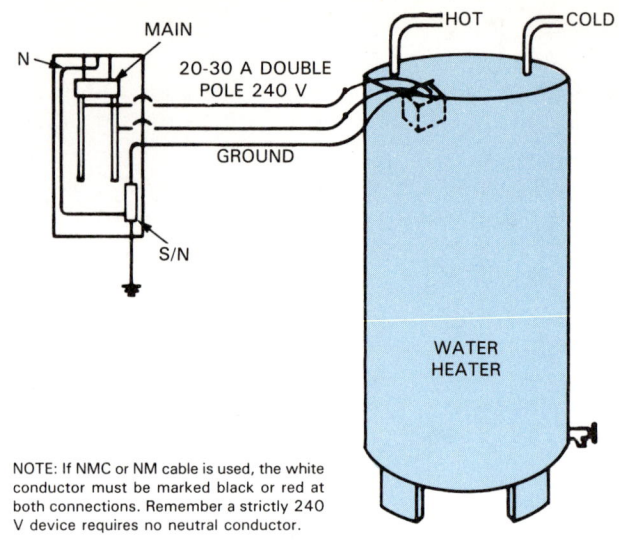

NOTE: If NMC or NM cable is used, the white conductor must be marked black or red at both connections. Remember a strictly 240 V device requires no neutral conductor.

Fig. 13-16. Water heater connection for 24-hour heating.

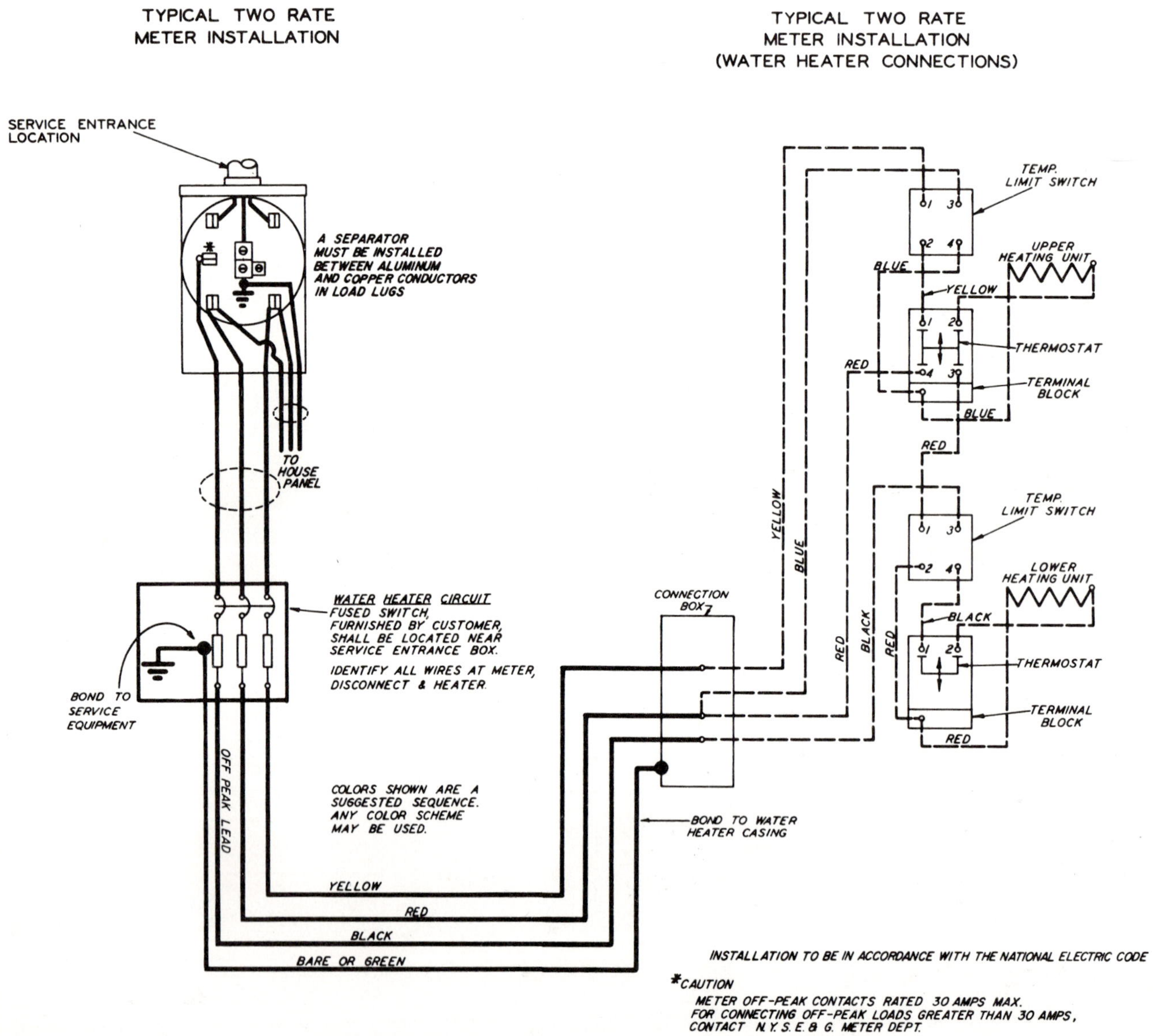

Fig. 13-17. Alternate method of connecting a water heater to provide off-peak heating. (NYSEG)

ROOM AIR-CONDITIONER UNITS

Air conditioners are connected for either 120 or 240 volts and generally are rated between 4 and 10 ampere. For example, a typical air conditioner might be rated 1.5 kVA (240 V x 6.25 A). This air conditioner could be placed on a 15 A double-pole breaker or plugged into a receptacle like the one shown in Fig. 13-18. Other air conditioners, which operate on 120 V, can be plugged into any regular duplex receptacle outlet.

Air conditioners can often be connected to the same circuit as heating units, since they would not be operating at the same time. In fact, when computing service requirement, only the larger load is considered.

GARBAGE DISPOSAL UNIT

Garbage disposal units, such as the one pictured in Fig. 13-19, may be controlled directly from the panel circuit breaker, from a plug and cord connection, or most often, with an on-off switch. They are rated at approximatley 5-9 A and 120 V. Since they are motor-operated, they must have overcurrent protection not exceeding 125 percent of their full-load current rating.

DISHWASHERS

Most dishwashers have a cord and plug which may be connected to one of the 20 A kitchen appliance receptacles. Like garbage disposal units, they carry a rating within the range of 5-10 A at 120 V. In fact if location permits, the dishwasher and garbage disposal unit can be connected to the same kitchen duplex outlet.

REFRIGERATOR AND FREEZERS

Refrigerators and freezers should have special outlets dedicated only to them. Both are motor-type appliances and, as mentioned earlier, should have overcurrent protection, fuses, or breakers rated at not more than 125 percent of their nameplate current. Connection is almost always by cord and plug.

COOKING TOPS AND WALL MOUNTED COOKING UNITS

Counter cooking tops and wall-mounted cooking units are each rated at approximately 7.5 kW (7.5

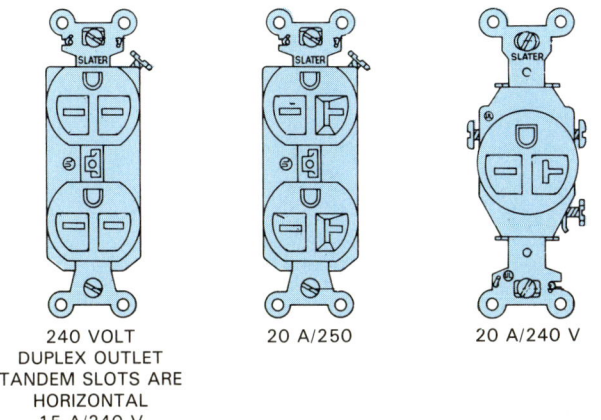

Fig. 13-18. 240 volt air conditioners, rated at 16 A or less, most often are simply plugged into special outlets such as these.

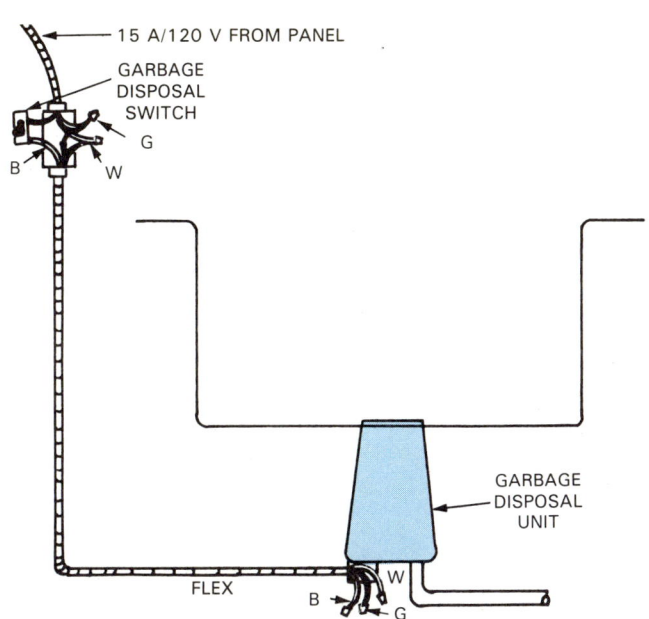

Fig. 13-19. Garbage disposal units are often installed with an "on-off" control nearby.

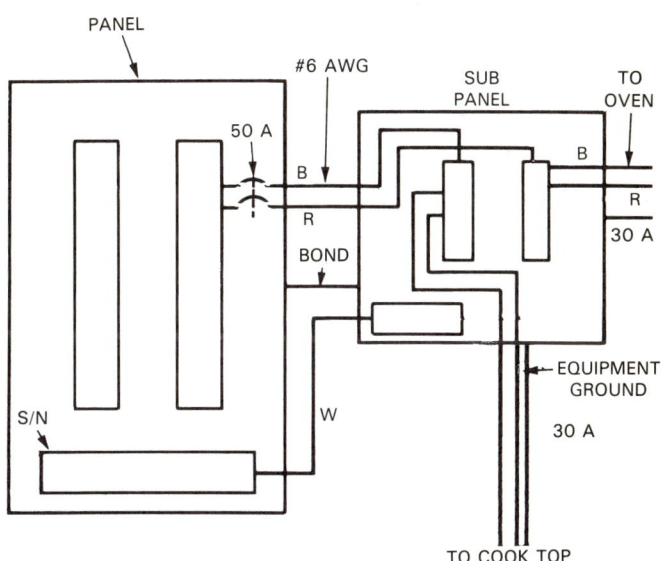

Fig. 13-20. Due to their close location, often two appliances or more may be placed on a single, but properly rated, circuit. Here, the circuit is shared by a pair of cooking units.

kVA) and 120/240 volts. These types of units require special circuits and overcurrent protection in the form of fuses or breakers, usually 30-40 A. They are connected in the same manner as ranges or dryers discussed earlier. Service entrance cable, No. 8-10 AWG, is often the method of connection, originating from the main panel or subpanel, directly to the unit. In instances where the units are close, it is permitted to connect the pair to a single circuit, providing the overcurrent protection is suitable for the combined load, Fig. 13-20.

MISCELLANEOUS SPECIAL OUTLET UNITS

There are several other devices which should have special outlets and circuits dedicated to them. They are merely listed since they are connected in a similar manner to large appliance outlets.
1. Central heating/cooling system.
2. Attic fans.
3. Well pump.
4. Sump pump.
5. Large shop tools.

ELECTRIC MOTORS, GENERAL PROVISIONS

Motors, like other electrically operated devices, must have a control. In addition, they should be protected against overloading, short-circuiting, or overcurrent surge.

Small motors (1/20, 1/10, 1/8, or 1/5 hp) can usually be controlled by a plug-in cord. To disconnect, simply unplug the electric cord. But larger, nonportable motors require more amperage and, thus, need more sophisticated disconnect means. This may be a circuit breaker, fuse block within the service panel, fused switch-box or motor-starter switch near the motor itself. See Fig. 13-21.

Regardless of the type of motor, it's horsepower, or amperage rating, make sure that it is properly grounded. This can be done through the conduit or metal armor of the conductor or by a separate conductor between the motor itself and a grounding electrode or water pipe. Do not overlook this important grounding method. Refer to grounding section of Chapter 5. Motors are discussed in depth in Chapter 22.

REVIEW QUESTIONS — CHAPTER 13

1. List five appliances which require special circuits or outlets.
2. Name the three methods by which heat is

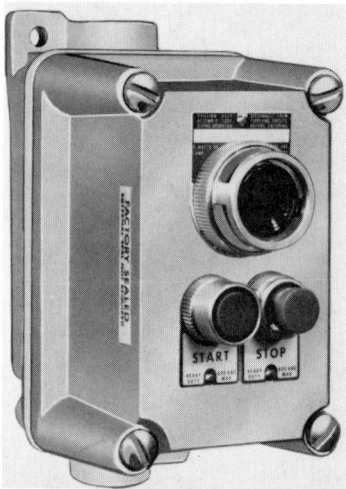

Fig. 13-21. Heavy-duty motor switches. These are used as the disconnecting means for motors of 1/4 hp or greater. (Killark Electric Mfg. Co.)

tranferred from place to place.
3. The method of heating water during the night-to-early morning hours is called _____ water heating.
4. Motor-operated appliances such as garbage disposal units should have overcurrent protection in the form of fuses or breakers which will not exceed _____ percent of their full load rating.
5. A counter cook top unit rated at 8 kVA and 120/240 V, should be on a separate circuit having a double-pole circuit breaker of _____ A.
6. Baseboard heaters are often on the same circuit as room air conditioner units. True or False?

Chapter 14

LIGHT COMMERCIAL WIRING

In this chapter, we will study two very different examples of commercial electrical wiring:
1. Light commercial structures.
2. Multi-family or multiple occupancy structures.

Light commercial wiring of the type required by a small store, is perhaps the least complex form of wiring an electrician will encounter. In general, these establishments need only the bare minimum of electrical power. Most stores, unless very special, require only a simple 100-ampere service, True, specialty stores such as appliance, electronics, or heavy-duty tool shops, would require much more capacity. These, however, are the exceptions rather than the rule.

FINDING LOADS

At the other extreme, multiple occupancy structures such as apartment complexes with three or more stories require careful considerations as to individual and overall load requirements. A 30-unit apartment building, which is certainly a rather small multiple occupancy, may easily require a service entrance of 1000 A or more.

In essence, the apartment building is a complex commercial structure. Wiring should be attempted only by an electrician experienced in this type of installation.

Refer to Article 220, of the NEC throughout this chapter. In addition, study the Examples Section of Chapter 9 in the NEC.

SMALL RETAIL STORE

Wiring a small retail store is one of the least complicated tasks for an electrician. A commercial structure such as this usually requires a minimum of electrical power. It has only a few lights, receptacle outlets, a modest heating system, air-conditioning, and, perhaps, a small water heater.

The structure represented by Fig. 14-1 is a small establishment of about 900 sq. ft. (45' x 20'). A show window extends 25 ft. along it's front and side walls.

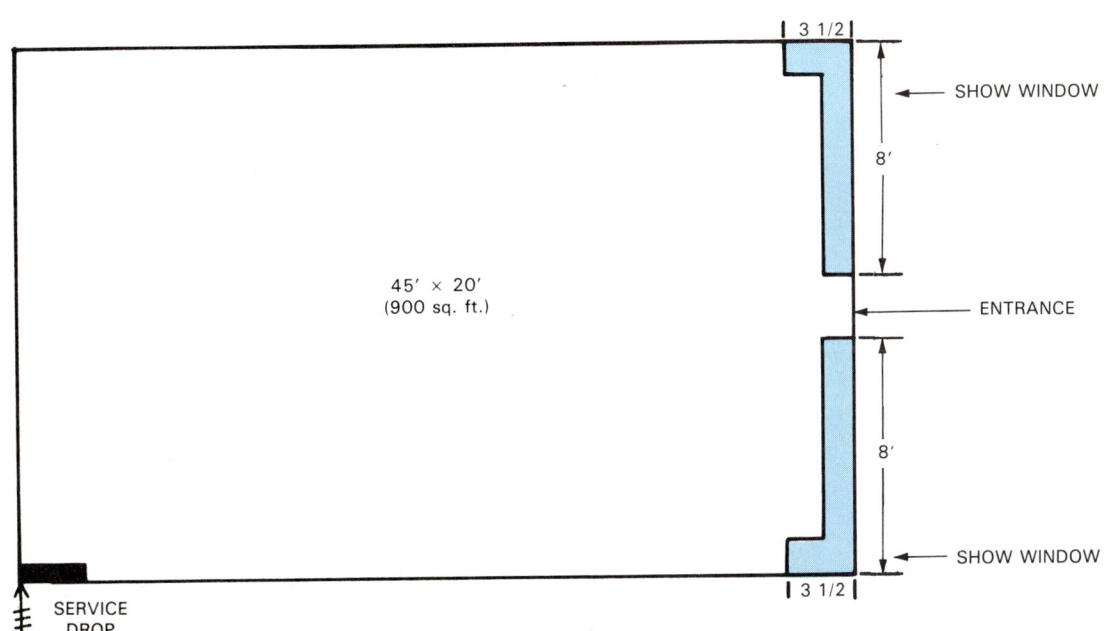

Fig. 14-1. A small retail store with 25 ft. of show window. Lighting requirement for show windows is 200 watts per lineal foot.

GENERAL LIGHTING

The NEC sets the general lighting load at 3 watts/sq. ft. for stores. Therefore, this store will require 2,700 VA (W) for general store lighting. This will translate into 22.5 A at 120 V which will require two 15 A circuits using No. 12 AWG wire. In addition, 24 duplex receptacles will be installed. (See Section 220-13, NEC.) Each is rated at 180 volt-amperes. A volt-ampere (VA) is a measure of apparent power, rather than actual wattage. Its values are the same as watts.

SHOW WINDOW LIGHTING

Show windows carry a special significance since they are to be well lighted. NEC 220-12 requires that a load of at least 200 VA be included for each running foot (305 mm) of show window.

In our example, you would multiply 25 feet by 200 volt amperes to find total volt amperes (5000 VA). Accordingly, three 15 A circuits are necessary. Again, No. 12 AWG conductors are recommended; No. 14 AGW is minimal.

HEATING, AIR CONDITIONING AND WATER HEATING

We shall assume the heating unit to be 8 kVA. The air conditioner is rated 5 kVA and a small water heater, 2.5 kVA.

SERVICE LOAD

Computing the service load for a small retail store, such as the one described here, is a simple matter. In fact, it is extremely easy when you consider the computations involved in wiring a large apartment building or a giant industrial complex.

We simply add up the factors:
```
General lighting . . . . . . . . . . . .   2700 VA
Show window . . . . . . . . . . . . . .   5000 VA
Receptacles (180 VA each) . . .   4320 VA
Heating . . . . . . . . . . . . . . . . . .   8000 VA
Air conditioning . . . . . . . . . . . . . . . . . . . . . .
   (omit, NEC 220-21 permits you to omit the
   smaller load of heating/cooling.)
Water heater . . . . . . . . . . . . . .   2500 VA
Total . . . . . . . . . . . . . . . . . . . .  22,520 VA
```
22,520 VA ÷ 230 V = 98 A

A 100 A single-phase, three-wire service entrance is adequate. If desired, this may be increased to 150 A to allow for future expansion.

SMALL MULTIFAMILY DWELLINGS

The wiring of multifamily dwellings is somewhat complex. This section will try to simplify the subject. We will consider only small multifamily dwellings rather than large apartment buildings. Once you have mastered the basics learned here you will be able to go on to more complex installations.

The term multifamily (or multiple occupancy, as it is sometimes called), refers to building which house more than two apartments or family occupancies. One and two-family dwellings are not considered multiple occupancy by the *Code*. This is an important distinction and should be remembered when interpreting various sections of the *Code*.

SERVICE DROPS

Except in rare instances, only one service drop is permitted for an entire building. This service drop feeds into separate meters for each apartment. See Fig. 14-2. In addition, one meter is usually used to record power consumed by common areas such as stairways, hallways, outside lighting, or security lights. Electrical requirements of common areas are known as the HOUSE LOAD.

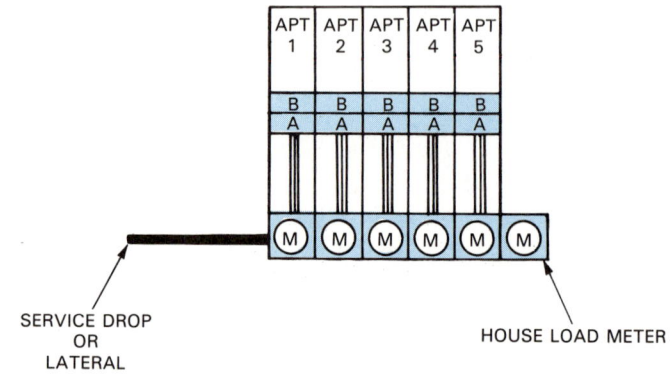

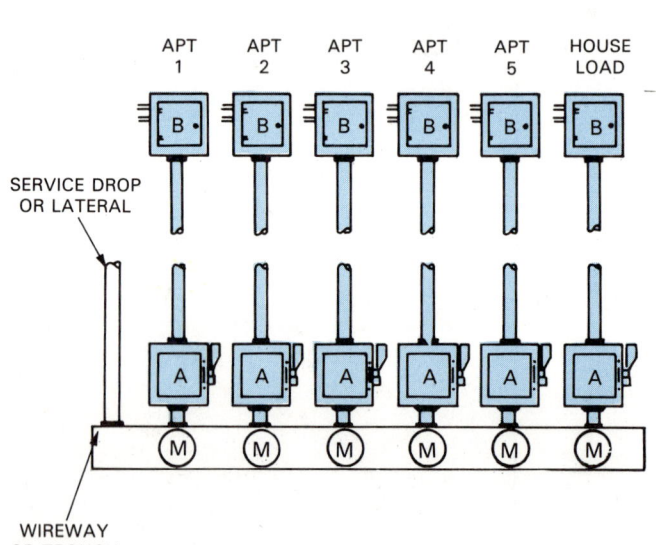

Fig. 14-2. For small building of one or two floors, the service is fed directly to separated metering equipment for each apartment. A—Main service disconnect. B—Branch circuits. M—Meter equipment.

Light Commercial Wiring

BUILDING CATEGORIES AND SERVICE SCHEMES

The *Code* generally separates multiple occupancy dwellings into several categories:
1. One or two-story structures.
2. Three-story-plus structures without occupants above the second floor.
3. Structures of three or more stories with occupancy on or above the third floor.

These categories are very important in determining the proper arrangement of the service equipment. The key difference is that multiple occupancy structures of one and two floors can be treated in much the same manner as individual houses. Each apartment will have a separate meter and panel with disconnecting means. Refer again to Fig. 14-2. It illustrates a five-apartment building of one or two stories having a meter for each unit plus a house load or common meter.

second floor must have all the service entrance equipment in a common accessible location. The branch circuit disconnects shall be in each apartment.

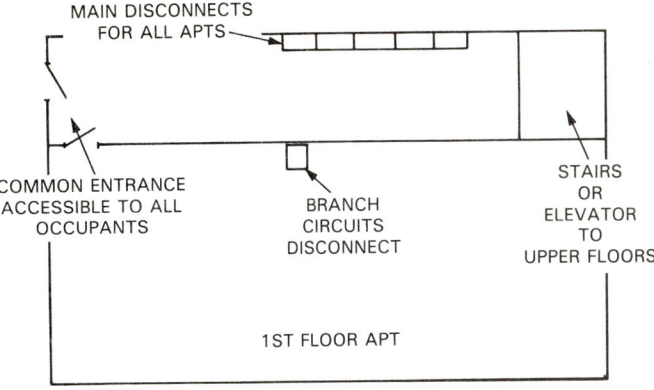

Fig. 14-4. Main floor of multiple occupancy dwelling, having occupancy above the second floor as well. Disconnects are in a main entrance area.

COMPUTING SERVICE LOADS FOR MULTIPLE OCCUPANCIES

A simple way to compute apartment building loads is to first treat each apartment as a separate load. Then, add to the total the load of the apartments' common loads such as central air conditioning, heating, laundry, etc. These are used by all the occupants.

As an example, we shall compute the service requirements for a multifamily dwelling having five apartments. Each apartment has:
1. An area of 600 sq. ft.
2. An electric range rated at 8 kVA.
3. An air conditioner rated at 5 kVA.
4. Six 1.5 kVA heaters.
5. A 5 kVA water heater.

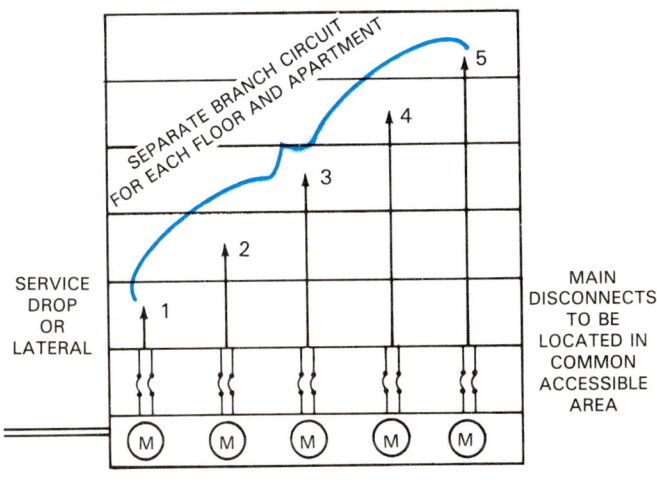

Fig. 14-3. In three-story or greater occupancies, metering equipment and main disconnects for all apartments must be in a common, accessible area.

When buildings are occupied on or above the third floor, the arrangement shown in Fig. 14-2 is not permitted. In these situations, the service equipment, meters, and main disconnects for each and every apartment must be in a common, accessible location. (This must be a place that all occupants can reach usually a hallway or basement.) The only pieces of service equipment placed in each apartment, are the lighting and branch circuit panel with individual circuit disconnect. Fig. 14-3 and Fig. 14-4 illustrate this type of layout.

There are alternatives, of course, to the arrangements illustrated in Figs. 14-2 through 14-4. The main idea is that dwellings of one or two stories have *main* disconnects in each apartment. Buildings of three or more stories with occupancies above the

First, compute the individual load for each apartment:
```
600 sq. ft. × 3 VA/sq. ft. . . . . . 1800 VA
8 kVA range @ 80%
   (NEC 220-19) . . . . . . . . . . . 6400 VA
Electric heating 6 × 1.5 kVA
   (NEC 220-15) . . . . . . . . . . . 9000 VA
Water heater 5 kVA . . . . . . . . . 5000 VA
Air conditioner 5 kVA . . . . . . . 5000 VA
```

Next, determine the number of branch circuits for each apartment:
General lighting . . . . 2-15 A or 20 A circuits, No. 12 AWG conductors
Required small appliances . . . 2-20 A circuits, No. 12 AWG conductors
Laundry circuit . . . (common to all apartments)

Range circuit ..... 1-30 A, No. 10 AWG conductors
Electric heating ..... 2-20 A, No. 12 AWG conductors
Air conditioner ..... 1-30 A, No. 10 AWG conductors
Water heater ..... 1-30 A, No. 10 AWG conductors

Next, find the size of the conductors needed to sub-feed these apartments from the service meter or main disconnect:

Computed load (see NEC 220):
```
General lighting load .......... 1800 VA
  (600 sq. ft. × 3 VA/sq. ft.)
Small appliance load .......... 3000 VA
  (2-1500 VA circuits)
Total ..................... 4800 VA
```

Demand factor (See sections under Article 220 NEC): Omit air conditioner load, NEC 220-21.
```
First 3000 VA at 100% ..... 3000 VA
Remainder at 35% ......... 630 VA
Subtotal ................. 3630 VA
+ range .................. 6400 VA
+ heating ................ 9000 VA
+ water heater ........... 5000 VA
Total .................. 24,030 VA
```

For a single-phase, three-wire system:
24,030 VA ÷ 230 V = 104 Amperes
No. 1 AWG TW conductors will be needed for the service feeder conductors.

Finally, compute the sub-feeder neutral for each apartment:
```
Lighting + Small appliance ..... 3630 VA
Range 8000 VA × .70
  (NEC 220-22) ............. 5600 VA
Heating and water heating (no neutral)
Total neutral load ......... 9230 VA
```

Therefore 9230 VA ÷ 230 V = 40 Amperes. Use No. 8 AWG TW conductors.

MAIN SERVICE ENTRANCE CONDUCTORS

Once individual apartment requirements are completed, you can the determine main service entrance conductor sizes needed for the entire building.

Computed load:
```
Lighting and small appliance load,
  5 × 4800 VA .......... 24,000 VA
Water and space heating × 5   70,000 VA
Ranges 5 × 8000 VA .....   40,000 VA
5 Apt., Net load ..........  134,000 VA
(See NEC table 220-32)
  134,000 VA × .45 .....  60,300 VA
60,300 VA ÷ 230 V = .....     262 A
```
Main neutral feeder size:
```
5 × 4800 VA ............. 24,000 VA
First 3000 at 100% .......  3000 VA
Remainder at 35% .........  7350 VA
+ 5 ranges = 20,000 VA × .70 = 14,000 VA
Total ...................  48,350 VA
```
43,350 VA ÷ 230 V = 210 Amperes

HOUSE LOADS

If they are of any consequence, house loads should be computed as you would a separate apartment. In reality, the house load is on its own meter and its cost is shared by all. Before making your calculations, and laying out the house circuit for a house load, carefully review Article 220 of the National Electrical Code regarding the finer points of branch circuit feeder calculations.

REVIEW QUESTIONS — CHAPTER 14

1. The general lighting load for retail stores is _____ VA/sq. ft.
2. Show window lighting should be computed at no less than _____ watts per linear foot.
3. List the general categories for multiple occupancy dwellings.
4. For buildings that have occupancy on or above the third floor, the service equipment, meters, and main disconnects for each and every apartment must be located in:
   a. The cellar.
   b. The individual apartments.
   c. A locked closet.
   d. A common accessible place.
   e. Any of the above locations.
5. Branch circuit equipment for each apartment must be located in _____.

Chapter 15

FARM WIRING

The sizes of farms and their production capabilities are increasing at a rapid rate. Accordingly, their electrical demands must be met with good quality and high capacity electrical systems.

Farms are complex operations requiring careful planning and layout considerations prior to the actual installation of electrical equipment. A farm should be thought of as a small industrial plant or business having several buildings, but involved in a single purpose, food production. Operating toward that end, each unit or building requires electrical power.

POWER DISTRIBUTION

Farm buildings may include: dairy or stock barn, equipment shed(s), hog house, poultry house, machine shed, granary, garage, and farmhouse.

The electrical power is brought to the farm by the power company. The service drop, in this case, ends not at the dwelling, but rather at a centrally located yard pole. Power is subsequently distributed from the pole to the individual buildings. Fig. 15-1 illustrates electrical power distribution on a typical farm.

FARM EQUIPMENT

As with any electrical installation, the total electrical demand should be calculated before beginning actual wiring. Farmers, by necessity, have great electrical needs and these must be accurately figured. For example, devices such as milking machines, coolers, silo unloaders, feed conveyors, barn cleaners, fans, feed mixers, and other motor-driven equipment are typically found on modern

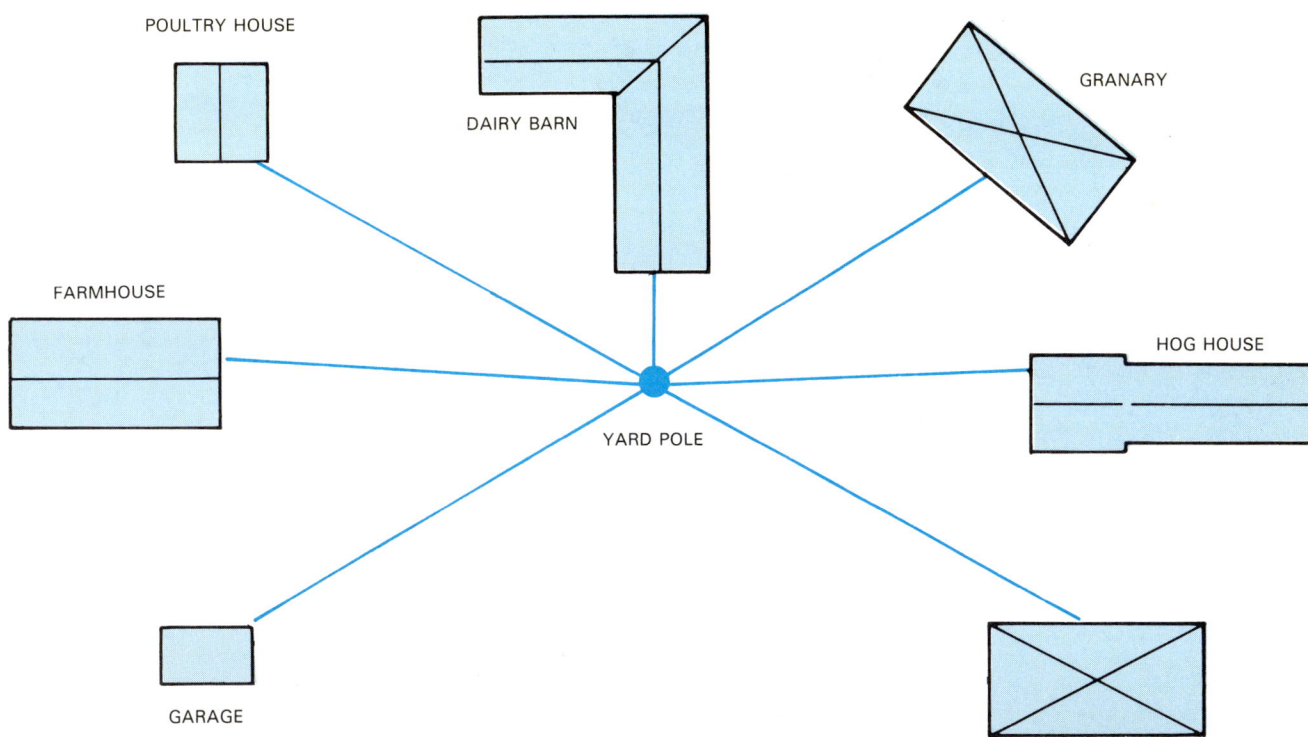

Fig. 15-1. The farm power central distribution layout. Power is directed to the various buildings from the farm yard pole.

farms, Fig. 15-2. Each of these devices, as well as many others, must be properly wired and connected for safe and efficient operation. Almost all modern farms require a 200 to 400 A service entrance.

THE YARD POLE SERVICE DROP

Although each building will have its own service panel containing fuses or breakers for overcurrent protection, the service drop and meter will be located at the yard pole. This will serve as a power center and overall disconnecting means for the farm. A yard pole is shown in Fig. 15-3.

GROUNDING

Because of the importance of grounding and particularly grounding of large motor-driven devices, farm electrical systems often have many grounds. This is crucial especially in farm building because of the dampness.

For this reason, metallic sheathed cable and/or metal conduit are not suitable choices for farm wiring systems. Too often the metal corrodes and weakens the ground. This creates hazardous situations.

To insure proper grounding and safe wiring in farm buildings, metal equipment is used as little as is practicable. Nonmetallic cable such as LTFMC, NMC, or UF, as well as nonmetallic outlet boxes, Fig. 15-4, are preferred. Boxes are made of bakelite, PVC, fiberglass, or similar materials. These materials are not affected by dampness and will provide long-lasting service.

The ground is established through a ground wire within the nonmetallic conduit or cable, as shown in Chapter 9. This ground wire is made continuous from the ground at the service entrance to all parts

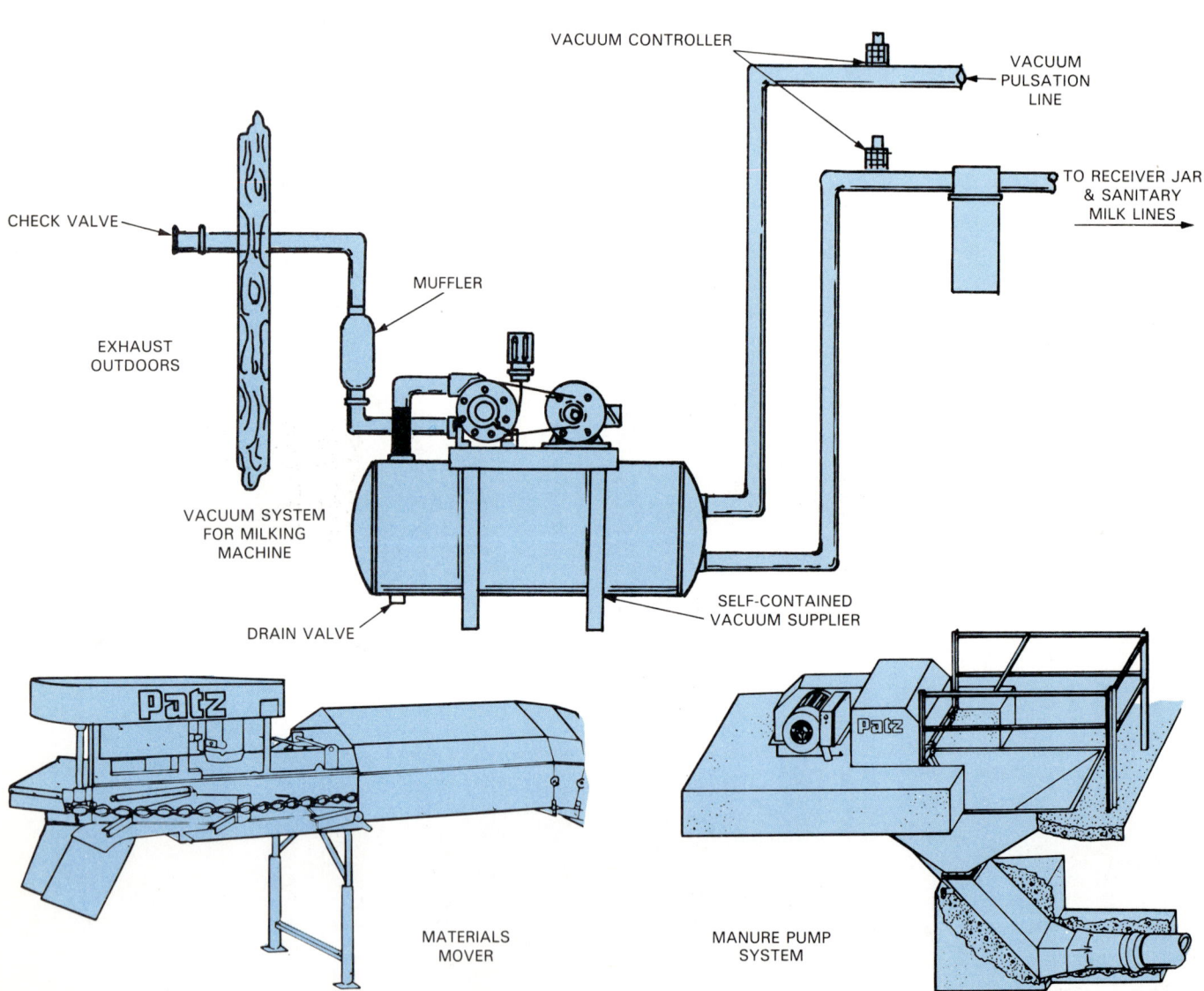

Fig. 15-2. Farms have great electrical demands which must be carefully considered. Many devices like the ones shown here are common farm necessities. (Patz Co. and DeLaval Agricultural Div.)

Farm Wiring

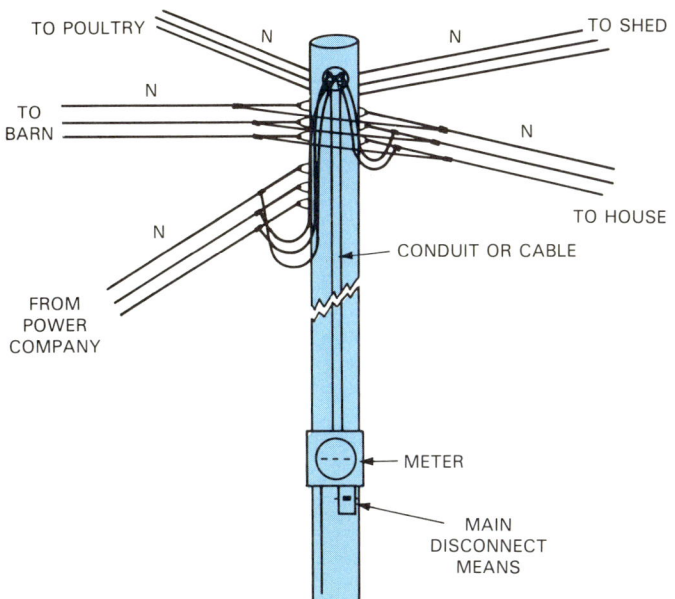

Fig. 15-3. The yard pole is the farm's central "hub" of power distribution.

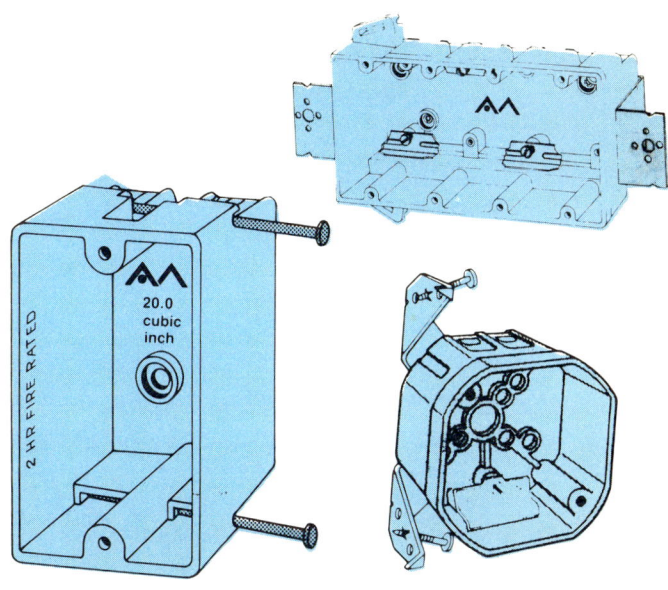

Fig. 15-4. Nonmetallic boxes are particularly useful in damp and wet locations.
(RACO Inc. and Allied Moulded Products, Inc.)

of the system. Sometimes, additional ground rods are located near and connected to motor-driven devices such as large fans and conveyors.

SPECIAL DEVICES

On many farms, as in other damp locations such as basements of homes, surface mounted nonmetallic units are used in lieu of metal outlet boxes. These units are self-contained, being combinations of box and switch, box and lamp holder, or box and receptacle, Fig. 15-5. They are simply interconnected with the NMC, LTFMC, or UF cable.

Other special devices and equipment used on farms may include: Vapor-proof light fixtures, timer-switches, waterproof (reflector) yard lights, light-

ELECTRIC EYE

SWITCH

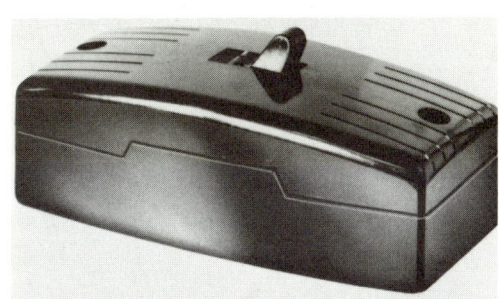
SWITCH

LAMPHOLDER

RECEPTACLE

LAMPHOLDER, PULL CHAIN

Fig. 15-5. Self-contained nonmetallic devices are often used on farms and other damp locations. (Bryant)

ning arrestors, weatherproof (outdoor) switches, and outlet boxes, Fig. 15-6 through 15-11.

DAIRY BARN

The dairy barn will undoubtedly require the most electrical power of any building on the farm. Indeed, it is the center of the farm.

RUNNING CABLE OR CONDUIT

Running of cable or raceways in barns and outbuilding should be done with the same care as house wiring. Make your parallel runs along the sides, rather than the bottoms of beams and joists. Make cross runs through drilled holes rather than over edges. If runs must be made across the surface, protect cable with 1 x 4 boards. Place wiring where it will be protected from weather wherever possible.

Electrical outlets should be plentiful. One outlet every 12 ft. or closer is best. These may be located along the walls behind the stanchions. The receptacles should be mounted no lower than 60 in. (1.5 m) above the floor. Higher is even better as it prevents their being damaged by animals or tools.

Lights should be located both to the front and rear

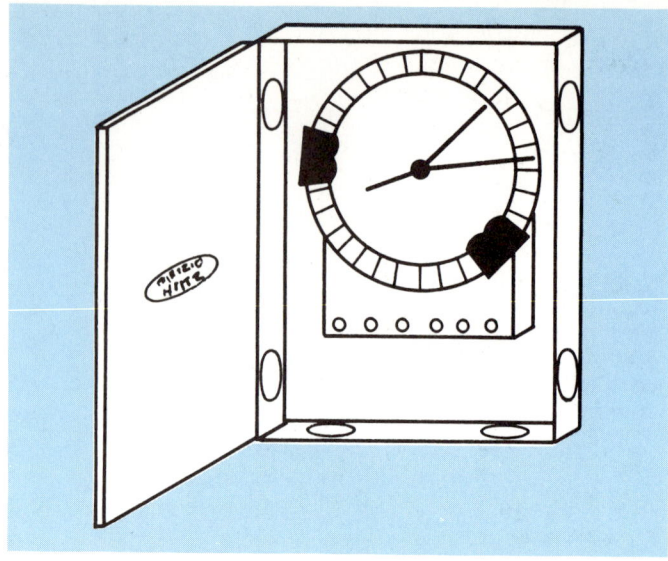

Fig. 15-7. Timer-switch is often used on the farm to automatically turn lights on and off.

of the stanchions. A 60 watt porcelain or plastic fixture each 15 ft. (4.6 m) is enough. Dustproof lights should be installed in the feed-storage, silo, and haymow areas. These lights should be controlled by a wall switch at every entrance to the area. Open feeding areas as well as calf pens should be provided with similar outlets and lighting circuits.

LOCATING LIGHTS

In planning locations for lights in the dairy barn consider putting them at all alley crossings as well as at stairs, ladders and chutes. If ceilings are dark it is advisable to use reflectors so light is directed downward rather than to the ceiling where much of it will be absorbed.

Fig. 15-6. Vapor-proof light fixture. Because of the excessive damp and dusty conditions in farm buildings, these units are installed as another safety measure. The bulb is encased in a glass dome, around which is a strong, protective cage. These units are essential in areas like the haymow.
(Appleton Electric Co.)

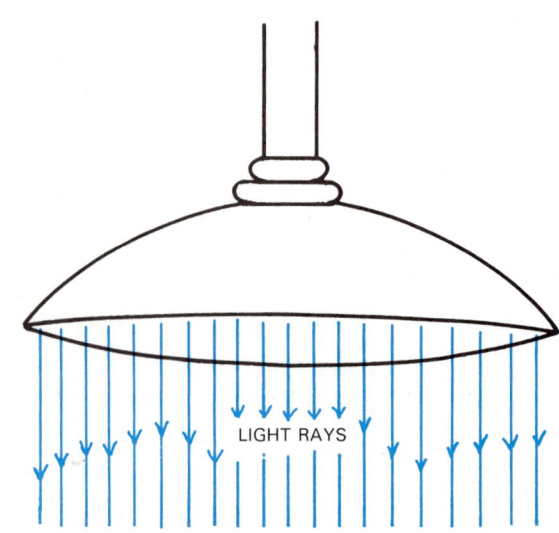

Fig. 15-8. Reflector lights. In barns and other farm buildings, reflector lights direct the light downward where needed.

Farm Wiring

Place the lights where joists, beams and posts will not obstruct the light from a fixture. Higher wattage bulbs are usually advisable especially since the lights are used only for a short time. Small bulbs represent only scant savings, at best.

Place switches so that lighting can be controlled at all entrances to the building. Use three way switches. As with house wiring, pilot lights are advisable when switches control lights that cannot be seen from the switch location.

The milkhouse or milking parlor, if there is one, would have similar lighting and receptacle arrangements as the dairy barn. Again, lighting should be plentiful and outlets mounted high enough to avoid damage. Special branch circuits will be required where the facility includes bulk tank storage, and/or motors for running vacuum pumps for milking machines.

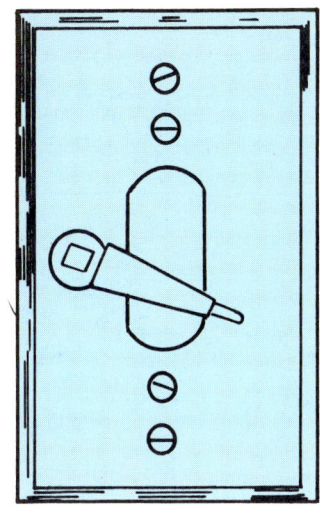

Fig. 15-10. Weatherproof switch.

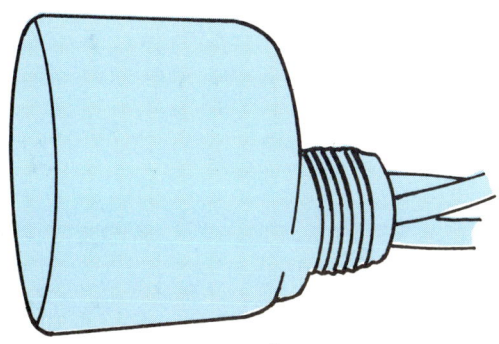

Fig. 15-9. Lightning arrestor.

POULTRY HOUSE

The poultry house must be well electrified. While the electrical demand for this building will be much less than that of the barn, wiring must be adequate to serve the functions of the building:
1. Egg laying.
2. Brooder or heat lamps.
3. Feed conveyors.
4. Water warmers.
5. Egg cleaners, etc.

Lighting should be plentiful. One light for every 150 sq. ft. of floor area is suitable. Wall-switched lights are advisable. Timer switches to increase egg production should be a part of the lighting circuits.

Receptacles should be located every 12 ft. to 15 ft. along the walls. These should be mounted no lower than 5 ft. above the floor.

HOG BARN, SHEEP HOUSE, HORSE BARN

To adequately provide electrical power for these buildings, follow the general provision outlined for the poultry house.

OTHER OUTBUILDINGS

These areas should have more than the minimum in lighting and receptacle provisions. Lights should be spaced 10 ft. or less apart and outlets every 8 to 10 ft.

The foregoing is merely a general plan for lighting and outlets. These buildings will each require additional circuits for specific loads. Thus, each building's electrical need must be computed carefully to help determine the total farm load.

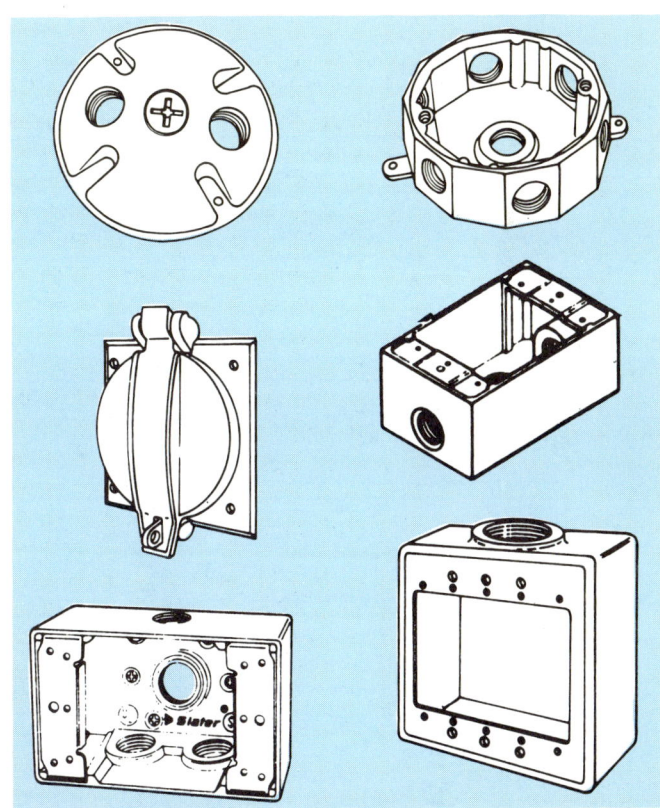

Fig. 15-11. Weatherproof surface-mount outlet boxes and covers. (Slater)

169

MOTOR HORSEPOWER	POWER RATING (IN WATTS)
1/4	250 W
1/2	500 W
1	1200 W
2	2000 W
3	3000 W
4	4500 W

Fig. 15-12. Power ratings of electric motors of different sizes (horsepower). These must be known to figure motor's load demand on an electrical circuit.

COMPUTING FARM POWER REQUIREMENTS

The farmhouse power requirements are figured the same as any residence. The other buildings—barns, sheds, and shops—are computed differently. The farm buildings, having more motors, heating devices, and lighting circuits, will generally, require as much, if not more, power than the farmhouse.

The procedure, briefly, for computing such demands begins with a list of the buildings. Include the farmhouse since it will also be connected to the yard pole service drop. The list might appear something like this:
1. Farmhouse.
2. Barn.
3. Equipment/tool shed.
4. Poultry house.
5. Hog house.

BARN ELECTRICAL REQUIREMENTS

Now, start with the barn. It probably will have the greatest electrical demand. List the number of lighting circuits, special appliances, motors, etc., that it will need. For example:

15 lighting circuits 120 V/15 A or 20 A.
6 motors @ 1/2, 1, 2, 4, 4, 4, hp., etc.
1 water heater.
1 sterilizer.

Next, determine volt-amperage requirements. A good rule of thumb for general lighting circuits

FARM BUILDING LOADS SUMMARY (Excluding Farmhouse)		
FARM BUILDING	SPECIAL EQUIPMENT — OTHER THAN LIGHTING AND RECEPTACLES	AVERAGE CONNECTED LOAD
Main barn (dairy barn) including milkhouse milking parlor, silos, haymows, feeding pens, stanchion area and calf pens	Gutter cleaners, milking machine pumps compressor motor, refrigeration, water heater(s), silo unloaders, feed conveyors, material movers, heat lamps, calf brooders, circulating fans, water pumps	79 kVA +
Poultry house including egg packing and feed storage areas	Feed conveyor, candling equipment, chick brooders, egg washers, water heater, fans	18 kVA
Hog barn	Hog brooders, water heater, fan, feed conveyor	10 kVA
Horse barn	Water heater, fan	8 kVA
Sheep barn	Feed conveyor, water heater, fan	8 kVA
Machinery shed, includes storage and repair areas	Compressor, heater units, welding equipment, lathe, hoists, lifts, water heater, gas pump, water pump	20 kVA
Garage Tool shed may be separate or as one	Compressor, heating, welder, table saws, drill stands, battery charger, water heater	12 kVA

Fig. 15-13. Chart like this is helpful in figuring electrical loads for the farm.

is about 400 VA each. For motors, see Chapter 22. For heaters, sterilizers, etc., use nameplate ratings stamped on the equipment housing. Also, consult Chapter 13 on appliances and various tables under Article 430 of the *Code.* See Fig. 15-12.

Add up the total power needed by the building:

lighting circuits	=	6000 VA
*motors, total × .50	=	8600 VA
heater	=	5000 VA
sterilizer	=	7000 VA
total	=	26,600 VA
		or 26.6 kVA.

The same procedure is followed for each of the other farm buildings. (The house is not included. It must be figured separately using requirements covered in Chapter 10.) Add up totals. The figures may look something like this:

farmhouse	=	25.6 kVA
barn	=	26.6 kVA
equip./tool shed	=	7.5 kVa
poultry house	=	6.3 kVA
hog house	=	3.0 kVA
total farm power needed	=	69 kVA

This total, 69 kVA, is called the CONNECTED LOAD. It represents the power that would be needed if everything was operating at one time.

A more conservative figure called the DEMAND LOAD, represents the amount of power which would most probably be needed at any given time. Generally, the minimum demand load is considered to be about 35 percent of the connected load. Based on this figure, the wire sizes and over-current protection at the yard pole are determined. In our example, 69 kVA × .35 = 24.15 kVA (24,150 VA). At 240 V, we arrive at 100 A as our current needed for the farm. Consider also, the

*The 50 percent factor may be used when not all motors will be operating at any given time.

future growth and increased power conditions. Thus, sizes of service conductors should be *AT LEAST* one size larger. In this case, No. 4 AWG would be minimum, but a wiser choice would be No. 2 or No. 1. Without a doubt, this would be a very small farm.

The wire sizes for the feeders interconnecting the various buildings to the yard pole should be determined by using this same methods.

Although there are no magic amperage ratings for specific farm buildings, Fig. 15-13 may serve as a general guide for average farm structures equipped with typical devices.

Code requirements for agricultural buildings are detailed in Article 547 of the *Code* book. You should be familiar with these requirements before attempting farm wiring.

REVIEW QUESTIONS — CHAPTER 15

1. Generally, the farm service drop ends at a centrally located distribution point called the _____.
2. List three common farm buildings, other than the farmhouse, which require electrical service equipment.
3. Most up-to-date farms require a _____ ampere service as a minimum.
4. Indicate which of the following wiring systems is recommended for farm buildings:
   a. Metal conduit.
   b. Nonmetallic conduit (PVC).
   c. BX or armored cable.
   d. Flexible metal conduit.
   e. Any of the above.
5. As a rule of thumb, the general lighting power factor is about _____ watts/circuit.

Chapter 16

MOBILE HOME WIRING

The *National Electrical Code* provides special wiring requirements to meet the needs of mobile homes and mobile home parks. The *Code* takes into account the nature of the homes which are easy to move and have no permanent foundation. While the service entrance provided is similar, in some respects, to a permanent residential installation, there are important differences.

NEC requirements (Article 550) apply to all mobile homes. It does not matter whether the unit is located in a mobile home park or on a private lot. See Fig. 16-1.

DEFINITIONS

The *Code* also defines certain terms which apply specifically to the installation of electrical service for mobile homes. These terms are used in the chapter and should be thoroughly understood. They include:

1. Distribution panelboard. This is an assembly which includes bus bars, automatic overcurrent devices, and, sometimes, switches. The devices are enclosed in a cabinet or cutout box that is attached to the interior wall of the mobile home, Fig. 16-2.
2. Feeder assembly. These are the conductors with their fittings and equipment, which carry electrical current from the mobile home service equipment to the distribution panelboard, Fig. 16-3. One of the conductors in the assembly must be a green colored, insulated, grounding conductor. This assembly may feed from overhead, from under the chassis, or it may be

A

B

C

D

Fig. 16-1. Typical mobile home electrical installation. A—Pad transformer. B—Pedestal for 100-ampere service. C—Main disconnect in pedestal. D—Under chassis feeder assembly is protected by conduit.

DISTRIBUTION PANELBOARD

The distribution panelboard corresponds to the service electrical panel in a permanent structure. It must be securely fastened to a structural member of the home either directly or by using a substan-

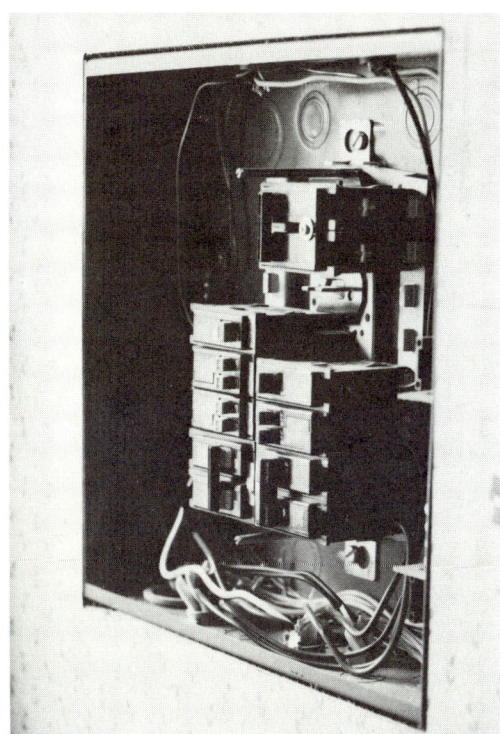

Fig. 16-2. NE Code requires that distribution panelboard be securely mounted inside the mobile home.

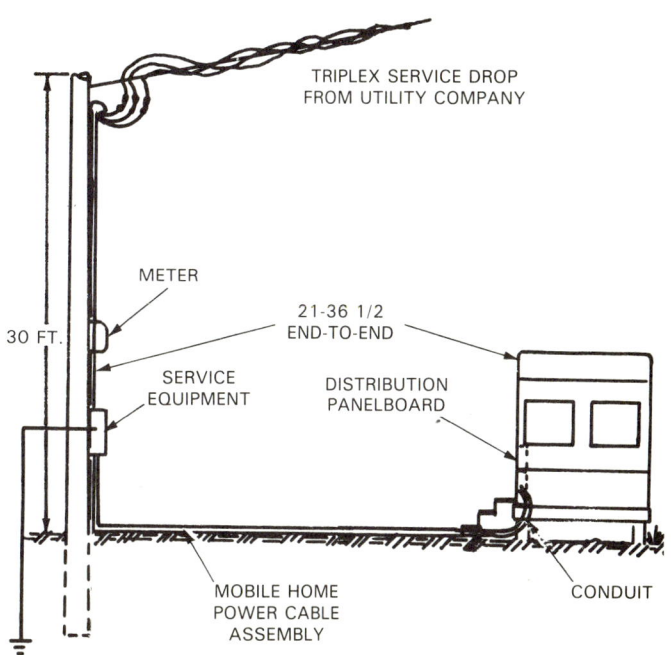

a power supply cord "approved for mobile home use."

3. Mobile home service equipment. All of the equipment, including a main disconnecting means, overcurrent protective devices, and receptacles or other arrangements for connecting the supply end of a mobile home feeder assembly. The meter may be regarded as part of the service equipment.

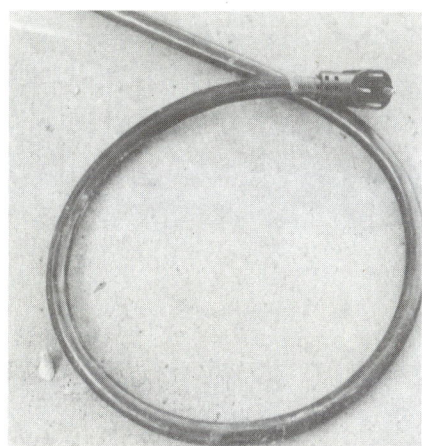

Fig. 16-3. A single power cord with a molded plug meets NE Code requirements for a mobile home when the electrical load does not exceed 50 amperes. There must be four conductors, all insulated.

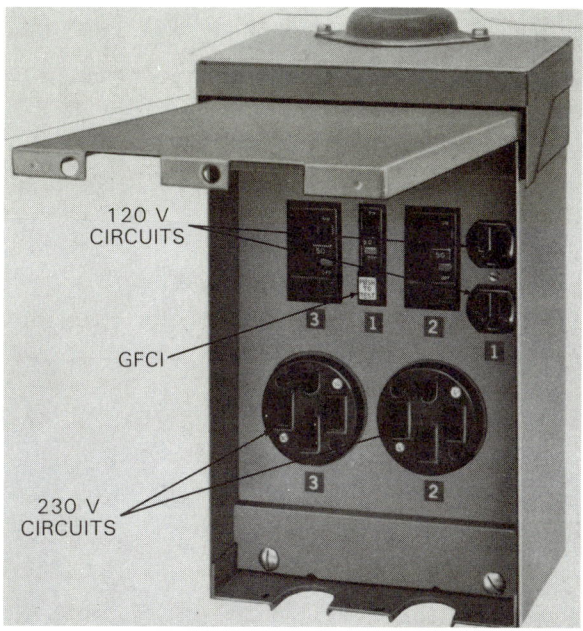

Fig. 16-4. One common hookup for mobile homes uses the power cord. One end of the cord is permanently wired into the distribution panelboard in the mobile unit. Other end will plug into 125/250 V grounding type receptacle. A—Sketch of complete hookup. B—A power panel designed to supply two mobile homes with 50 A circuit to each. A 20 A circuit is also available for other purposes. Circuits are protected by a GFCI.

173

tial brace. It receives its power supply through the feeder assembly. The electrician will rarely be called upon to install a mobile home distribution panelboard since it is almost always supplied by the manufacturer of the mobile home.

FEEDER ASSEMBLY

The feeder assembly must have four conductors. Three must be insulated circuit conductors (two ungrounded and one grounded). The fourth is the insulated green grounding conductor. Depending primarily upon the calculated electrical load of the mobile home and any additional requests of the owner, the electrical feed can be supplied by any of three different ways:

1. A mobile home power supply cord. This is permitted if the load is 50 ampere or less. However, if the mobile home has factory-equipped gas or oil heating and cooking equipment, a 40 ampere cord is permitted. Fig. 16-4 shows a typical installation.

   The cord, which must always be approved and labeled for mobile home use, must be no shorter than 21 ft. and no longer than 36 1/2 ft. from end to end. The plug must be of the molded type. The opposite end of the cord must be permanently attached to the distribution panelboard in the mobile home. A clamp should be provided at the panelboard knockout to prevent strain on the cord from damaging the cord or its connections to the terminals, Fig. 16-5. Only one cord per mobile home unit is permitted.

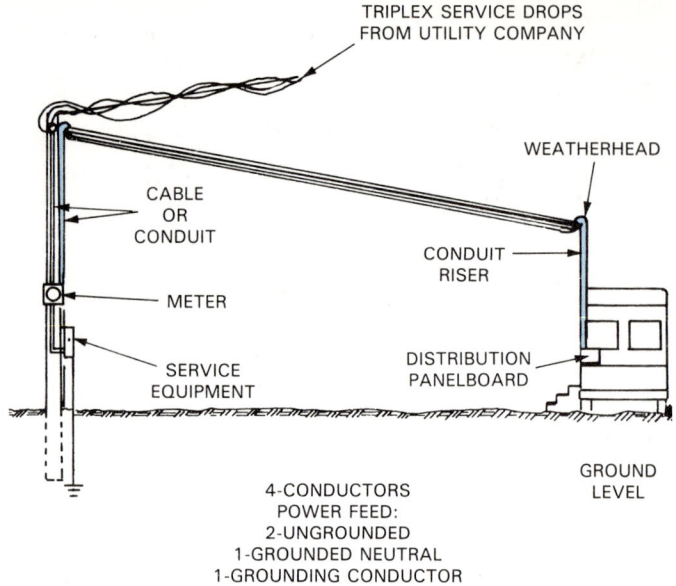

Fig. 16-6. Overhead feeder hookup. Mast and weatherhead can be attached to the mobile home, but main disconnect, meter and breakers are to be located on pole no more than 30 ft. away.

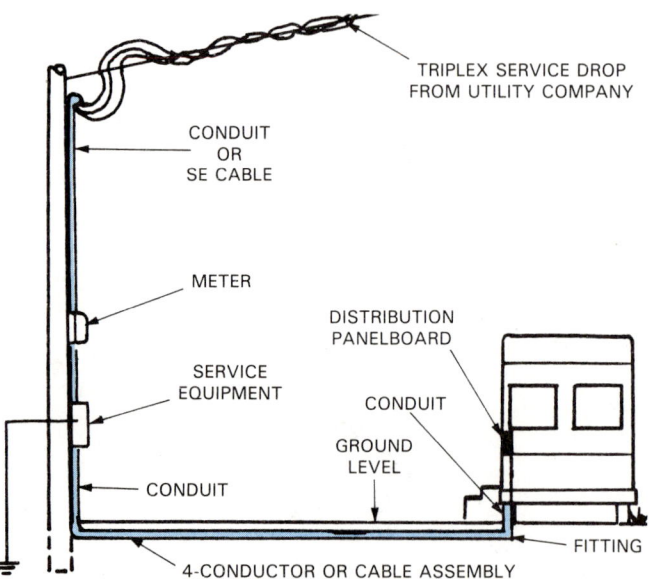

Fig. 16-7. Underground feeder hookup for mobile home is also permitted for permanent installation. This illustration shows cable being used; however, metallic or nonmetallic (PVC) conduit protection can be used as well.

2. An overhead installation consisting of a mast and a weatherhead. This type is for permanent installations or where the load exceeds 50 ampere, Fig. 16-6.
3. An underground lateral service using cable or conduit. This is an alternative to an overhead permanent installation. It is probably the most common method used. See Figs. 16-7 and 16-8.

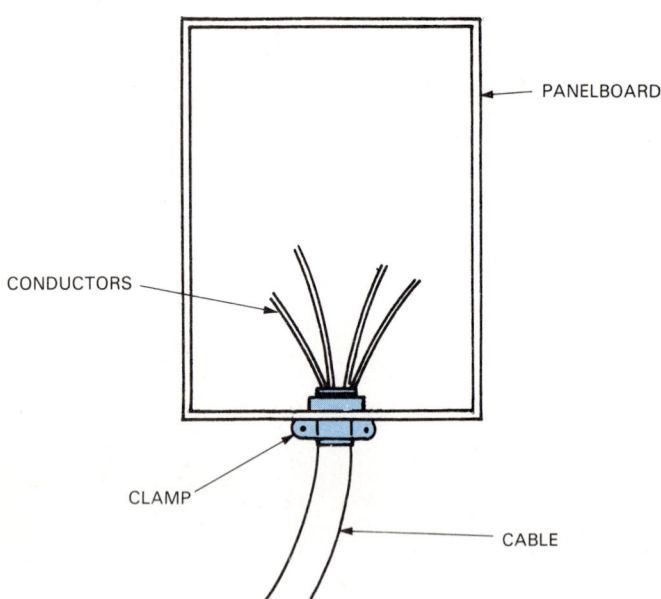

Fig. 16-5. When a feeder cord is used to supply electricity to mobile home a clamp must be provided at the panelboard knockout to prevent strain on the cord connections inside the cabinet.

MOBILE HOME SERVICE EQUIPMENT

Service equipment must not be mounted on the mobile home, itself. The *Code* requires that it be within sight and located a specific distance from the unit—usually 20 but not more than 30 ft. away. It must be supported by a pole or a pedestal arrangement. Sometimes the main disconnect, breaker, and meter are mounted together. In other cases, the meter is mounted on a pole and the overcurrent protective devices are on a pedestal.

OVERHEAD INSTALLATION

The mast and weatherhead installation must have four continuous conductors which are color coded. One of the conductors must be an equipment ground. Refer to Article 320 of the *Code* for the kinds and sizes of conductors to be used. See also Article 550-3L(1).

SERVICE LATERAL

For a lateral installation, the mobile home must be fitted with a metal raceway, such as a conduit, which connects to the distribution panel inside the unit and extends to the underside of the chassis. The raceway must have provision for attaching a junction box or fitting.

The manufacturer may or may not have provided the raceway with conductors. Refer to NEC, Section 550-3L(1 & 2).

The electric utility company usually installs the necessary service drop and transformer. The customer must supply the service pole and components as well as the meter pedestal (for underground supply). Both pole and pedestal are permanent installations. The pedestal must be rigidly mounted so that it cannot be easily knocked over. Fig. 16-9 is a sketch showing a pedestal arrangement.

DEVICES AND OUTLETS

The various electrical devices and equipment are almost always installed by the manufacturer and must meet minimum NEC requirements. These are then checked by the electrician and the electrical inspector for proper polarity and grounding.

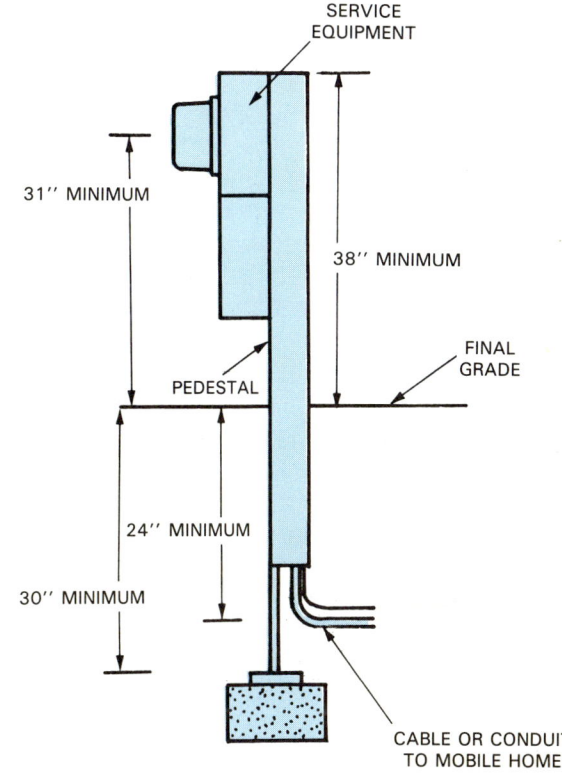

Fig. 16-9. A pedestal arrangement for supporting the service equipment.

DISTRIBUTION PANEL HOOKUP

The distribution panel, along with a main breaker for branch circuit disconnect purposes, is usually installed by the manufacturer. To connect the power-supply feeder assembly to this panel, carefully study Fig. 16-10.

Note, particularly, that the neutral conductor is completely *insulated and isolated* from the grounding conductor as well as the equipment enclosure. In addition, the mobile home chassis must be securely bonded to the grounding bus or grounding conductor, as must all metal, non-current-carrying equipment inside the mobile home. This grounding continuity is essential and will be checked by the inspector.

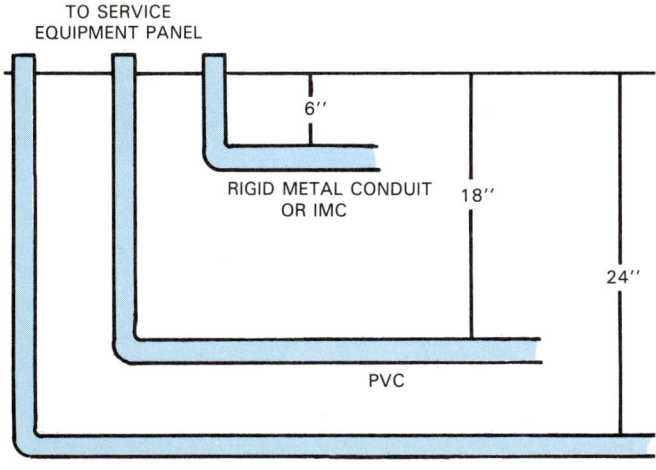

Fig. 16-8. Depths are prescribed for direct buried cable and conduit.

CALCULATING PANELBOARD LOAD

The distribution panel load of the mobile home is computed somewhat like a standard residential load, with some minor differences.

Lighting and small appliances requirements are determined similarly to other one or two-family residences. The 3 volt-ampere per square foot rule is used plus the two small appliance and one laundry circuit consideration. To this, is added:
1. Nameplate ratings for motors.
2. Loads for heater, garbage disposal, dishwasher, water heater, clothes dryer, wall-mounted oven, cooking units and the like.

Free-standing ranges, unlike wall-mounted ovens or cooking tops, are rated at 80 percent of their volt-amperage up to 10,000 VA. Beyond that, figure 8000 VA for those loads between 10,000 and 12,500. Add 400 VA for every 1000 VA increase above 12,500. The smaller of the heating and cooling loads can be omitted from the load circulation.

For example, find the total load for a 65 by 12 ft. mobile home. It has two small appliance circuits, a laundry circuit, a 2500 VA space heater, a 100 VA fan, a 500 VA dishwasher, an 800 VA air-conditioner, 8000 VA range, and a 5 kVA water heater.

Find the general lighting and small appliance load:

Area = 65 × 12 = 780 × 3 VA/sq. ft.
            = 2340 VA
2 small appliance circuits at 1500 VA
            = 3000 VA
1 laundry circuit   = 1500 VA
                          6840 VA
First 3000 VA at 100%   = 3000 VA
Remainder (3840), × 35%
                        = 1344 VA
Net lighting and small appliance = 4344
            ÷ 230 V
            = 18.9 A
2500 VA space heater ÷ 230 = 10.9 A
100 VA fan ÷ 115   = .9 A
500 VA dishwasher ÷ 115 = 4.3 A
Omit air conditioner (only heating or air conditioning is computed, whichever is larger)
8000 VA range ÷ 230 × 80% = 27.8 A
5000 VA water heater ÷ 230 × 100%
            = 21.7 A

Note: We divide by 230 V rather than 240 V. This provides a higher ampacity figure and a greater buffer for future loading.

Total load is about 84 A. Therefore, a 100 A service is needed. This is considered the minimum.

MOBILE HOME PARKS

The NEC, in Article 550, part B, outlines the various requirements for mobile home parks. The general demand factors to be considered are shown in Fig. 16-11.

The application of these demand factors may best be illustrated through the use of an example: If a particular mobile home park is to accommodate 25 mobile homes, we would use the demand factor of

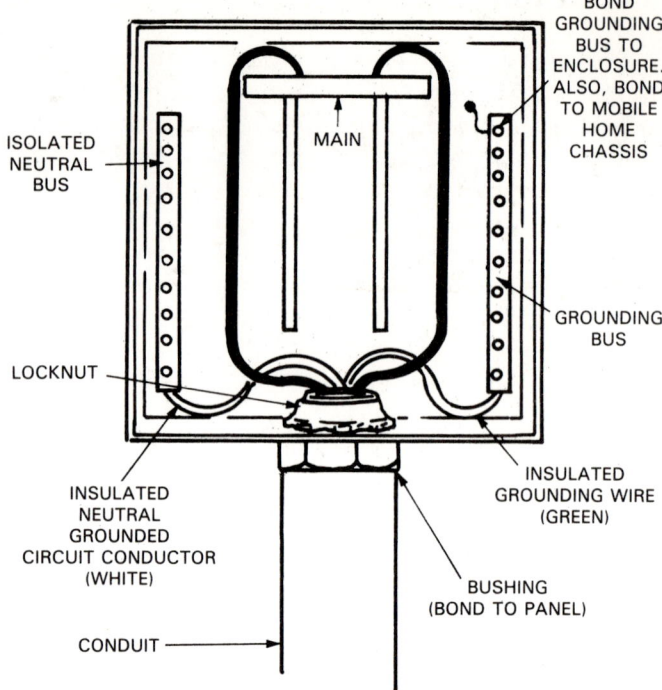

Fig. 16-10. Proper connections at the mobile home's distribution panelboard are crucial to the equipment grounding integrity.

NEC TABLE 550-22
DEMAND FACTORS AND WATTS PER MOBILE HOME SITE (MINIMUM) FOR FEEDERS AND SERVICE ENTRANCE CONDUCTORS

Number of Mobile Homes	Demand Factor (Percent)	Volt-Amperes/Mobile Home Site (Min.)
1	100	16,000
2	55	8800
3	44	7040
4	39	6240
5	33	5280
6	29	4640
7-9	28	4480
10-12	27	4320
13-15	26	4160
16-21	25	4000
22-40	24	3840
41-60	23	3680
61 and over	22	3520

Fig. 16-11. Size of service needed for a mobile home park can be based on this chart. (National Fire Protection Assoc.)

24 percent or 3840 VA per mobile home for a total of 417 A for the park's main service entrance.

$$\frac{16{,}000 \text{ VA} \times 25 \times 0.24}{230 \text{ V}} = 417 \text{ A}$$

or, more simply, $\dfrac{3840 \text{ VA} \times 25}{230 \text{ V}} = 417 \text{ A}$

In addition, common park facilities and equipment such as security lighting must be considered in the overall total service demand.

REVIEW QUESTIONS — CHAPTER 16

1. List two factors which distinguish mobile homes from ordinary one or two-family residences.
2. For mobile homes, the service equipment (main disconnect and meter) must be placed:
   a. On the mobile home.
   b. In the mobile home.
   c. On a pole or pedestal separated from the mobile home.
   d. At any of the places indicated above.
3. Mobile home power cords must be no less than _____ and no more than _____ ft. long.
4. The grounded neutral conductor must be _____ and _____ from the grounding conductor within the mobile home distribution panel.
5. The lighting load of a mobile home is based upon the _____ volt-amperes/sq. ft. rule.
6. The minimum service rating of a 20-unit mobile home park would be _____ A.

Typical electrical service serving six mobile homes in a mobile home park. Meters and service equipment are securely mounted on a power pole.

Modern Residential Wiring

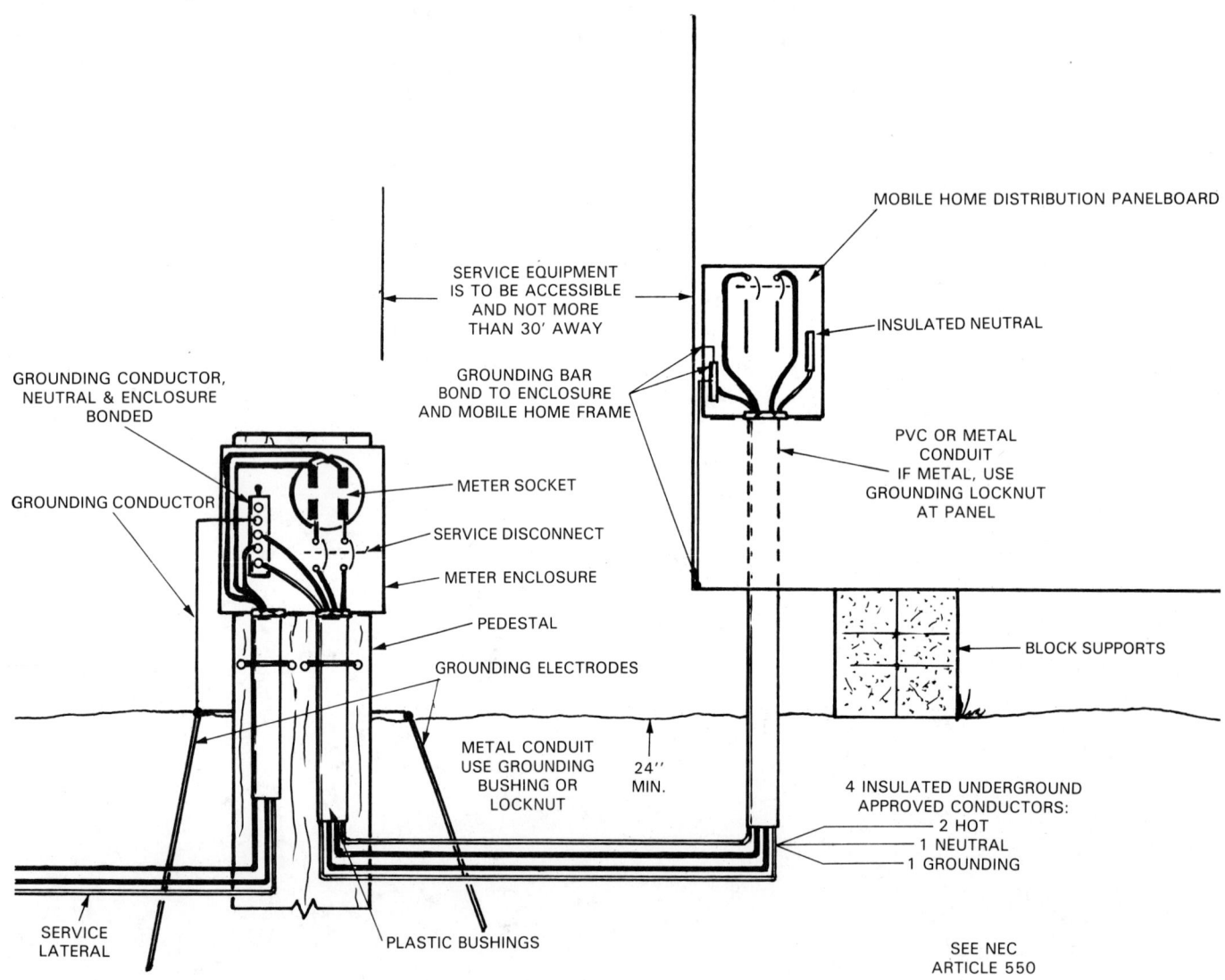

Typical mobile home hookup shows proper *Code* hookup both at meter socket and mobile home distribution panelboard.

Chapter 17

LOW-VOLTAGE CIRCUITS

Low-voltage circuits are those which operate at voltages much less than 120 V. They are found in residences chiefly as a method of controlling devices that will then operate on higher voltage. When low-voltage is used for this purpose it is known as remote control wiring. It is an important use of low-voltage that is growing rapidly because it is flexible, convenient, and easy to install. In this chapter, you will explore the meanings of remote control and study how the components operate. Moveover, you will learn how to connect remote-control switching systems.

BASIC REMOTE CONTROL OPERATION

In a remote control wiring system, pressing of a switch at one point can operate a switch at a second distant point. The motion or force needed to activate the second switch is provided by a relay. A relay is nothing more than a small electromagnet. It is wired into a point along the circuit. As long as there is no current in the circuit, the relay has no magnetic properties. However, if a current is introduced in the circuit, the relay becomes a small magnet. The current must flow only enough to let the relay operate a switch. See Fig. 17-1.

Since the second switch will stay on once it is activated, the current flow through the remote, low-voltage circuit is needed only for a brief moment. A normally-off, rocker switch is all that is needed to energize the remote control circuit.

When the rocker switch is pressed, low-voltage current flows through the coil surrounding the iron core of the relay. The core becomes magnetized and attracts the metal in the switch the relay is controlling. The rocker switch moves to the on position and current begins to flow through that circuit.

The first part of the remote circuit is only capable of turning the switch on. A second part of the circuit must be used to turn the switch off again. Fig. 17-2 shows a simplified arrangement of a complete remote control circuit capable of on-off switching operations.

The foregoing is a simplified example of a relay in action. Actual design of the relay is much more

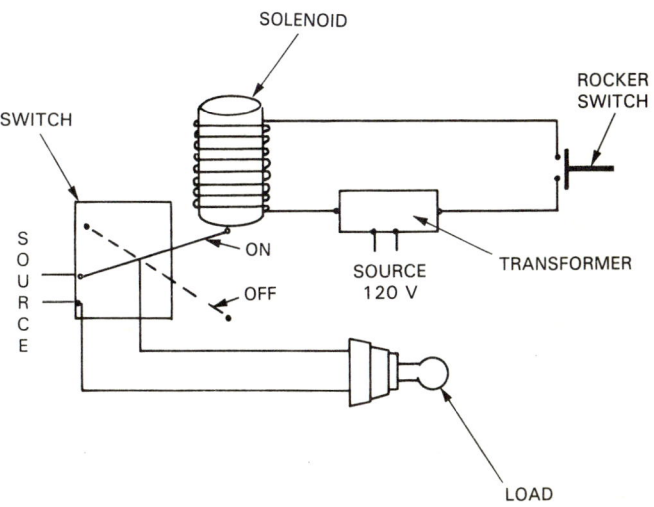

Fig. 17-1. The job of a remote control, low-voltage circuit is simply to open or close a switch to start current flow in a higher voltage circuit. Only the part of the circuit which turns the switch on is shown.

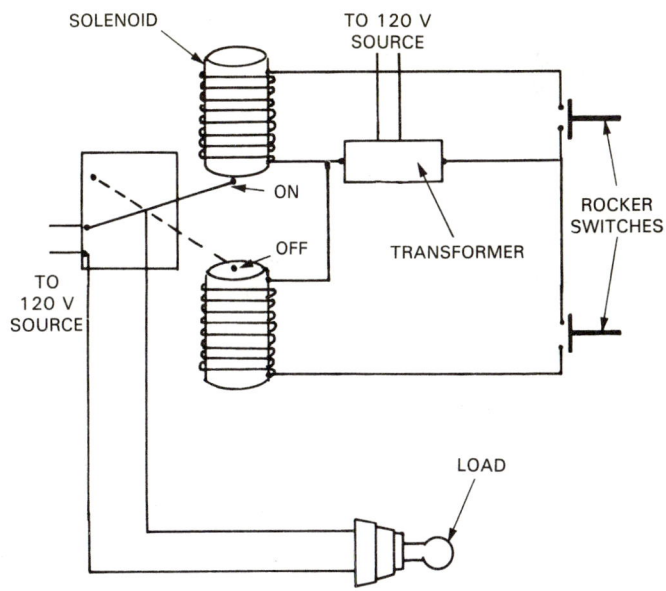

Fig. 17-2. A complete low-voltage circuit which will turn a switch on or off to control a 120 V circuit which has a light in it. In the actual design, the same housing encloses both the relay mechanism and the 120 V switch. Refer to Fig. 17-5.

sophisticated. The principle is the same, however. An electromagnet is energized when a pushbutton is pressed. It attracts a switch which is activated causing current to flow or to stop flowing.

A relay has two basic parts. One is the electromagnet which is operated by low voltage current. The other part is the switch which controls 120 V current.

Low-voltage circuits can be used to perform various functions around a residence. Their design will vary somewhat depending on what they are required to do.

TYPES OF LOW-VOLTAGE CIRCUITS

All the circuits covered in this chapter are alike in that they operate at voltages less than conventional or standard power circuits. Included are:
REMOTE CONTROL CIRCUITS which control other circuits or parts of circuits through devices such as relays. NEC classifies these as Class 1 circuits.
SIGNALING CIRCUITS which energize devices specifically designed to give visual or audible signals. They are classified by NEC as Class 1 circuits.
LOW-VOLTAGE CIRCUITS which usually operate at 30 V or less. Low-voltage switching, controlled by relay devices, is considered the same as remote control.
POWER-LIMITED CIRCUITS which have current and voltage limitation in accordance with the NEC. Table 725-31, of the *Code* outlines the power regulations. These circuits are also classified as Class 2 and Class 3.

APPLICATION

These circuits are used for many special purposes, including the following:
1. Remote control (relay switching) and low-voltage.
    a. Motor-load circuits.
    b. Lighting systems.
    c. Heating systems.
2. Signaling.
    a. Chimes.
    b. Buzzers.
    c. Bells.
    d. Intercom systems.
    e. Burglar alarms.
    f. Smoke detectors.
    g. Fire alarms.
3. Power-Limited circuits (Classes 2 and 3).
    a. Circuits operating at 30 to 150 V.
    b. Intercom systems.
    c. Electric door opener circuits.
    d. Battery-operated annunciator (bell or buzzer) systems.

LOW-VOLTAGE WIRING ADVANTAGES

Remote control, low-voltage switching systems have certain advantages:
1. Ease of installation.
2. Less possibility of shock hazard.
3. More flexibility in control of lighting.
4. Lower installation costs.
5. Lower maintenance costs.
6. Low operating costs.

LOW-VOLTAGE CABLE AND DEVICES

A low-voltage switching system operates at 24 volts of alternating current. There is little or no shock hazard from operating these low-voltage devices. In addition, the circuit is made up of less expensive 18 AWG two, three, and four wire cable. This size can easily handle the load, Fig. 17-3. Cable with 19 or 26 conductors is also available. Such cable is used for making runs to master control switches or motor master control units. Some cable is designed for indoor use, other for outdoor use in any kind of climate or weather.

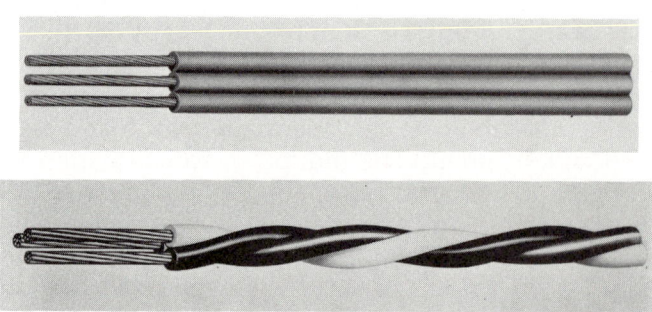

Fig. 17-3. Two styles of No. 18 low-voltage cable. A—Flat three-conductor cable. B—Standard three-conductor cable. (General Electric Co.)

TRANSFORMERS

The heart of the remote control switching system is the step-down transformer. One transformer usually supplies voltage for the entire system. The transformer reduces the 120 V supply to a nominal 24 V on the remote control side. A transformer such as the one shown in Fig. 17-4 may supply voltage to 30 or more relays at the same time.

RELAYS

The relay, shown in Fig. 17-5, is essentially a solenoid. It controls the 120 V lighting circuit by opening and closing points or contacts. See circuit diagramed in Fig. 17-6.

Fig. 17-4. The transformer is the heart of the remote control switching system. It converts 120 V voltage to 24 volts. (Acme Electric Corp.)

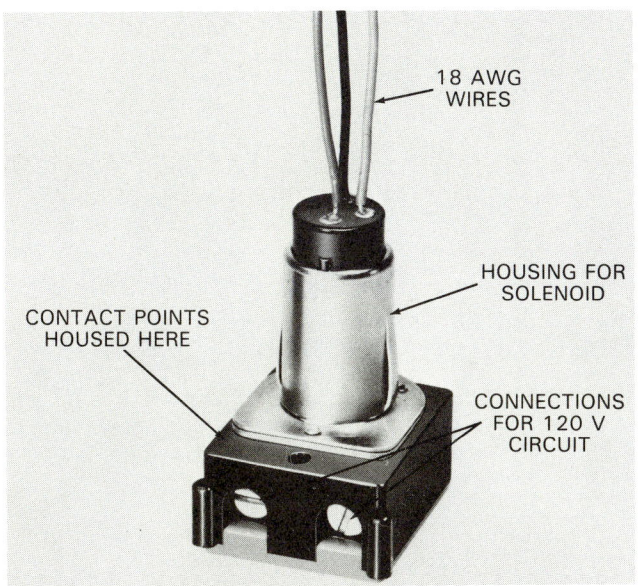

Fig. 17-5. The relay device combines the solenoid of the 24 V system and the switch of the 120 V system. (General Electric Co.)

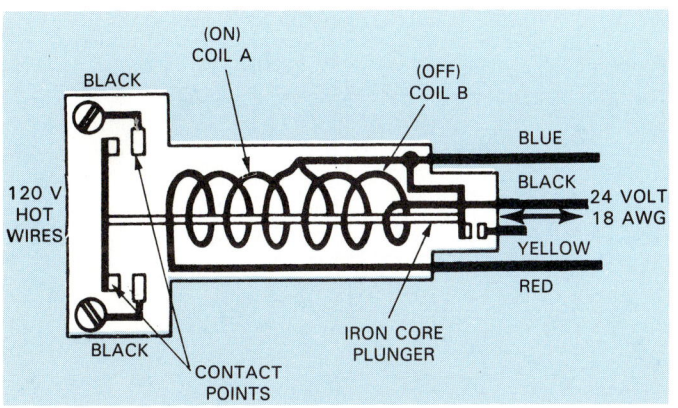

Fig. 17-6. The inside of a relay. It is actually a double-acting electromagnetic coil. The iron core can move in either direction to open or close the load side of the switching circuit. Motion in one direction closes the contact points. Opposite movement opens them.

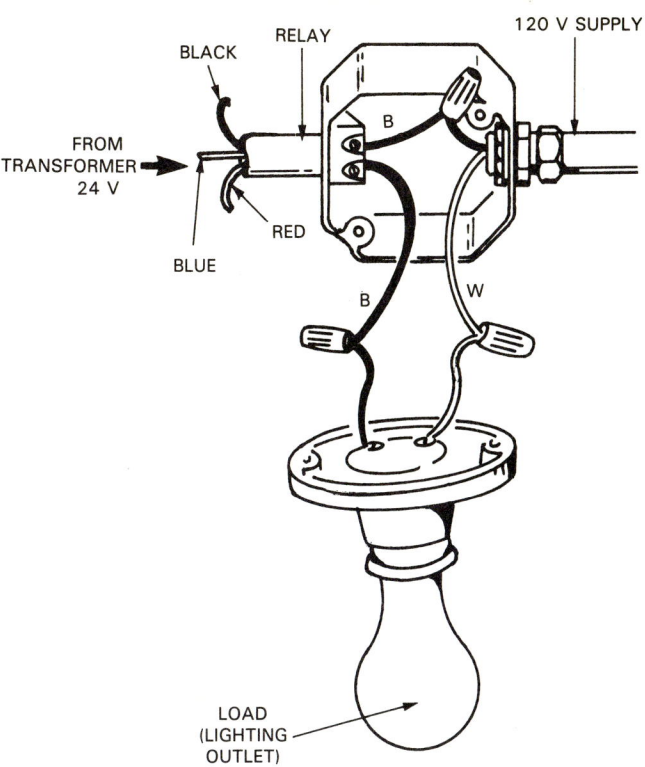

Fig. 17-7. The relay installed in an outlet box. Note the proper conductor connections inside the box. Red and black wires run between the relay and the low-voltage switch.

The relay is usually installed in the fixture outlet box. The power operating the relay comes from the 24 V side of the transformer. The black leads are terminated in the fixture outlet box. The installation and connection of conductors is shown in Fig. 17-7. Remember, low-voltage leads are OUTSIDE the box, the 120 V leads, inside. Note the proper connection of leads to the load.

LOW-VOLTAGE SWITCH

The low-voltage switches used in a remote control system are single-pole, double-throw types.

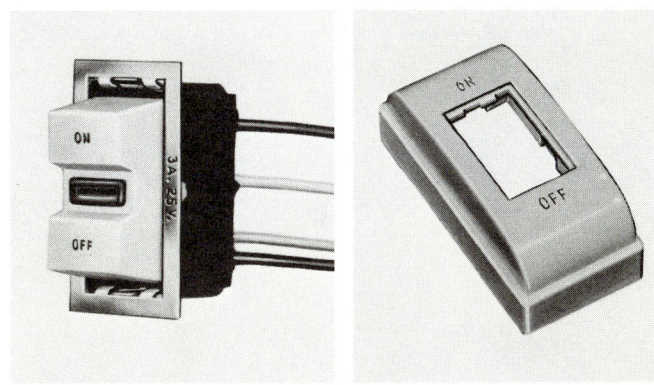

Fig. 17-8. Left. Rocker type low-voltage switch. It activates the relay in the low-voltage circuit. Scores of these may be used in as many locations to control a light. Right. Typical low-voltage switch plate. (General Electric Co.)

Modern Residential Wiring

They are about half the size of those normally used in standard switching. They require switch plates of special design. Because of their small size and low voltage, low-voltage switches can be mounted either in standard switch boxes or simply on raised plaster rings. A typical low-voltage switch and switch plate are shown in Fig. 17-8. Fig. 17-9 is a simplified sketch of how the L-V switch works.

Using low-voltage switches, a light may be controlled from numerous locations without the three-way and four-way types needed in standard wiring systems.

MASTER SWITCH

A particularly convenient and versatile method of employing low-voltage switching, is through the use of a master selector switch. See Fig. 17-10.

The master selector switch can activate or deactivate all or any of the various low-voltage switches at remote locations. By glancing at the pilot light indicators you can tell which lights are one and which are off. This is of immense benefit in large homes and for outdoor security lighting.

Master selector switches are available with motorized timing mechanisms. The timer can activate or deactivate lights at varying intervals.

Wiring of a master selector switch is a relatively simple matter, as illustrated by the schematic diagrams of Fig. 17-11 and 17-12. For simplicity, only circuit Nos. 1, 2, and 3 are shown.

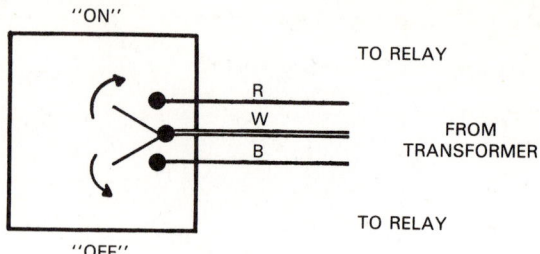

Fig. 17-9. Simple diagram of how a rocker type low-voltage switch is arranged to make contact with "on" and "off" terminals.

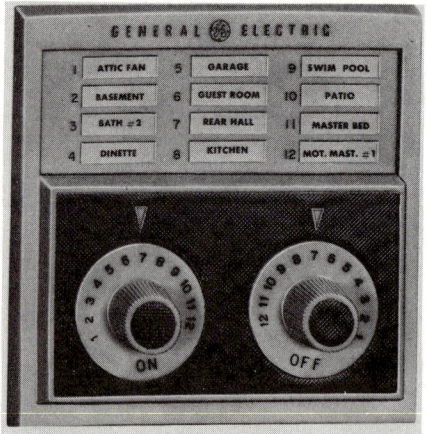

Fig. 17-10. A master selector switch houses a number of remote switches so any or all switches can be activated or deactivated from one location. (General Electric Co.)

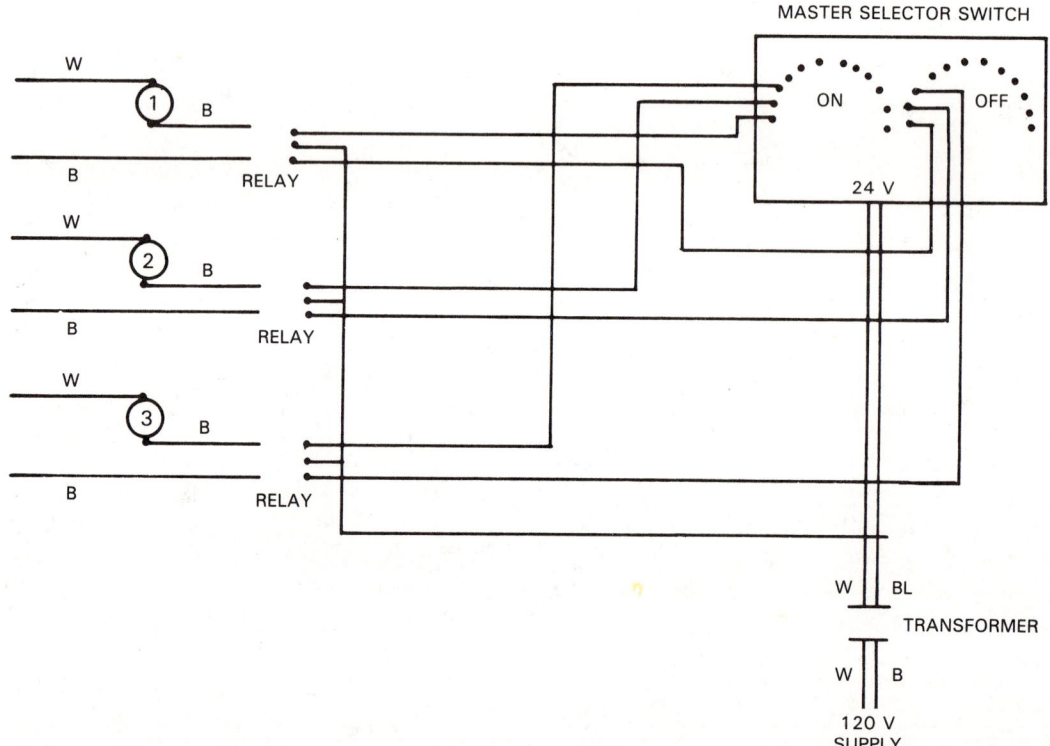

Fig. 17-11. Schematic diagram of the master selector switch. It may be manually operated or motorized.

182

Low-Voltage Circuits

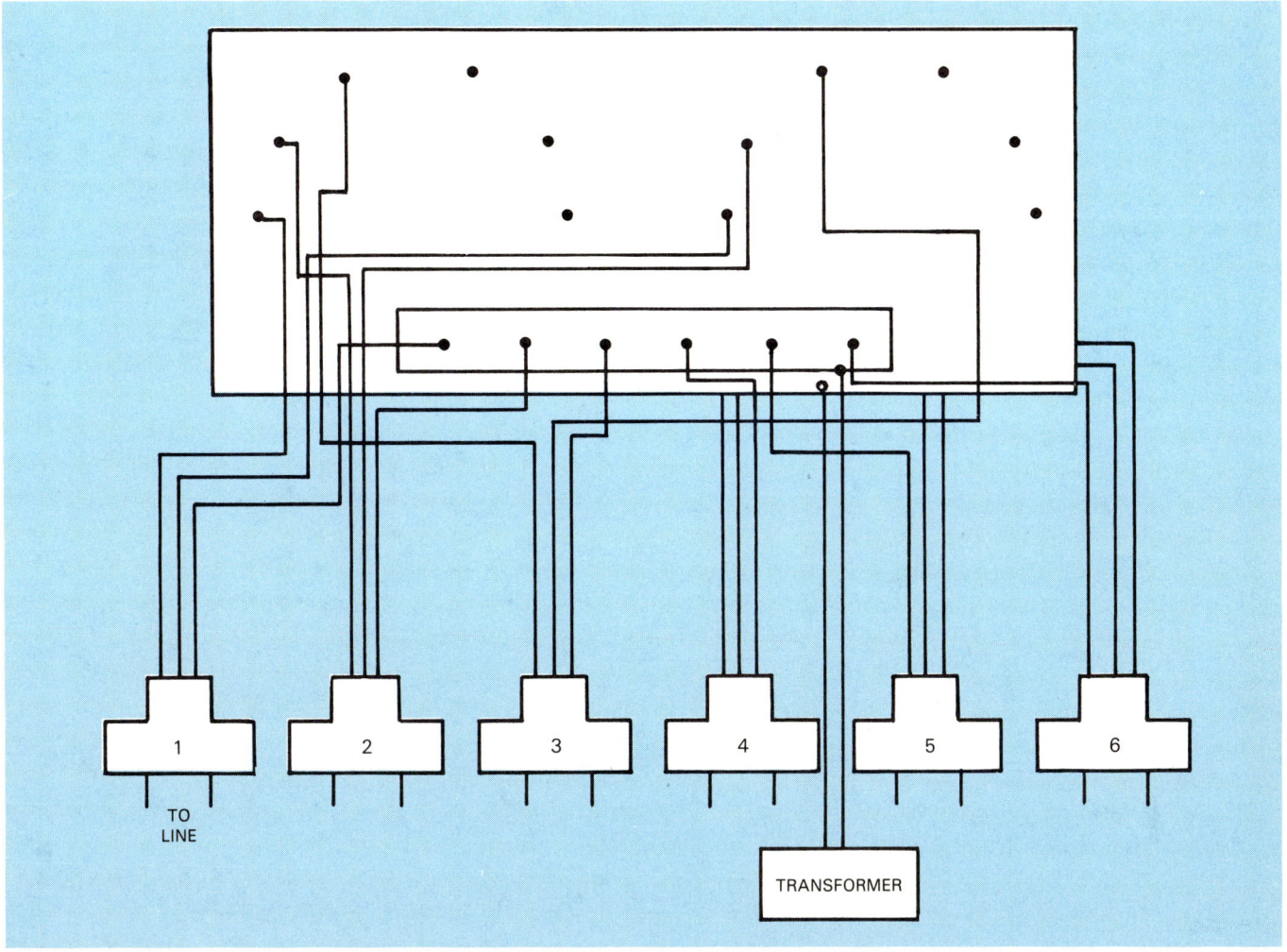

Fig. 17-12. Routing low-voltage cable between relays and master selector switch.

INSTALLING REMOTE CONTROL SYSTEMS

You can think of a low-voltage circuit as a second circuit operating with 120 V circuits. It supplies switching capabilities to 120 V or 240 V circuits. This section of Chapter 17 will show the steps for installing devices and cable for a 24 V system.

1. Install the 24 V step-down transformer. It can be in a central location for the circuits it will supply. Less wire will be needed. The transformer reduces supply to 24 V.
    a. Attach the transformer to a junction box that is supplied with 120 V line power.
    b. Connect line side leads to neutral and hot wires on the transformer. *Be sure that power feed to the circuit is deenergized.*
2. Install a low-voltage relay at each location where remote control is desired.
    a. Remove a knockout in the outlet box where a light is to be controlled.
    b. Insert the neck of the relay through the knockout. High side (120 V) of the relay *must* be inside the outlet box.
    c. Inside the box, connect white wire from source to load. Then connect one black wire from relay to black wire from source. Connect second black wire from relay to black wire to load.
    d. Terminals on the part of the relay outside of the box are for connecting low-voltage cable to switches and transformers.
3. Install low-voltage cable. This cable is very light and has little insulation since the voltage is so low. The cable can be stapled to surfaces or carried between the walls with no additional protection. Drill holes through studs and joists and run cables through them. If strung out-of-doors between buildings support the cable on a stronger cable called a "messenger." This will reduce stress on the lighter low-voltage cable.
    a. Run cable from the transformer low-voltage side to the relay (blue colored conductor).
    b. Run more cable from the relay to each switch position (red and black conductors).
    c. For hookup of a master switch refer to the schematic in Fig. 17-11 and the layout in

Modern Residential Wiring

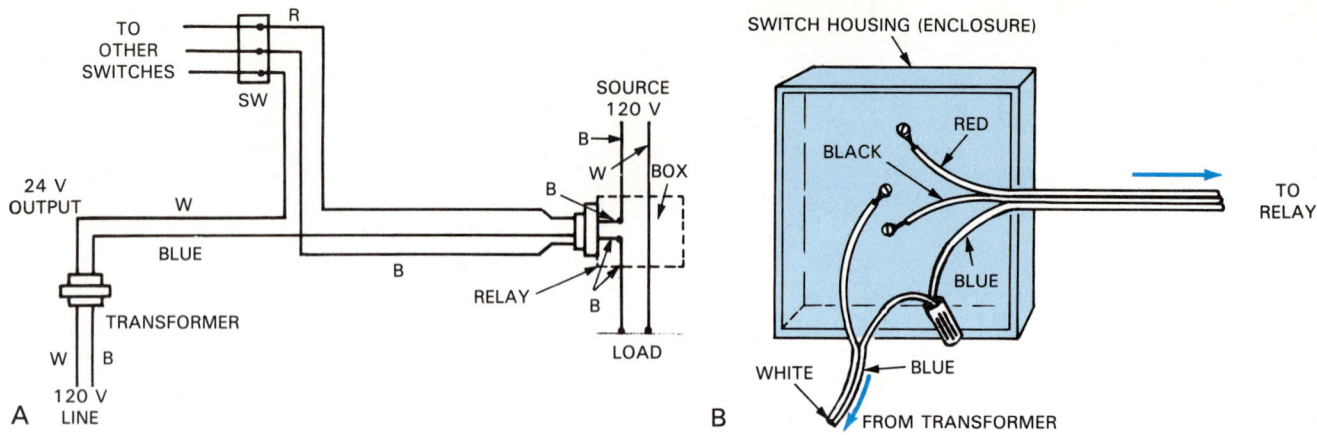

Fig. 17-13. A—Wiring schematic for low-voltage connections of conductors between transformer, relay, and switches. B—Sketch shows how to wire up the circuit in "A." You must run a two-wire cable from switch to relay and a three-wire cable from switch to relay.

Fig. 17-12. You will need to run a two-conductor cable to the master selector switch and three-wire cables to the individual relays. Fig. 17-13 shows a schematic for a much simpler installation involving one relay and several switches.

d. You may find it helpful to prepare a cable layout on an electrical plan. Refer to layouts described in Chapter 11.

4. Install the low-voltage switches at locations where you previously terminated the cable. The switches can be mounted in plaster rings or standard boxes, as was noted earlier in the chapter.
5. Make conductor connections to switches.
6. Activate the system by energizing the circuit feeding the transformer.

Fig. 17-14 shows another circuit schematic for a motor master control.

LOW-VOLTAGE SYMBOLS

Special symbols will be used on electrical plans to indicate low-voltage cable and devices. Refer to the chart in Fig. 17-15 for the symbols and what they mean.

L-V CODE RULES

Low-voltage wiring is not governed by the same rules as 120 V wiring. The *Code* does, however, discuss regulations for low-voltage, remote control

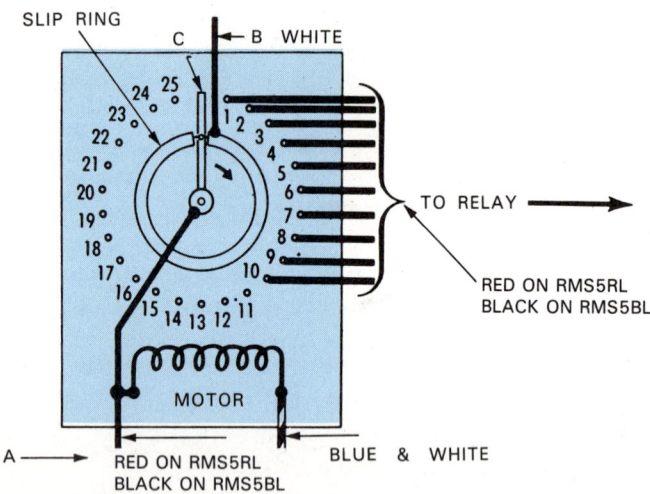

Fig. 17-14. Internal wiring of motor master control unit. Starting switch is connected to lead A and white lead B. When starting switch is activated, contact arm C rotates clockwise connecting white lead to contact points 1 through 25. It picks up the slip ring connected to the white lead and moves it through one revolution. Lead A may be either red or black. (General Electric Co.)

SYMBOL	DEVICE
— — —	Low voltage wiring
T	Transformer
R₁ Rp	Relay 1 = Number of relay Rp = Relay/pilot switch
Sp St	Standard switch Sp = Switch/pilot light St = Interchangeable
MS 12	Master selector panel (Number indicates number of positions)
MMB MMR	Motorized master panel B = Black for off R = Red for on

Fig. 17-15. Low-voltage symbols.

Low-Voltage Circuits

systems under Article 725-Class 1, Class 2, and Class 3 Remote-Control Signaling, and Power-Limited Circuits.

NEC sections 725-31 and 725-42, should be understood thoroughly before you make any remote control or low-voltage installations. Also, as with all wiring, check with local code enforcement agencies, the utility company and the inspector if there are questions or problems regarding regulations and procedures. See Fig. 17-16.

Table 725-31(a). Power Limitations for Alternating Current (Class 2 and Class 3 Circuits)

Circuit	Inherently Limited Power Source (Overcurrent protection not required)				Not Inherently Limited Power Source (Overcurrent protection required)			
	Class 2			Class 3	Class 2			Class 3
Circuit Voltage V_{max} (Note 1)	0-20†	Over 20-30†	Over 30-150	Over 30-100	0-20†	Over 20-30†	Over 30-100	Over 100-150
Power Limitation $(VA)_{max}$ (Note 1) (Volt-Amps)	—	—	—	—	250 (see Note 3)	250	250	N.A.
Current Limitation I_{max} (Note 1) (Amps)	8.0	8.0	0.005	$150/V_{max}$	$1000/V_{max}$	$1000/V_{max}$	$1000/V_{max}$	1.0
Maximum Overcurrent Protection (Amps)	—	—	—	—	5.0	$100/V_{max}$	$100/V_{max}$	1.0
Power Source Maximum Nameplate Ratings — VA (Volt-Amps)	$5.0 \times V_{max}$	100	$0.005 \times V_{max}$	100	$5.0 \times V_{max}$	100	100	100
Power Source Maximum Nameplate Ratings — Current (Amps)	5.0	$100/V_{max}$	0.005	$100/V_{max}$	5.0	$100/V_{max}$	$100/V_{max}$	$100/V_{max}$
Supply Conductors and Cables	See Section 725-37							
Circuit Conductors and Cables	See Section 725-40							

† Voltage ranges shown are for sinusoidal ac in indoor locations or where wet contact is not likely to occur. For nonsinusoidal or wet contact conditions, see Note 2.

Table 725-31(b). Power Limitations for Direct Current (Class 2 and Class 3 Circuits)

Circuit	Inherently Limited Power Source (Note 4) (Overcurrent protection not required)					Not Inherently Limited Power Source (Overcurrent protection required)			
	Class 2				Class 3	Class 2			Class 3
Circuit Voltage V_{max} (Note 1)	0-20††	Over 20-30††	Over 30-60††	Over 60-150	Over 60-100	0-20††	Over 20-60††	Over 60-100	Over 100-150
Power Limitation $(VA)_{max}$ (Note 1) (Volt-Amps)	—	—	—	—	—	250 (see Note 3)	250	250	N.A.
Current Limitation I_{max} (Note 1) (Amps)	8.0	8.0	$150/V_{max}$	0.005	$150/V_{max}$	$1000/V_{max}$	$1000/V_{max}$	$1000/V_{max}$	1.0
Maximum Overcurrent Protection (Amps)	—	—	—	—	—	5.0	$100/V_{max}$	$100/V_{max}$	1.0
Power Source Maximum Nameplate Ratings — VA (Volt-Amps)	$5.0 \times V_{max}$	100	100	$0.005 \times V_{max}$	100	$5.0 \times V_{max}$	100	100	100
Power Source Maximum Nameplate Ratings — Current (Amps)	5.0	$100/V_{max}$	$100/V_{max}$	0.005	$100/V_{max}$	5.0	$100/V_{max}$	$100/V_{max}$	$100/V_{max}$
Supply Conductors and Cables	See Section 725-37				Circuit Conductors and Cables	See Section 725-40			

†† Voltage ranges shown are for continuous dc in indoor locations or where wet contact is not likely to occur. For interrupted dc or wet contact conditions, see Note 5.

Notes for Tables 725-31(a) and (b)

Note 1. V_{max}: Maximum output voltage regardless of load with rated input applied. I_{max}: Maximum output after 1 minute of operation under any noncapacitive load, including short circuit, and with overcurrent protection bypassed if used. VA_{max}: Maximum volt-ampere output regardless of load and overcurrent protection bypassed if used. **Note 2.** For nonsinusoidal ac, V_{max} shall be not greater than 42.4 volts peak. Where wet contact (immersion not included) is likely to occur, Class 3 wiring methods shall be used or V_{max} shall be not greater than: 15 volts for sinusoidal ac; 21.2 volts peak for nonsinusoidal ac. **Note 3.** If the power source is a transformer, $(VA)_{max}$ is 350 or less when V_{max} is 15 or less. **Note 4.** A dry cell battery shall be considered an inherently limited power source provided the voltage is 30 volts or less and the capacity is equal to or less than that available from series connected No. 6 carbon zinc cells. **Note 5.** For dc interrupted at a rate of 10 to 200 Hz, V_{max} shall be not greater than 24.8 volts. Where wet contact (immersion not included) is likely to occur, Class 3 wiring methods shall be used or V_{max} shall be not greater than: 30 volts for continuous dc; 12.4 volts for dc that is interrupted at a rate of 10 to 200 Hz.

Fig. 17-16. Continued.

Fig. 17-16. Check power limitation charts before making low-voltage installations. (National Fire Protection Assoc.)

Modern Residential Wiring

Table 760-21(a). Power Limitations for Alternating-Current Fire Protective Signaling Circuits

	Inherently Limited Power Source (Overcurrent protection not required)			Not Inherently Limited Power Source (Overcurrent protection required)		
Circuit Voltage V_{max} (Note 1)	0-20	Over 20-30	Over 30-100	0-20	Over 20-100	Over 100-150
Power Limitation $(VA)_{max}$ (Note 1) (Volt-Amps)	—	—	—	250 (see Note 2)	250	N.A.
Current Limitation I_{max} (Note 1) (Amps)	8.0	8.0	$150/V_{max}$	$1000/V_{max}$	$1000/V_{max}$	1.0
Maximum Overcurrent Protection (Amps)	—	—	—	5.0	$100/V_{max}$	1.0
Power Source Maximum Nameplate Ratings — VA (Volt-Amps)	$5.0 \times V_{max}$	100	100	$5.0 \times V_{max}$	100	100
Power Source Maximum Nameplate Ratings — Current (Amps)	5.0	$100/V_{max}$	$100/V_{max}$	5.0	$100/V_{max}$	$100/V_{max}$
Supply Conductors and Cables	See Section 760-27					
Circuit Conductors and Cables	See Section 760-30					

REVIEW QUESTIONS — CHAPTER 17

1. List three major components of a low-voltage switching system.
2. Low-voltage switching systems operate at _____ volts.
3. One transformer is capable of serving about _____ or more relays.
4. The relay is installed at each lighting outlet, with the low-voltage leads (inside, outside) the junction box.
5. Low-voltage switching systems are installed using No. _____ AWG conductors.
6. Complete the following diagram:

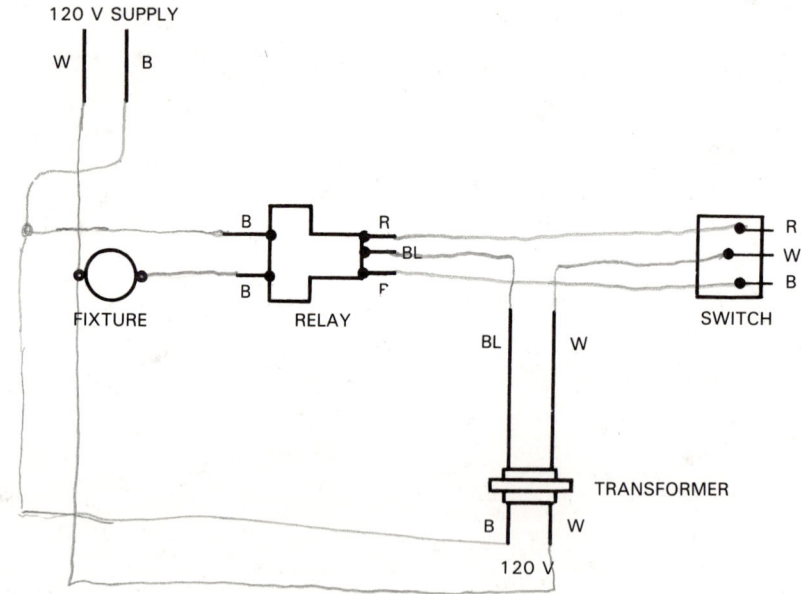

Chapter 18
ELECTRICAL REMODELING

Updating and extending existing wiring systems should only be undertaken with careful planning and attention to the layout. Familiarity with the building's construction is equally important.

This chapter will describe the procedures which seem to work best in most conditions. However, each finished building has special problems which will require special handling. Techniques for specific problems must be handled on a one-to-one basis as they arise.

BASIC CONSIDERATIONS

Remodeling or modernizing an electrical installation falls under the category known as *old work*. It is, perhaps, the most difficult kind of wiring to do. Considerable time, effort, and special techniques are needed to bring about a satisfying result.

Portions of finished walls or ceilings may have to be removed or altered during a rewiring job. Wires must often be "fished" or pulled through blind wall, floor, ceiling, or attic spaces to get from one place to another.

Great care must be used in routing wire cables through walls and similar spaces so as not to damage them. Also, special effort has to be taken to insure the continuity and integrity of the grounding conductor when extending or adding to an existing system. *The grounding conductor must be continuous from the service panel to all outlets.*

There are no set rules concerning the procedures to follow for old work. There is no substitute for experience. However, some basic steps will help you resolve certain difficulties.

SAFETY

When remodeling wiring, follow the same safety rules outlined in Chapter 5. Never work on circuits which are electrically "hot". Turn off the power to the device or circuit on which you are working. If in doubt whether power is off test with a neon tester.

Plan a safe installation. Check local codes for your own protection. Use fuses or circuit breakers rated for the amperage in the circuit. Fire or damage to the wiring can result from improper overcurrent ratings.

Never touch electrical fixtures when you are wet, standing in water, or on a wet surface.

Use tools with insulated handles. Current in less likely to pass thorough insulating materials and cause shock.

SPECIAL TOOLS

Certain special tools are neded for old work:
1. Chisels and crowbars are needed to pry up floor boards, see Fig. 18-1.
2. A keyhole saw is used to cut through floor boards and wall materials, Fig. 18-2.
3. Extension drill bits are necessary to drill deep holes through walls, floors, and beams. The length of these bits may be extended through the use of steel shanks with special ends to attach to other steel shanks of similar construction. For example, a simple 2 in. brace bit can

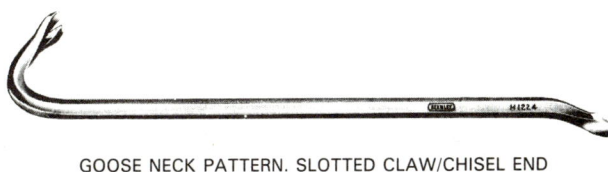
GOOSE NECK PATTERN. SLOTTED CLAW/CHISEL END

SOLID STEEL WITH ENAMEL FINISH/POLISHED BIT

WOOD CHISEL

Fig. 18-1. Chisels and crowbars are useful for cutting into and prying up floor boards. This allows access to old wiring or to space where new wiring must be placed.

Modern Residential Wiring

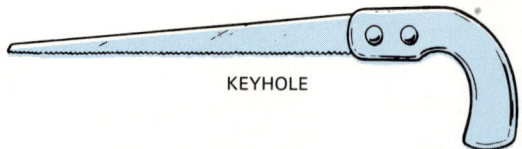

Fig. 18-2. Keyhole saw is designed for use in tight openings. It is used to cut openings into ceilings and walls where new outlets must be installed. It may also be used to cut flooring. (GE Wiring Devices Dept.)

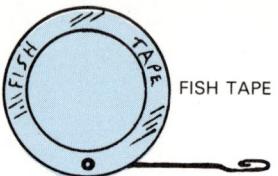

Fig. 18-4. Old work requiring running of cable through walls is impossible to do without fish tape for pulling the cable. (GE Wiring Devices Dept.)

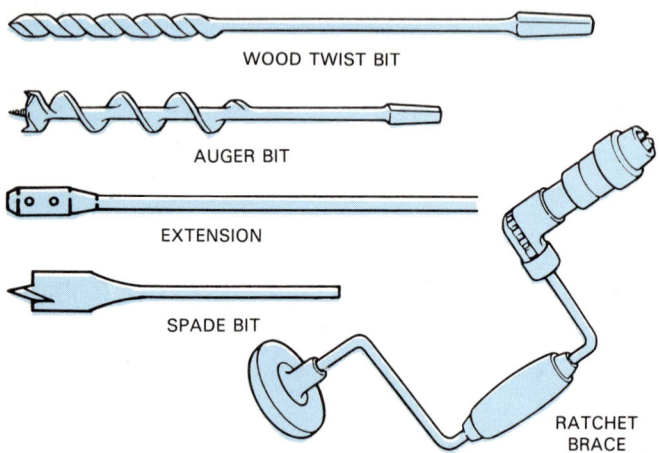

Fig. 18-3. Drills and brace. Extension bits are essential in boring from one floor to the next or through thick beams.

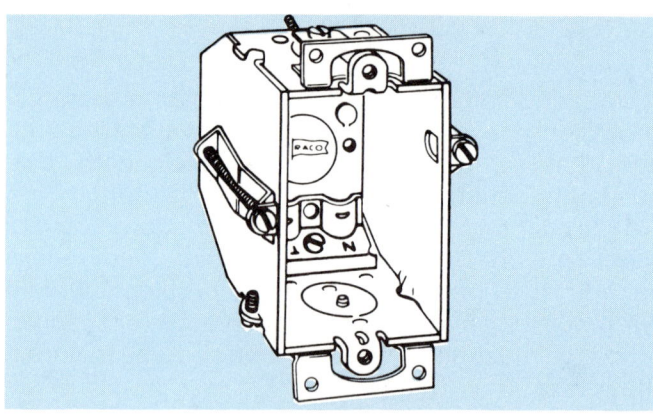

Fig. 18-5. Special old-work boxes like this one have expansion devices which grip the wall. Turning the screws on either side will draw a flange against the back side of the wall. (RACO, Inc.)

be extended to lengths of 20 ft. or more using this method. All this may be necessary to bore a hole continuously, through the first and second floor of a structure, Fig. 18-3.

4. A wire snake or fish tape pulls wires through bored holes and wall spaces, Fig. 18-4.

MATERIALS

The materials used in old work are essentially the same as those used in new work. However, there are special boxes, box hangers, and extenders made to ease installation. See Figs. 18-5 and 18-6.

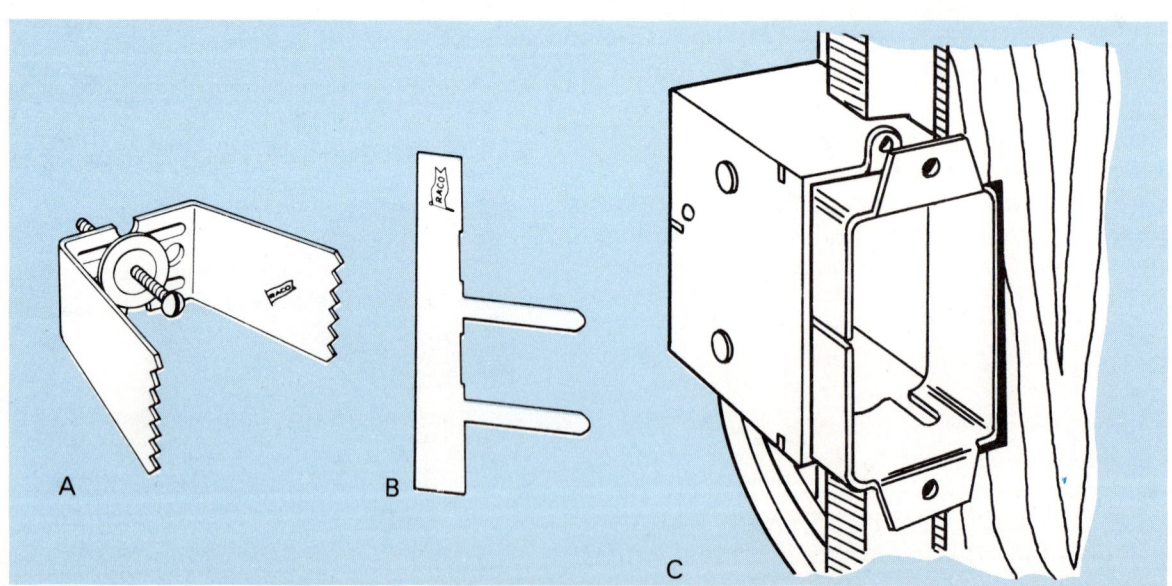

Fig. 18-6. Other devices for adapting and holding boxes in old work A—Snap-in bracket. B—Metal support for standard electrical box. C—Adapter brings old box flush to new wall surface. (RACO, Inc.)

188

Conduit is out of the question. Installation would be virtually impossible without removing large sections of finished walls. Therefore, nonmetallic sheathed cable, Greenfield, or armored cable (BX) are used. These are more readily fished through walls and floors.

BUILDING CONSTRUCTION

The type of building will often dictate the tools and materials used. Walls may be constructed of sheetrock (drywall), wallboard, plaster on lathe, brick, or concrete blocks. The type of surface will make enormous differences in the wiring method used.

WALL OPENINGS

Wall openings for switch boxes outlet boxes as well as for wire-pulling should be carefully cut. This cannot be overemphasized especially where the wall is plaster on lathe. A switch box opening can easily balloon into a nightmare without careful tracing and cutting.

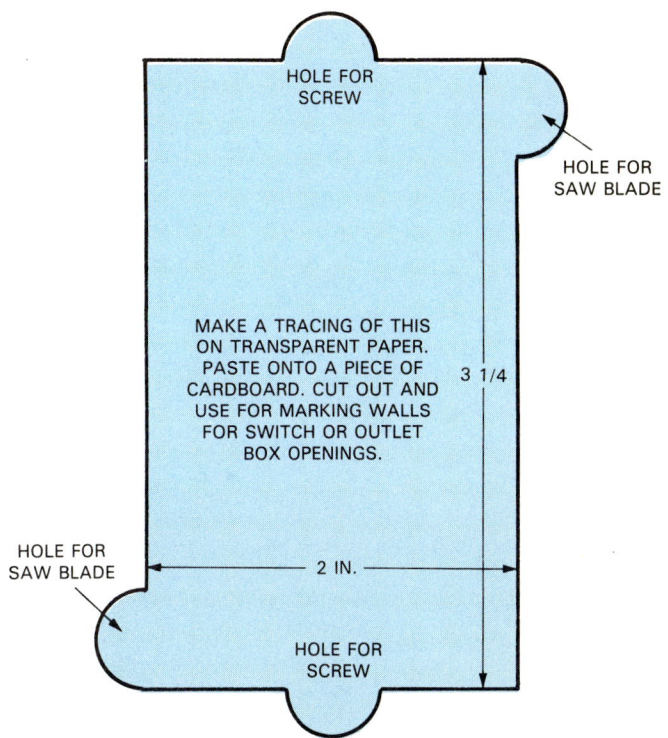

Fig. 18-8. Actual size template can be cut out of cardboard to mark wall openings.

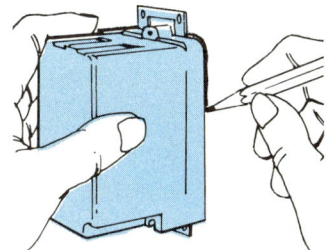

Fig. 18-7. The old-work box can be used as a pattern for cutting the opening. Place the box, open side to wall over the spot where the opening is to be cut. Carefully trace around it. Be sure that the box is held plumb.
(Allied Moulded Products, Inc.)

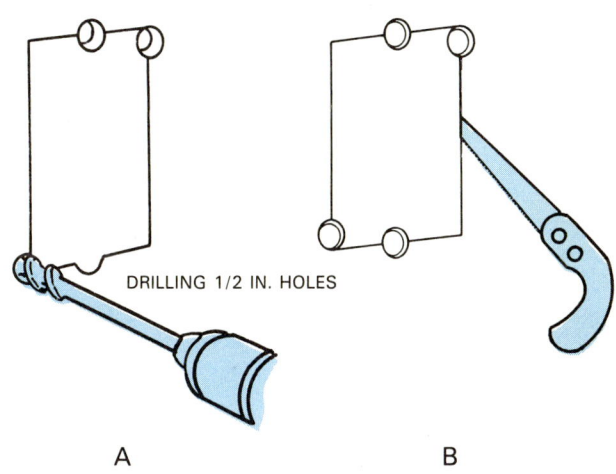

Fig. 18-9. Steps for cutting box openings. A—Drill holes at two corners and at holes for screws. B—Working through holes saw along all lines. Use a keyhole saw.
(GE Wiring Devices Dept.)

Using the box, Fig. 18-7, or a template like the one shown in Fig. 18-8, trace the box shape on the wall where new boxes are to be located. Make sure that no studs are present at that location. Use a stud finder or tap with a hammer to locate the stud. The half circles at two corners are for drilling holes where you can start saw cuts with a keyhole saw.

Next, drill the holes as indicated in Fig. 18-9. Cut out the opening and slip the outlet box in to check the fit. Trim as needed to adjust fit. Keep opening as small as possible.

Boring holes for other openings, primarily fishing cables, require less skill. However, these, too, must be repaired in the end; so, the smaller the opening the better. Methods of routing cable through structural timbers and walls are shown in Figs. 18-10, 18-11, 18-12, and 18-13.

INSTALLING CABLE

Installing cable in old work involves pushing or pulling it through wall openings you have prepared. Sometimes, this is easily accomplished by hand, at other times fish tape must be used to pull the wires through.

For some wire pulling jobs, a length of ordinary steel wire can be used. Simply bend a small loop into one end. This can be used to attach cable and

Modern Residential Wiring

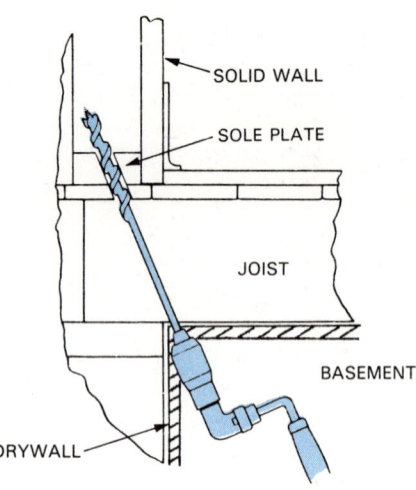

Fig. 18-10. When boring for access with fish tape or cable keep openings small and as neat as possible. Here a brace and bit are used to drill from one level to the next. The opening will be large enough to fish cable through it but small enough to easily repair.

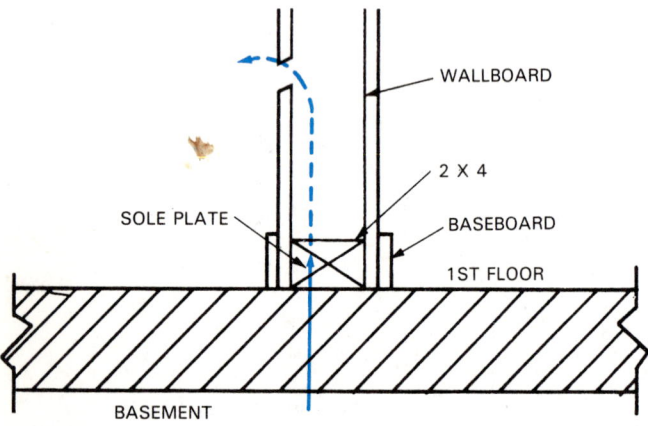

Fig. 18-11. To go from a basement to a wall on the first floor, locate the wall and drill through rough flooring and sole plate (2 x 4) which supports the studs. Arrow shows path of fish tape.

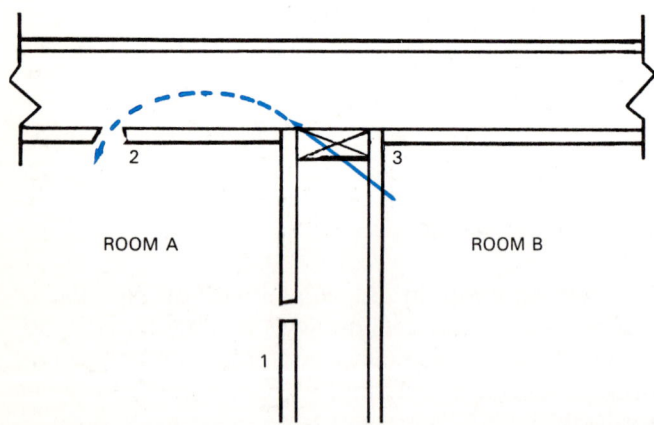

Fig. 18-12. Going from a wall location to a ceiling location in the same room. Black arrow shows location and direction of drilled hole.

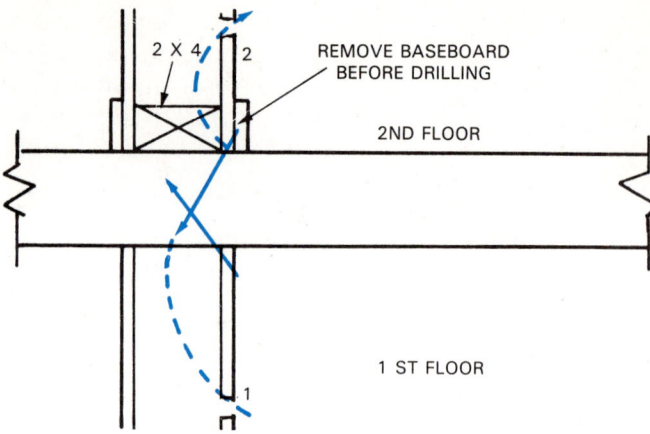

Fig. 18-13. Going from a first floor to a second floor. Drilling from two locations is required and holes must intersect.

prevent it from snagging. See Fig. 18-14. When completed you have a "homemade" fish tape.

Commercially sold fish tape comes in reels of 25, 50, or 100 feet and is essential for conduit work. Refer again to Fig. 18-4.

USING THE FISH TAPE

Fish tapes, whether home made or purchased, are a practical necessity for running cable in old work. They are flexible enough to be worked through 90-degree bends. Yet, they are stiff enough to be worked up through narrow openings and between walls without buckling.

To use them:

1. Feed the tape from top down, if you can. Usually, you will need to work the tape from one box opening, through the wall or a partition, to a second opening. See Fig. 18-15.
2. Twist and turn the tape until it enters the hidden opening (such as a hole drilled in a plate). Continue to push it until it can be seen or felt at the second position to which cable must be fed.
3. If bends prevent a tape from reaching the second opening, a second tape may be inserted

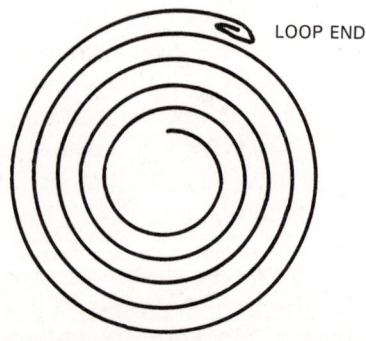

Fig. 18-14. Making fish tape from ordinary steel wire. Suit length to the job at hand.

190

Electrical Remodeling

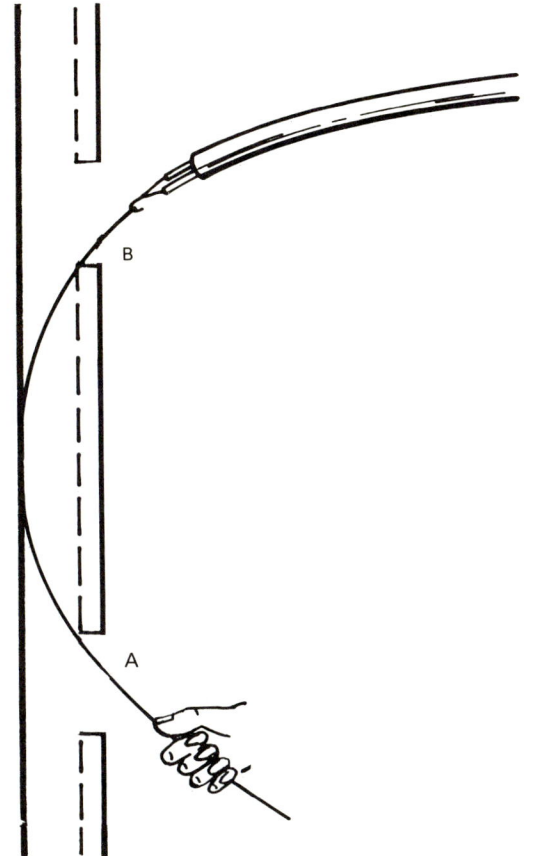

Fig. 18-15. Using the fish tape to go from one opening to another in a wall. Tape is fed in at A and is pushed upward until it can be pulled through the opening at B. (GE Wiring Devices Dept.)

through the second hole. Twist and pull on the tapes until they catch each other. Pull on one of the tapes until the other is drawn through the second opening.

4. If a box from old work cannot be removed, insert the fish tape through a knockout as shown in Fig. 18-16.
5. Secure the cable to the fish tape, Fig. 18-17, and pull the cable through the wall to the first opening.
6. Once the cable is brought to the appropriate wall opening, strip back the insulation about 8 in. and install the cable connector.
7. Attach the cable to the box, Fig. 18-18, and install the box in the wall.

USING NONMETALLIC BOXES

Nonmetallic boxes have also been designed for use on old work. Locate the correct knockout (in direction of the cable) and tap out the thin section as in Fig. 18-19. Slide the conductors through the knockout and the box clamp. Push through at least

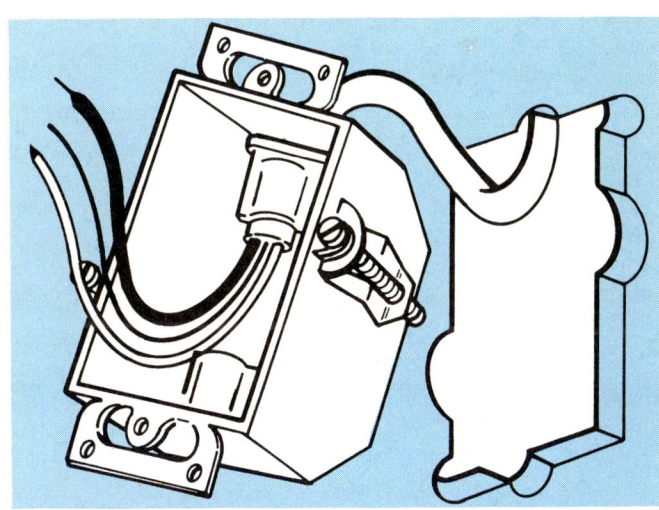

Fig. 18-18. Cable must be attached to box before the box is installed. Some connectors are installed inside the box. Others are attached to the cable before being pulled into the box. (GE Wiring Devices Dept.)

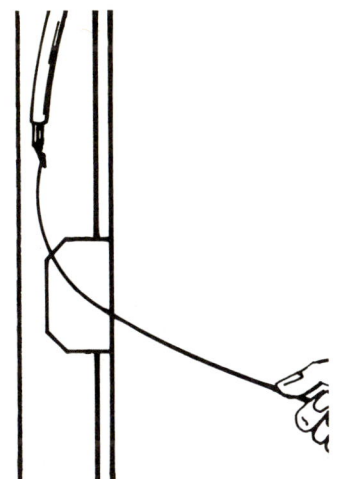

Fig. 18-16. Fish tape can be fed through a knockout in an electrical box when going from an existing outlet to a new one.

Fig. 18-17. Connect cable securely to end of fish tape in preparation for pulling through blind space.

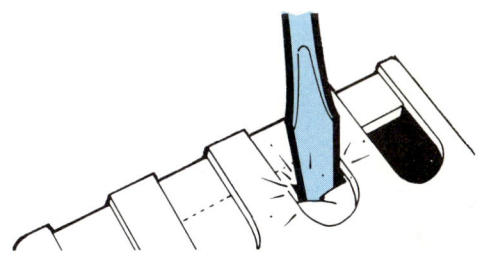

Fig. 18-19. Opening up a knockout on a fiberglass or plastic electrical box. Thin section is removed by a sharp tap of a screwdriver. (Allied Moulded Products, Inc.)

Modern Residential Wiring

1/4 in. of the cable jacket (covering). See Fig. 18-20. Using a screwdriver, tighten the cable clamp until the cable is securely fastened in the box.

To install the box, gently guide it with the snap-in bracket attached, through the cutout. Push it into the opening as far as the "ears." Secure the box by tightening the screw or screws in the bottom of the box, Fig. 18-21. This will draw the snap-in bracket against the back side of the wall.

EXTENDING A CIRCUIT

Often, old work will consist of extending an existing circuit. To do this you need to run cable from an existing outlet to the point where a new outlet is needed. Routing of the new cable should be done to create the least work or destruction of plaster or drywall.

As shown in Fig. 18-22, the new work can be routed through an attic, behind a baseboard, through a basement or crawl space. Which route is best depends upon:
1. Where the new outlet is located.
2. What obstructions lie in the way.

If you are installing a ceiling fixture, going through an attic space may be the obvious route. If a new receptacle is being added along a wall and power is being taken from another receptacle 6 ft. away, a baseboard installation would seem best. However, if there is a door in the way you would likely choose a different route.

WIRING BEHIND A BASEBOARD

Baseboard installations can be done without use of a fish tape. Before opening up the live outlet trip the circuit breaker to the off position to deenergize the outlet. Check with a neon tester to make sure the power is off.
1. Look at Fig. 18-23. Remove receptacle plate and receptacle at A.
2. With a screwdriver remove a knockout facing in the direction the new wiring will go, in this case, toward the floor.
3. Locate the new outlet. Use a stud finder or tap on the wall to find where studs are located. Mark the spot where the outlet will be.
4. Put the new box against the wall or use a template to mark the outline of the new opening.
5. Drill starter holes and cut out wall material with a keyhole saw.
6. Remove the baseboard carefully to avoid damage to either the baseboard or the wall.
7. Drill holes below both A and B in area that will be concealed by the baseboard.
8. Cut a groove in the wall and the studs from A to B. Make it deep enough so a piece of conduit or cable will fit behind the baseboard. Select cable or conductor size according to the circuit ampacity.
9. Install conduit or cable. (Cable can be installed in conduit if desired. This would protect it from nails driven through the baseboard. If cable only is used protective steel plate must be placed over the notch at every stud.)
10. Make connections inside the boxes.
11. Install the new box in the wall.
12. Install receptacles and cover plates. Test the circuit extension.
13. Replace the baseboard.

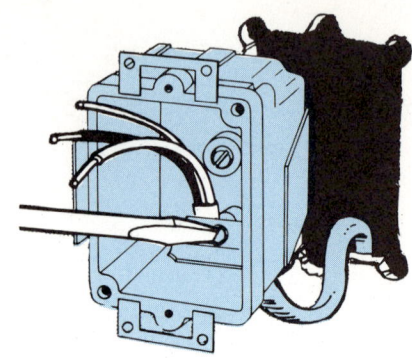

Fig. 18-20. Fastening cable inside a nonmetallic box. Pull at least 1/4 in. of cable jacket through the clamp.

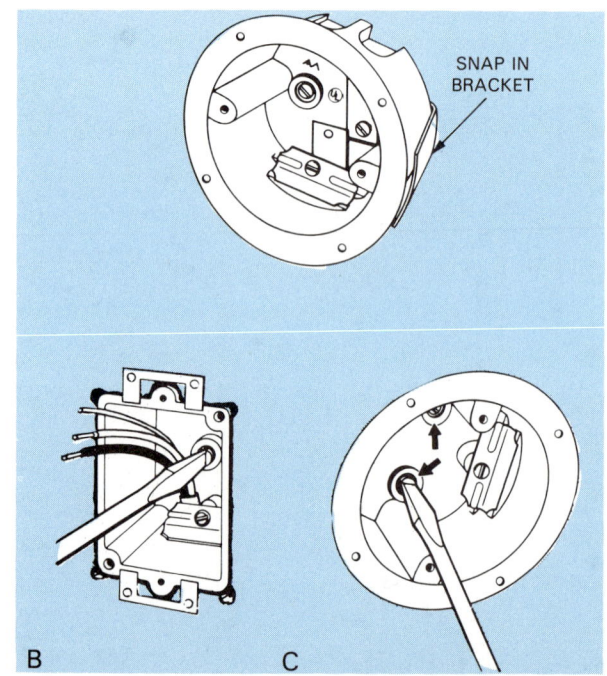

Fig. 18-21. Securing box in the wall. A—Snap in bracket is attached to back of box with adjusting screw. B—When box is in place with ears against surface of wall, bring bracket against the reverse side by tightening the screw. C—Some boxes have two screws for adjusting snap-in bracket. (Allied Moulded Products, Inc.)

Electrical Remodeling

BOX INSTALLATION

In old work, as in new work, box installation is a simple procedure. However, in old work the technique is different since the boxes are usually not attached to the building studs.

For this reason, special boxes and mounting devices are available. See Figs. 18-24 and 18-25.

Whatever devices are used, the box must be prepared and the cable attached to it *before* it is installed. Simply knock out the appropriate holes and attach the cable using the connector(s) you've already installed, Fig. 18-26. Then slip the box into the wall opening. Secure it, and connect the load device.

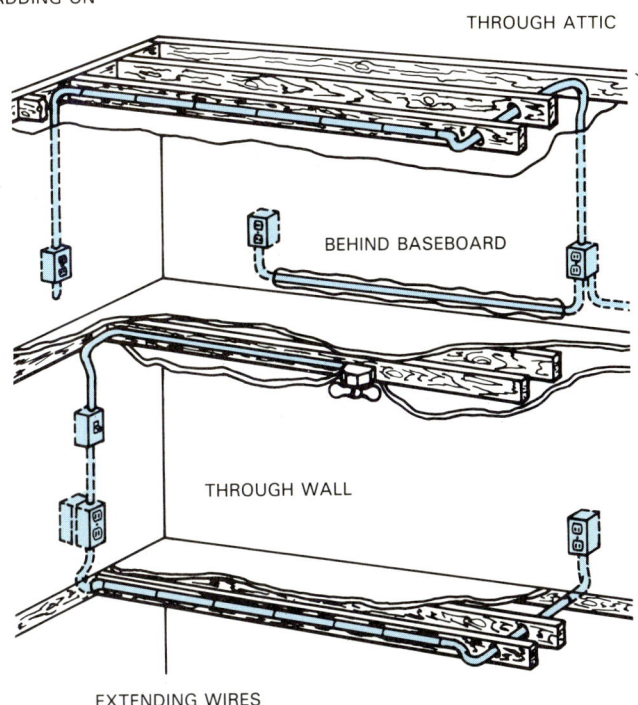

Fig. 18-22. Methods of routing wiring during remodeling. (GE Wiring Devices Dept.)

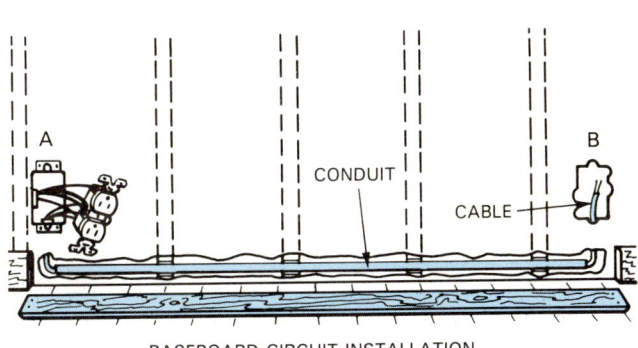

Fig. 18-23. Routing wiring where it will be concealed by a baseboard.

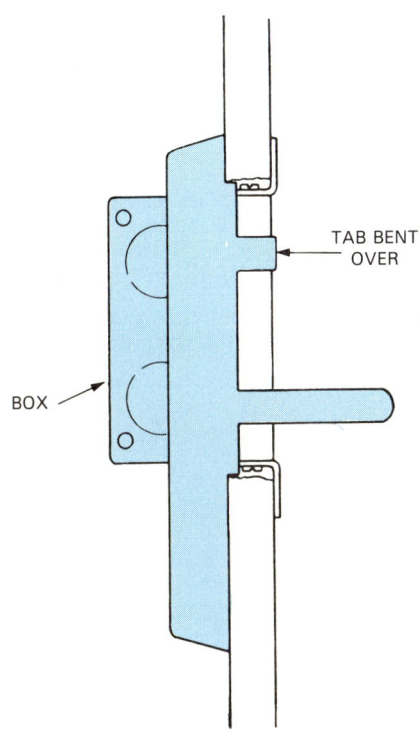

Fig. 18-25. This style of hanger strip is designed to be used with standard electrical boxes.

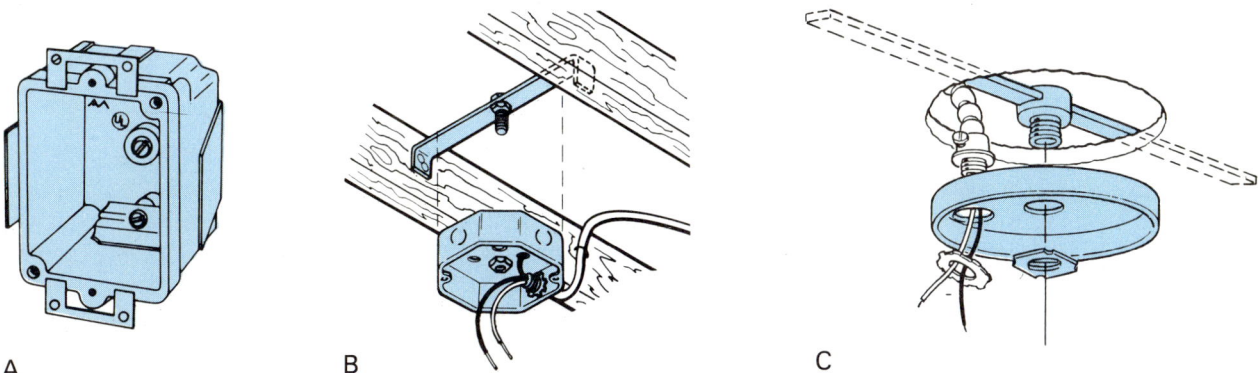

Fig. 18-24. Special boxes and devices for clamping boxes in old work. A—Snap-in bracket is part of the box. B—Hanger mount for mounting from ceiling joists in an attic. C—Hanger designed to bridge hole in plaster or drywall.

Modern Residential Wiring

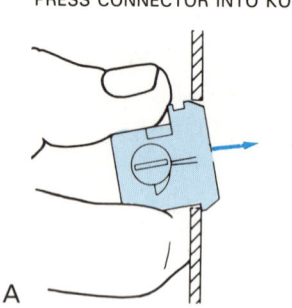
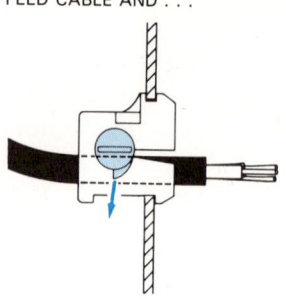
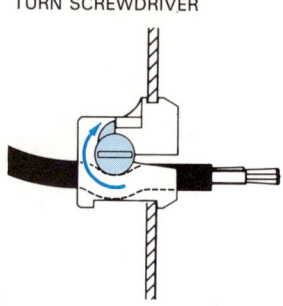
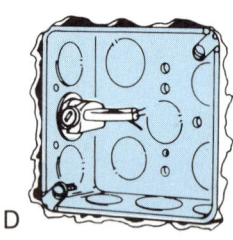
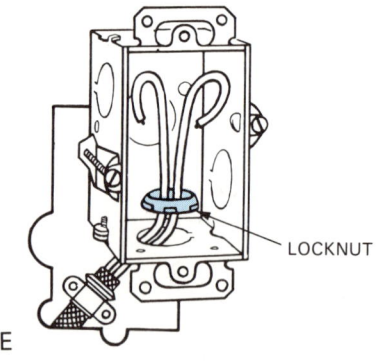

Fig. 18-26. Cable connectors for old work. A, B, and C—Method of installing special connector which needs no locknut. D—Completed installation with all steps completed inside the box. E—Connector designed for nonmetallic cable. It is fastened to cable before box connection is made. Note locknut. (RACO Inc.)

MODERNIZING SERVICE ENTRANCES

More often than not, the electrical demands made on a house built as recently as 15 or 20 years ago, exceed its capacity, Fig. 18-27. Sometimes a larger service entrance is the best solution.

There are several ways to update a service entrance. Which is best will be based on:
1. Calculating present loads.
2. Planning for future loads.

In any case, a minimum of 100 A will be required. In a home where there is electric heating, air conditioning, ranges, clothes dryers, and other 240 volt appliances, a 200 A service is advisable. The updated service entrance must be installed as described for new service entrances. See Chapter 12. Make sure to provide a breaker panel large enough for all the existing circuits, new circuits, and some spares for future expansion.

CONTACT POWER COMPANY

Inform the power company before work starts. Certain information, forms, and metering equipment must be obtained before the job begins. In addition you will need to arrange for a disconnect of the old service and a reconnect of the new or updated one.

After the new service entrance is installed, the old circuits must be connected to it. There are several possibilities:
1. If the new panel is to be placed in the same location as the old one, then the old circuits may be directly connected to it.
2. If the new panel cannot be located in the same position then you must junction the old circuits

Fig. 18-27. Old, outdated electrical equipment is best replaced. Overloaded circuits and extension cords are fire hazards. (Scott Harke)

Electrical Remodeling

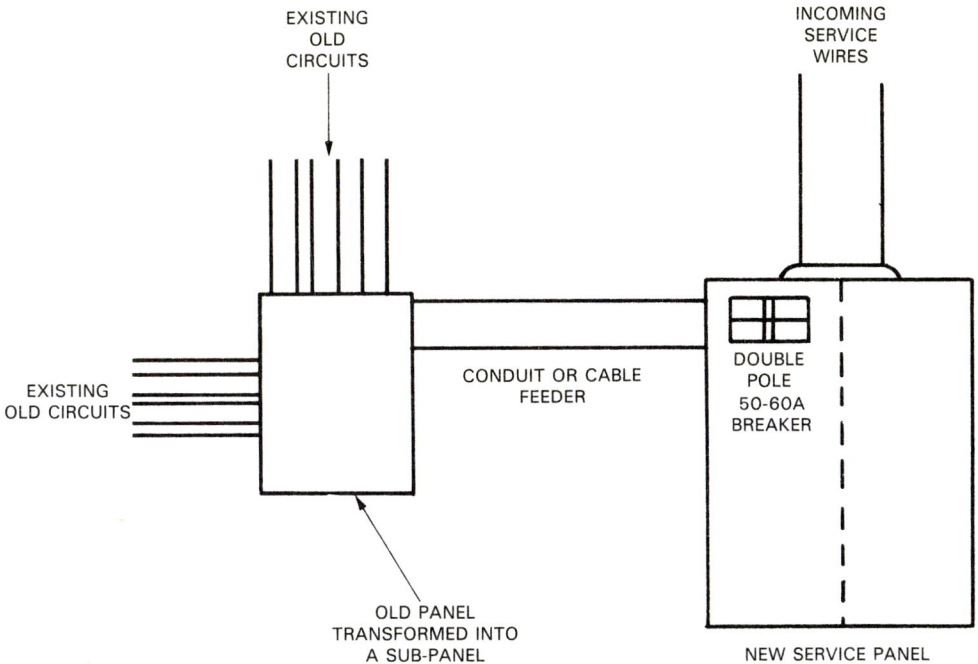

Fig. 18-28. Incorporating the old service panel into a new one. The old panel is transformed into a sub-panel.

to it. One way of doing this is to use the old service panel or fuse box as a sub-feed panel, Fig. 18-28. Use a 50 or 60 A breaker in the new panel to feed it.
3. Another possibility is to replace the old fuse box or outdated panel with a new lighting panel and circuit breakers. Then connect old circuits directly to the new panel.
4. Still another alternative is to replace the old box or panel with an approved junction box of "J" box, as shown in Fig. 18-29. Again, connect the old circuits to the new main panel, giving each its own circuit breakers.

ADDING A SUB-PANEL

The cable feeding the sub-panel should match the panel's ampacity. To determine the ampacity needed, refer to the panel rating and the manufacturer's instructions. *Be sure to check local Codes as well.* Also, be sure that the sub-panel has sufficient capacity for the number of circuits you wish to add. To install a sub-panel use two circuit breaker spaces in the main panel. See Fig. 18-30.
To make the installation:
1. Switch the main circuit breaker to the "off" position or remove the main fuses. Be careful.

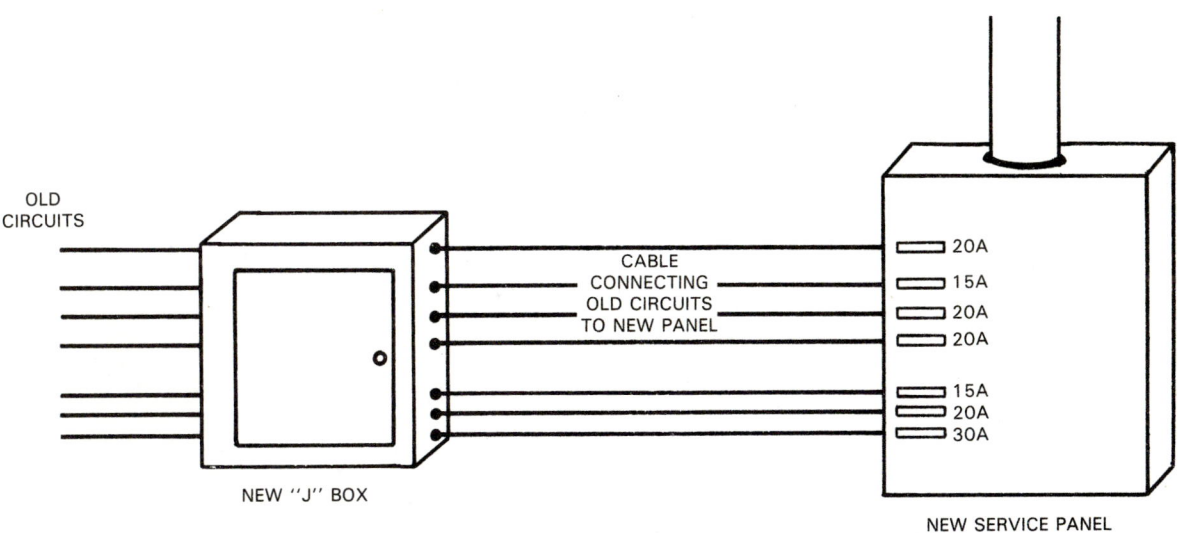

Fig. 18-29. Often a junction box is installed to replace the old panel. It offers a connecting means to make the home run to the new service panel.

195

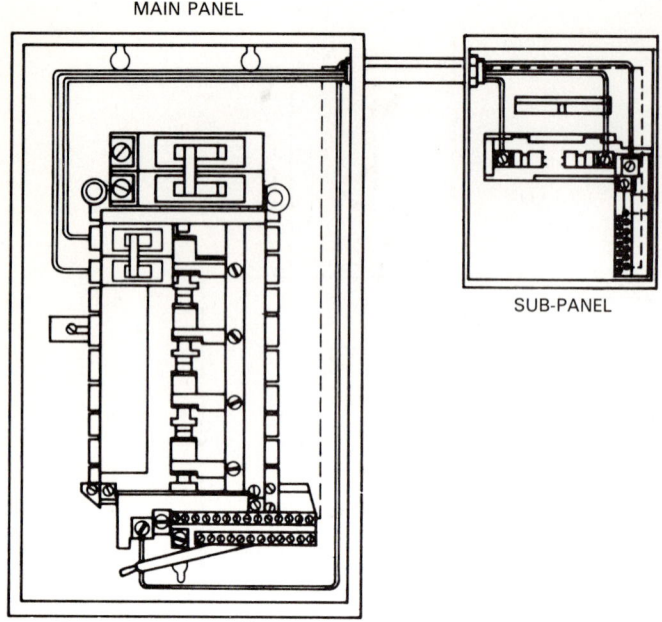

Fig. 18-30. Hookup for adding a sub-panel to a service entrance main panel. It is similar to Fig. 18-28 and the hookup is made in the same way. (GE Wiring Devices Dept.)

5. The two ungrounded conductors (red and black wires) connect to the screw terminal at the top of the new panel. The white wire must be connected to the neutral bar. The green insulated or bare wire goes to the equipment ground bus bar.
6. Make connections inside the main panel. Cut the wires to length allowing enough to make connections to the sub-feed overcurrent device. As in the sub-panel, connect the white wire to the neutral bus bar, and the green or bare wire to the equipment ground bus bar.
7. This completes the wiring of the sub-panel. The next step is to add branch circuits and circuit breakers.

GROUNDING REMODELED SYSTEMS

When installing new panels, or "J" boxes as substitutes for the old fuse box, or when using the old fuse box or panel as a "J" box, the neutral and the ground bus bars must be isolated. That is, ALL neutral or white wires must go to a neutral bus bar and ALL ground wires must go to a separate ground bus bar. This bus bar must be insulated and isolated from the neutral bus bar. (See Section 250-61 of the NEC.) The ground bus must be bonded to the enclosure; the neutral bus must be insulated from the enclosure. This is especially important when using nonmetallic cable. Fig. 18-31 shows the wiring diagram when installing any sub-panel.

SURFACE WIRING

A discussion of surface wiring will include: flat conductor cables, wireways, and multi-outlet

Remember, the incoming power cable and line-side lugs are still live even though the rest of the panel has been disconnected.
2. The sub-panel is usually mounted as near to the main panel as possible.
3. Run cable from the main panel to the sub-panel. Make sure that there is extra cable at each end to make the connections.
4. Cut insulated wires long enough to make connections to screw terminals, neutral bar, and equipment ground bar.

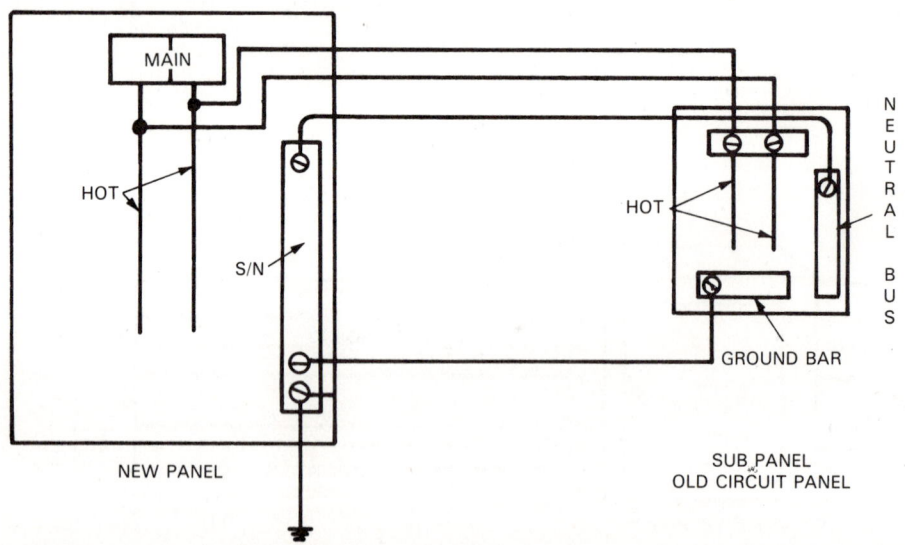

Fig. 18-31. Feeder from main panel to sub-panel will contain four conductors. Two are hot, one is neutral and one is the grounding conductor. In the sub-panel, the grounded and grounding conductors must be insulated from each other. Rules regarding this procedure may be found in Section 250-23 of the NEC.

Electrical Remodeling

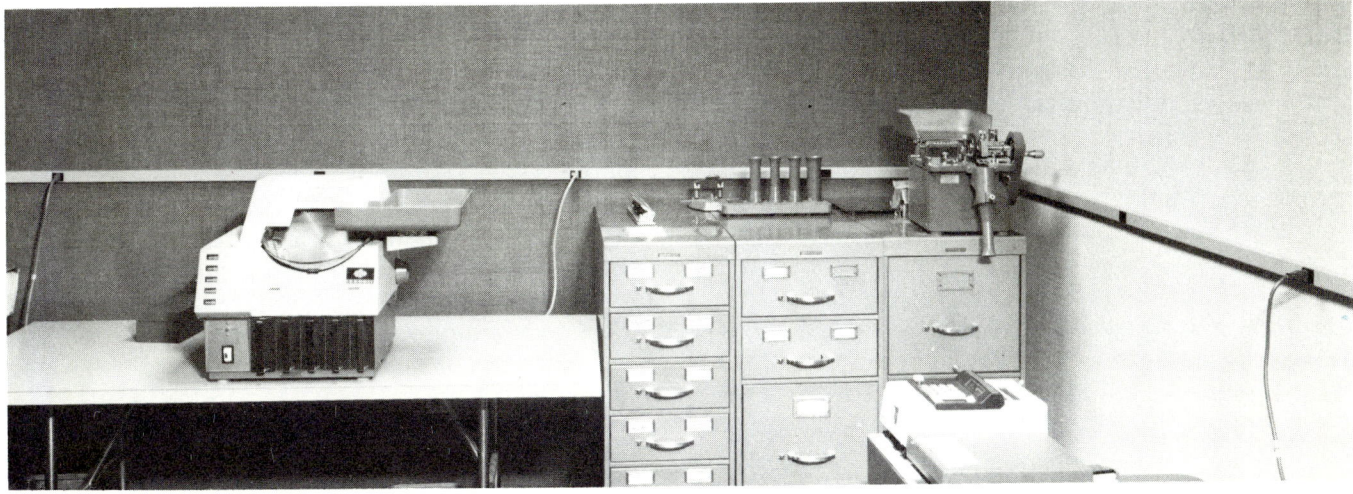

Fig. 18-32. A surface wireway is a simple, inexpensive way to extend a circuit to get more receptacles. Instructions are supplied by the manufacturer. (The Wiremold Co.)

assemblies or raceways. See Fig. 18-32. The *Code* defines such systems under Article 100 as "A type of surface or flush raceway designed to hold conductors and attachment-plug receptacles, assembled in the field or at the factory." The uses and provisions of such systems are spelled out in Code Articles 352, 353, and 363. You should become familiar with these articles of the *Code* before making any installation of this nature.

The main benefit of surface wiring systems is that walls, ceilings, and floors do not have to be opened up for installation. This saves time, effort, and money.

INSTALLING SURFACE WIRING

Manufacturers of surface wiring provide detailed instructions on installation of their systems. A typical installation begins with bringing the line side conductors into the base of the surface assembly. Remove an entrance knockout from the base and attach a suitable connector. (If connector is already attached to feeder cable, pull connector and cable through the base.) Install attaching locknut, Fig. 18-33, and attach base to the wall with mounting screws provided by the manufacturer.

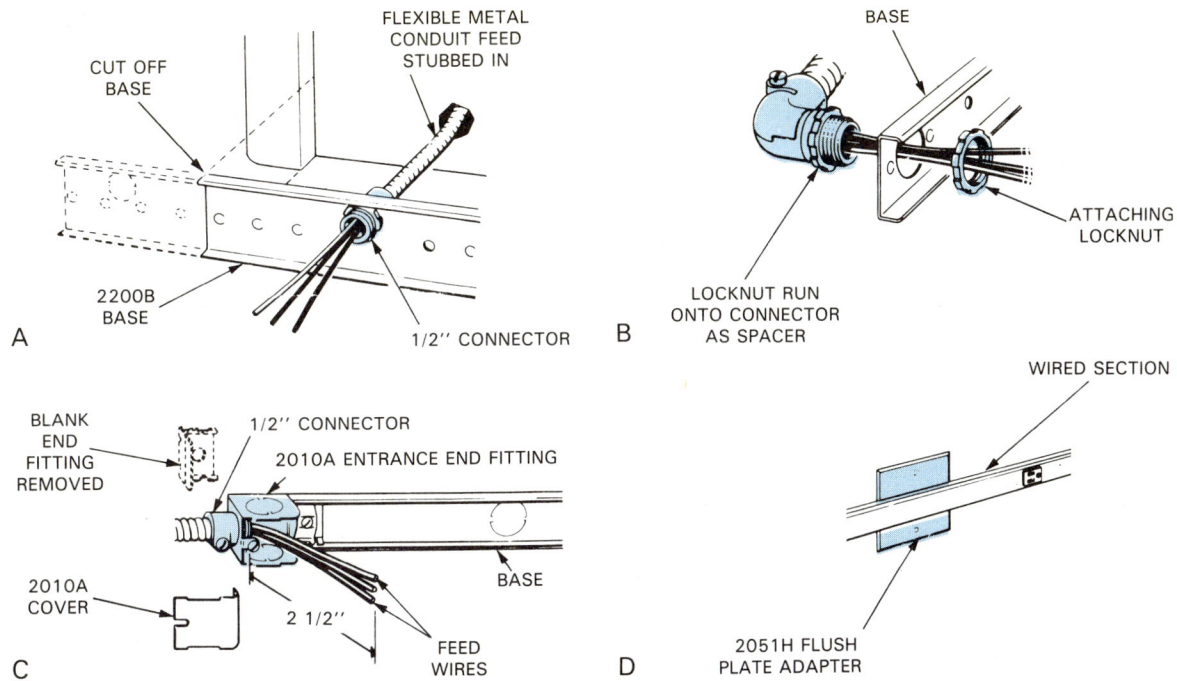

Fig. 18-33. Methods of bring in line conductors to a wireway. A—Cable can be brought in through a knockout in the base. No box is required in the wall. B—Using a 90-degree connector. C—Special closure for end feeding. D—Wireway fed from an outlet box using a special cover adaptor.

Receptacles are normally mounted in the cover section of the assembly. Detach one or two receptacles from the cover. Align cover and receptacles with the base to determine where connection should be made.

Cut and strip black and neutral wires on both harness and feed as shown in Fig. 18-34. Connect black and neutral feed wires to black and neutral harness wires. Use wire nuts (solderless connectors) or special connectors provided by the manufacturer. Strip away enough insulation from the green insulated grounding conductor in the harness to make an approved connection with the green grounding wire from feed. Fig. 18-35 shows completed connections.

If nonmetallic sheathed cable is used for feeder, connect the bare ground wire as shown in Fig. 18-36. Use solderless connectors to cap unused leads at the end of the harness, Fig. 18-37.

When connections are completed, replace the receptacles and harness in the cover. Snap the cover into the base to complete the installation.

To extend surface wiring, attach the new base section to the wall being careful to align and butt the end to the previous section. Remove blank end fittings from sections. Make an electrical connection between the harnesses in the two sections.

Reinstall the covers. Figs. 18-38 through 18-41 show installation of nonmetallic type surface raceway.

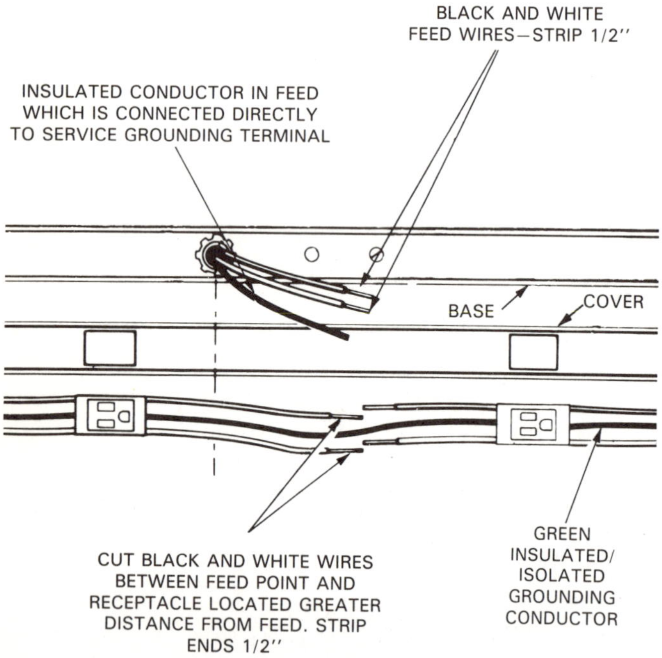

Fig. 18-34. Cutting of wires in harness prior to making connections to feed. Green insulated wire is not cut. (The Wiremold Co.)

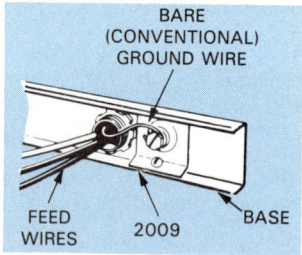

Fig. 18-36. Method of making ground connection when nonmetallic sheathed cable is used. (The Wiremold Co.)

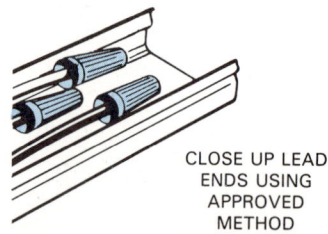

Fig. 18-37. Bare leads at the end of the harness must be capped to prevent short circuits.

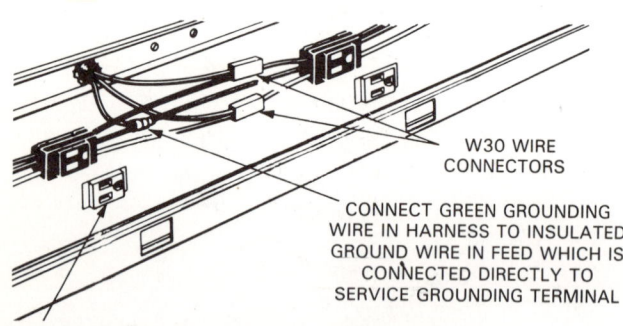

Fig. 18-35. Making connections from feed to wire harness. Solderless connectors or special connectors supplied by manufacturer can be used.

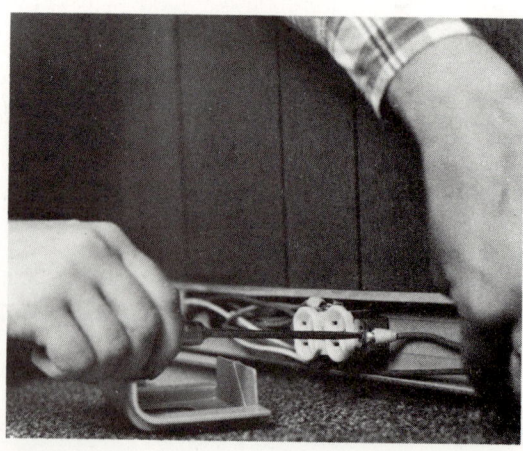

Fig. 18-38. Installing nonmetallic molded raceway. Branch circuit conductors are brought in through hole cut in base. (Carlon, an Indian Head Co.)

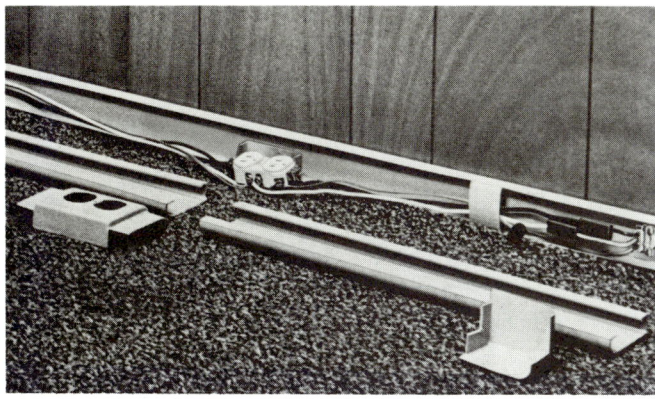

Fig. 18-39. Outlets are connected to appropriate circuit conductors and positioned at desired intervals. Outlet mounting brackets hold outlet securely. Wire retaining clips help support conductors in raceway.

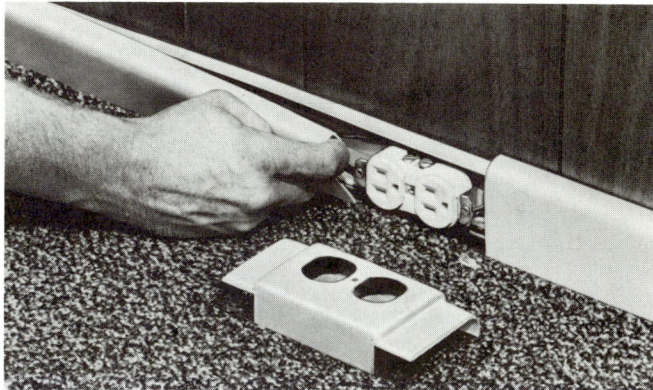

Fig. 18-41. With base cover properly installed, outlet plate is ready to be attached.

Fig. 18-40. Base cover has been cut to length and is being snapped over the base. (Carlon, an Indian Head Co.)

REVIEW QUESTIONS — CHAPTER 18

1. Discuss the major differences between old work and new work.
2. List five special tools needed for remodeling wiring.
3. When extending or remodeling a wiring system, the grounding conductor must be _____ from the service panel to all outlets.
4. The method(s) of wiring used in old work most often is (are):
    a. Nonmetallic sheathed cable.
    b. EMT.
    c. Rigid conduit.
    d. Armored cable (BX).
    e. Greenfield.
    f. None of these.
5. A _____ or the _____, itself, is used to trace the shape of an outlet box on a wall when doing old work.
6. Explain the use of fish tape in electrical remodeling.
7. In old work, boxes are usually attached to studs or joists. True or False?
8. Give two common ways that old circuits can be connected to a new service entrance panel.
9. List several benefits of surface wiring.
10. Explain what is meant by insulating the neutral grounded conductor from the grounding conductor.

Chapter 19

ELECTRICAL METERS

In electricity, as in other technical fields, it is important to be accurate. To make accurate measurements of electrical quantities, we use electrical METERS. They are to electrical measure what rules are to length measure. Without the meter we could only be approximate in our measurements of electrical quantities.

In this section you will become familiar with technical measuring devices. They give us an accurate insight into electrical circuitry and are an invaluable aid to electrical design and troubleshooting.

Many types of electrical meters have been developed. Some are much more useful than others. We will study meters designed to measure voltage, current, resistance, and power.

METER DESIGN

Most meters used for electrical measurement employ a "moving-coil meter movement." This design, illustrated in Fig. 19-1, was first produced by the French physicist, Jacques Arsene D'Arsonval, to detect small electrical currents. This early instrument, called a GALVANOMETER, has been modified for different purposes, but the basic construction is still about the same. In fact, the coil movement is called the D'Arsonval movement, honoring its inventor.

As shown, the basis of a D'Arsonval movement is a permanent horseshoe magnet made of soft iron. A cylindrical or circular soft iron core is placed within the magnetic field created by the horseshoe magnet. Very fine wire is wrapped around this core and makes up the coil. Opposite ends of the coil are fastened to springs, which are permitted to pivot on highly polished bearings to reduce friction. This allows the coil to move freely between the poles of the magnet.

The two springs are each connected to a terminal. These serve as the meter terminals. Leads are attached to the terminals. Further, a pointer registers coil movement. Stops are provided at the left and right extremes to prevent the pointer from moving too far in either direction. A *calibrated* (numbered) scale, placed in back of the pointer, is used to in-

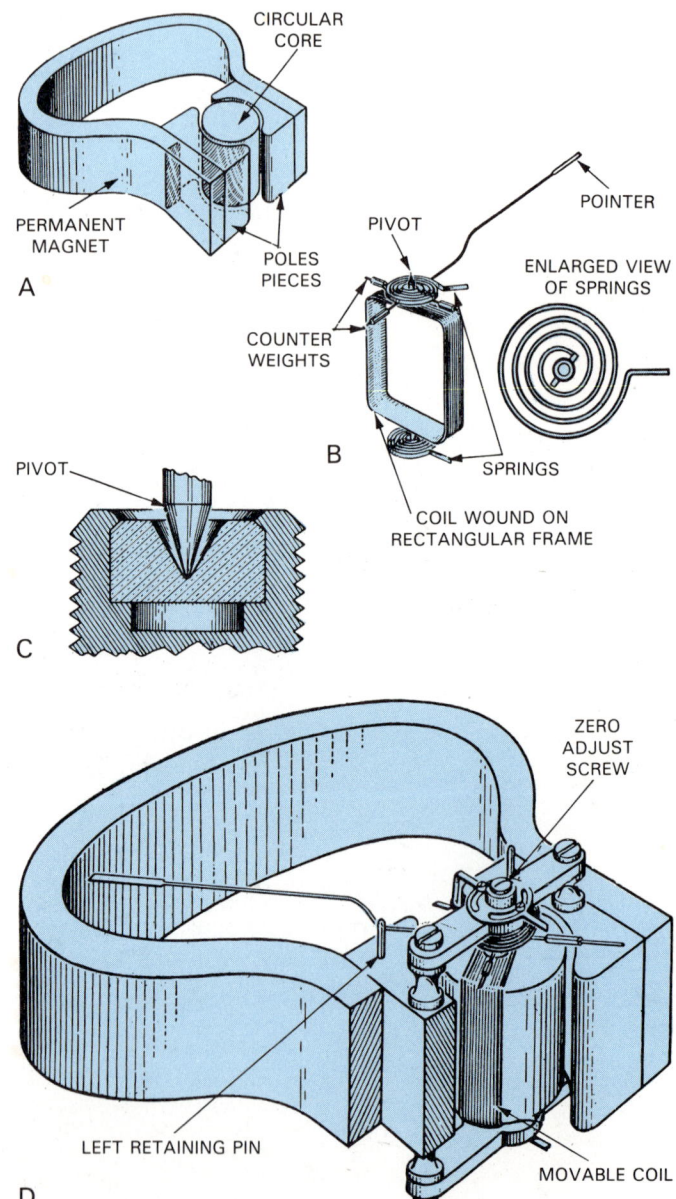

Fig. 19-1. The basic mechanism of electrical meters like the ammeter and voltmeter is the D'Arsonval movement illustrated above. A—A permanent magnet surrounding a circular core is at the heart of the movement. B—A coil wound on a frame with sensitive springs envelopes the core. C—A carefully machined pivot allows for movement of the pointer. D-The fully assembled electrical measuring device.

Electrical Meters

dicate the appropriate readings of the electrical values being measured, Fig. 19-2.

METER FUNCTION

A meter operates in a predictable way. When connected to a circuit across its terminals, the meter reacts to current flow into the meter coil. Current sets up a magnetic field around the coil. The magnetic field formed by the current interacts with the magnetic field of the permanent horseshoe magnet. This reaction creates torque, making the coil pivot. The greater the current, the greater the torque and, therefore, the more the pointer will be moved, Fig. 19-3.

Fig. 19-2. A calibrated or numbered scale indicates quantity of units being measured.

TYPES OF METERS

An electrician will only have need for six different types of meter. This group will include:
1. Galvanometer. *milliamp meter*
2. Ammeter. *current*
3. Voltmeter. *voltage*
4. Ohmmeter. *resistance*
5. Multimeter.
6. Wattmeter and Watt/hour meter. *power*

GALVANOMETER

The galvanometer, as you read earlier, is an electrical instrument used to detect and measure very small amounts of current. Its construction is shown in Fig. 19-1. When current is sent through the coil, it becomes an electromagnet. The north of this coil is then attracted to the south of the permanent magnet and deflection occurs.

Galvanometers measure current in thousandths of an ampere, and, for practical purposes, are not often used by electricians. However, because other instruments, such as the voltmeter, ammeter and ohmmeter, are special adaptations of the galvanometer, we have included it in our discussion. See Fig. 19-4.

AMMETER

This instrument is employed to measure the current flow in an electric circuit. It, too, is a moving-coil galvanometer, but it has been provided with a SHUNT. A shunt is a low resistance. It is connected in parallel with the coil as illustrated in Fig. 19-5.

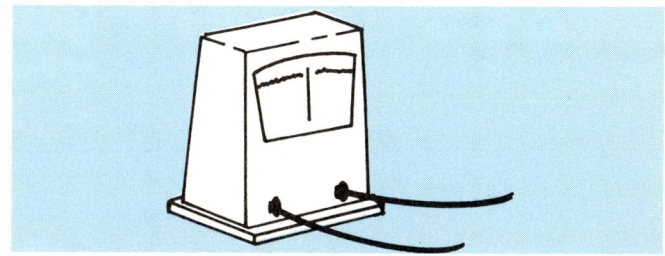

Fig. 19-4. A galvanometer is used to measure very small electrical currents. It is identical in construction to the mechanism shown in Fig. 19-1.

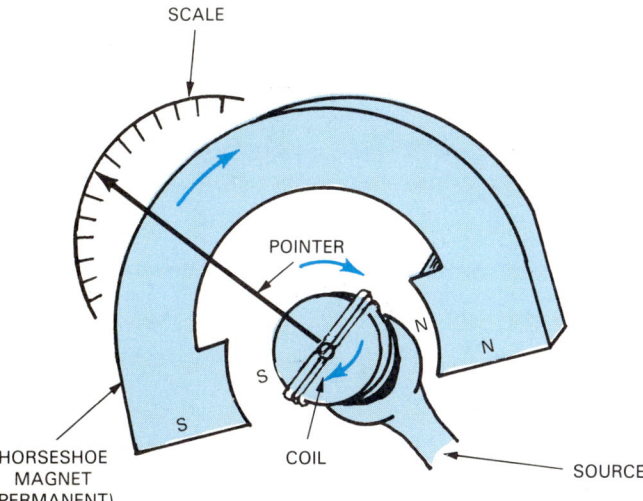

Fig. 19-3. Interaction of magnetic field, created in coil, with that of the permanent horseshoe magnet causes coil to pivot, moving the pointer across the scale.

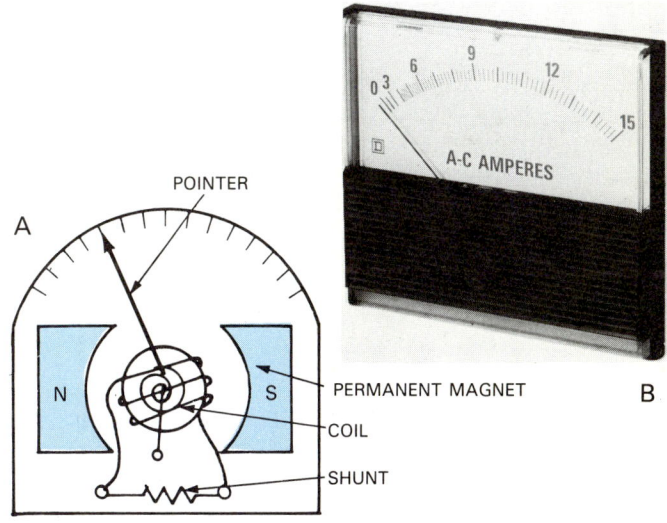

Fig. 19-5. A—Schematic shows how galvanometer principle is modified for ammeter. B—Modern ammeter.

Modern Residential Wiring

The ammeter works like the galvanometer. However, most of the current goes through the shunt rather than through the coil. The coil would burn up without it, Fig. 19-6. The less the resistance in the shunt, the more current will pass through it. The fraction of the total current that flows through the coil is thus determined by the resistance of the shunt. This comparison is used in the calibration of the ammeter.

Ammeters may have one or more scales, depending on the number of shunts used and their resistances. Because ammeters have a small internal resistance, *they must be placed in series with the circuit to be measured.* If placed in parallel, they will burn out. Fig. 19-7 shows proper connection of an ammeter in a circuit.

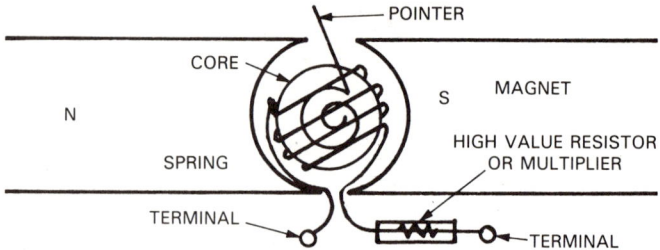

Fig. 19-8. The voltmeter is used to measure the electromotive force (EMF), potential difference, or, as commonly termed, the voltage of an electric circuit.

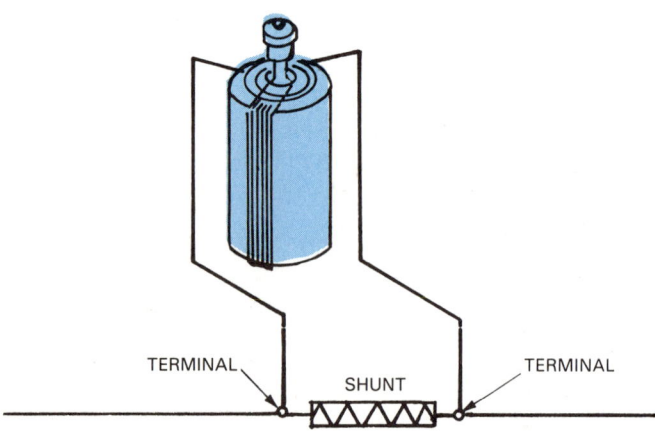

Fig. 19-6. A shunt permits a small amount of current through the meter coil, while a large amount passes through the shunt.

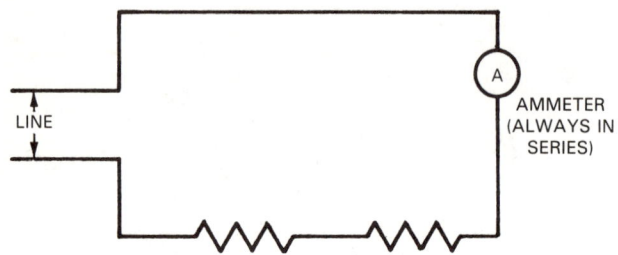

Fig. 19-7. Ammeter measures the current flow of a circuit. They must always be placed in series with the circuit components.

VOLTMETER

The voltmeter is an electrical instrument used to measure difference in potential or the voltage across two points along a circuit. The voltmeter is not unlike the galvanometer and ammeter. The major difference is that the voltmeter has a MULTIPLIER placed in series with the coil. A multiplier is a high-value resistor. See Fig. 19-8.

The scale of a voltmeter is merely calibrated to indicate the potential energy needed to push the small amount of current through the coil.

Voltmeters are *always placed in parallel* to the circuit or with any component part of it. Fig. 19-9 shows the proper connection of a voltmeter in a circuit.

Like all meters, the voltmeter is a delicate instrument which must be handled with care. Moreover, the meter must be placed in the circuit correctly. Failure to do so will surely result in severe damage to the instrument.

OHMMETER

The ohmmeter is used to measure circuit resistance. It utilizes the same basic meter movement as the previously mentioned meters. The probes of an ohmmeter are *placed in series or across*

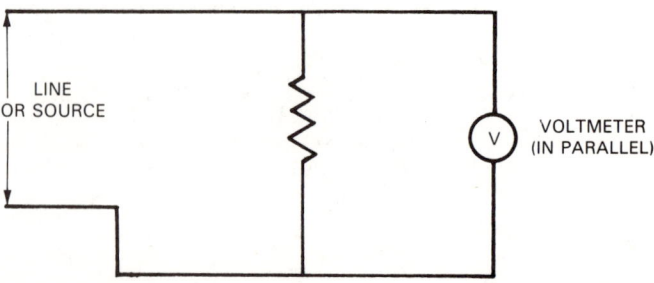

Fig. 19-9. Voltmeters are always placed in parallel with circuit components to be measured.

202

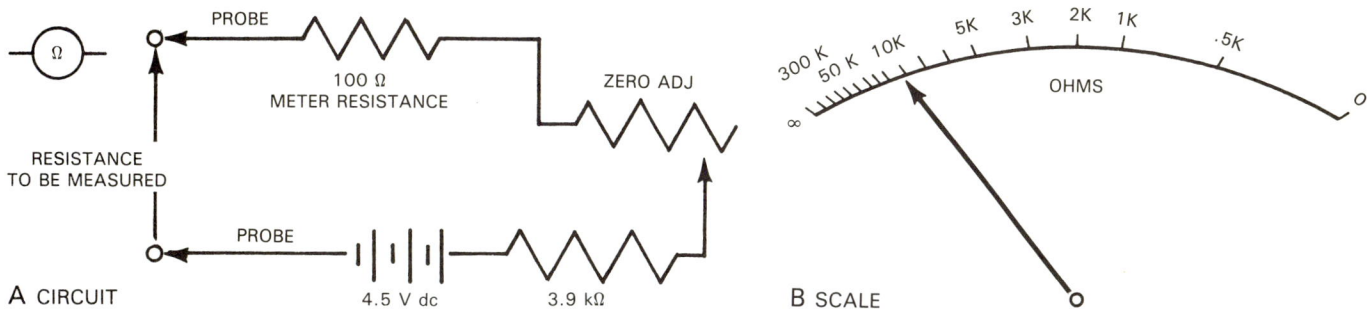

Fig. 19-10. A—The internal schematic of an ohmmeter. B— The ohmmeter scale is not uniform in graduations. The far left of the scale begins with infinity while the far right ends with 0 resistance.

the resistance to be measured. The value is read directly from the meter scale. The scale ranges from infinite resistance to a low of zero resistance.

The ohmmeter contains its own source of electricity, usually one or several dry cell batteries, several resistors, and an adjusting device (zero adjust) to compensate for battery wear. A simple series ohmmeter ciruit is shown in Fig. 19-10, along with a typical ohmmeter scale.

Care must be exercised when using an ohmmeter. Be absolutely sure that *all current has been shut off* and that the portion of the circuit to be analyzed is disconnected. Voltage in the circuit can severely damage the ohmmeter.

Another way of determining the resistance of a circuit is through the application of Ohm's Law. As long as we know the amperage and voltage, we can find the resistance:

$$\frac{\text{Voltage}}{\text{Amperage}} = \text{Resistance}$$

MULTIMETER

Most commonly, a multitester or multimeter instrument is used to measure electrical circuit quantities. They are far more handy than the individual ohmmeter, voltmeter and ammeter since they combine all of these functions in one inexpensive unit. A picture of the multimeter, often called the VOM or volt-ohm-milliammeter, is shown in Fig. 19-11. Note the front panel which provides a switch for selecting function and range. Figs. 19-12 through

Fig. 19-11. The multimeter or VOM is a truly versatile electrical measuring instrument combining in one unit the attributes of a voltmeter, ohmmeter, and milliammeter. (Simpson Electric Co.)

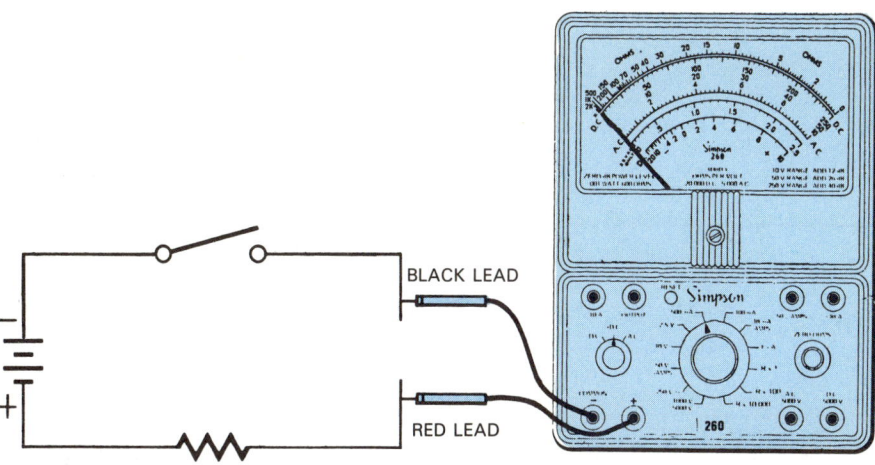

Fig. 19-12. Measuring amperage of a simple circuit using a VOM. Note the positioning and polarity of the probes.

Modern Residential Wiring

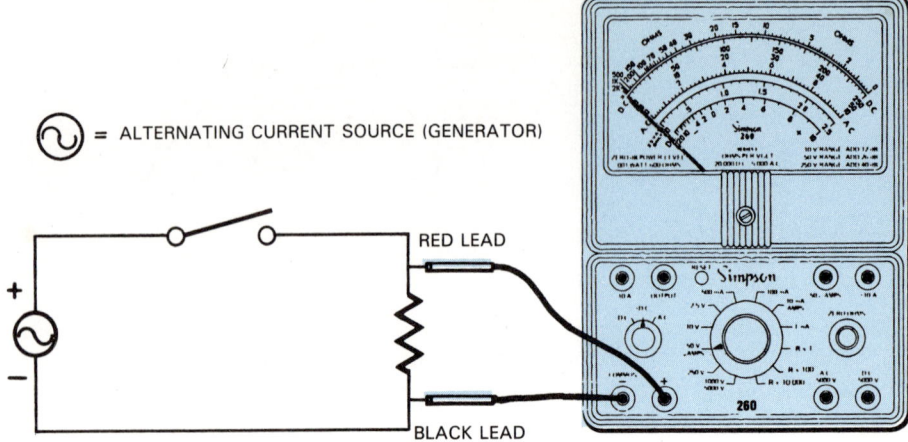

Fig. 19-13. Measuring ac voltage with the VOM is frequently done by electricians. Again, take special note of the hookup and polarity detail.

19-14 show the multimeter hookups for measuring amperage and resistance of a circuit. Complete instructions and suggested procedures are supplied by the manufacturers of the instruments. It is very important to study these instructions carefully before you use the instrument. This precaution will help prevent injury to you and damage to the meter.

ADAPTING METERS TO AC MEASUREMENT

The meters discussed this far are those used for dc measurement. However, with minor alterations, these same meters can be adapted to ac measurement. The main procedure that is used is called rectification. By adding one or several rectifiers into the meter circuit, we can convert a dc meter into one which will measure ac electricity.

THE WATTMETER AND WATT/HOUR METER

The main object of any electric circuit is to provide power to the individual devices within the circuit. If you recall, power is measured in watts which is the product of the electromotive force (voltage) and the rate of current flow (amperage). Thus, voltage × amperage = power (in watts or VA.)

A meter designed for measuring power in an electric circuit is called a WATTMETER. The internal structure of this type of meter is shown in Fig. 19-15. In essence, it is a combination voltmeter-ammeter.

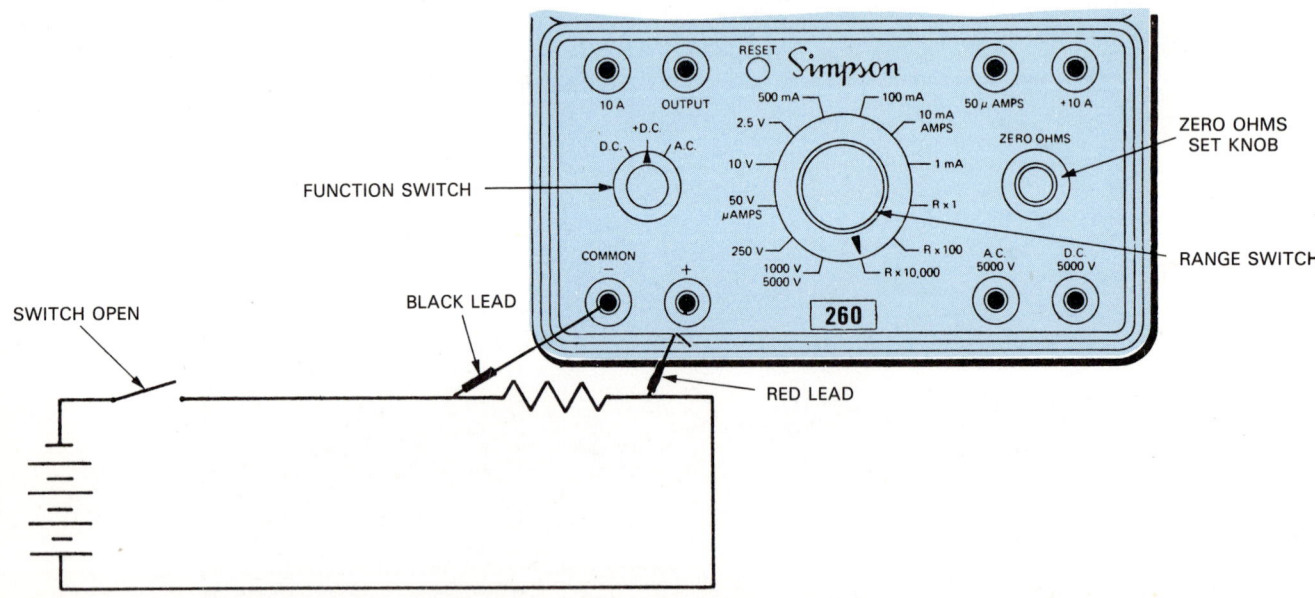

Fig. 19-14. In order for resistance measurements to be accurately taken, the range switch must be set on the highest rating for resistance readings; the zero adjust knob must be set to bring the pointer (not shown here) to zero; the function switch must be on dc, the circuit must be de-energized and the polarity must be correct.

Electrical Meters

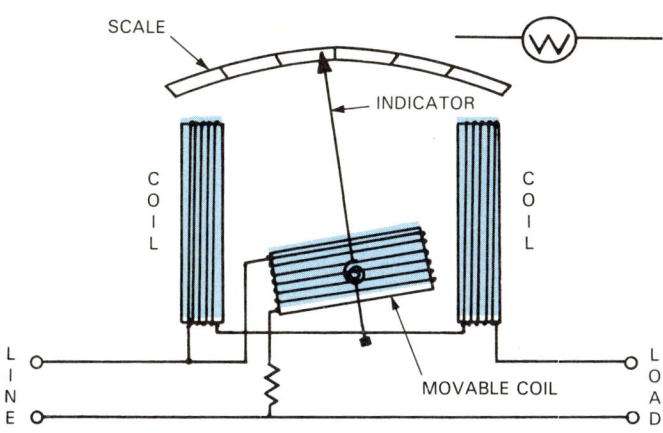

Fig. 19-15. Wattmeter symbol and design.

WATT/HOUR METER

A modification of the wattmeter which measures the amount of electrical energy used by the consumer, is called the watt/hour meter. Utility companies require all of there customers to have one as part of the service entrance. This watt/hour meter records the power used in units of *kilowatt hours (kW/hr).* One kilowatt hour is equivalent to 1000 watts of power being used for one hour of time, two kW/hr would be equivalent to 1000 watts of power being used for a two hour period of time, and so on. An average family of four may use 1000 kW/hr of power each month. The circuitry of the watt/hour meter is illustrated in Fig. 19-16.

Fig. 19-17, shows the dial markings on a typical watt/hour meter. To read the meter, simply write down the number that the pointer *has just passed* on each dial going from left to right. For example, the dial markings illustrated, would be read 1559 kW/hr.

METER CARE

Electricians will rely upon their meters to a greater or lesser degree. How long the meter will serve them will, in turn, depend on the amount of use and care it receives.

The main thing to remember is that meters are delicate measuring instruments. They *must* be handled with care. Magnetic type meters should never be placed near strong magnetic fields. Terminals or probes of a meter must be placed properly in the circuit to be tested. Direct current polarity must be observed. Maximum current or voltage must not be exceeded. Keep the meter clean and store it in a protected case or cover.

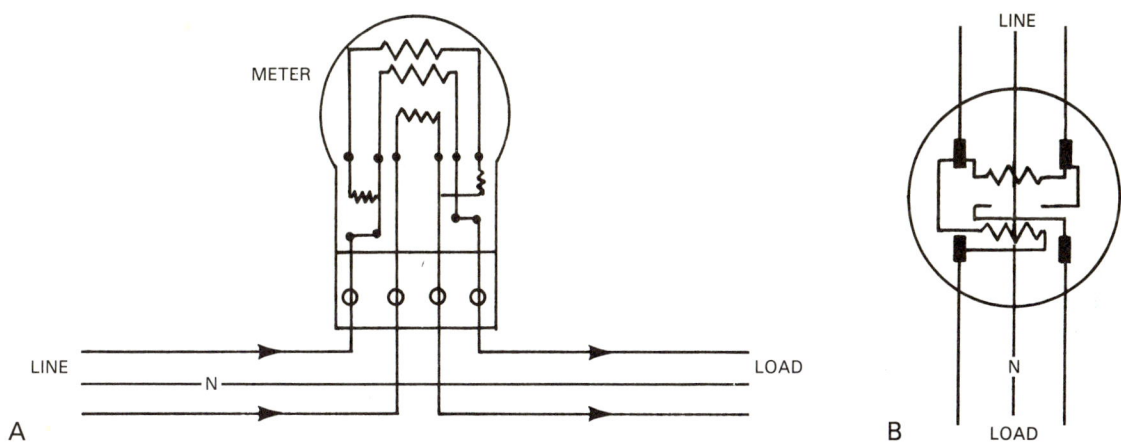

Fig. 19-16. The watt/hour meter. A—Bottom-connected type. B—Socket type.

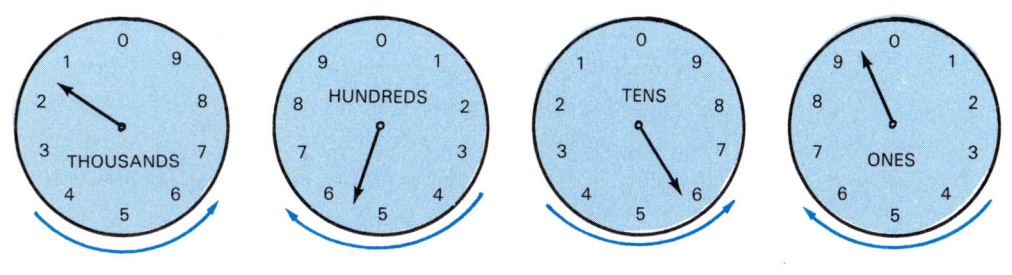

KILOWATT-HOURS

Fig. 19-17. Reading a kilowatt-hour meter. Note the direction of the pointer rotation. The numbers you read are the ones which the pointer has just passed.

Modern Residential Wiring

REVIEW QUESTIONS — CHAPTER 19

1. Complete the following chart:

METER:	MEASURES:	SYMBOL:
	Small amounts of electrical current	G
	Electrical current	
Voltmeter		
	Resistance	Ω or OHM

2. The basic meter component, called the _____ movement, consists of a cylindrical iron core placed within a magnetic field created by a horseshoe magnet.
3. A _____ is a low resistance connected in parallel with the coil of a(n) _____.
4. A _____ is a high resistance connected in series with the coil of a(n) _____.
5. A most versatile measuring meter called the _____ combines the capabilities of several individual meters.
6. Rectifiers are added to a meter circuit to convert it from direct current to alternating current. True or False?
7. Voltage x Amperage =
   a. Potential difference.
   b. Power.
   c. Rectification.
   d. Current.
8. The meter dials below would be read as _____ kW/hr.

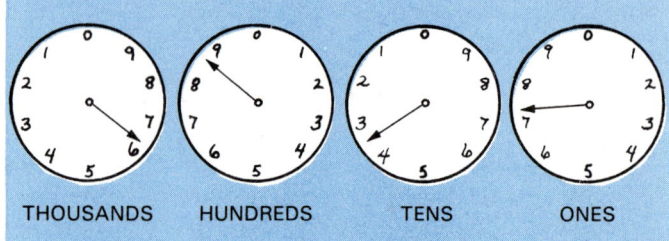

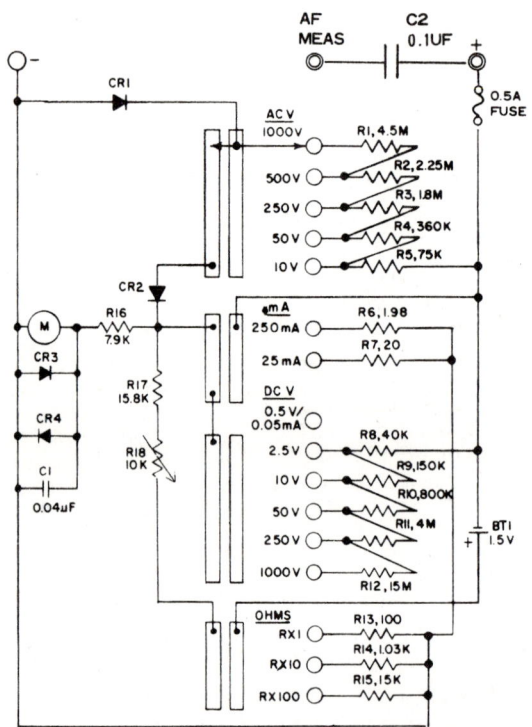

NOTES:
1. UNLESS OTHERWISE SPECIFIED:
   A- ALL RESISTOR VALUES ARE IN OHMS ±1%, 1/4W
   B- ALL CAPACITOR VALUES ARE IN MFD.

Schematic diagram of typical VOM.

Chapter 20

ELECTRICAL TROUBLESHOOTING

The electrician is often called upon to correct or repair a malfunction in an existing electrical system. Although there are no hard and fast rules regarding the steps in such a process, we will discuss those procedures which are most helpful in tracing common troubles.

SAFETY CONSIDERATIONS

Safety must be a continual concern of all electricians. Safety must be uppermost in the troubleshooter's mind since this aspect of electrical work can be dangerous. In most instances, you will de-energize the device, the circuit, or the entire system in order to correct or repair a malfunction. However, at certain times, it is necessary that the system remain energized in order to track down the problem. It is during these times, that alertness and safety consciousness is essential.

TROUBLESHOOTING TOOLS

Some of the tools of particular importance to troubleshooting are:
1. The *VOM.* See Chapter 19 and Fig. 20-1.

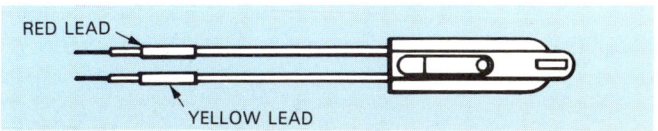

Fig. 20-2. A neon tester will indicate the presence of 120 or 240 V by how brightly it glows when placed across the load. (GE Wiring Devices Dept.)

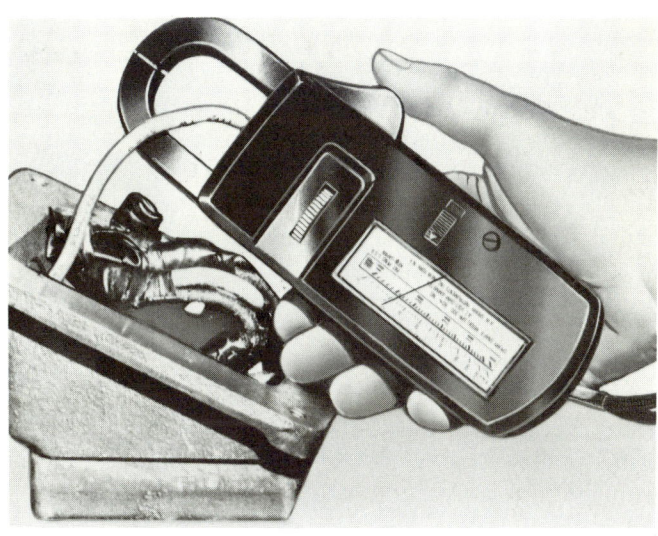

Fig. 20-3. The amp clamp is indispensable when the current reading of a live conductor is needed. (A.W. Sperry Instruments Inc.)

Fig. 20-1. A truly handy testing instrument is the VOM or volt-ohm-milliameter. It can be used to measure most characteristics of an electric circuit.

2. The *Neon Tester,* Fig. 20-2. The neon tester is a simple device which will indicate 120 V or 240 V by the brightness of its glow when connected across the hot and neutral wires or two hot wires of a circuit. The greater the voltage, the brighter it will glow.
3. The VOM will indicate the same information. However, it will be in numerical form on a calibrated scale.

Other more precise instruments can be used when more exact information regarding circuit characteristics is needed. For example, an AMP CLAMP, shown in Fig. 20-3, is essential when you

Modern Residential Wiring

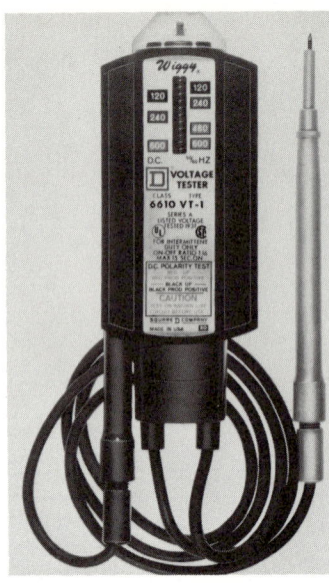

Fig. 20-4. A good quality voltage tester will indicate voltage as well as polarity. This model is lightweight, compact, and inexpensive. (Square D Co.)

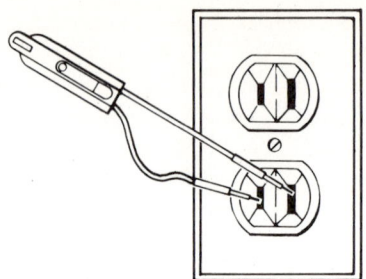

Fig. 20-5. Checking for power at outlet. Insert leads of neon tester into slots of receptacle. If tester light glows, circuit and receptacle are all right.

need to know the exact current flow in a given conductor. A compact VOLTAGE TESTER like the one in Fig. 20-4 will indicate voltage and polarity through a series of vertical neon lamps.

DIAGNOSING PROBLEMS

Electrical troubles show up as a wide range of symptoms. The common simple problems usually involve faulty receptacles, switches, lighting fixtures, fuses, breakers, and the like. More complex situations may include open circuits, broken conductors, voltage fluctuations, and ground-faults, or current leakages. We will limit our discussion to the more commonly encountered and readily solvable troubles.

TESTING RECEPTACLES

When receptacles are not performing properly, chances are either the circuit or the receptacle itself is at fault. If no voltage is indicated when the tester leads are inserted in the receptacle contact slots, Fig. 20-5, check the line conductors for voltage to the receptacle. This is accomplished by removing the wall plate and checking for voltage at the terminal screws, Fig. 20-6. If voltage is present at the terminal screws, the receptacle is defective and must be replaced.

Sometimes, an open neutral conductor in the receptacle is the cause of the problem. If no voltage is present at the receptacle slots as in Fig. 20-5, but a positive test lead reading is obtained as shown in Fig. 20-7, the chances are very good that the neutral is open. If there is no voltage at either the

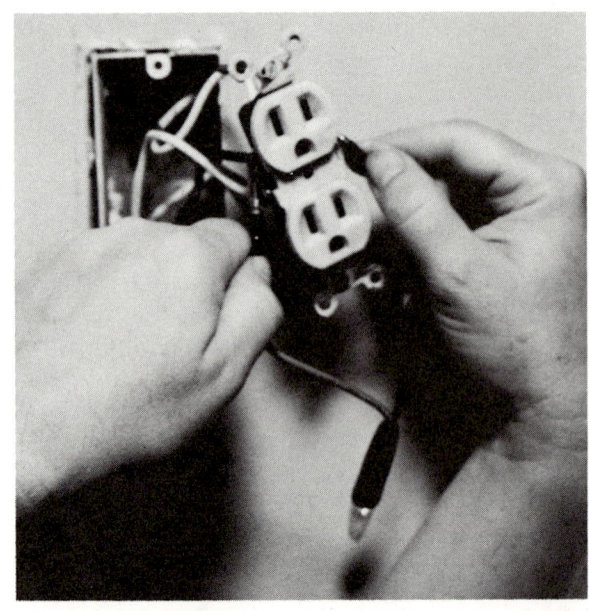

Fig. 20-6. If the tester does not glow when leads are in receptacle slots (Fig. 20-5), pull off plate and try this test. If test light goes on, voltage is present up to the receptacle. It indicates that the receptacle is not working.

Fig. 20-7. If the tester does not light during the test shown in Fig. 20-5 try this test. If you get a positive reading between the "hot" conductor and the ground it indicates that the neutral is open.

receptacle slots or the terminal screws, then an open circuit is the problem. This may have occurred at the distribution panel or somewhere along the circuit wires. Possibly it happened at a previous receptacle.

CONTINUOUS GROUND CHECK

Too often, improper grounding can result in shocks which cause discomfort to the user of an outlet or fixture. In some cases, the shock may be serious enough to cause death. We have stressed the importance and methods of grounding throughout this text. Checking the grounding system should be routine. Each time a system comes under scrutiny or repair for any reason a check should be made. The method of checking circuit grounding at the receptacles is the same one shown in Fig. 20-7.

Older type receptacles do not have a ground slot. To check the ground continuity, insert one lead of the tester into the slot for the hot wire. Touch the other lead to the metal mounting screw. Tester will glow brightly if the box is properly grounded. See Fig. 20-8.

TESTING SWITCHES

To test switches, remove cover plate and determine if there is power to the switch by touching one lead of neon tester to the metal box and the other lead to the line side terminal, Fig. 20-9. If tester lights, circuit is working. Now turn switch to "on" and touch one lead to load terminal of switch while other lead is touching the metal box. If test light glows, switch is alright and fixture or wiring to it is faulty. Where nonmetallic boxes are used, one lead of tester must contact a neutral wire.

TESTING FIXTURES

Fixture condition can be determined by checking whether power reaches the supply conductors at the fixture outlet, Fig. 20-10. If there is voltage indicated at this point, then the fixture is not functioning. It needs repair or replacement.

OVERLOADED NEUTRALS AND UNBALANCED CURRENTS

It is often more convenient and economical to use one three-wire cable rather than two two-wire circuits. This is particularly true when both circuits would follow the same general direction. Installing two two-wire cables takes more material and time.

The substitution of a single three-wire cable for two two-wire cables constitutes a multiwire branch circuit. The rules regarding such circuits are

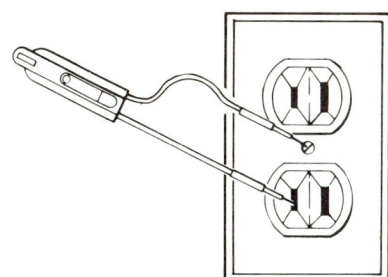

Fig. 20-8. Testing a two-wire outlet for proper ground. Tester should glow just as brightly as when the tester leads are in both slots of the outlet. (GE Wiring Devices Dept.)

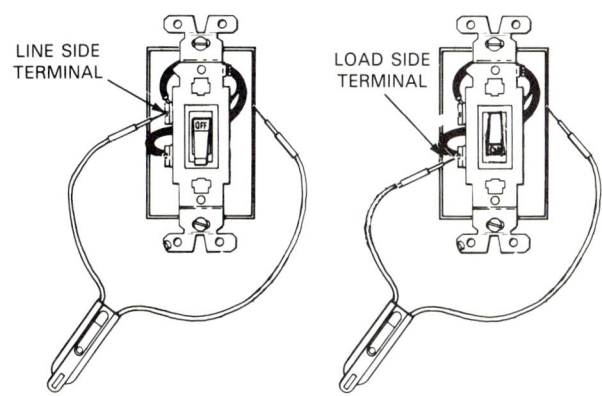

Fig. 20-9. With switch off as at left, the tester will indicate voltage at the line side terminal screw if circuit is "live." If tester does not light, circuit is faulty. Right. With the switch on, neon tester should indicate voltage between load side switch terminal and box or neutral conductor. If tester fails to glow, switch is faulty. Replace switch and repeat test.

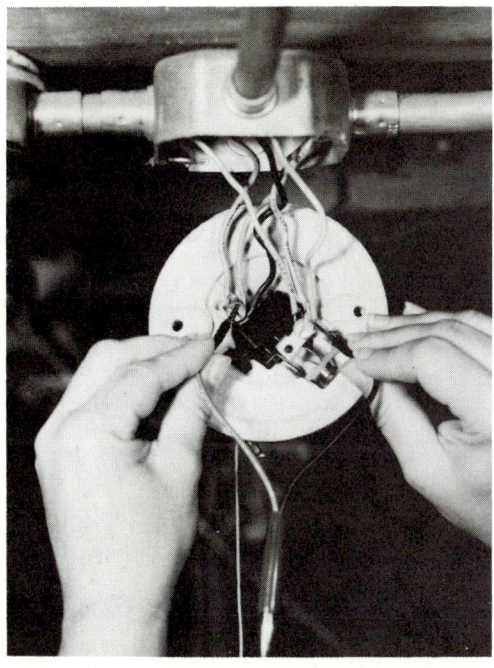

Fig. 20-10. If fixture will not function properly, but test light indicates voltage is present at circuit conductors, then the fixture must be repaired or replaced.

referenced in *Code* section 210-4. Such a circuit may be used when split-wiring receptacles or when a circuit run consists of many light fixtures. The two circuits formed this way use a common neutral (the third white, grounded conductor in the cable). That is, both "hot" wires use the same neutral. See Fig. 20-11.

Current will travel one direction in one hot wire while it is traveling the other direction in the other hot wire. The neutral wire will carry any unbalanced current resulting from differences between the hot wire loads. For example, if one conductor in Fig. 20-11 is carrying 5 ampere of electricity, and the other conductor is carrying 7 ampere, the common neutral will conduct the 2 ampere difference.

IMPROPERLY CONNECTED CIRCUITS

Some problems do occur with multiwire circuits under certain conditions. With proper troubleshooting, the electrician can detect and correct the problems.

By far the most common problem occurring with split or multiwired circuits is the improper connection of the circuit conductors at the panel. Too often the hot conductors are placed on the same hot busbar. This overloads the neutral because the current load it carries is the sum of the two hot wire currents. The overcurrent frequently overheats the conductor and insulation, causing the neutral conductor to short circuit. The insulation may actually melt away from the conductor, Fig. 20-12.

Further, should the neutral open, excessive voltage surge may occur along one of the hot conductors. High voltage could damage the loads connected to the circuit.

It is important that the electrician insure that the loads on each hot conductor are closely matched. They need not be exactly the same, but the closer the better. Remember, the neutral will carry the imbalance or difference. In addition the NEC requires that the two hot conductors have a common disconnect which acts simultaneously in case of overcurrent. This rule is important and must not be disregarded. An ordinary double-pole, single throw breaker will satisfy this requirement.

FUSES

Whenever fuses "blow," the immediate concern is to determine why they did so, Fig. 20-13. Only when the cause is corrected, should the fuses be replaced. In addition, never replace a fuse with one having a higher rating. This will only defeat the purpose of the fuse in the first place.

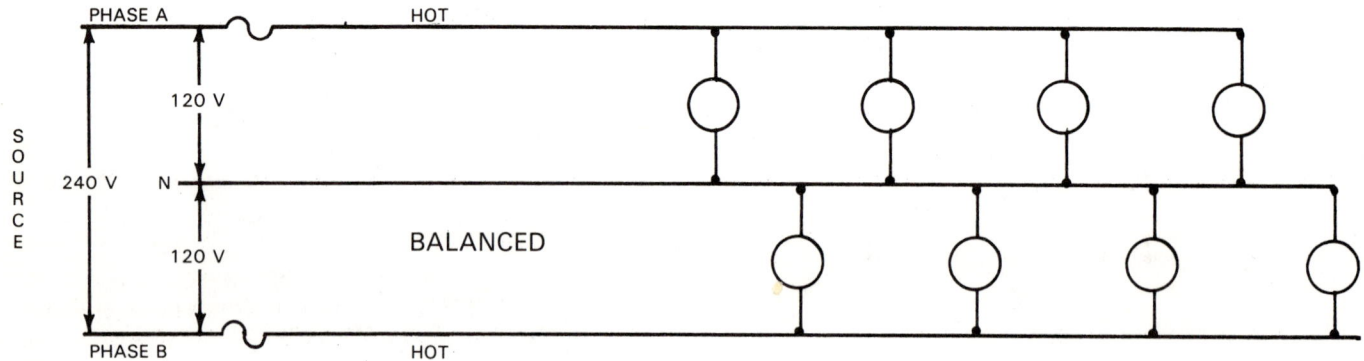

Fig. 20-11. Properly constructed multiwire or split wired circuit. Neutral will safely carry unbalanced portion of load.

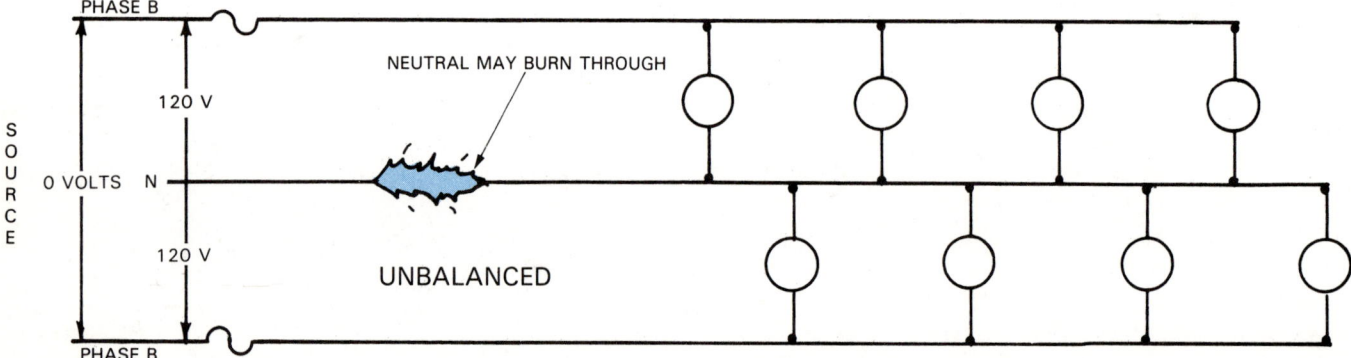

Fig. 20-12. One of the most common problems with multiwire or split-circuits is the improper phase division at the lighting panel. In this situation, a serious imbalance can cause overloading and probable damage to the neutral, as well as to the loads.

Electrical Troubleshooting

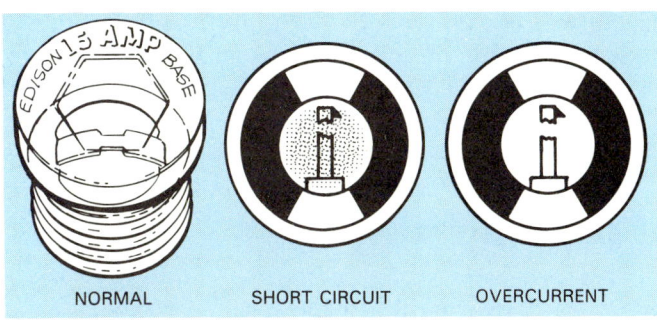

Fig. 20-13. When fuses "blow" the first concern is to identify the cause. The pattern of a blown fuse is the first clue. Note the normal, short-circuit, and overcurrent patterns of the common Edison-base fuses. (GE Wiring Devices Dept.)

When plug fuses blow they are easily located and replaced. When cartridge fuses blow we must first identify the blown fuse, since there is often no visible evidence.

You can identify the blown cartridge fuse in one of two ways:
1. Remove the cartridge fuses from the panel and test them for electrical continuity. Use the VOM.
2. With the cartridges still installed, test them using the neon tester. The second method, Fig. 20-14, is easier and saves time.

BREAKERS

Breakers are treated in much the same way as fuses. Should a breaker "trip," there is undoubtedly a cause. Find it before resetting the breaker.

Occasionally, breakers malfunction and will not reset. This is determined using the same method

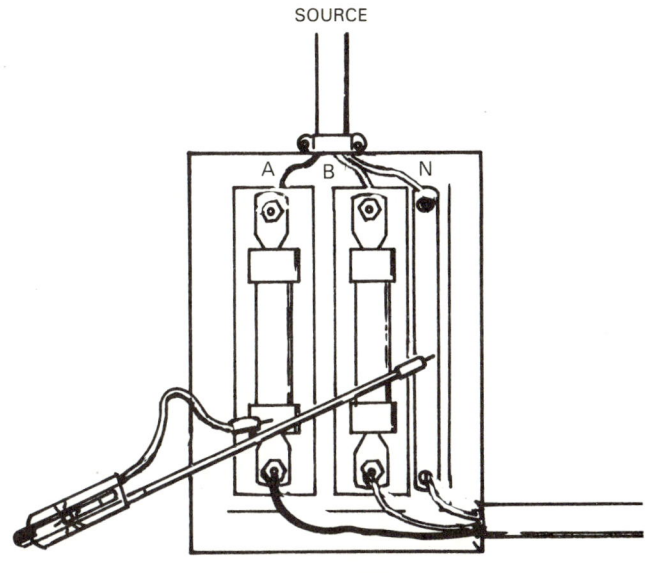

Fig. 20-14. Method of checking out a fuse. If fuse is good, light on neon tester will light.

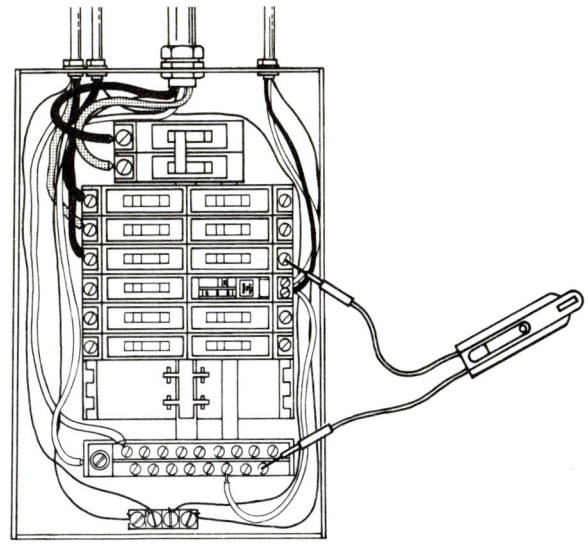

Fig. 20-15. Method of testing circuit breaker. If light in tester does not go on, reset and test again. If test is still negative replace the breaker with one of the same rating. If breaker trips when reset, the problem lies somewhere else in the circuit. (GE Wiring Devices Dept.)

as with a blown cartridge fuse. With the breaker reset, place the neon tester between its terminal and the neutral busbar. If the results are negative the breaker is not functioning, as shown in Fig. 20-15.

MISCELLANEOUS PROBLEMS

Obviously, there are other problems which can arise in electrical systems. Those originating with electrical equipment can range from intermittent shorts (those which occur only at certain times and under certain conditions) to entire system shutdowns because of service equipment failures. These problems, however, are rare in comparison to those just discussed. They are beyond the scope and purpose of this text. However, the key point to keep in mind is that all electrical problems are traceable through careful analysis of the system components and perseverance on the part of the troubleshooting electrician.

REVIEW QUESTIONS — CHAPTER 20

1. List three common testing tools used in troubleshooting an electrical system.
2. Describe the procedure for troubleshooting each of the following:
   a. Receptacle outlets.
   b. Switches.
   c. Fixtures.
3. In a multiwire branch circuit two different circuits will use a _____ _____.
4. List the two ways of identifying a blown cartridge fuse.

Chapter 21

SPECIALIZED WIRING

The electrical wiring installations covered in this chapter are performed by the electrician on a less frequent basis than those mentioned in previous chapters. Nonetheless, many of these slightly "offbeat" installations require a high level of skill, competence, and *Code* knowledge. Also included in this chapter is a section on telephone wiring.

SIGNALING CIRCUITS

Signaling circuits supply electrical energy to doorbells, buzzers, signal lights, and other warning devices. In addition, remote-control low voltage circuits, covered in Chapter 17, are included in this category.

The NEC distinguishes signaling circuits by dividing them into classes:
1. Class 1 circuits are those in which power is limited to no more than 30 V and not over 1000 VA. Remote motor-switching systems are in this classification.
2. Class 2 and Class 3 circuits are power limited not to exceed 30 V and 100 VA. Such circuits are used for chimes, doorbells, humidistats, thermostats, and the like.

You will find detailed information on all three classes of remote control and signaling circuits in Article 725 of the *Code*. We shall discuss only *signaling circuits,* such as the household buzzer or bell systems.

CONDUCTORS

Buzzers, bells, and chimes operate at a nominal 6–20 volts. They require wire ranging in size from 16 to 20 AWG, depending on the device rating and length of run. See Fig. 21-1.

Conductors serving buzzers, bells, or chimes must never be run with regular or full-power circuits that operate on 120 or 240 V. They must never pass through or terminate in boxes or enclosures containing full-voltage conductors, unless separated by a firmly fixed partition.

TRANSFORMER

Doorbell type circuits may be battery-operated as shown in Figs. 21-2 and 21-3. However, a more efficient system requiring less maintenance uses a transformer to supply power, Fig. 21-4.

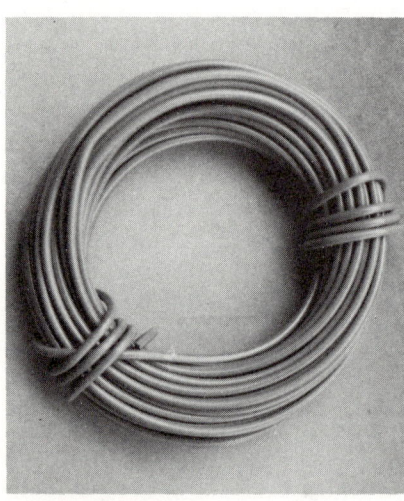

Fig. 21-1. Bell wire, 16-20 AWG, is used in circuits running between transformers and bells, buzzers, chimes, and the like.

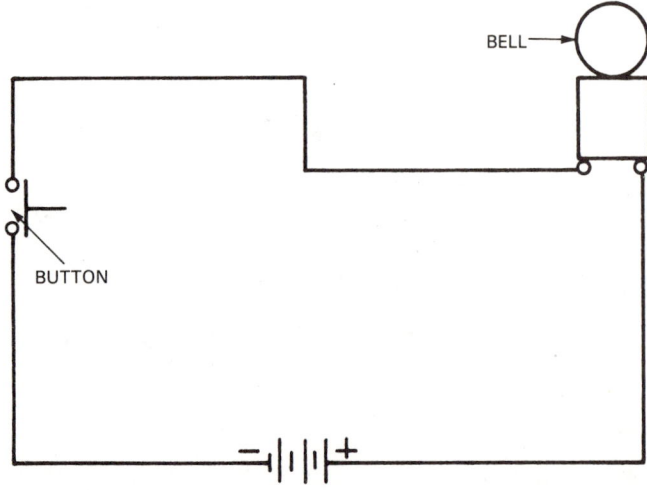

Fig. 21-2. A single bell operated by a 6 V battery is the simplest type of signaling circuit.

212

Specialized Wiring

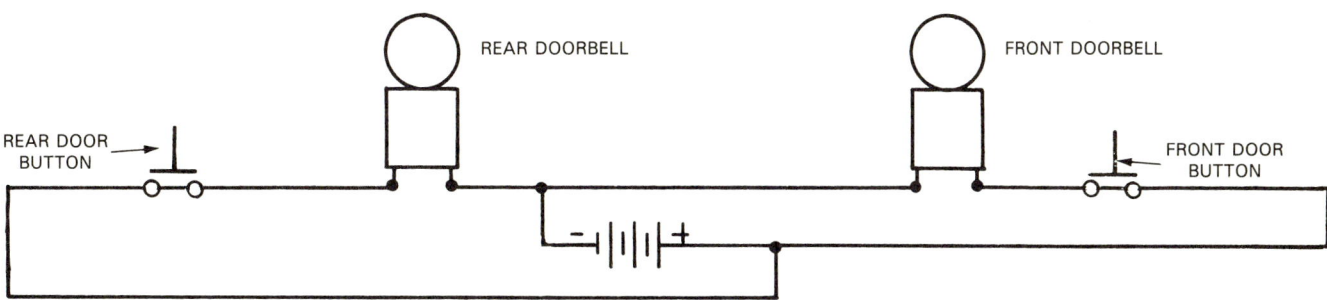

Fig. 21-3. A double-bell system energized by a battery. This system is more flexible, but is limited by battery charge.

Fig. 21-4. Doorbell, chime, or buzzer transformers operate at 6 to 20 volts.

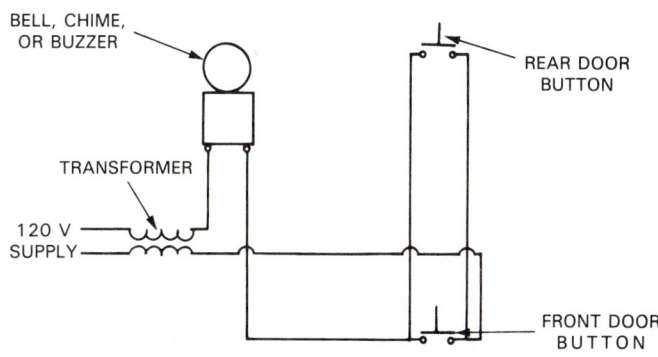

Fig. 21-6. A typical, yet simple, bell system operated by transformer supply.

The transformer is placed at an outlet box from which the 120 V supply will be tapped. The lead wires must be connected to the 120 V line conductors. The connection must be made within an enclosure.

To accomplish this, the transformer is generally placed on its own outlet box. The 120 volt supply conductors are brought into this box, Fig. 21-5. From the transformer load side, the bell or buzzer wire is directed where needed.

A simple schematic for a transformer operated bell system is shown in Fig. 21-6. A similar two-button chime system is sketched in Fig. 21-7.

HAZARDOUS LOCATIONS

The NEC classifies locations where explosive, ignitable, combustible, or otherwise dangerous materials are produced, stored, or handled as HAZARDOUS. These hazardous locations are

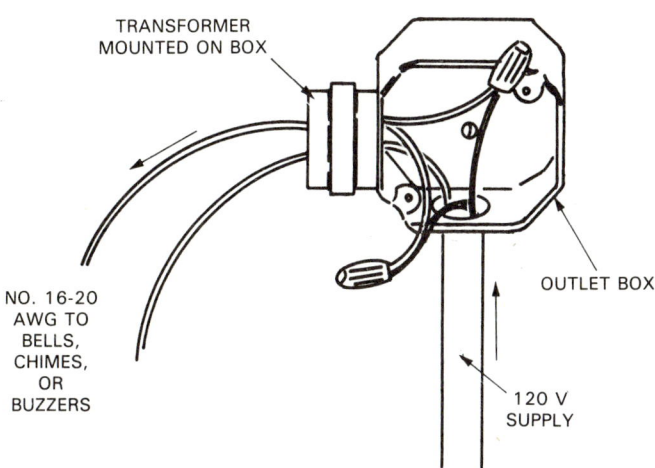

Fig. 21-5. Proper mounting and connection of the doorbell, buzzer, or chime transformer. Note that all full voltage wires are housed in an outlet box.

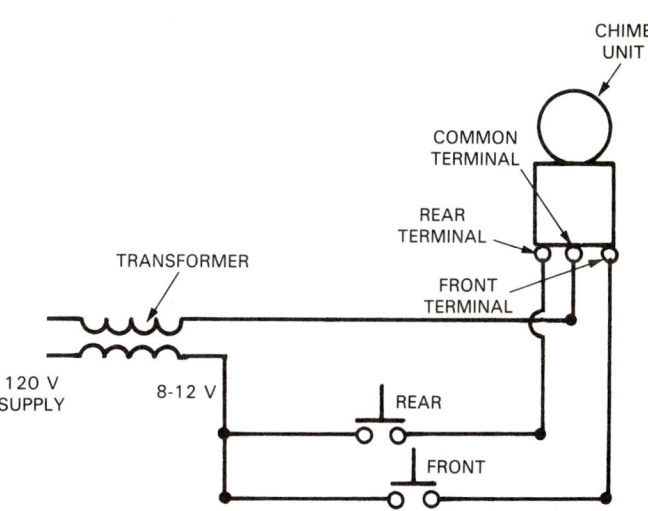

Fig. 21-7. Double-button (front, rear) chime systems.

213

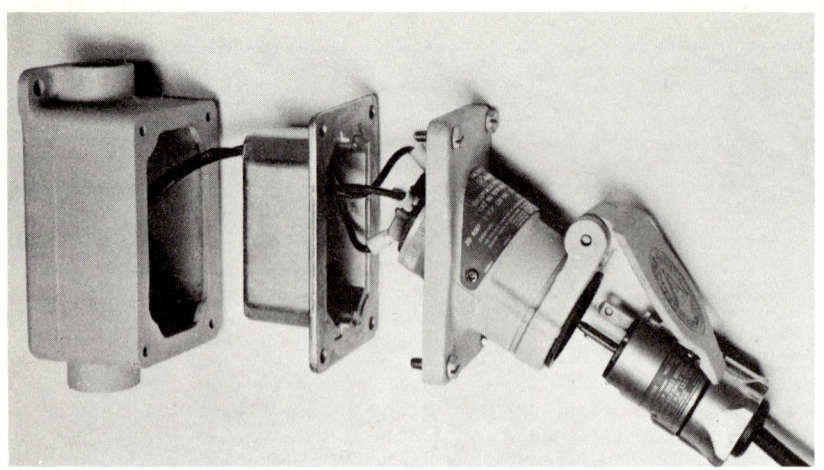

Fig. 21-8. Explosion proof receptacles such as these are used in hazardous locations. Left. Exploded view. No outside seals are needed. Right. Fully assembled. Plug is also watertight. (Appleton Electric Co.)

divided into three classes based on the environment and type of hazardous material. Refer to articles 500-503 of NEC.

1. Class I—Locations in which flammable gases or vapors are or may be present in the air in quantities sufficient to produce explosive or ignitable mixtures.
2. Class II—Locations which are hazardous due to the presence of combustible dust.
3. Class III—Locations hazardous because of the presence of ignitable flyings or fibers, but where such fibers will probably not be present in the air in sufficient quantities to produce ignitable mixtures.

Each class is divided into:

Divison 1 — The hazardous material is usually present in the air because of the manufacturing, storing, or handling process.

Division 2—The hazardous material is usually confined in cartons, containers, etc., and, therefore, will not produce explosive or ignitable mixtures in the air unless by accident or through some failure of the normal operating conditions.

WIRING OF HAZARDOUS LOCATIONS

Installing electrical wiring in hazardous locations require special attention and consideration:

1. For Class I, Division 1, hazardous locations, the wiring must be enclosed in rigid or intermediate metal conduit or with type MI cable. Circuits must be terminated in specially constructed explosion-proof enclosures. See Figs. 21-8, 21-9, and 21-10. Hazardous locations in this classification would include operating rooms of hospitals, dry cleaning plants, factories which handle or transfer flammable or explosive gases, paint spraying booths, and the like.
2. For Class I, Division 2, hazardous locations, the wiring should be protected by flexible metal conduit, a flexible metal raceway, or special heavy duty cord approved for this purpose. Terminations must be in explosion-proof enclosures. Examples of such locations would be flammable gas or liquid storage areas (service stations, propane or bottled-gas storage facilities, and similar places).
3. In Class II, Division 1, hazardous locations, the wiring may be in threaded conduit, MI cable, flexible metal conduit, or type S heavy duty cord. All connections, equipment, and devices must be housed in dustproof enclosures, or wireways. Examples would be coal and grain elevators.
4. Class II, Division 2, hazardous locations such as flour processing plants, grain storage facilities, and carbon-black holding areas should

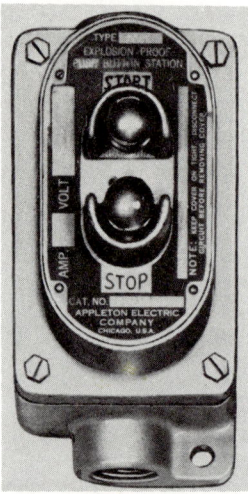

Fig. 21-9. Explosion-proof push button station is suitable for use in Class I, II, and III hazardous locations. (Appleton Electric Co.)

Fig. 21-10. Hazardous location lighting fixtures. (Killark Electric Mfg. Co.)

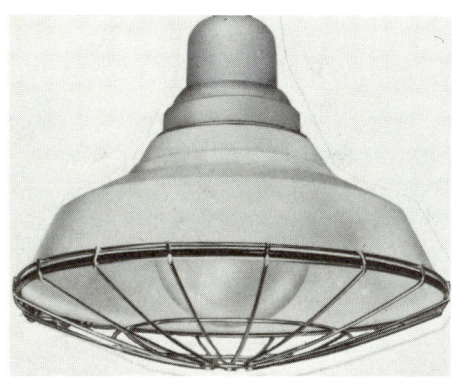

Fig. 21-11. Lighting fixture typically used in areas where high humidity and dust are present. (Killark Electric Mfg. Co.)

be wired in either the manner described for Class II, Division 1, or with EMT.
5. Class III, Division 1, hazardous locations require the same wiring methods as do Class II, Division 2. Examples would include textile or fabric mills as well as enclosed woodworking or lumber mills.
6. Class III, Division 2, hazardous locations should be wired by methods which exclude dust, lint, and fibers. That is, the enclosures, fittings, conduit, cable, and cords should be sealed to keep out dust particles, lint, etc. See Fig. 21-11. Examples of these type of locations would be textile or fabric storage plants.

ADDITIONAL CONSIDERATIONS

It is strongly recommended that those wishing to install electrical systems in these types of locations

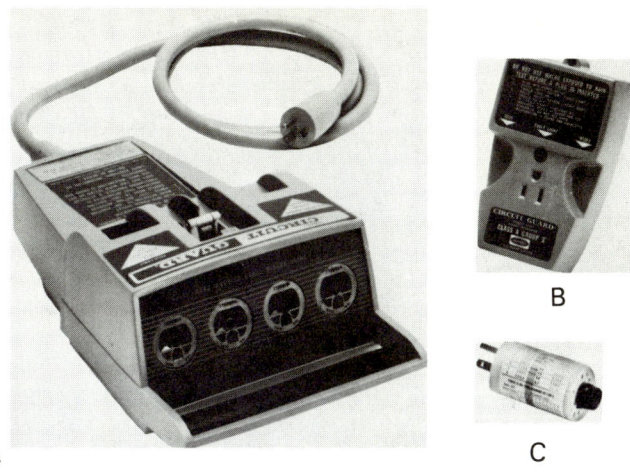

Fig. 21-12. Portable, plug-in GFCI units are particularly useful in areas where GFCIs are not part of the permanent wiring. These units are easily carried from one location to the next. A—Four-outlet plug-in GFCI. B—Single outlet portable GFCI. C—Handy GFCI tester. (Harvey Hubbell Inc.)

consult the *National Electrical Code,* power authority, and local inspector for more complete information and suggestions. More often than not, certain facilities have areas which fall under more than one *class* and *division* of the hazardous locations classifying scheme. Each portion of such a facility must be wired accordingly. Usually commercial garages, aircraft hangers, and service stations will fall under several classifications.

GARAGES AND OUTBUILDINGS

Garages and outbuildings will usually require special wiring for two reasons:
1. These structures are often built below grade. This factor entitles these structures to special electrical consideration as they are often damp, if not wet.
2. Garages and outbuildings are not often contiguous (attached) to the main portion of the residential or commercial structure. Therefore, they must have separate (sub-feed) power panels.

CIRCUITS, FIXTURES, AND RECEPTACLES

In most cases, the garage or outbuilding should have circuits which are *separate* from the main structure. This remains true whether the garage or outbuilding is physically attached to it or not.

Each structure should have at least one lighting and one receptacle outlet circuit. The lighting fixtures should be overhead, switched, and in sufficient number to clearly light the entire area.

The receptacle outlets should be spaced, as in the main dwelling, no more than 6 ft. from any point along the wall. These should be installed at about 50 in. from the floor. Each should have ground fault circuit interrupter protection. GFCIs are required for

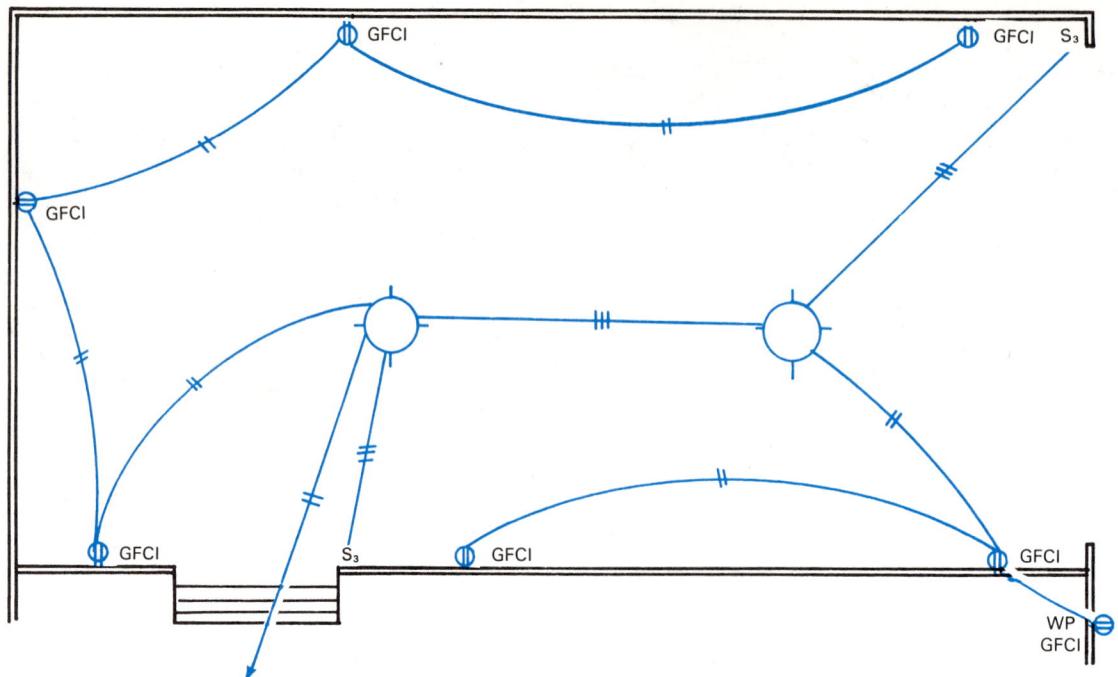

Fig. 21-13. Garage receptacles and lighting outlet layout.

garage receptables.* They would be advisable in outbuildings too, Fig. 21-12.

A garage layout might look like the one in Fig. 21-13. Note the use of GFCIs and cable routes.

SUB-PANEL POWER SOURCE

As stated earlier in the text, sub-panels fed from the main service equipment will be the four-wire type. Two are hot wires; a third is neutral, and a fourth is a grounding conductor. The latter must be isolated from the neutral at the sub-panel. Sub-panels will usually have provisions for four to six circuits.

EMERGENCY AND STANDBY SYSTEMS

Certain public structures are required to have a second wiring system which takes over when the regular electrical power source is not functioning. The *National Electrical Code* identifies two separate systems: emergency systems and standby systems. Requirements for the systems are outlined in Articles 700-702.

EMERGENCY SYSTEMS

Emergency systems are those legally required for emergency power by a municipal, state, or federal governmental agency. Emergency systems are intended to *automatically* supply enough power to assure the safety of a building's occupants during a power failure, Fig. 21-14.

*Except for those receptacles dedicated to a freezer or similar permanent appliance, or one which serves a garage door opener unit.

LOCATIONS

Emergency systems are required mostly in places such as health care facilities (like hospitals) for fire detection and fire alarm systems, elevator operation, fire pump units, public safety communication systems, essential ventilating systems, and any other building services whose failure to operate could produce a serious threat to life.

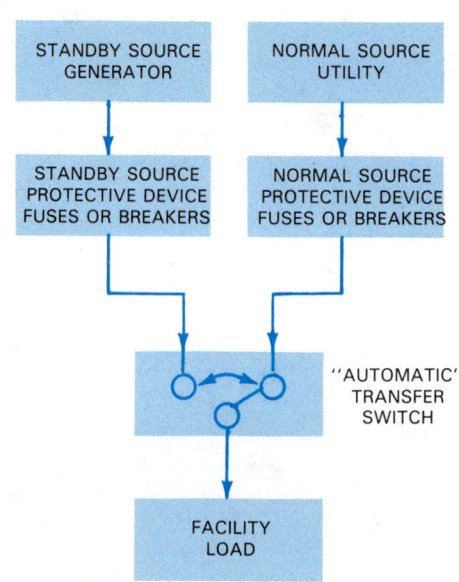

Fig. 21-14. Schematic of emergency power systems which automatically energize when power is interrupted. Note that the generator unit, as well as the regular source, has overcurrent protection in the form of fuses or breakers.

Specialized Wiring

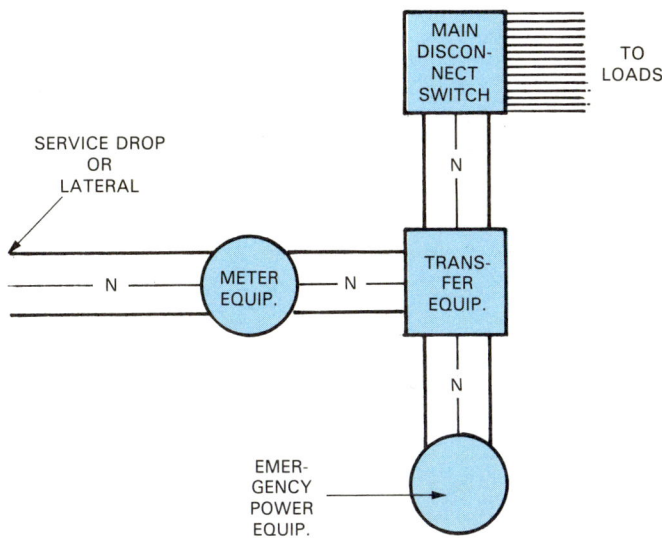

Fig. 21-15. General overall arrangement of emergency power supply equipment in relation to service disconnect and metering equipment.

MAINTENANCE AND TESTING

Emergency systems must be monitored and maintained. They will require periodic testing by the authority in charge. What this means is that, every month or two, the system(s) will be activated for several minutes to assure proper functioning in the event of an actual current interruption.

CURRENT CAPACITY

The amperage rating or capacity of the system should be at least enough to operate those devices (such as lights, elevators, etc.), which are a must during a power failure. Emergency systems are available, which can supply upwards of 500,000 kVA.

EMERGENCY SWITCHING LOCATION AND ACCESSIBILITY

The transfer switch for emergency systems, *should be located ahead of the main service disconnect,* Fig. 21-15, unless contrary to the *Code* authority in charge. Further, the transfer equipment for these systems, should be located in a place which is easily accessible to those persons in charge of controlling such systems. At the same time, some type of provision should be made to keep unauthorized persons out.

CIRCUIT WIRING FOR EMERGENCY SYSTEMS

The *National Electrical Code* requires that wiring from an emergency source and overcurrent protection to emergency loads be kept completely separate from all other wiring and equipment. It is not permitted to carry parts of the emergency system in the same raceway, cable, box, or cabinet with other wiring. However, Article 700-9 of the NEC allows several exceptions to the regulation.

LEGALLY REQUIRED STANDBY SYSTEMS

These systems, are discussed in Article 701 of the *Code.* They serve loads necessary to the normal operation of vital building systems. Sewage disposal, heating, refrigeration, communication, ventilation and smoke removal, lighting, and other systems, if stopped, could cause great hazards to occupants and could also hamper fire-fighting efforts. In most other respects, the standby system is arranged, controlled, tested, and maintained in a fashion similar to the emergency system.

OPTIONAL STANDBY POWER

Many individuals want their residences and commercial properties to have some form of standby power in the event of a power outage. These systems are not legally required since a power failure at the residence or private business location does not represent a life-endangering situation.

Nonetheless, there are situations where loss of heating, refrigeration, data processing, and commercial industrial processing capability could result in great losses. This is particularly true with businesses such as computer processing centers, bakeries, ice cream manufacture or distribution facilities, frozen food packing, or preparation plants and dairy farms.

GENERAL TYPES OF EMERGENCY POWER EQUIPMENT

Emergency power equipment is almost always tied to a gasoline or diesel engine powered generator. These are rated from 2 kW-100 kW, or higher for large structures. See Fig. 21-16. They may be manually or automatically activated, Fig. 21-17. The generator may be self contained and either fixed or portable. On farms, the generators are often driven by a tractor with a PTO (power take-off). In addition, these generators may be used to supply single or three-phase power.

GENERAL CONNECTION PROCEDURES

Emergency and standby power systems must be connected to the electrical installation in such a manner that their power is isolated from the main power supply and vice versa. That is, it must be arranged so that an inadvertent interconnection to the normal source of supply cannot occur.

Modern Residential Wiring

A

B

Fig. 21-16. Emergency generating units may be rated from 2 kW to 50 kW. A—Gasoline, 30 kW. B—Diesel 8.5 kW. (Kohler Co.)

This is accomplished by using a double pole, double throw switch similar to the automatic one shown in Fig. 21-17. The nonautomatic switch in Fig. 21-18 does the same job and complies with *Code*.

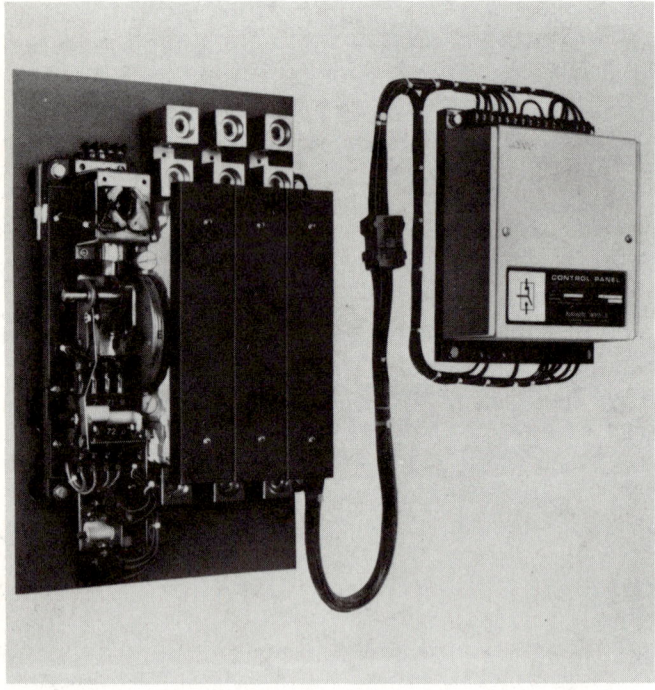

Fig. 21-17. Automatic transfer switch reacts almost instantly to regular power failure. It is easily installed and reliably activates the emergency system. This one has solid-state control panel. (Automatic Switch Co.)

TELEPHONE WIRING

A Federal Communications Commission ruling in 1977 makes it possible for persons, other than telephone company employees, to install telephone equipment. At the same time, the equipment, once available only at the phone company, is manufactured and sold by many different commercial firms.

BASIC TELEPHONE SYSTEM

Telephone service receives its power from a local telephone exchange called a central office. The telephone connection from the building to the central office is called the LOCAL LOOP. The telephone conductors or "line" are actually two wires called a CABLE PAIR. At the central office the wires are connected to and powered by a battery which supplies a direct current to operate the telephone equipment.

As shown in Fig. 21-19, distribution lines can be run overhead on poles or buried in the ground. Many cable pairs are bundled together in large cables to carry telephone service to each customer.

A CABLE TERMINATION BOX is the enclosure where connections are made for the cable pair coming from each building or dwelling. It is found in both above-ground and below-ground installation.

Cable pairs from the building actually terminate at a PROTECTOR. This is a device which protects the cable pair from high voltage should lightning strike the wires.

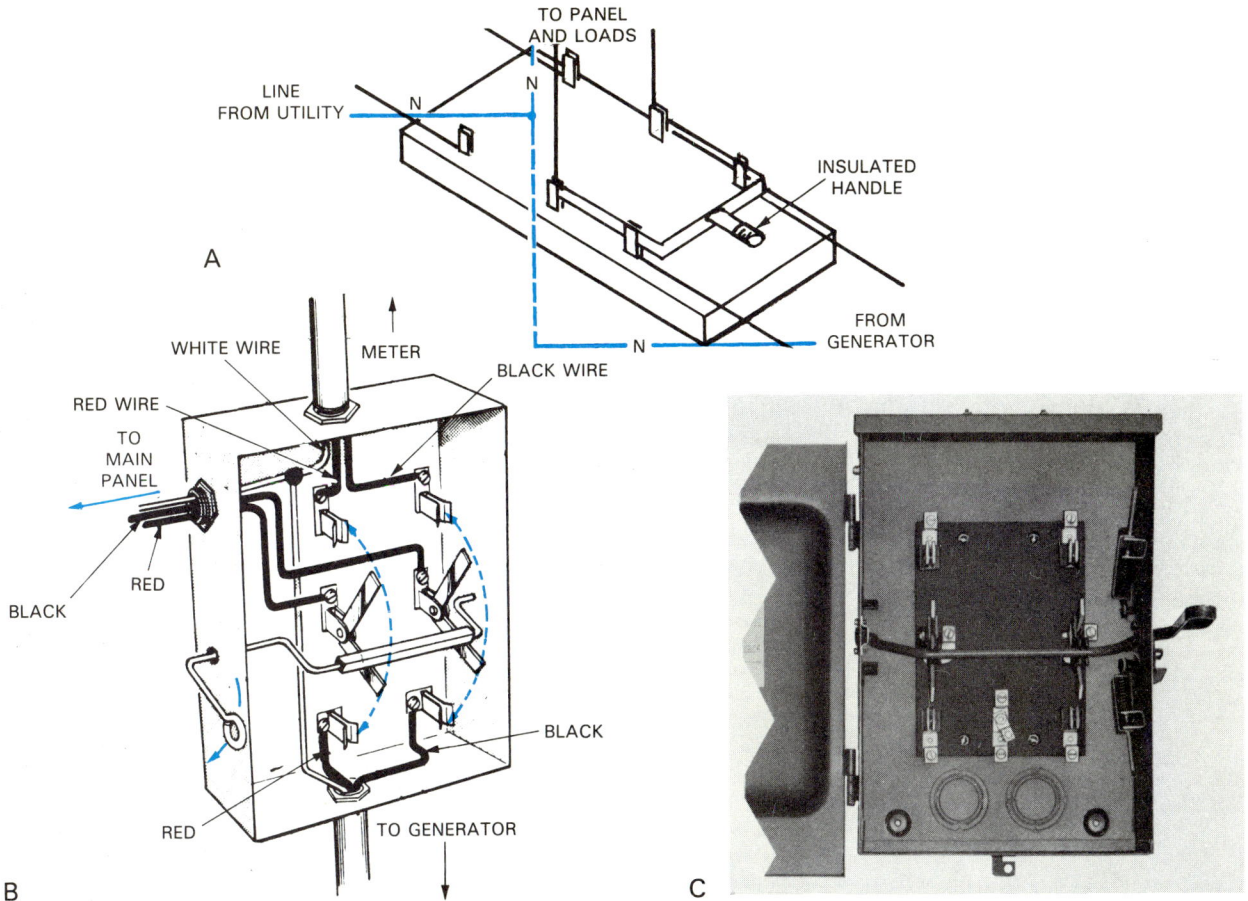

Fig. 21-18. A simple double-pole, double-throw switch will accomplish the transfer from standard to emergency supply. Note the continuous, unswitched, neutral conductor. A—Schematic. B—Simplified drawing of a transfer switch mechanism. C—Actual switch. (Wadsworth Electric Mfg. Co., Inc.)

From the single protector, the cable pair enters the building. They may go:
1. Directly through the wall.
2. Feed through a roof overhang (soffit or rake) into an attic.
3. Through a wall into a basement.

The cable pair usually connects to a terminal block or notework interface inside the building. This terminal is also called a 42A block, a modular interface, or a modular outlet. It is the junction point for all the telephones in the building. All wire pairs for each phone can connect to this terminal.

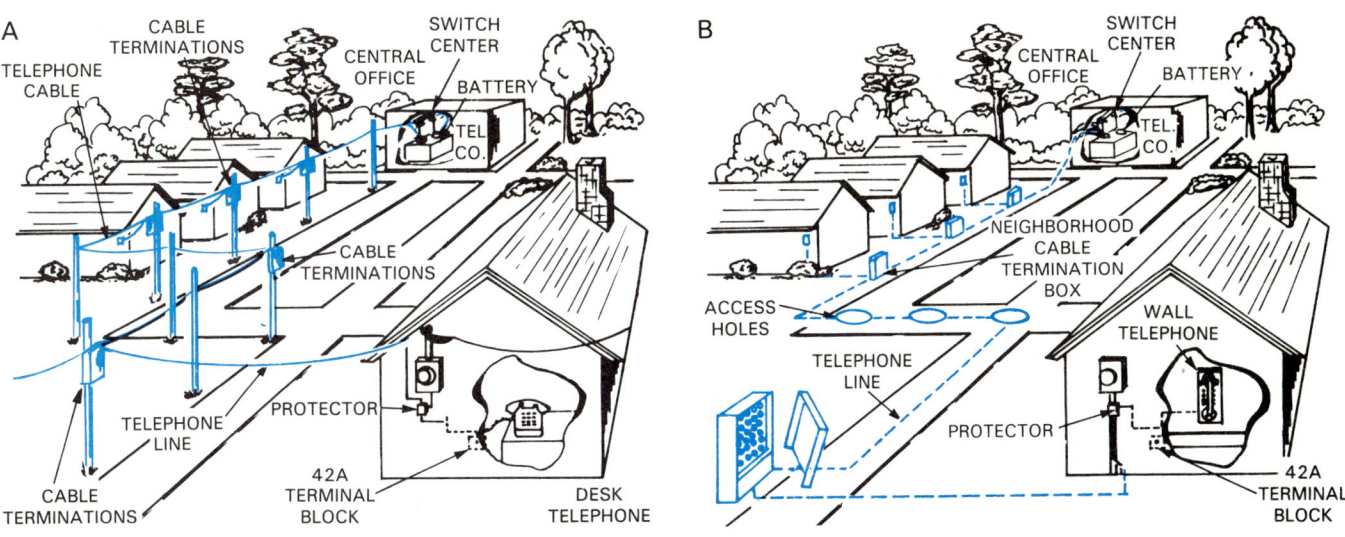

Fig. 21-19. Typical telephone systems. Phones are powered by central office of phone company. A—Overhead system. B—Underground system.

OPERATION OF EQUIPMENT

Two or more telephones can function on a circuit as easily as one. Additional telephones may be connected in one of the patterns shown in Fig. 21-20. Phones are always connected in parallel with each other.

Safety

There is little danger of electrical shock from the telephone's own voltage. When the unit is not ringing, 48 V dc is present. Operating the phone signal requires from 85 to 90 V ac. This voltage, ac and dc, is supplied by the telephone company.

The higher alternating current voltage is unpleasant but not harmful. To avoid it, take the receiver off the hook when working on the system.

Regardless of the lower voltages in telephone systems, it is advisable to avoid the possibility of shock by taking precautions:
1. Be sure that your inside wiring is not hooked up to the network interface. This ensures there is no power to the system.
2. If, for some reason this cannot be done, then use screwdrivers, cutters, and strippers with insulated handles. Avoid touching screw terminals or bare conductors with your hands.

Use care in drilling through walls, floors, or ceilings. Avoid areas where there might be electrical wiring, gas pipes, steam pipes, or water pipes.

When working on existing telephone wiring disconnect any transformer which might be supplying step-down voltage for dial lighting.

Keep telephone wiring at least 2 to 6 in. away from other wiring in the building. Keep jacks away from hazardous locations such as tubs, showers, swimming pools, and laundry areas. Telephones should not be used where the body will be in touch with water.

Never remove protectors or grounding means placed by the telephone company. Do not modify these installations or make any connection to them.

To avoid high voltage shock, keep telephone cable pairs away from accidental contact with 120 or 240 V line current. Carefully read Article 800 of the *National Electrical Code.*

MATERIALS AND TOOLS

Materials for installing telephone equipment include special types of cable, small terminal blocks, modular plugs, and jacks. Tools include fish tapes, test meters, and machines for installing cable underground.

Wire types

Telephone wire and cable is designed for either indoor or outdoor use. Outdoor cable avoids the use of any fibrous insulation material that is not weatherproof.

Indoor wiring is of three kinds:
1. Solid wire.
2. Stranded wire.
3. Spiral ribbon stranded wire.

The spiral ribbon type is difficult to work with because its strands are fragile. It requires a tool that pierces the insulation without removing it. Fig. 21-21 shows both wire and tool.

Solid and stranded wire can be attached to terminal screws in a manner similar to conductors for 120 and 240 V electrical systems. Solder is not recommended as it makes later disassembly difficult and slow.

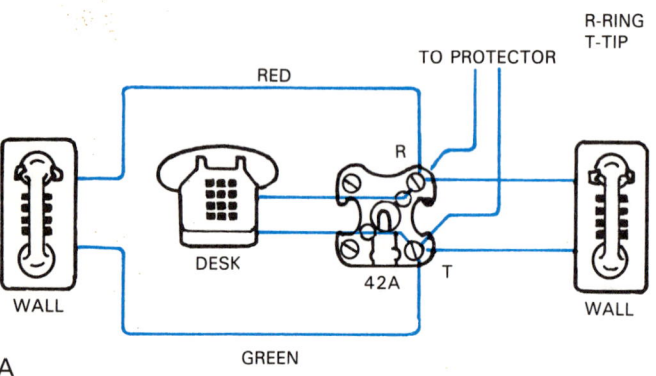

A

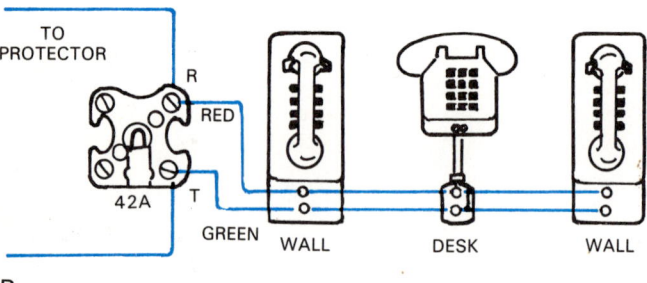

B

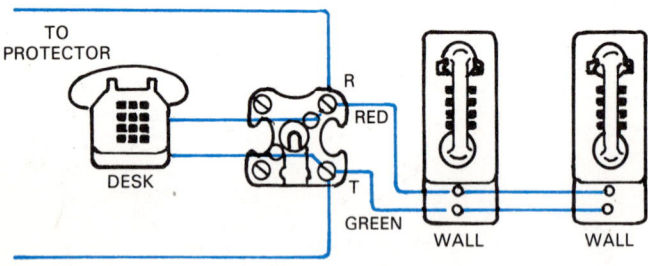

C

Fig. 21-20. Phones are always hooked up in parallel in these patterns. A—All cable pairs start at the same terminal block. B—Additional phones are added by connecting them to the terminals of another phone. C—A combination of A and B.

Specialized Wiring

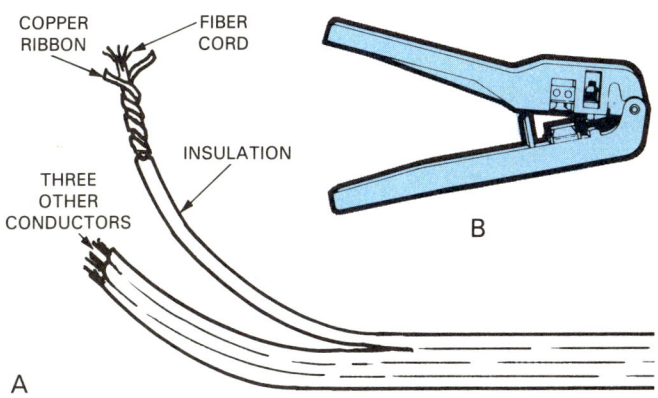

Fig. 21-21. A—Cable containing four spiral ribbon stranded wires is used for indoor extensions. B—Terminals cannot be attached without a special holding and crimping tool. Ribbons tend to unravel when insulation is removed. (Gemini Industries Inc.)

Terminal blocks

Modern terminal blocks are easy to use. Refer to Fig. 21-22. Sometimes no wiring work is needed other than to plug in the block.

Blocks can be used to convert the older terminal screw type of blocks to the modular system. Blocks should be attached to ceiling or floor joists in attic or basement.

When attaching cable pairs leave some slack in the wires so wires will not pull loose from the block.

Color coding

Terminal and cable pairs are color coded. In four-wire cable the wire colors are usually black, green, red, and yellow. Terminal screws are coded accordingly. Most blocks allow you to run cable pairs for as many as four phones.

Boxes

Some installers may use standard electrical boxes for termination points. See Fig. 21-23. These may be convenient to use in new construction but are not necessary. Voltages are low so that terminal blocks are commonly used instead. Because of the lower voltages found in telephone systems, boxes are not required at junctions.

Wall plates

Wall plates may be either metal or plastic. Usually the plate supports a wall phone. See Fig. 21-24. The wall plate has two pins with special heads designed to fit the slots on the phone's mounting base. The pins may be used as ground. If you change plates, you may need to connect a ground wire to the pins.

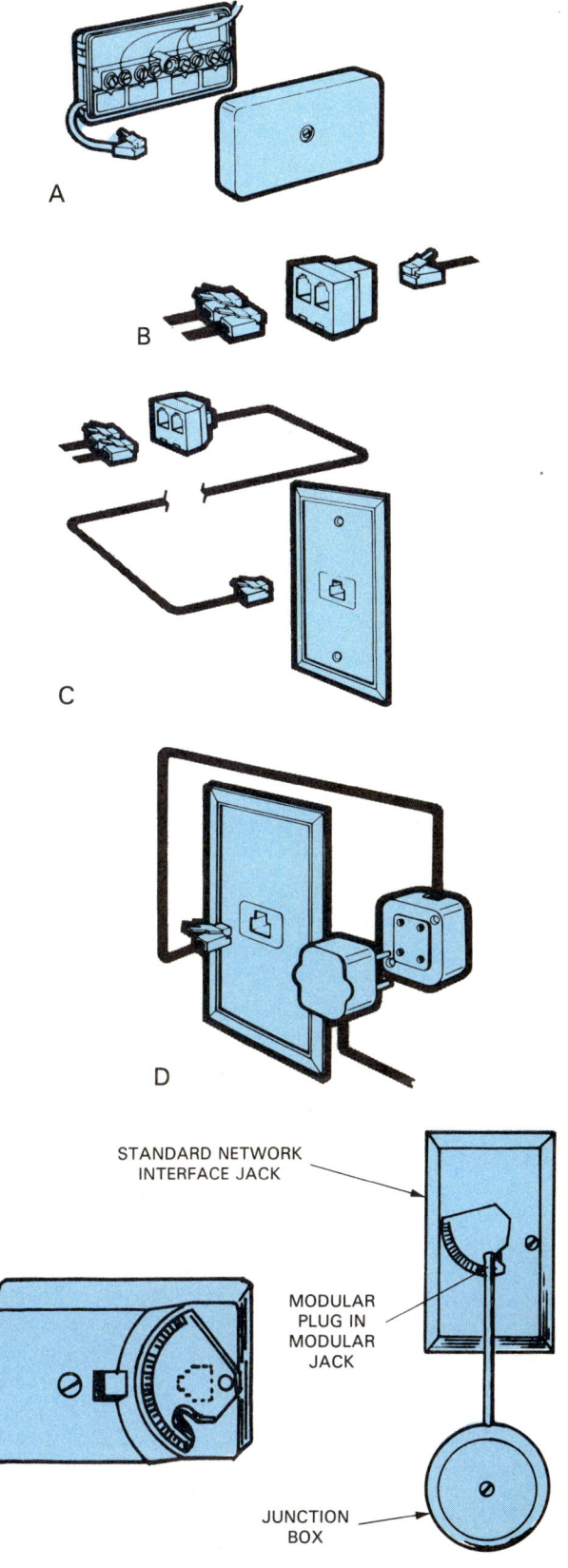

Fig. 21-22. Typical telephone connectors. A—Screw terminals accept lines for up to four phones. B—A three-way or Y modular jack. It allows three modular plugs to be connected together. C—A three way Y with a long extension cord. D—Adapter which converts an old style four-prong jack to fit a modular terminal. (Gemini Industries Inc.) E—Modular interface. It has pivoting cover to keep out dirt.

221

Modern Residential Wiring

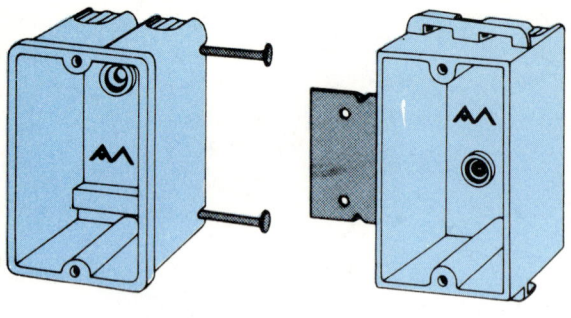

NEW WORK

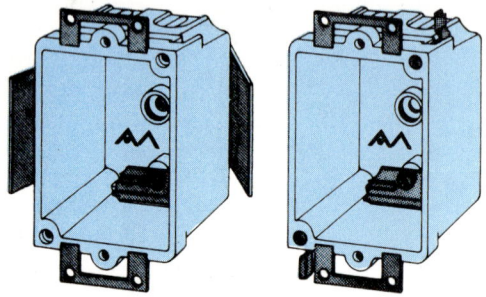

OLD WORK

Fig. 21-23. In new construction, standard electrical boxes are often used for concealed telephone wiring. Boxes support wall phones or special plates are used with provision for modular connections. (Allied Moulded Products, Inc.)

Tools

A useful tool for telephone installation is the fish tape. This is illustrated in Chapter 4. It makes concealed telephone wiring possible in old work. Two or more tapes can be used to fish wires through hollow walls. If a second tape is not available, a weighted string can be lowered inside the wall. The fish tape can then be used to hook the string through the hole, Fig. 21-25.

A drill bit with a hole drilled near the tip is another time-saver, Fig. 21-26. Before the bit is removed from a drilled hole, thread a fine wire or string through the hole in the bit. When the bit is removed from the hole the wire or string is also pulled through.

A multimeter, pictured in Fig. 20-1, will help trace wires and find loose connections. It will also help check a transformer for dial lighting.

INSTALLING INDOOR TELEPHONE WIRING

The indoor telephone wiring system is that part of the system that starts from the network interface and provides service to all the phones in the building. The network interface is the jack or terminal where the telephone company installation terminates. It is usually conveniently located where it can be connected to the house phone wiring. See Fig. 21-27.

When indoor wiring is completed a final connection is made between the network interface and the customer's jack. You are permitted to make modular plug connections between the two boxes. However, only the telephone company can make connections inside the company protector or network interface. See Fig. 21-28.

PHONE WIRING PLAN

Not all telephone installations will have or require a plan. It is customary on new installations. If there is one, Fig. 21-29, check it carefully. No more than five phones can be installed without overloading the

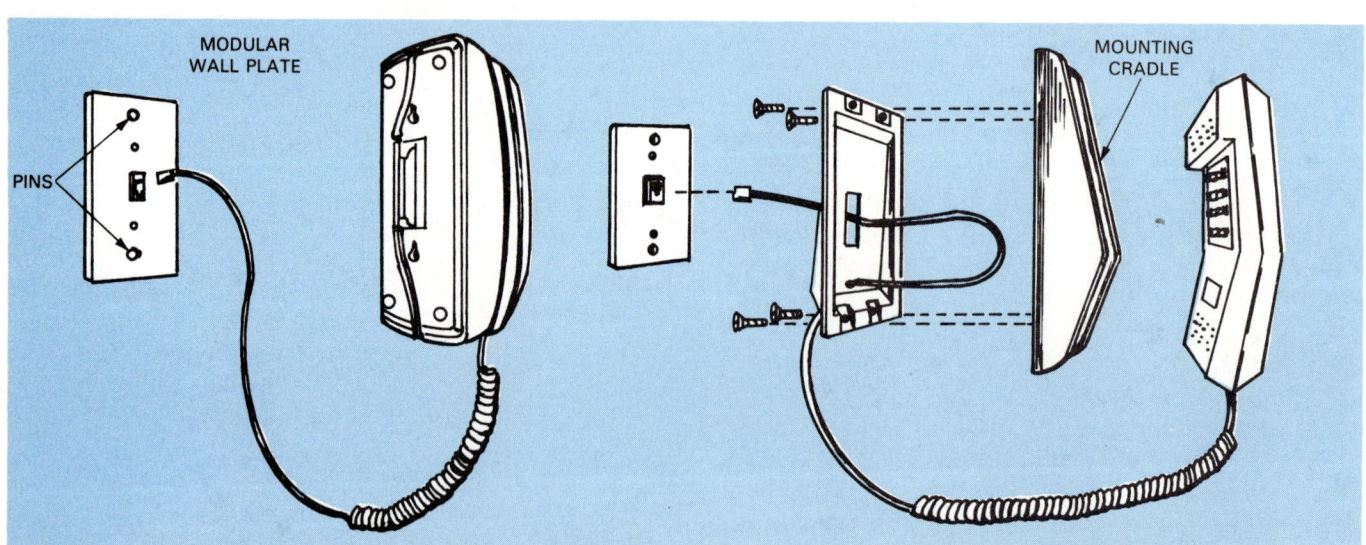

Fig. 21-24. Type of wall plate which can be attached to a box. Note two different styles of plate adapters for wall mounting of phones.

Specialized Wiring

incoming phone circuit. However, it is best to install a jack in each room in any new construction. The savings in time and convenience is well worth the extra initial cost to the customer.

NEW CONSTRUCTION

In new construction, the telephone wiring should be installed before application of drywall or paneling. Start the wiring at the jack/junction installed near the network interface. If outlet boxes are required by local code, be sure to allow them to extend beyond the stud the thickness of the wall covering materials. It is not necessary to attach the telephone wire to the stud. This will make it easier to replace or remove during possible later relocation. Allow about 6 in. of wire inside the outlet for your connections.

Be sure to check local codes. Some may require conduit for concealed wiring.

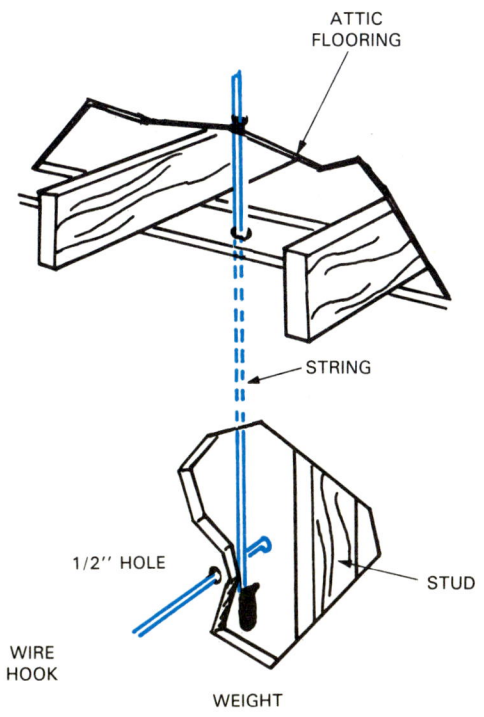

Fig. 21-25. Pulling wires through inaccessible places with aid of string and fish tape. Note weight on string.

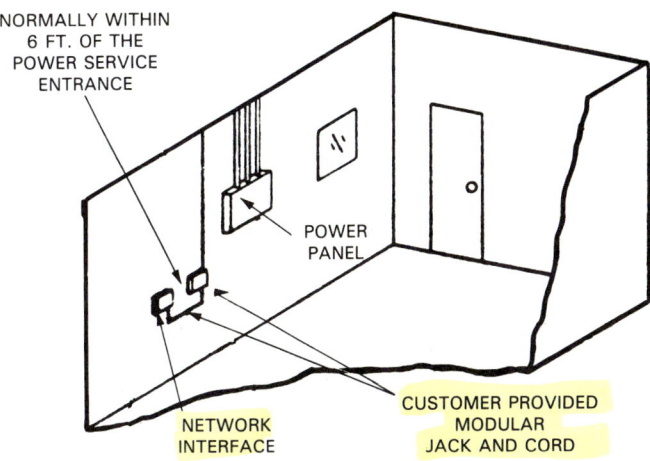

Fig. 21-27. Usually telephone workers will bring phone service into the building. The network interface is the junction point for outside and inside wiring. Normally, the network interface will be located in a garage, basement, or other indoor service area.

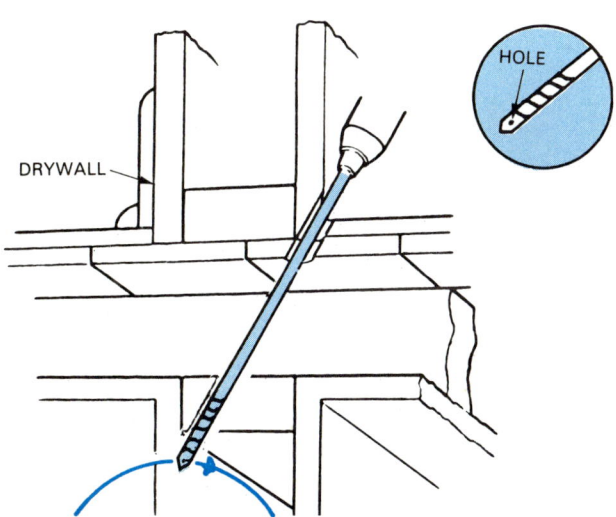

Fig. 21-26. Drilling holes to conceal wiring. Note use of bit with hole drilled near its point. Thread light wire or string through hole in drill and pull drill back through the drilled hole.

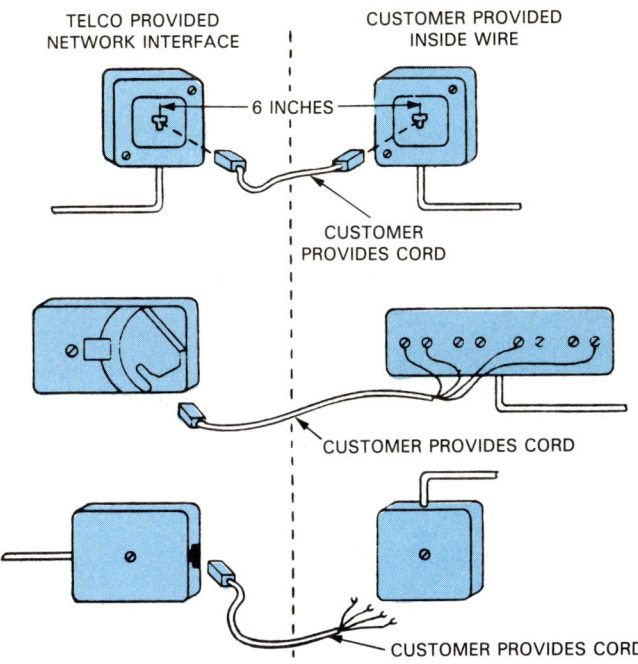

Fig. 21-28. Typical connections between the network interface and indoor telephone system. This connection should be made last to avoid the discomfort of shock while working on the system.

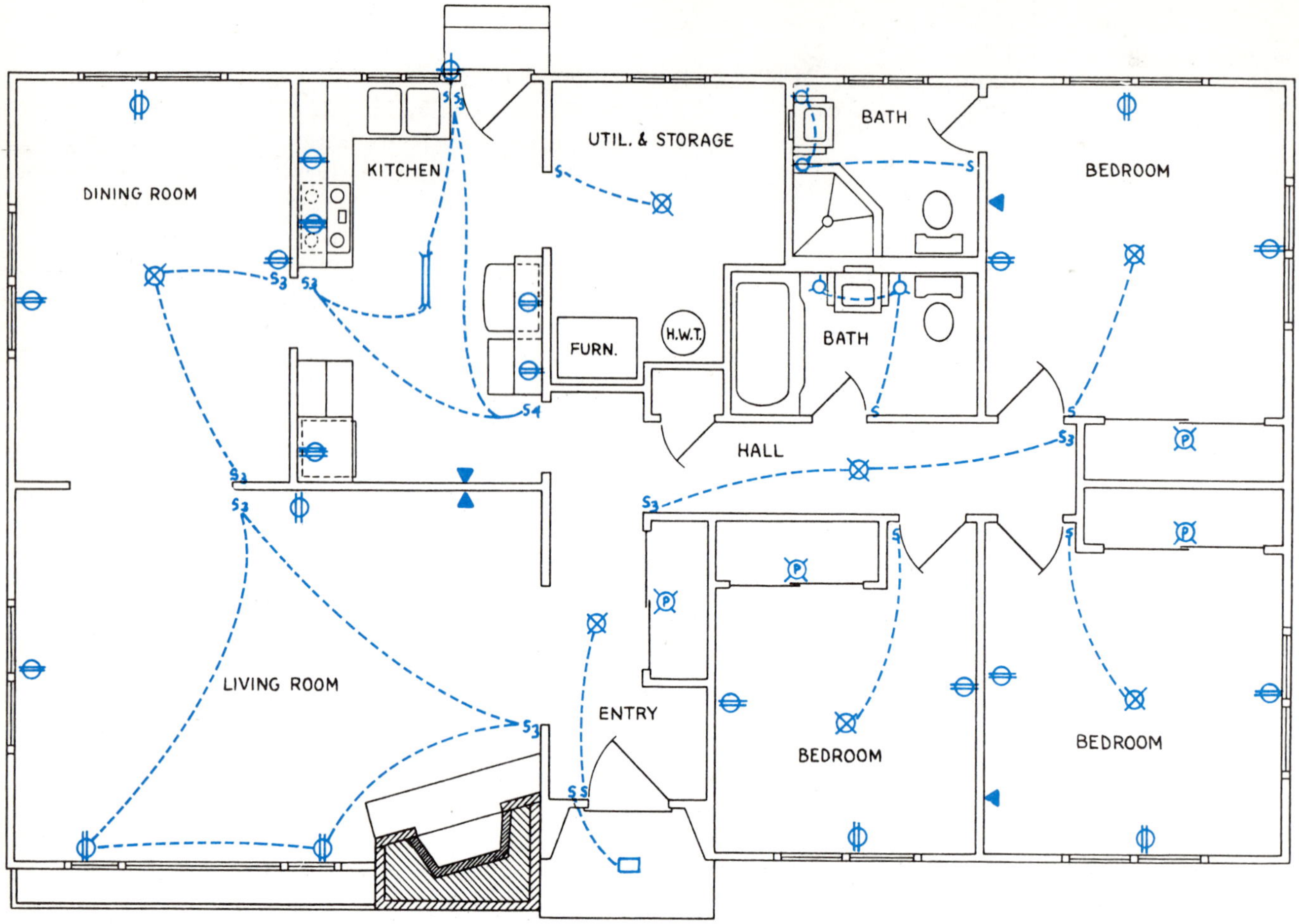

Modern Residential Wiring

Fig. 21-29. Residential floor plan. Electrical and telephone locations are marked. Care should be taken to keep telephone wiring away from electrical wires and electrical devices.

Desk phones require that outlets be placed nearer the floor at the same height as the electrical outlets. See Fig. 21-30.

Wall phone outlets should be about 56 in. from the bottom of the outlet to the floor. Over counters, allow 10 in. between the bottom of the outlet and the counter top.

OLD WORK

Installing telephone systems after wall coverings are in place entails more care and, sometimes, more work. This is especially true when the wiring is to be concealed. In these cases use of fish tape and string will help pull wires through wall cavities. Refer to Figs. 21-25 and 21-26 as well as Chapter 18, Electrical Remodeling, for information on pulling wires through existing construction.

CONCEALING SURFACE WIRING

Wires cannot always be run inside walls. Often they are mounted on surfaces. Refer to Fig. 21-31. Such wire must be supported. Use 1/4 in. wide staples made for supporting telephone wire. Insulated staples work well and protect the wire. Ordinary staples may be used if they are carefully attached. Place staples every 3 ft. Do not run the wire under carpet or near traffic areas.

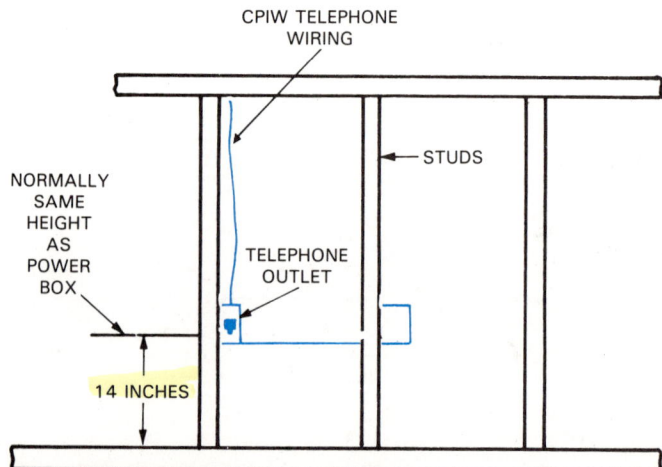

Fig. 21-30. Proper location of telephone outlet intended for a desk phone. Follow same standards as for receptacle outlets.

Specialized Wiring

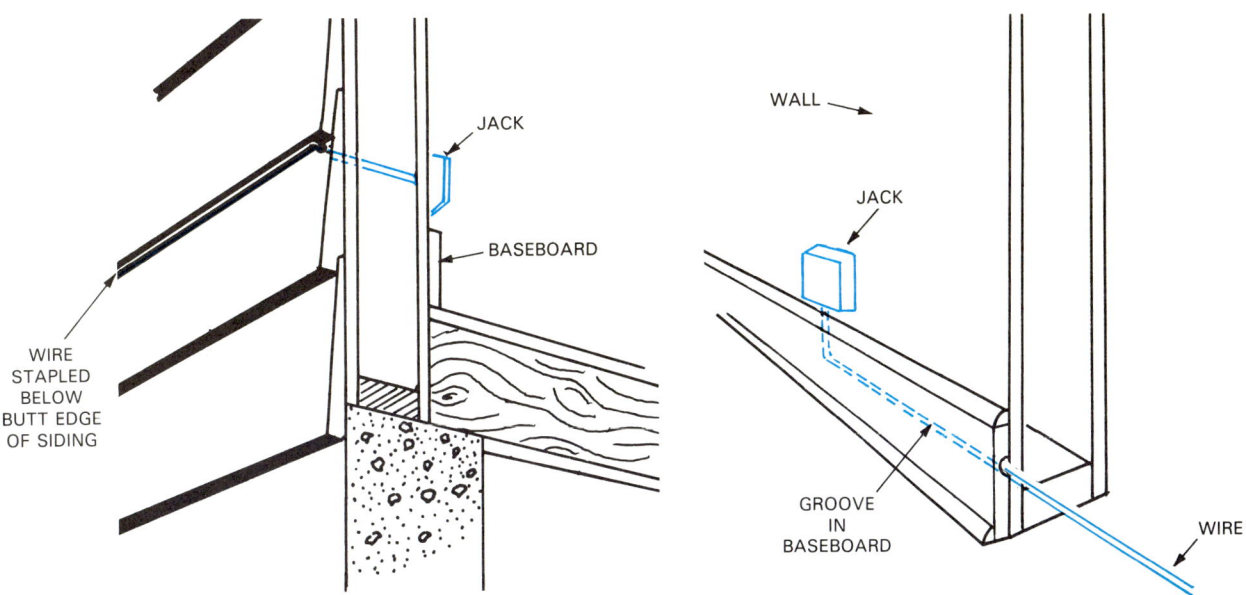

Fig. 21-31. Method of hiding wires. Left. Outside wire is in shadowline of siding. Right. Inside wire routed through a groove cut in hidden side of baseboard.

Wires can be made less noticeable if they are placed in areas where moldings, casing, grooves in paneling or coverings partially or wholly conceal them. Wiring can also be run inside cabinets.

DIAL LIGHT TRANSFORMER

Telephone cable usually has four wires: red, green, yellow, and black. The red and green carry the message as well as the 48 V dc and 85 V ac signals. The yellow and black can be used for an accessory like a lighted dial. Connect a 6 volt step-down transformer to the yellow and black lines feeding all of the phones, Fig. 21-32. One transformer will power all of the dials.

REPAIRING AND TROUBLESHOOTING

Suppose that a telephone cable is broken and that the break is visible. If the cable is the spiral type, it can only be spliced with two more modular terminals, Fig. 21-33. If the break cannot be located, replace the entire length of wire.

If you suspect an electrical short in a phone system, use an ohmmeter to find the problem. See Fig. 21-34.

An open circuit is most easily checked by plugging in a telephone known to be working. Dial tone indicates proper wiring. You can also make a shorting loop on a modular plug to use in place of a telephone. The ohmmeter will show low resistance when loop is placed in any good outlet. It will not respond when loop is placed in an open outlet.

Fig. 21-32. Use four-wire cable to accommodate lighted phone dials now or later. Connect 6 V transformer to yellow and black wires.

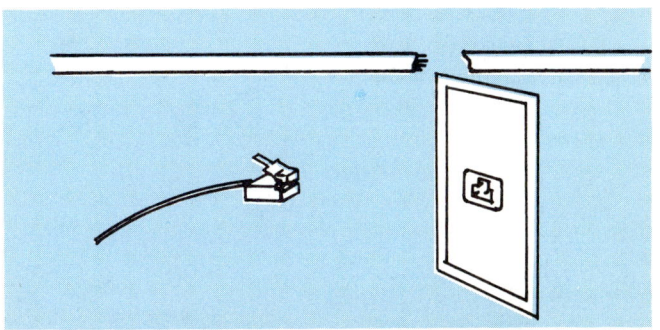

Fig. 21-33. Splicing a spiral ribbon cable requires a pair of modular connections.

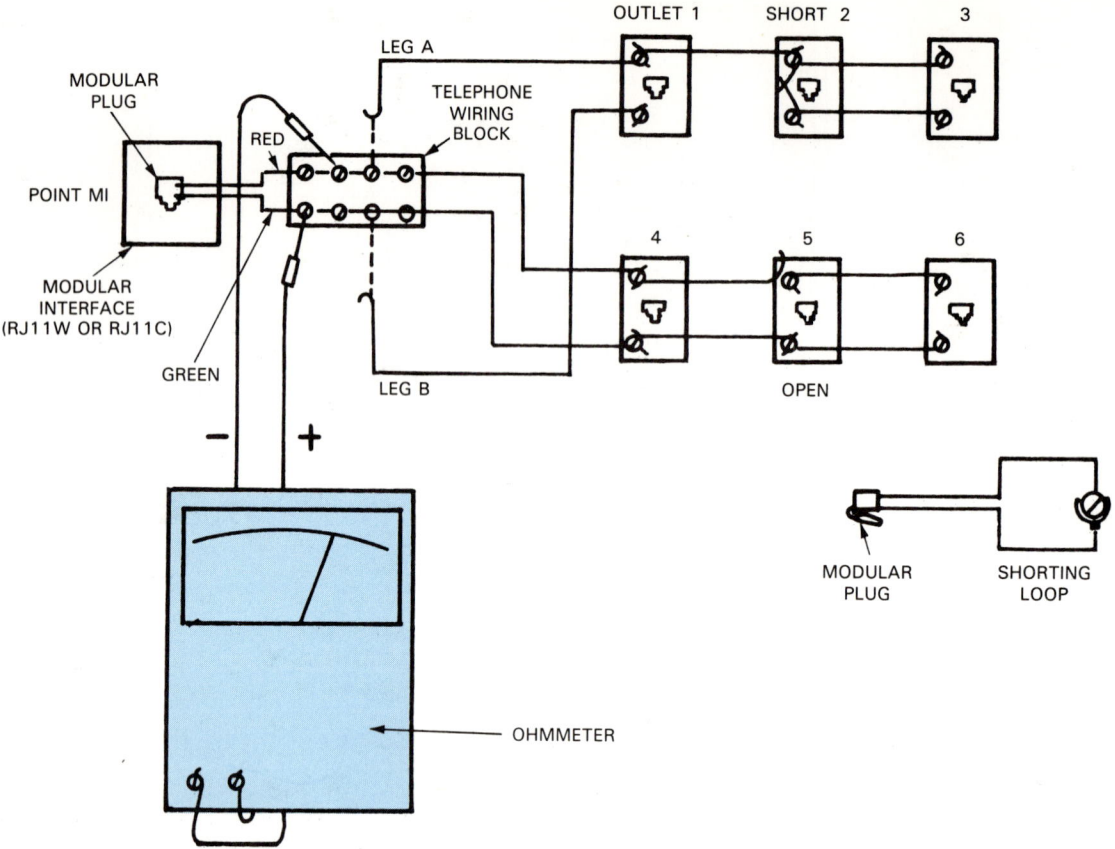

Fig. 21-34. Using an ohmmeter to check out shorted or open branch circuits. With open circuit (Leg B) connected to wiring block, ohmmeter reading will show infinite resistance. When Leg A is connected and Leg B disconnected, resistance will drop to zero because of the short.

REVIEW QUESTIONS — CHAPTER 21

1. Doorbell, buzzers, and chimes fall under the category of _____ circuits.
2. Conductors which serve the above circuits may be run with the full-power circuit conductors. True or False?
3. A Class I hazardous location is one in which:
   a. Ignitable fibers or flyings are present.
   b. Flammable gases or vapors are present.
   c. Combustible dust is present.
4. Explain what distinguishes Division 1, Hazardous locations from Division 2 areas.
5. Explosion-proof enclosures for wiring devices are required for:
   a. Class II, Division 2. Hazardous locations.
   b. Class II, Division 1. Hazardous locations.
   c. Class III, Division 1. Hazardous locations.
   d. All of the above.
   e. None of the above.
6. Explain the difference between emergency, legally required standby, and optional standby power systems.
7. By what means is an emergency or standby power system isolated from the regular source of power supply?
8. Electrical power for the telephone comes from (list all correct answers):
   a. 120 and 240 volt service in the building.
   b. Local telephone exchange, called a central office.
   c. Batteries at the central office of the telephone company.
   d. Step-down transformer in building (for dial lighting only).
9. A _____ is a device which protects incoming cable pairs from accidental high voltage.
10. What is the telephone network interface and who installs it?
11. The ringing voltage for a telephone is:
    a. 48 V ac.
    b. 85-90 V ac.
    c. 48 V dc.
    d. 85-90 V dc.
12. *Code* requirements for telephone installation are covered under Article _____ of the *National Electrical Code*.
13. What is included in the indoor telephone wiring system?
14. No more than _____ telephones can be used on one circuit at the same time.

Chapter 22

MOTORS AND MOTOR CIRCUITS

Electricians must be familiar with the wiring requirements for electric motors. They also need to know how to order the right replacement when an old motor is no longer serviceable.

Motors operating at 600 V or less are encountered frequently. Motors with higher voltage, found in larger commercial and industrial operations, should be handled only by maintenance electricians having considerable experience with these motors.

Articles 430 and 440 of the *National Electrical Code* should be reviewed carefully before you attempt to install motors and their associated circuit components. These *Code* articles deal with the many concerns and variations of motor circuits.

MOTOR NAMEPLATE

One of the most important aids to installing and wiring a motor correctly is the information found on the motor itself. The nameplate, Fig. 22-1, provides a wealth of data regarding the characteristics of the motor.

NEC and NEMA (National Electrical Manufacturers Association) requirements state that all motors are to have nameplates providing the following information: name of manufacturer, horsepower rating, type, frame number, time rating, rpm rating, design letter, frequency (in hertz), number of phases, insulation class, load rating in amperes, locked rotor code letter, voltage, duty, ambient temperature, and service factor. Other information, such as serial number, model number, bearing numbers, and efficiency ratings are often included. This nameplate data should be carefully reviewed before setting and wiring a motor.

NAMEPLATE DATA DEFINITIONS

Frame and type—The NEMA designation for frame designation and type.

Horsepower—The power rating of the motor.

Motor code—Designated by a letter indicating the starting current required. The higher the locked-rotor kilovolt-ampere (kVa), the higher the starting current surge. Table 4, Fig. 25-11, in the Technical Information section shows the most common letter designations and the locked rotor kVa they represent.

Cycles, or hertz—The frequency at which the motor is designed to be operated.

Phase—The number of phases on which the motor operates.

Revolutions per minute (rpm)—The speed of the motor at full load.

Voltage—The voltage or voltages of operation.

Thermal protection—An indication of thermal protection provided for the motor, if it has any.

AMPS (or A)—The rated current (in amperes) at full load.

Time—Time rating of the motor showing the duty rating as continuous or as specific period of time the motor can be operated.

Ambient temperature, or temperature rise—The maximum ambient temperature at which the motor should be operated, or the permissible temperature rise of the motor above the ambient air at rated load.

Service factor—The amount of overload that the motor can tolerate on a continuous basis at rated voltage and frequency.

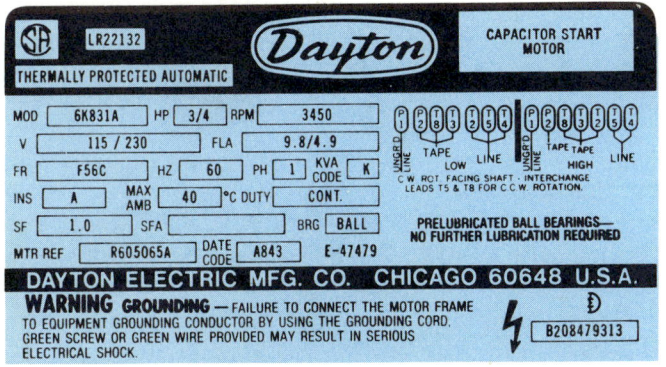

Fig. 22-1. The motor nameplate gives motor characteristics. The code designation, service factor, time rating, and temperature rise are important considerations in selecting a motor for a given job. Refer to the text for definitions of each entry. (Dayton Electric Mfg. Co.)

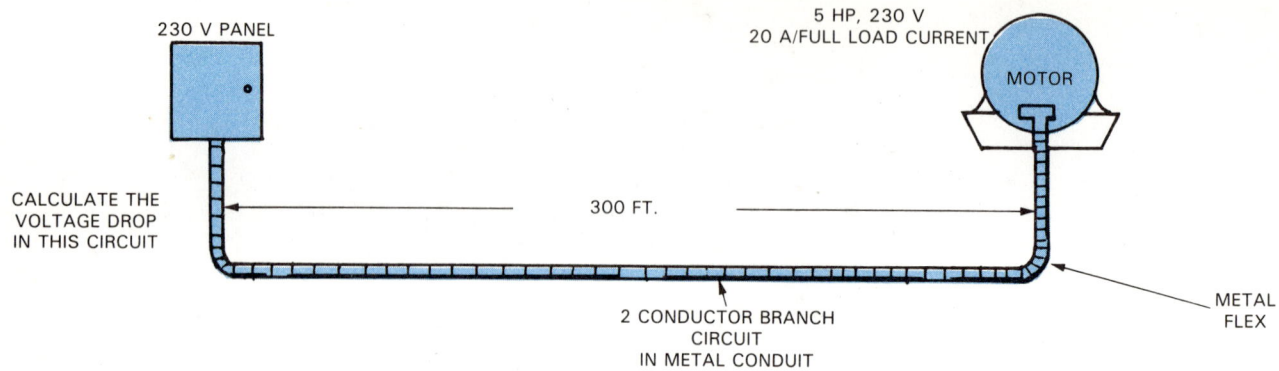

Fig. 22-2. Calculation of voltage drop in a circuit is related to the speed and torque of the motor, the length of the branch circuit and the size of the conductors.

Insulation class—A designation of the insulation system used, primarily for convenience in rewinding.

NEMA design—A letter designation for integral horsepower motors specifying the motor characteristics.

FRAME NUMBER

Motors of a given horsepower rating are built in a certain size of frame or housing. For standardization, NEMA has assigned the frame size to be used for each integral horsepower motor so that shaft heights and dimensions will be the same to allow motors to be interchanged.

PROPER OPERATION FACTORS

For proper operation of an electric motor, the supply voltage and frequency (cycles per second or hertz) at the motor terminals must match the values specified by the manufacturer as closely as possible. Performance is best over a range of plus or minus 10 percent of rated voltage and 5 percent above or below rated frequency. Frequency is usually no problem since it rarely varies.

However, if applied voltage varies too much from the nameplate specifications, it will produce noticeable changes in the motor torque (turning force). Low voltage from inadequate wiring or any other cause, can create severe problems. Motor starting torque may be too low to start the load and keep it moving. The sections following, "Voltage Drop" and "Size and Protection of Motor Feeder Conductors," describe how to size wire properly.

VOLTAGE DROP

The greatest allowable voltage drop on a motor circuit is basically related to the speed and torque of a motor. If, for example, a 230 V motor requires a minimum of 224 V for adequate performance, then a 6 V drop is the most that is permissible. This is easy to check before the actual wire is run. Checking is advisable, as it will avoid incorrect sizing of conductors. Look at Fig. 22-2, for instance.

The NEC requires this motor to be supplied by branch circuit conductors having a capacity of 125 percent of the full-load current (20 A). Thus: 1.25 × 20 = 25 A. No. 10 AWG (copper) conductors would be required to handle the amperage.

The resistance of No. 10 copper wire is 1.018 ohms per 1000 ft. at 36°F (25°C). To find the total resistance, multiply the resistance per foot (1.018 divided by 1000) by the number of feet (300):

1.018 ÷ 1000 = 0.001018 × 300 = 0.30540 ohms resistance per conductor. Multiplied by 2 = 0.6108 ohms = total resistance in two conductors.

When the full-load current of 20 A is flowing, the voltage drop is 20 A × 0.6108 ohms or 12.216 V. Thus, actual voltage to the motor is only 230 − 12.216 = 217.784 V. This is inadequate. A larger conductor must be used.

What is needed is a conductor with about half the resistance of a No. 10. If you consult Table 8, Chapter 9 of the NEC, you will find that a No. 6 copper conductor has a resistance of 0.410 ohm/1000 ft. Calculate the voltage drop for the circuit using No. 6 wire:

0.410 ohms × 0.3 = 0.123 ohm/300 ft. for each conductor

0.123 × 2 = 0.246 ohms for two circuit conductors

Therefore, 20 A × 0.246 Ω = 4.92 V. Subtract this voltage drop from the supply voltage: 230 − 4.92 = 225.08 V.

Since this size of conductor can carry more than the minimum voltage of 224 V, the No. 6 circuit conductors are acceptable and should be used.

One important note: the resistances used in the foregoing example are based on an ambient (surrounding) temperature of 77°F (25°C). Higher temperatures will create higher conductor resistance and, consequently, higher voltage drops. The formula used to adjust resistance values in copper wire is:

$R_x = R[1 + 0.00385(t_x - 25)]$

where R_x = the resistance, in ohms, at the ambient temperature t_x (°C)
R = the resistance, in ohms, at 25°C

SIZE AND PROTECTION OF MOTOR FEEDER CONDUCTORS

The sizing and protection of motor feeder conductors rests primarily on the type of motor circuit involved. For practical purposes, there are four common motor circuits.

1. "Single motor" circuit (circuit serving only one motor).
2. "Multiple motor" circuit (circuit serving two or more motors).
3. "Motor and other loads" circuit (circuit serving one or more motors *plus* other loads such as lighting and appliances).
4. Hermetic compressor motor circuit.

The simplest, of course, is the motor circuit which serves only one motor. For this circuit the NEC requires that the feeder capacity be 125 percent of the full-load current rating of the motor. This rating may be obtained directly from the motor nameplate or from Table 430-150 of the NEC.

For example, the circuit conductors supplying a 30 hp 460 V three-phase wound rotor motor is found as follows:

1. From the NEC Table 430-150, the full-load current rating is shown to be 40 A.
2. The branch-circuit conductors for continuous safe operation must be 125 percent of this rating or 50 A.
3. Thus, the conductors for this circuit must be no smaller than No. 6 AWG copper (THW, RHW, THHN, RHH or TW) as shown in Table 310-16 of the *National Electrical Code*.

When aluminum circuit conductors are used, then No. 4 AWG types TW, THW, RHH, THHN or XHHW would be suitable. (Again, refer to NEC Table 310-16.)

CIRCUIT PROTECTION

Protection for motor circuits is provided by an overload relay within the motor starter, itself, and with adequate fuses or circuit breakers at the panel. These, too, must be properly sized following NEC guidelines. Accordingly, running overload protection in the starter must have a rating or setting of not more than 125 percent of the motor full-load current. Fusing or circuit breaker requirements are found in Section 430-52 and Table 430-152 of the *Code*. The maximum fuse or breaker size permitted for short-circuit or ground fault protection is 150 percent of the full-load current rating. Thus, 60 A fuses or breakers are permitted.

For example, look at Fig. 22-3. Note that the procedure for determining conductor size and protection is the same as for the larger motor used in our previous example. It makes no difference that the motor type and supply voltage are different.

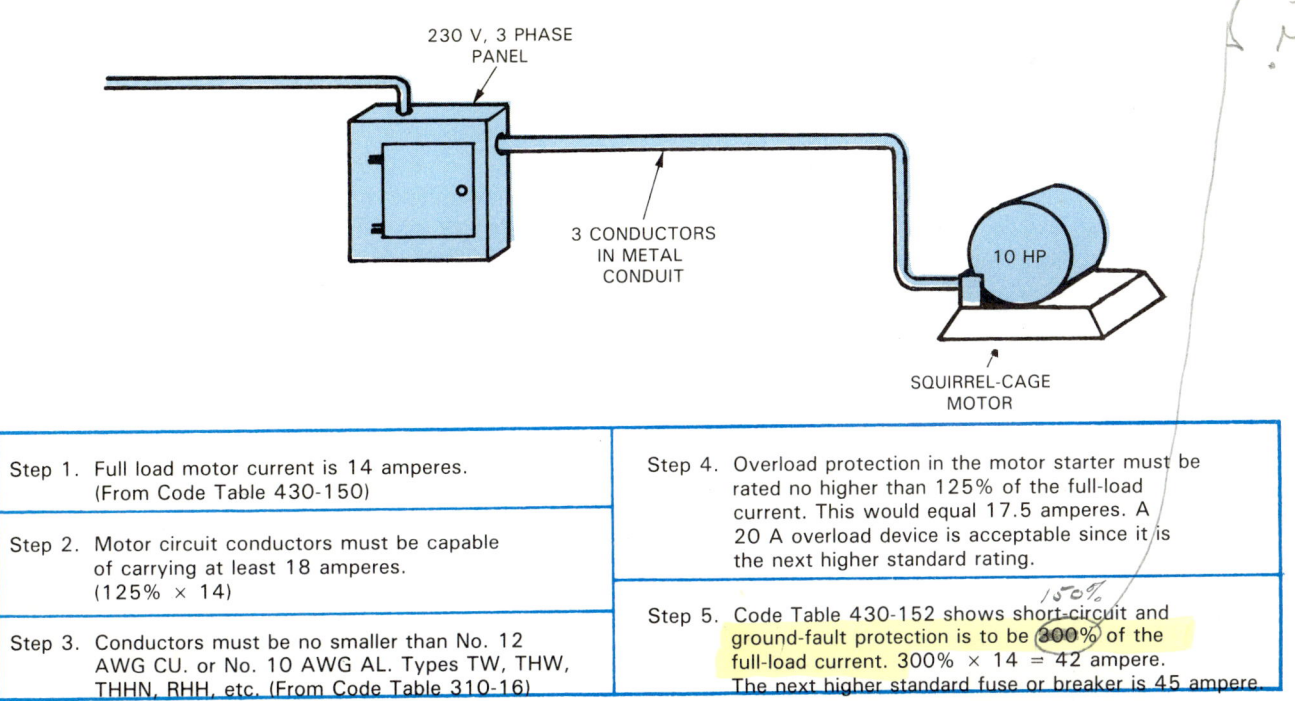

Fig. 22-3. Steps for determining adequate overcurrent protection and ground-fault protection for a circuit.

Modern Residential Wiring

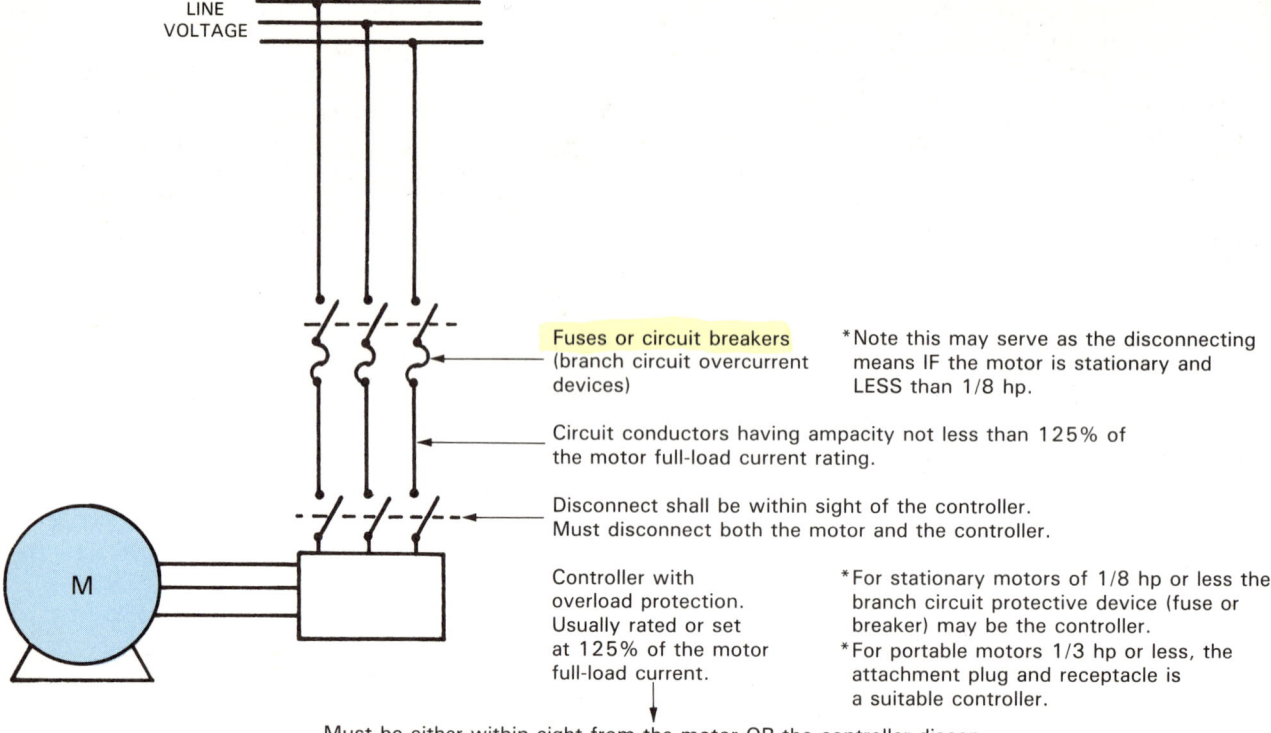

Fig. 22-4. This layout provides all NEC requirements for motor circuit components.

GENERAL LAYOUT

Fig. 22-4 shows the general layout acceptable for a single motor branch circuit. This circuit must provide:
1. Conductors rated at 125 percent of the motor full-load current.
2. Circuit overcurrent protection in the form of fuses or breakers.
3. A motor controller with overload protection.
4. A disconnect means.

GENERAL PURPOSE CIRCUIT GUIDE

Follow these general guidelines:
1. When each motor is 1 hp or 6 A or less, the circuit will require 20 A/125 V or smaller fuse or breaker; or 10 A/250 V. See Fig. 22-5.
2. Larger motors require a controller and overcurrent protection "approved for group installation."
3. Cord and plug connection is permitted. A 15 A circuit with fuse or circuit breaker protection is suitable for motors of 1/3 hp or less, Fig. 22-6.

MOTOR BRANCH CIRCUIT

1. Each motor needs motor running overcurrent protection and this protection must not exceed the amperage stamped on "approved for group installation" overcurrent device of the smallest motor on the branch circuit.
2. Each motor may have individual disconnect means.
3. However, a single disconnect is alright on several motors if:
   a. All the motors are part of one machine.
   b. Disconnect is in sight of all motors (motors all in same room).
   c. All motors are 6 A or less and less than 300 V.

Conductors supplying two or more motors must have a current rating not less than 125 percent of the full-load current rating of the largest motor plus the sum of the other motors on the circuit.

CONTROLLER REQUIREMENTS

Motors must be provided with some type of control. The NEC defines a motor controller as a switch or other device normally used for starting and stopping a motor. The controller must be protected against short-circuit current damage. There are numerous options available to meet the requirement. The method used depends primarily on the size and type of motor.

CONTROL METHODS

On small, single-phase motors, the overcurrent device (fuse or circuit breaker) may serve as the

Motors and Motor Circuits

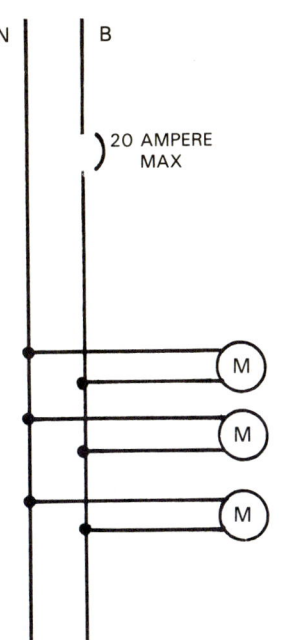

Fig. 22-5. When each motor on the circuit is 1 hp or 6 A or less the circuit shall be protected by a 20 A fuse.

controller for motors rated not more than 1/8 hp which are stationary and normally left running. See Fig. 22-7. A good example would be a clock motor.

For motors up to and including 2 hp, the controller may be a switch. This switch should be of the general-use type and have a current capacity of at least twice the full-load current rating of the motor.

The switch may also serve as the disconnect. Fig. 22-8 shows a suitable switch.

The circuit breaker for a branch circuit may serve as the controller. However, the longer delay prior to opening may permit a fault current to damage the controller or the motor. The circuit breaker may also serve as the disconnect.

A cord and plug arrangement may serve as the controller for motors at or less than 1/3 hp, Fig. 22-9. The same arrangement may also serve as the disconnect.

SEALED HERMETIC MOTORS

Motors which are hermetically sealed inside a refrigerating or air conditioning system have low operating temperatures and stay cool during normal operation. They can handle heavier loads than general purpose motors because of this temperature advantage. Such motors are not rated in horsepower but in terms of full-load current and/or locked-rotor current.

Controller selection is often determined from NEC Tables 430-148, 149, or 150 by checking the horsepower equivalent. Table 430-151 will give the horsepower equivalent for locked-rotor current ratings. The larger of the two values derived from the tables should be used to find the controller rating. (The nameplate provides full-load and locked-rotor current ratings.) If exact current values are not given in the charts, then go to the next higher current in the table to find the horsepower rating.

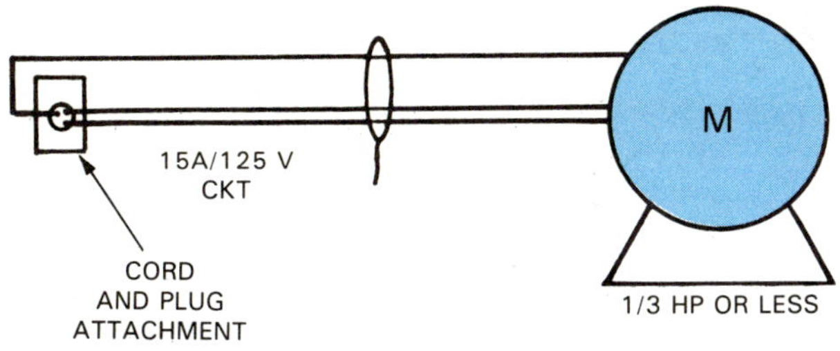

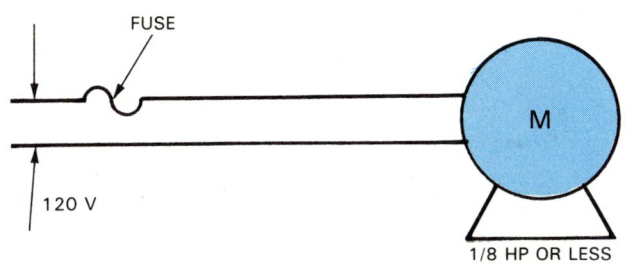

Fig. 22-6. General purpose circuit. Cord and plug are permitted. 15 A fuse or breaker protection is acceptable for motors of 1/3 hp or less.

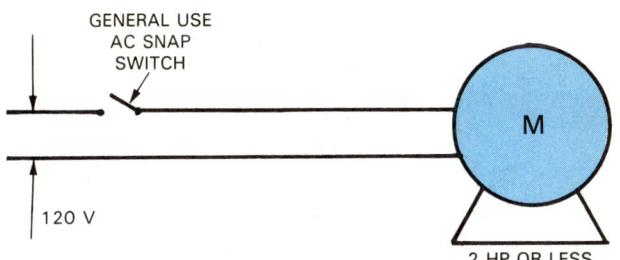

Fig. 22-8. A switch may serve as the controller for motors up to 2 hp. Switch should be of the general-use type.

Fig. 22-7. On small, single-phase motors, overcurrent device may serve as the controller for motors of 1/8 hp or less.

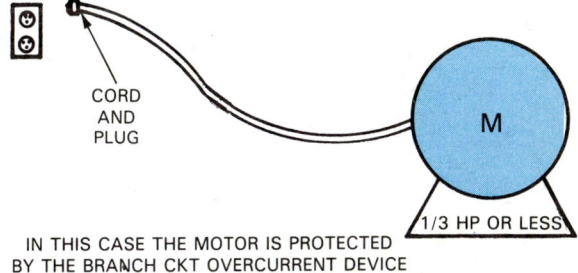

Fig. 22-9. Cord and plug may serve as controller of motors having less than 1/3 hp.

Some controllers are rated in full-load current or locked-rotor current. In such cases, the horsepower equivalent is not needed.

SAMPLE PROBLEM

For example, what is the correct size of controller for a 230 V three-phase wound-rotor motor which has a nameplate showing 90 A locked rotor and full-load current of 25.3 A? Find the full-load current and 90 A locked-rotor current of this motor:
1. From Table 430-150 find the full-load amperage. The next full-load amperage above 25.3 is 28. This indicates a horsepower equivalent of 10.
2. From Table 430-151 the next higher locked-rotor amperage shown is 91. This indicates a 5 hp equivalent. Since the higher hp value is to be used, the controller must be rated for 10 hp.

MOTOR DUTY

Motor design has an influence on the capacities of the controller and fuses used with a motor. All electric motors are designed for either:
1. Continuous duty.
2. Limited duty.

Those designed for continuous duty will deliver the rated horespower for an indefinite period without overheating. General purpose motors should always be the continuous-duty type.

Limited-duty motors will deliver rated horsepower for a specified period of time but cannot be operated continuously at the rated load. A limited-duty motor is one used to operate valves, pumps, or elevators. If its operating period is extended, a limited-duty motor will overheat and may burn out prematurely.

SELECTING THE PROPER SIZE DISCONNECT

Before discussing proper disconnect sizing, we must first deal with several definitions:
RATED-LOAD CURRENT—The current resulting when a motor-compressor is operated at the rated load, rated voltage, and rated frequency of the equipment it serves. The rated-load current is indicated on the nameplate of the motor.
BRANCH-CIRCUIT SELECTION CURRENT—The current value, in amperes, is to be used instead of the rated-load current to determine branch circuit conductor, disconnecting means, and controller and fault device ratings. This value is always greater than the rated-load current.
The disconnecting means rating for a hermetic motor shall be 115 percent of the nameplate rated-load current or the branch-circuit selection current, whichever is greater. The equivalent horsepower is determined in the same way as described earlier for determining controller horsepower. Again, the larger value is the one used to select the proper disconnect.

SIZING CIRCUIT COMPONENTS FOR COMBINATION LOADS

The circuit shown in Fig. 22-10 includes two 1 1/2 hp, one 7 1/2 hp, and one 10 hp squirrel-cage motors plus a lighting load of 26,000 VA. The supply consists of a four-wire, three phase, 120/240 V delta transformer. In order to properly install this circuit we must determine:
1. Conductor ampacities.
2. Conductor types and sizes.
3. Conduit sizes and fill.
4. Switch requirements and sizes.
5. Overcurrent protection—fuses or circuit breaker capacities.

Step-by-step method

The procedures which follow illustrate how to find the values needed. This step-by-step way is only an example of one method. It should be mastered by every electrician, however, as it is quick and practical. Furthermore, the method can be applied to many power/lighting combination circuits.

Continue referring to Fig. 22-10 as you work through the following steps:
1. Determine the full-load current for each motor. From NEC Table 430-150:
   - 1 1/2 hp = 5.2 amperes
   - 7 1/2 hp = 22 amperes
   - 10 hp = 28 amperes
2. Calculate the main motor feeder capacity and conductor size.
   (Refer to NEC Section 430-24.)
   (2) 1 1/2 hp @ 5.2 amperes = 10.4 amperes
   (1) 7 1/2 hp @ 22 amperes = 22 amperes
   (1) 10 hp @ 28 amperes × 125%
   = 35 amperes (This is the largest motor)
   TOTAL = 67.4 amperes
   Three No. 4 AWG (copper) conductors, types THHN, TW, or THWN (NEC Table 310-16) will work well. These should be installed in 1 in. conduit. (See NEC Chapter 9, Tables 3A and 3B.)
3. Find motor feeder protection rating.
   (Refer to NEC Sections 430-62, 430-24, and 430-52.)
   (2) 1 1/2 hp @ 5.2 amperes each
   = 10.4 amperes
   (1) 7 1/2 hp @ 22 amperes
   = 22 amperes
   (1) 10 hp @ 28 amperes × 250%
   = 70 amperes
   TOTAL = 102.4 amperes

Motors and Motor Circuits

Step 1. Find full-load current for each motor.
Step 2. Calculate main motor feeder capacity & conductor size.
Step 3. Find motor feeder protection rating.
Step 4. Determine lighting load.
Step 5. Find combined feeder load.
Step 6. Size combined feeder switch and overcurrent device rating.

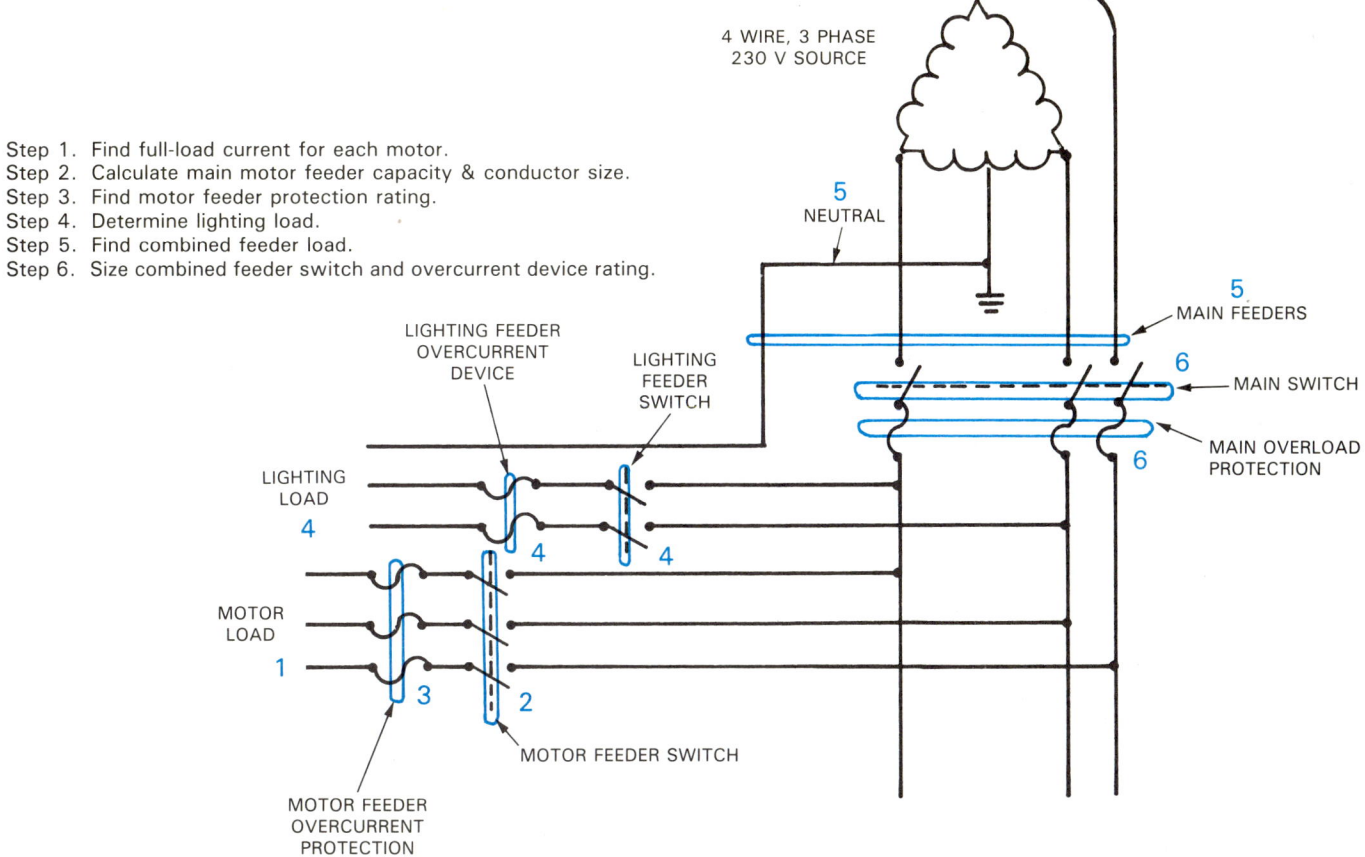

Fig. 22-10. Sizing circuit for combination loads.

Therefore, the main motor feeder protection should consist of a 100 ampere switch having 100 ampere fuses. These fuses should be the time-delay type.

4. Determine the lighting load.

$$\frac{26,000 \text{ VA}}{240 \text{ VA}} = 108.7 \text{ amperes}$$

Use three No. 2 AWG (copper) TW, THHN, or THW conductors in 1 1/4 in. conduit and provide feeder protection consisting of a 200 ampere switch with 110 amperes fuses in each hot leg.

5. Find combined feeder load.

Note that, of the three main feeder conductors, two will carry both the motor and lighting loads while one conductor will carry only a motor load. The two phase conductors carrying the combined motor and lighting load will need a minimum current capacity consisting of:

67.4 amperes for motor load, plus
108.7 amperes for lighting load
= 176.1 amperes, TOTAL

Each of these phase conductors must be adequately sized. No. 2/0 AWG, THHN, RHH, or THW are best suited. The remaining phase conductor serves only the motor load and will need a capacity equal to 67.4 amperes. A No. 4 AWG TW, RHH, THW, or THHN will be adequate.

The neutral (for the lighting load) must be rated to carry the maximum unbalance between it and the hot leg (108.7 amperes). Use a No. 2 AWG THW or THWN conductor (Refer to NEC Table 310-16).

The four main feeder conductors (two No. 2/0 + one No. 4 + one No. 2) should be run in 2 in. conduit. This is well within the NEC Chapter 9, Table 1 guidelines.

6. Size the combined feeder switch and overcurrent devices. Since two of the phase conductors will carry a 176.1 ampere load and the other a 67.4 ampere load, the main feeder switch should be rated at 400 amperes.

Overcurrent protection for the phase conductors carrying both the lighting and motor loads will be sized at 225 amperes (NEC, 430-63):

motor current draw = 102.4 amperes
light current draw = 108.7 amperes
Total draw = 211.1 amperes
(next higher fuse rating is 225 A)

A fuse or breaker supplying 110 amperes of overcurrent protection must be used for the hot leg carrying only the motor load (102.4 amperes).

INSULATION SYSTEMS FOR SMALL MOTORS

Four insulation systems are available for small induction motors. They are as follows:

System	Maximum (Hot Spot) Continuous Temperature
Class A	105°C (221°F)
Class B	130°C (266°F)
Class F	155°C (311°F)
Class H	180°C (356°F)

Temperature limits are established by two agencies:
1. Underwriters Laboratories—to protect against fire hazards.
2. National Electrical Manufacturers Association (NEMA)—to assure adequate motor life.

Nameplate data generally give the allowable or maximum temperature rise above the ambient (surrounding) air for motor operation. If these are observed, the hot spot temperature of the motor should remain within the specified value for the insulation system used. Normal maximum ambient temperature is 40°C (104°F) for most motor ratings.

MOTOR INSTALLATION TIPS

Proper installation of an electrical motor is essential for satisfactory operation, maximum service, and personal safety. The installation and wiring should conform to the recommendations of the *National Electrical Code,* and to any local code.

CAUSES OF MOTOR FAILURE

A motor properly selected, installed, and used can give many years of satisfactory service. Failures are most often due to overheating, moisture, bearing failure, or starting mechanism failure. *Preventive maintenance and proper motor loading are the best insurance against motor failure.* Motor life is prolonged by keeping the motor cool, dry, clean, and lubricated.

Overheating

Heat is one of the most destructive agents and can cause premature motor failure. Overheating occurs because of motor overloading, low voltage at the motor terminals, excessive ambient temperatures, or poor cooling caused by accumulation

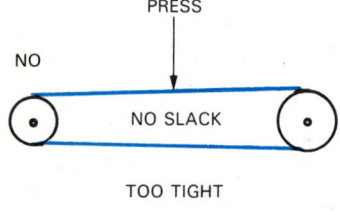

NO — NO SLACK — TOO TIGHT

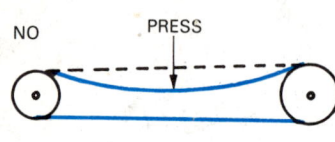

NO — VERY LOOSE BELT NOT SEATED IN VEE

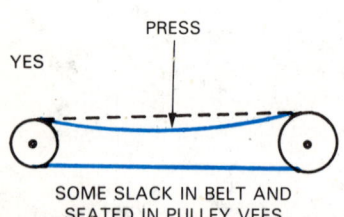

YES — SOME SLACK IN BELT AND SEATED IN PULLEY VEES

V-BELT CROSS SECTION	SMALL SHEAVE DIAMETER RANGE	SMALL SHEAVE RPM RANGE	SPEED RATIO RANGE	DEFLECTION FORCE	
				Minimum	Maximum
type	inches			pounds	pounds
A	3.0 to 3.2		2.0 to 4.0	2.3	3.2
	3.4 to 3.6		2.0 to 4.0	2.5	3.6
	3.8 to 4.2		2.0 to 4.0	2.9	4.2
	4.6 to 7.0		2.0 to 4.0	3.5	5.1
B	4.6		2.0 to 4.0	4.0	5.9
	5.0 to 5.4		2.0 to 4.0	4.5	6.7
	5.6 to 6.4		2.0 to 4.0	5.0	7.4
	6.8 to 9.4		2.0 to 4.0	5.8	8.6
C	7.0		2.0 to 4.0	7.1	10.0
	7.5 to 8.0		2.0 to 4.0	7.9	11.0
	8.5 to 10.0		2.0 to 4.0	9.3	13.0
	10.5 to 16.0		2.0 to 4.0	11.0	16.0
D	12.0 to 13.0		2.0 to 4.0	16.0	24.0
	13.5 to 15.5		2.0 to 4.0	18.0	27.0
	16.0 to 22.0		2.0 to 4.0	21.0	31.0
E	21.6 to 24.0		2.0 to 4.0	33.0	47.0
3V	2.5 to 3.5	1200 to 3600	2.0 to 4.0	3.0	4.3
	3.51 to 4.50	900 to 1800	2.0 to 4.0	3.5	5.3
	4.51 to 6.0	900 to 1800	2.0 to 4.0	4.3	6.0
5V	7.0 to 9.0	600 to 1500	2.0 to 4.0	8.8	13.0
	9.1 to 12.0	600 to 1200	2.0 to 4.0	9.5	14.0
	12.1 to 16.0	400 to 900	2.0 to 4.0	11.0	15.0
8V	12.5 to 17.0	400 to 900	2.0 to 4.0	22.0	31.0
	17.1 to 24.0	200 to 700	2.0 to 4.0	23.0	34.0

Fig. 22-11. Left. Proper procedure for adjusting V-belt tension. Apply pressure at right angle to the belt midway between pulleys. Right. Recommended force for 1/64 in. deflection in V-belt for each inch of span.

of dirt or lack of ventilation. If heat is not dissipated, insulation failure can result, ruining the motor.

Moisture

Moisture should be kept from entering a motor. Cover it to protect it from the weather, particularly during periods when it is not used.

Bearing failure

Bearings should be kept properly lubricated. Bearings may fail (seize) in unused motors that are not rotated for extended periods. Special care in lubrication may be required for these motors.

Starting mechanism failure

Choosing a well-built motor will help solve this problem. Also, the starting mechanism must be kept free of dirt and moisture. The same goes for bearings and motor windings.

Mounting

Secure mounting and correct alignment with the load are essential for proper motor performance. The motor should be positioned where it is readily accessible, but not in the way. If possible, the motor should be located so that it will not be exposed to excessive moisture, dust, or abrasive material.

Mount the motor on a smooth, solid foundation. Fasten the mounting bolts tightly. If mounted on an uneven base or fastened insecurely, the motor may become misaligned with the load during operation. This will throw unnecessary strain on the frame and bearings, causing rapid wear and overheating. Loose mounting also causes vibration and noise.

Connecting to load

Motors may be connected to load by:
1. Direct drive.
2. Belt and pulley.
3. Chain and sprocket.

Direct drive can be used only when the motor and the driven equipment operate at the same speed. A flexible coupling should be used. The motor shaft and driven shaft should be in near-perfect alignment. This prevents excessive wear of the shaft bearing.

Using a V-belt is the easiest and most common way of connecting a motor to the load. High-speed chain drives are used when a positive drive is necessary or when the torque required is more than a V-belt drive can transmit.

Proper belt tension must be maintained. If a belt

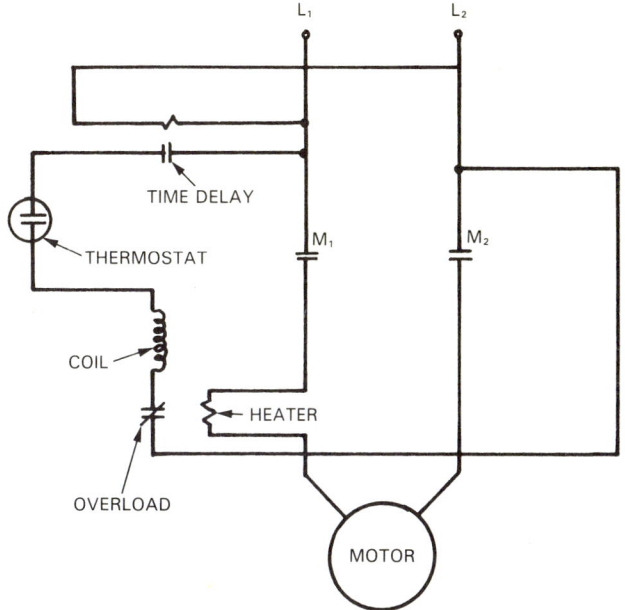

Fig. 22-12. Circuit for random restarting of motors under automatic control.

is too loose, it will slip on the drive pulley, overheat, and wear out quickly. If it is too tight, it will cause the belt and bearings to wear excessively.

To properly tension a V-belt drive, measure the span between shafts as shown in Fig. 22-11. Also, measure the force required to deflect the belt 1/64 in. for each inch of span. The force required should be within the values shown in Fig. 22-11 for the type of belt used.

SERVICE AND REPAIR OF MOTORS

Generally, motors should not be allowed to restart automatically after a loss of power. If automatic operation is necessary, provision should be made for random restarting to prevent the excessive voltage drop in the wiring that would occur if all motors came on at one time. This can be accomplished by including a low-cost time-delay relay in the magnetic motor starter control as shown in Fig. 22-12. This random restart feature is especially desirable for large horsepower motors.

Figs. 22-13 through 22-15 provide further information on the proper wiring of ac magnetic motor starters. Varioius types of starter enclosures and their applications are found in Fig. 22-16.

A well-made and properly installed electric motor requires less maintenance than many other types of electrical equipment. However, for the best and most economical performance, periodic servicing is required.

The service operations listed should be performed once a year or more often if the motor operates under conditions of severe heat, cold, or dust.

WIRING DIAGRAM

A WIRING DIAGRAM shows, as closely as possible, the actual location of all of the component parts of the device. The open terminals (marked by an open circle) and arrows represent connections made by the user.

Since wiring connections and terminal markings are shown, this type of diagram is helpful when wiring the device, or tracing wires when troubleshooting. Note that bold lines denote the power circuit, and thin lines are used to show the control circuit. Conventionally, in ac magnetic equipment, black wires are used in power circuits and red wiring is used for control circuits.

A wiring diagram, however, is limited in its ability to convey a clear picture of the sequence of operation of a controller. Where an illustration of the circuit in its simplest form is desired, the elementary diagram is used.

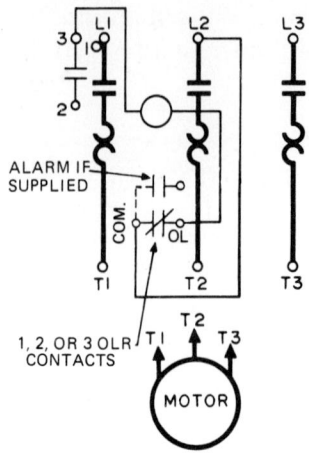

ELEMENTARY DIAGRAM

The elementary diagram gives a fast, easily understood picture of the circuit. The devices and components are not shown in their actual positions. All the control circuit components are shown as directly as possible, between a pair of vertical lines, representing the control power supply. The arrangement of the components is designed to show the sequence of operation of the devices, and helps in understanding how the circuit operates. The effect of operating various interlocks, control devices etc. can be readily seen — this helps in trouble shooting, particularly with the more complex controllers. This form of electrical diagram is sometimes referred to as a "schematic" or "line" diagram.

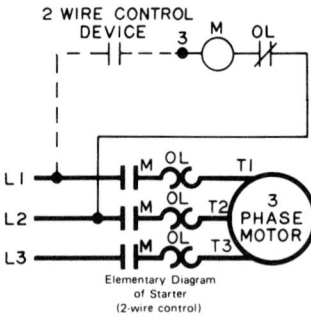

Fig. 22-13. These two diagrams will help you understand the circuit and wiring for an ac magnetic motor starter. (Square D Co.)

MOTOR SERVICE OPERATIONS

1. Remove dust and dirt from the air passages and cooling surfaces of the motor to insure proper cooling. Plugged air passages of an open motor, or a coating of dust on a totally enclosed motor, will cause the motor to overheat under normal operation.
2. Check bearings for wear. Excessive side or end play may cause the motor to draw higher-than-normal starting current, develop less starting torque, and, therefore, damage the motor.
3. Make sure the motor shaft turns freely. Tight or misaligned bearing will cause the motor to overheat.
4. Lubricate the motor according to the manufacturer's specifications. *Do not overlubricate. Too much lubricant is as bad as too little.*
5. Check all wiring for frayed or bare spots. Repair or replace as needed.

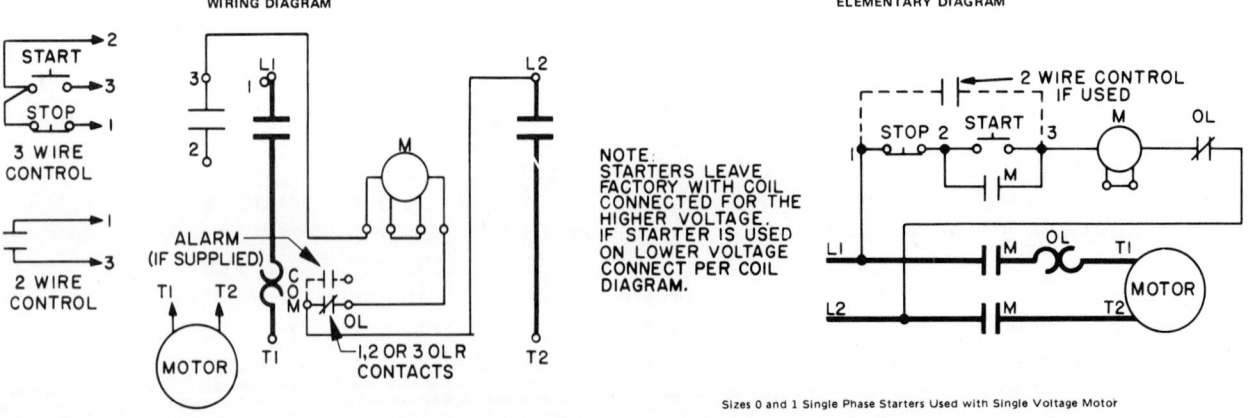

Fig. 22-14. Wiring diagrams for starter for single phase voltage. (Square D Co.)

Motors and Motor Circuits

6. Clean the starting switch contacts of split-phase and capacitor motors and the commutator and brushes in would-rotor (repulsion-type) motors. Use *very* fine sand paper, *not emery cloth.*
7. Replace worn brushes, and make sure the brush-lifting and shorting-ring action works smoothly in wound-rotor motors.
8. Check belt pulleys to be sure they are secure on their shafts. Align the belts and pulleys carefully. Improper alignment causes excessive wear on belts and pulleys. Check and adjust belt tension. Replace belts that are badly worn.

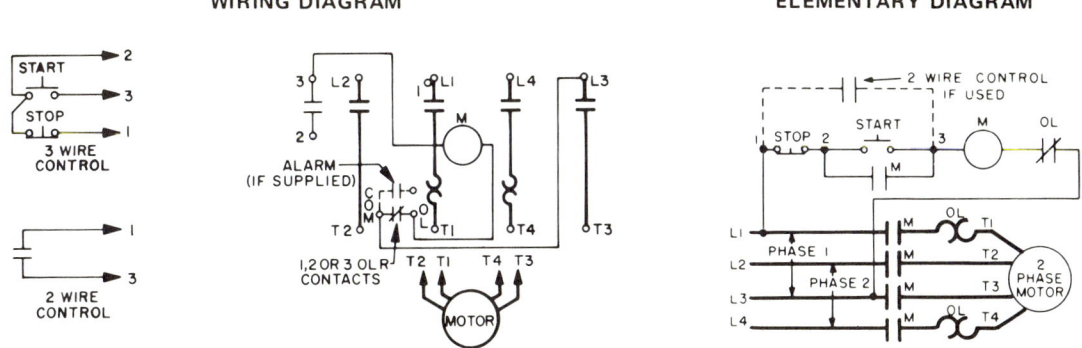

Sizes 3 and 4, 4 Pole, 2 Phase, 4 Wire Starters with External 2 or 3 Wire Control

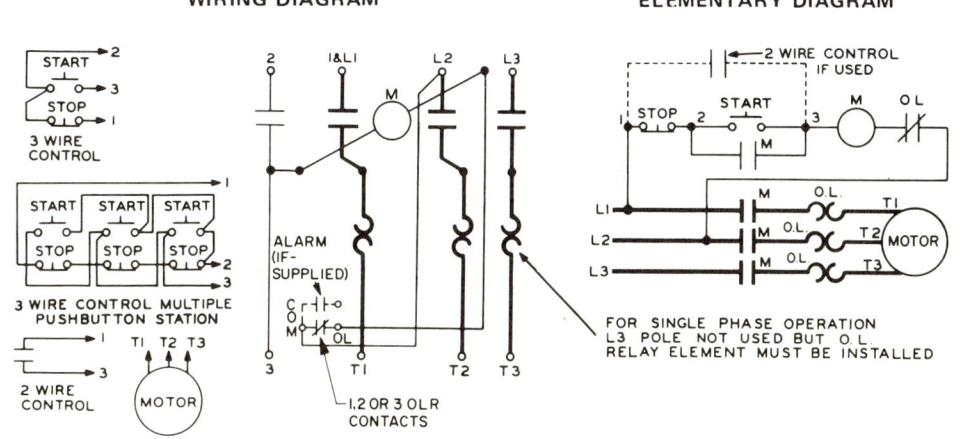

Size 00, 3 Pole, 3 Phase Starter with External 2 or 3 Wire Control

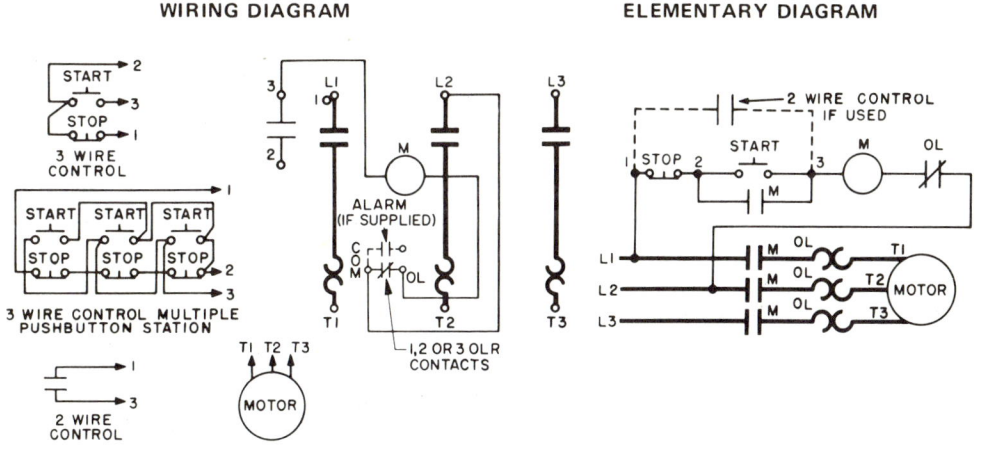

Sizes 0-4, 3 Pole, 3 Phase Starters with External 2 or 3 Wire Control

Fig. 22-15. Diagrams for wiring of two-phase and three-phase magnetic starters.

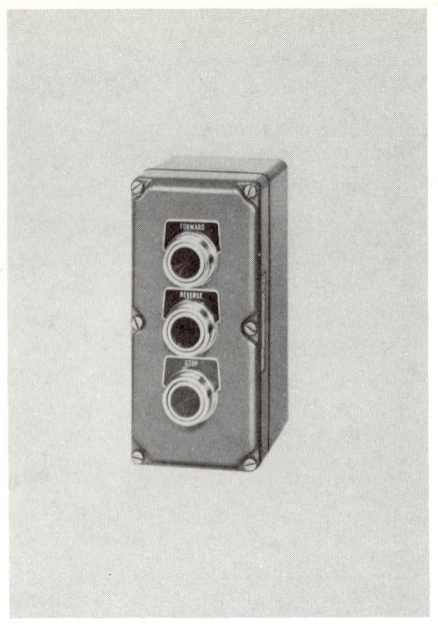  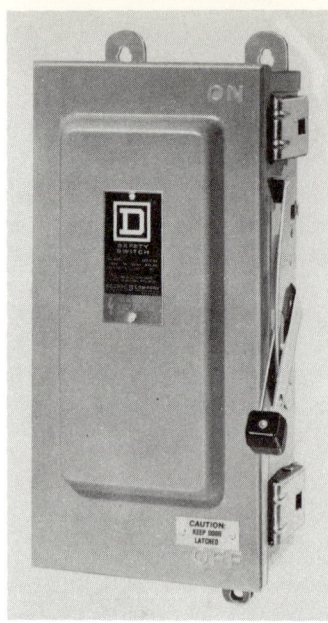

Fig. 22-16. Motor control units. A—Heavy duty motor control unit is used primarily in foundries, steel mills, etc. Unit is oil and dust tight. B—Watertight, dust tight and corrosion resistant manual motor starter enclosure. C—Single motor branch circuit safety switch. (Square D Co.)

REVIEW QUESTIONS — CHAPTER 22

1. For proper performance, the _____ voltage and frequency at the _____ _____ must match the values specified by the manufacturer as closely as possible.
2. Low voltage from inadequate wiring (can, cannot) cause problems in an electric motor.
3. List the information which usually appears on the nameplate attached to an electric motor.
4. Explain what frame number of an electric motor has to do with standardization of electric motors.
5. List the four common motor circuits.
6. NEC requires that the feeder capacity of a single motor circuit be _____ percent of the full-load _____ rating of the motor.
7. What is the full-load current of each of the following squirrel-cage motors? (Refer to tables in Article 430 of NEC.)

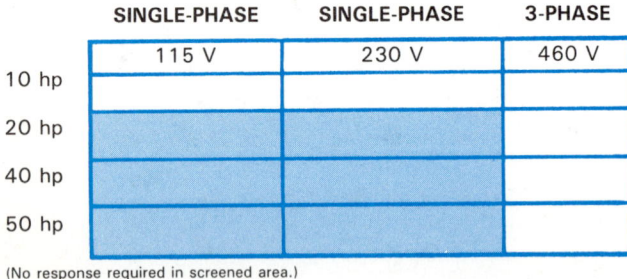

(No response required in screened area.)

8. A _____ _____ is a switch or other device normally used for starting and stopping a motor.
9. Study the illustration and determine:
   a. Full-load current of the electric motor. _____
   b. Correct size of conductors to keep voltage drop within NEC-set limits. _____

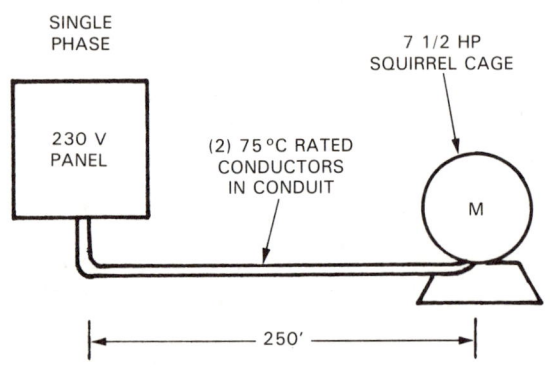

NOTE: MAX VOLT. DROP AT MOTOR TO BE 3% or 7V. I.E. 223 V. MIN. AT MOTOR TERMINALS

10. List all of the following which are causes for overheating in an electric motor.
    a. Lack of insulation.
    b. Excessive ambient temperatures.
    c. Low voltage at motor terminals.
    d. Too high voltage at motor terminals.
    e. Poor cooling from dirt or lack of ventilation.
    f. Overloading of motor.
    g. Lack of lubricant.
11. During periodic motor service, bearings (should, should not) be lubricated.
12. Emery cloth (should, should not) be used to clean contacts of split-phase and capacitor motors and the commutator and brushes of wound-rotor motors.

Motors and Motor Circuits

PROBLEM	CAUSE(S)	REMEDY
Motor overheats	Rotor rubbing on stator —Bent shaft —Worn bearings	Replace shaft or bearings
	Overloading —Binding load	Correct load factor Check current input
	Poor ventilation	Keep motor clean Clear vent holes or venting system
Motor won't start	Power failure —Insufficient voltage	Check the circuit breakers or fuses Replace if necessary
	Improper connections	Check wiring diagram and make necessary corrections
	Overload	Reduce load or use larger motor
	Worn or incorrectly set brushes	Replace brushes or reset them
	Open circuit	Check all connections Clean starting switch contacts Check for short-circuit or ground faults Reset thermal overload
	Too much end play	Add washers to shaft
Rotor or stator burns out	Moisture Corrosive chemicals Dust	Keep motor clean and dry. Shield motor if necessary
Brush sparking	Short-circuit or open circuit	Clean, repair or replace armature
	Sticking or worn brushes	Replace brushes
	Dust	Clean the motor
	Overload	Decrease load or use more powerful motor
	Improperly fit brushes	Refit the brushes to match commutator
	Loose commutator	Replace commutator
Noisy motor	Unbalanced load	Balance the load and pulley
	Loose parts	Tighten motor components and retighten motor mounts
	Faulty alignment	Properly align motor with load
	Bent shaft	Straighten shaft and realign Make sure load is balanced
	Worn bearings	Lubricate bearings; replace bearings.
	Dust	Clean motor

Motors are expensive and will last a long time with proper maintenance. This guide is useful should problems arise. Good maintenance is important; without it a motor may have a short life.

Chapter 23
ELECTRICAL CAREERS

If you have more than just a casual interest in electrical wiring, perhaps a career in electricity is for you. This chapter will briefly list and explain the various career categories open to you.

CAREER CATEGORIES

The manual and mental skills required for electrical jobs vary greatly. However, all of them require knowledge of electricity and electrical wiring techniques. Following is a partial list of jobs which fall under the general category of electrical construction or maintenance work.
1. Power line installers and repairers.
2. Power line troubleshooters.
3. Cable splicers.
4. Construction electricians.
5. Maintenance electricians.

Fig. 23-1. Powerline installers are shown working on new primary distribution lines. (New York State Electric and Gas Corp.)

Fig. 23-2. Line repairer-installer is correcting damaged transmission line.

Fig. 23-3. Cable splicers-line installers are positioning pad transformer. (New York State Electric and Gas Corp.)

TRANSMISSION AND DISTRIBUTION ELECTRICIANS

One-fourth of the workers in the electric power industry are involved in transmission and distribution jobs. The principal workers in these occupations are those who control the flow of electricity. They are the load dispatchers and substation operators. Others in this group construct and maintain power lines. Job titles in this category include: line installers and repairers, cable splicers, troubleshooters, ground helpers, and laborers. Line installers and repairers make up the largest single occupation in the industry.

Line installers and repairers construct and maintain the network of power lines that carries electricity from generating plants to consumers, Fig. 23-1. Their work consists of installation, equipment replacement, repairs and routine maintenance. When wires, cable, or poles break, it means an emergency call for a line crew. Line repairers splice or replace broken wires and cables, and replace broken insulators or other damaged equipment.

Most installers and repairers work from "bucket" trucks with pneumatic lifts that take them to the top of the pole at the touch of a lever, Fig. 23-2. In some power companies, line crew employees specialize in particular types of work.

Troubleshooters are experienced line installers and repairers who are assigned to special crews that handle emergency calls. They move from one job to another, as directed by a central service office which receives reports of line trouble. Often, troubleshooters receive their orders by direct radio communication with the central service office.

These workers must have a thorough knowledge of the company's transmission and distribution network. They first locate and report the source of trouble and then attempt to restore service by making the necessary repairs.

Depending on the nature and extent of the problem, troubleshooters may restore service, or simply disconnect and remove the damaged equipment. They must be familiar with all the circuits and switching points so that they can safely disconnect live circuits.

Cable splicers install and repair insulated cables on utility poles and towers, as well as those buried underground or those installed in underground conduits, Fig. 23-3. When cables are installed, the cable splicers pull the cable through the conduit and then join the cables at connecting points in the transmission and distribution systems. At each connection in the cable, they wrap insulation around the wirings. They splice the conductors leading away from each junction of the main cable, insulate the splices, and connect the cable sheathing. Most of the physical work in placing new cables or replacing old ones is done by cable splicers and their assistants.

Cable splicers spend most of their time repairing and maintaining cables and changing the layout of the cable systems. They must know the arrangement of the wiring systems, where the circuits are connected, and where they lead to and come from. When making repairs, they must make sure that the conductors do not become mixed up between the substation and the customer's premises. Cable splicers also periodically check insulation on cables to make sure it is in good condition.

MAINTENANCE ELECTRICIANS

Maintenance electricians keep lighting systems, transformers, generators, and other electrical equipment in good working order. They also may install new electrical equipment.

Duties vary greatly depending on where the electrician is employed. Electricians who work in large factories may repair items such as motors and welding machines. Those in office buildings and small plants usually fix all kinds of commercial or industrial electrical equipment. Regardless of location, electricians spend much of their time doing *preventive maintenance,* and making periodic inspection of equipment to locate and correct defects before breakdowns occur. See Fig. 23-4.

When trouble occurs, they must find the cause and make repairs quickly to prevent costly "downtime" or production losses. In emergencies, they advise management whether continued operation of equipment would be hazardous.

Maintenance electricians make repairs by replacing items such as fuses, circuit breakers, or switches. When installing new or replacing existing wiring, they splice wires and cut or bend conduit through which the wires are run.

Maintenance electricians sometimes work from blueprints, wiring diagrams, or other specifications. They use meters and other testing devices to locate faulty equipment. To make repairs, they use an array of specialized devices besides pliers, screwdrivers, wirecutters, drills, and other basic tools.

TRAINING, OTHER QUALIFICATIONS, AND ADVANCEMENT

Most maintenance electricians learn their trade on the job or through formal apprenticeship programs. A relatively small number learn the trade in the military service. Training authorities generally agree that apprenticeship gives trainees more thorough knowledge of the trade and improves job opportunities during their working life. Because the

Fig. 23-4. Maintenance electrician checks attic insulation. (New York State Electric and Gas Corp.)

Fig. 23-5. Technician uses X-ray machine to rate liquid and solid flame retardant. (Underwriters Laboratories)

training is comprehensive, people who complete apprenticeship programs qualify either as maintenance, construction electricians, or technicians. See Figs. 23-5 and 23-6.

Apprenticeship usually lasts four years, and consists of on-the-job training and related classroom instruction in subjects such as mathematics, electrical and electronic theory, and blueprint reading. Training may include electric motor repair, wire splicing, installation and repair of electronic controls and circuits, and welding and brazing.

INFORMAL TRAINING

Although formal apprenticeship is the preferred method of training, many people learn the trade informally on-the-job by serving as helpers to skilled maintenance electricians. Helpers begin by doing simple jobs, such as replacing circuit breakers and switches. With experience, they advance to more complicated jobs such as splicing cable and connecting circuit wires. They eventually get enough experience to qualify as electricians.

SELECTING HELPFUL COURSES

Persons interested in becoming maintenance electricians can obtain a good background by taking high school or vocational school courses in subjects such as electricity, electronics, physics, algebra, mechanical drawing, shop, and physical sciences. To qualify for an apprenticeship program, an applicant should be at least 18 years old and usually must be a high school or vocational school graduate.

Because the electrician's craft is subject to constant technological change, experienced electricians must continue to learn new skills.

All maintenance electricians must be familiar with the *National Electrical Code* and local building codes. Many cities and counties require maintenance electricians to be licensed. Electricians can obtain a license by passing an examination that tests their knowledge of electrical theory and its application.

ELECTRICIANS (CONSTRUCTION)

Heat, light, power, air conditioning, and refrigeration components all operate through electrical systems that are assembled, installed, and wired by construction electricians. These workers also install electrical machinery, electronic equipment, controls, and signal and communications systems. Construction electricians follow blueprints and specifications for most installations.

Electricians, for safety reasons, must follow *National Electrical Code* regulations and, in addition, must fulfill all requirements of state, county, and municipal electrical codes.

TRAINING, OTHER QUALIFICATIONS, AND ADVANCEMENT

Most training authorities recommend the completion of a four-year apprenticeship program as the preferred way to learn the trade. As in the electrical maintenance trade, some individuals learn informally by working for many years as electrician's helpers.

Modern Residential Wiring

Fig. 23-6. Technician carefully checks resistance of product submitted for testing. (Canadian Standards Assn.)

Many helpers gain additional knowledge through trade school or correspondence courses, or through special training in the military.

Apprenticeship programs are often sponsored through and supervised by local union-management committees. These programs provide 144+ hours of classroom instruction each year in addition to comprehensive on-the-job training.

In the classroom, apprentices learn blueprint reading, electrical theory, electronics, mathematics, safety, and first-aid practices. On the job, under supervision of experienced electricians, apprentices must demonstrate mastery of electrical wiring principles.

At first, apprentices drill holes, set anchors, and set up conduit. In time and with experience, they measure, bend, and install conduit, as well as install, connect, and test wiring. They also learn to set up and draw diagrams for entire electrical systems.

To obtain a license, which is necessary for employment in most cities, construction electricians must also pass an examination which requires a thorough knowledge of electrical wiring and of state and local electrical codes.

For more information concerning careers in the electrical trades write to the following organizations:

International Brotherhood of Electrical Workers (IBEW)
1125 15th Street N.W.
Washington, DC 20005

National Electrical Contractors Association (NECA)
7315 Wisconsin Avenue
Washington, DC 20014

National Fire Protection Association (NFPA)
Batterymarch Park
Quincy, MA 02269

International Association of Electrical Inspectors (IAEI)
930 Busse Highway
Park Ridge, IL 60068

Do not overlook the local power suppliers, trade schools, neighborhood electricians, and electrical inspectors. They are potential employers as well as sources of information.

REVIEW QUESTIONS — CHAPTER 23

1. List five categories of jobs available to qualified electricians.
2. _____ are experienced line installers and repairers who are assigned to special crews that handle emergency calls.
3. Electricians who keep lighting systems, transformers, generators, and other electrical equipment in good working order are called _____ electricians.
4. Most electricians learn their trade on the job or through formal apprenticeship programs. (True or False?)
5. What must a construction electrician do to get a license to work in most cities?

Modern Residential Wiring

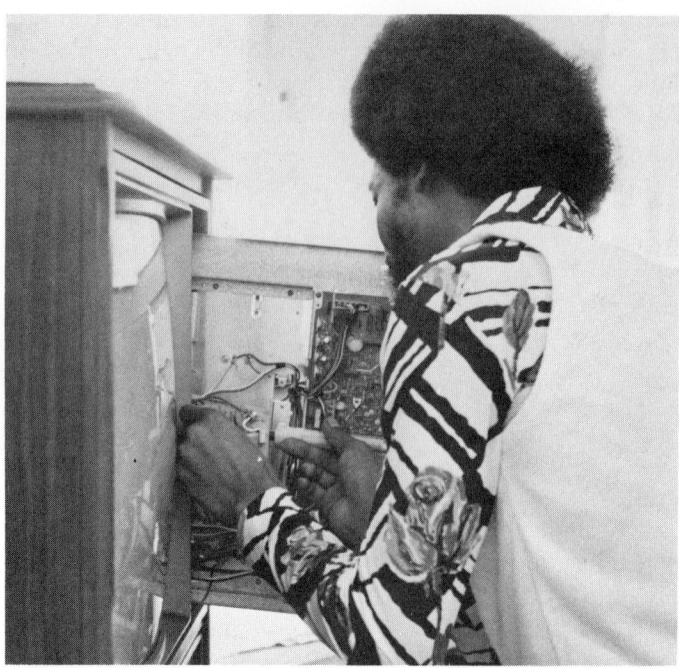

Some electrical/electronics specialists work at testing of new products to rate their safety and operation for public use. This technician is investigating the wiring in a television set. (Underwriters Laboratories, Inc.)

Powerline installers and repairers work mostly from "bucket" trucks with pneumatic lifts. In some power companies, line crew employees will specialize in particular types of work.

Chapter 24

MATH REVIEW

Every electrician must have a good background in math. This section will review some of the basic math operations needed by the electrician.

Above and beyond the simple addition, subtraction, division, and multiplication skills, an electrician must also be competent in computing or performing the following operations:
1. Converting fractions to decimals.
2. Calculating percentage.
3. Converting fractions to percentages.
4. Formula organization.
5. Basic trigonometry.
6. Circle area and circumference.
7. Rectangle or square area and circumference.

Each of these procedures will be used often by the electrician in such tasks as: figuring service entrance capacity, conduit fill, conduit bends, branch and feeder circuits, and wire ampacity. In addition there are routine business operations to perform on a day-to-day basis. For these reasons, math must be considered an important electrical tool.

Converting Fractions to Decimals and Vice Versa

On many occasions, it is necessary to change a fraction into its equivalent decimal. The electrician is constantly making measurements and reading specifications. Many of these figures are in fractions or decimals.

To change a fraction to a decimal, divide the numerator (top part of fraction), by the denominator (lower part of fraction).
Example:
Change 7/16 to a decimal.
Solution:

```
       .4375
16/7.0000
     64
     ──
      60
      48
      ──
      120
      112
      ───
       80
       80
```

To change a decimal to a fraction, place the decimal number over 10 or 100 or 1000 or 10,000 depending on the number of digits in the decimal. Then reduce to the smallest fraction.
Example:
Change 0.4375 to a fraction.
Solution:
Since the decimal is to the ten thousandths place, we will divide as follows;

$$\frac{4375}{10,000} = \frac{875}{2000} = \frac{175}{400} = \frac{35}{80} = \frac{7}{16}$$

Finding percentage

It is often necessary to compute the percentage of a given number. This is especially useful when computing conduit fill or discounts on supply items.

To find the percentage one number is of another, simply divide the larger number into the smaller one. Multiply the result by 100%.
What percentage of 87 is 14?
Solution:

```
      .16
87/14.00
   87
   ──
   530
   522
```
and, .16 × 100% = 16%

A practical application of this follows:
You have just purchased $240 worth of electrical fittings. If you pay the supplier's bill within 10 days, you can take a 5 percent discount. How much can you save by paying early? *Solution:* Multiply the total by 5% (.05) and subtract this result from the total.

```
  $ 240.00              $ 240.00
  ×    .05       and   − $  12.00
  ─────────             ─────────
  $  12.0000            $ 228.00
```

Converting fraction to percentage

To convert a fraction to a percentage, divide the numerator by the denominator and multiply the result by 100%.

Example:
What percentage is the fraction 1/6?
Solution:
$$\frac{1}{6} = 6\overline{)1.000}^{.166} \text{ and } .166 \times 100\% = 16.6\%$$

Formula Organization

Many times the particular unknown in a formula is not the one which is isolated. Thus, the formula must be reorganized to separate the unknown from the rest.

Example:
Ohm's law, in formula, is written, $E = I \times R$. Suppose we wish to find R, when E and I are known. We must rewrite the formula to isolate R.
Solution:
Divide both sides of the equation by I.
$E = I \times R$
to:
$$\frac{E}{I} = \frac{I \times R}{I}$$
result: $\frac{E}{I} = R$
Then, just solve for the unknown (R).

Basic trigonometry

Trigonometry relates to various parts of the triangle. In electrical work, trigonometry is used to find unknown sides or angles of triangular areas. A fundamental knowledge of trigonometry is helpful in working with conduit and for general measurement.

The basic six trigonometric functions are listed below along with the illustration of a right triangle for reference.

Sine A $= \frac{a}{c}$ Cosecant A $= \frac{c}{a}$

Cosine A $= \frac{b}{c}$ Secant A $= \frac{c}{b}$

Tangent A $= \frac{a}{b}$ Cotangent A $= \frac{b}{a}$

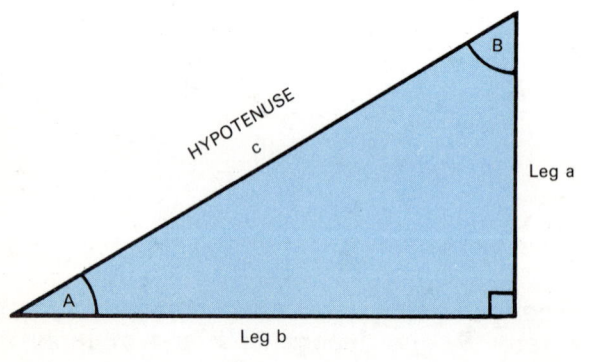

Useful right triangle facts:
1. In a 30° - 60° - 90° triangle, the hypotenuse of the triangle is twice as long as the side opposite the 30° angle.
2. In a 45° - 45° - 90° triangle, the sides oppostie the 45° angles are equal.
3. For any right angle triangle, the lengths of the sides and the hypotenuse are related by the relationship:
$$c^2 = a^2 + b^2$$

Facts about other than right triangles:
For any triangle, like the one shown below, we can compute the size of any two sides and an angle, if one side and the other two angles are known. This can be done by using the Sine Law or Cosine Law.

Sine law: $= \frac{a}{\text{Sine A}} = \frac{b}{\text{Sine B}} = \frac{c}{\text{Sine C}}$

Cosine Law: $a = \sqrt{b^2 + c^2 - 2bc \cos A}$
or $b = \sqrt{a^2 + c^2 - 2ac \cos B}$
or $c = \sqrt{a^2 + b^2 - 2ab \cos C}$

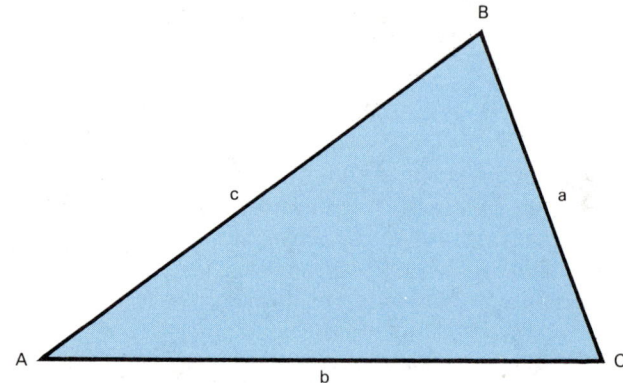

A table of trigonometric functions can be helpful in solving problems.

Circle area and circumference

To obtain the area of a circle, simply apply the formula $A = \pi r^2$. Here, A is the area, π is 3.14 and R is the radius of the circle.

For circle circumference, the formula is $C = \pi d$. C equals circumference, π is 3.14 and d represents the diameter of a circle.

Rectangle, area, square area, and circumference

1. Area of a rectangle is equal to its base times its height.
2. Area of a square is any side times itself.
3. Circumference of a rectangle is found by adding its four sides.
4. Circumference of a square is four times any side.

REVIEW QUESTIONS — CHAPTER 24

1. The decimal equivalent of 7/8 in. is _____.
2. The decimal, .5625, is equal to the fraction, _____.
3. If resistance in a circuit is 4 ohm, and current is 30 ampere, the voltage is _____.
4. A conduit is bent as shown. What is the angle of the bend at A?

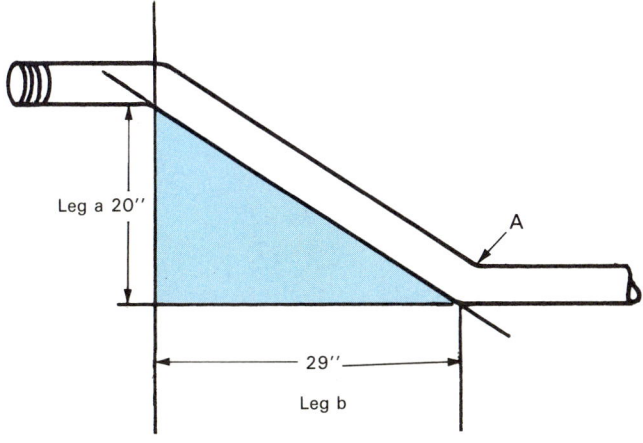

5. As shown, a guy wire is to be attached to the top of the 30 ft. utility pole, so that it makes an angle of 60° with the ground. How much guy wire is needed? How far from the base of the pole must it be anchored?

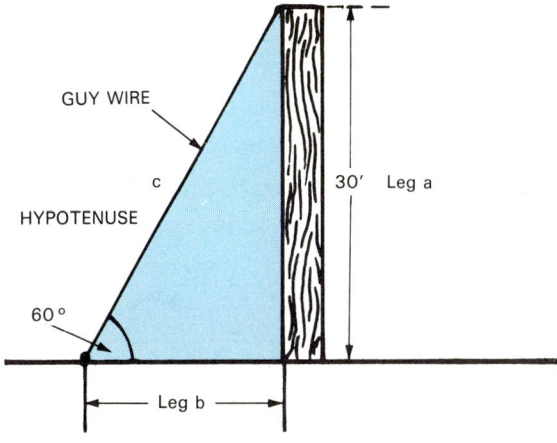

6. What is the area of a circle whose radius if 6 ft.?
7. What is the circumference of the circle discussed in Question 6?

Chapter 25
TECHNICAL INFORMATION

MINIMUM SIZE COPPER WIRE REQUIRED TO ALLOW NOT MORE THAN 2% VOLTAGE DROP

ON 120-VOLT CIRCUITS FOR DISTANCES SHOWN

Length of Circuit in Feet (One Way)

Cap. Amps.	20	30	40	50	60	70	80	90	100	120	140	160	180	200	240	280	320	360
1	....	....	....	....	....	....	....	....	....	....	....	....	....	....	14	14	14	14
1.5	....	....	....	....	....	....	....	....	....	....	14	14	14	14	14	14	12	12
2	....	....	....	....	....	....	....	....	....	14	14	14	14	14	12	12	12	10
3	....	....	....	....	....	14	14	14	14	14	14	12	12	12	10	10	10	8
4	....	....	....	14	14	14	14	14	12	12	12	10	10	10	8	8	8	8
5	....	....	....	14	14	14	14	12	12	12	10	10	10	10	8	8	8	6
6	....	....	14	14	14	14	12	12	12	10	10	8	8	8	8	8	6	6
7	....	14	14	14	14	12	12	12	10	10	10	8	8	8	6	6	6	6
8	....	14	14	14	12	12	12	10	10	10	8	8	8	8	6	6	6	5
9	....	14	14	12	12	12	10	10	10	8	8	8	6	6	6	6	5	5
10	14	14	14	12	12	10	10	10	10	8	8	8	6	6	6	5	5	4
12	14	14	12	12	10	10	10	8	8	8	8	6	6	6	5	5	4	4
14	14	14	12	12	10	10	10	8	8	8	8	6	6	6	5	5	4	4
16	14	12	12	10	10	8	8	8	8	6	6	6	5	5	4	3	3	2
18	12	12	10	10	8	8	8	8	6	6	6	5	5	4	4	3	2	2
20	12	12	10	10	8	8	8	6	6	6	5	5	4	4	3	2	2	1
25	10	10	10	8	8	6	6	6	6	5	4	4	3	3	2	1	1	0
30	8	8	8	8	6	6	6	6	5	4	4	3	3	2	1	1	0	0
35	8	8	8	6	6	6	5	5	4	4	3	2	2	1	1	0	00	00
40	6	6	6	6	6	5	5	4	4	3	2	2	1	1	0	00	00	000
45	6	6	6	6	6	5	4	4	3	3	2	1	1	0	00	000	000	000
50	6	6	6	6	5	4	4	3	3	2	1	1	0	0	00	000	000	0000
60	4	4	4	4	4	4	3	3	2	1	1	0	0	00	000	000	0000	0000
70	4	4	4	4	4	3	2	2	1	1	0	00	00	000	000	0000	0000	....
80	3	3	3	3	3	2	2	1	1	0	00	00	000	000	0000	0000	....	....
90	2	2	2	2	2	2	1	1	0	00	00	000	000	0000	0000	....	....	....
100	1	1	1	1	1	1	1	0	0	00	000	000	0000	0000	....	....	....	....
120	0	0	0	0	0	0	0	0	00	00	000	0000	0000	....	....	....	....	....

ON 240-VOLT CIRCUITS FOR DISTANCES SHOWN

Length of Circuit in Feet (One Way)

Cap. Amps.	20	30	40	50	60	70	80	90	100	120	140	160	180	200	240	280	320	360
1	....	....	....	....	....	....	....	....	....	....	....	....	....	....	....	....	....	....
1.5	....	....	....	....	....	....	....	....	....	....	....	....	....	....	....	....	14	14
2	....	....	....	....	....	....	....	....	....	....	....	....	....	....	14	14	14	12
3	....	....	....	....	....	....	....	....	....	....	....	14	14	14	14	14	12	12
4	....	....	....	....	....	....	....	....	....	14	14	14	14	14	12	12	12	12
5	....	....	....	....	....	....	....	....	14	14	14	14	12	12	12	10	12	10
6	....	....	....	....	....	....	14	14	14	14	14	12	12	12	10	10	10	8
7	....	....	....	....	....	14	14	14	14	14	12	12	12	10	10	10	8	8
8	....	....	....	14	14	14	14	14	12	12	12	10	10	10	8	8	8	8
9	....	....	....	14	14	14	14	14	12	12	12	10	10	10	8	8	8	8
10	....	....	14	14	14	14	12	12	12	10	10	10	10	8	8	8	8	6
12	....	....	14	14	14	14	12	12	12	10	10	10	8	8	8	8	6	6
14	....	14	14	14	14	12	12	12	10	10	10	8	8	8	6	6	6	6
16	....	12	12	12	12	12	12	10	10	10	8	8	8	8	6	6	6	6
18	....	12	12	12	12	12	10	10	10	8	8	8	6	6	6	6	5	5
20	12	12	12	12	12	12	10	10	10	8	8	8	6	6	6	5	5	4
25	10	10	10	10	10	10	10	8	8	8	6	6	6	6	5	4	4	3
30	8	8	8	8	8	8	8	8	8	6	6	6	6	5	4	4	3	3
35	8	8	8	8	8	8	8	8	6	6	6	5	5	4	4	3	2	2
40	6	6	6	6	6	6	6	6	6	6	5	5	4	4	3	2	2	1
45	6	6	6	6	6	6	6	6	6	6	5	4	4	3	3	2	1	1
50	6	6	6	6	6	6	6	6	6	5	4	4	3	3	2	1	1	0
60	4	4	4	4	4	4	4	4	4	4	3	3	2	2	1	1	0	0
70	4	4	4	4	4	4	4	4	4	3	2	2	1	1	0	0	00	00
80	3	3	3	3	3	3	3	3	3	2	2	1	1	0	00	00	000	000
90	2	2	2	2	2	2	2	2	2	2	1	1	0	0	00	000	000	000
100	1	1	1	1	1	1	1	1	1	1	1	0	0	00	000	000	0000	0000
120	0	0	0	0	0	0	0	0	0	0	0	00	00	000	000	0000	0000	....

Fig. 25-1. Voltage drop tables for 120 and 240-volt circuits using copper wire. (General Electric Co.)

Technical Information

TEMPERATURE CONVERSION TABLE

°C.	°F.	°C.	°F.
-80	-112.	47	116.6
-70	-94.	48	118.4
-60	-65.	49	120.2
-50	-58.0	50	122.0
-45	-49.1	51	123.8
-40	-40.0	52	125.6
-35	-31.0	53	127.4
-30	-22.0	54	129.2
-25	-13.0	55	131.0
-20	-4.0	56	132.8
-19	-2.2	57	134.6
-18	-.4	58	136.4
-17	1.4	59	138.2
-16	3.2	60	140.0
-15	5.0	61	141.8
-14	6.8	62	143.6
-13	8.6	63	145.4
-12	10.4	64	147.2
-11	12.2	65	149.0
-10	14.0	66	150.8
-9	15.8	67	152.6
-8	17.6	68	154.4
-7	19.4	69	156.2
-6	21.2	70	158.0
-5	23.0	71	159.8
-4	24.8	72	161.6
-3	26.6	73	163.4
-2	28.4	74	165.2
-1	30.2	75	167.0
0	32.0	76	168.8
1	33.8	77	170.6
2	35.6	78	172.4
3	37.4	79	174.2
4	39.2	80	176.0
5	41.0	81	177.8
6	42.8	82	179.6
7	44.6	83	181.4
8	46.4	84	183.2
9	48.2	85	185.0
10	50.0	86	186.8
11	51.8	87	188.6
12	53.6	88	190.4
13	55.4	89	192.2
14	57.2	90	194.0
15	59.0	91	195.8
16	60.8	92	197.6
17	62.6	93	199.4
18	64.4	94	201.2
19	66.2	95	203.0
20	68.0	96	204.8
21	69.8	97	206.6
22	71.6	98	208.4
23	73.4	99	210.2
24	75.2	100	212.0
25	77.0	105	221.
26	78.8	110	230.
27	80.6	115	239.
28	82.4	120	248.
29	84.2	130	266.
30	86.0	140	284.
31	87.8	150	302.
32	89.6	160	320.
33	91.4	170	338.
34	93.2	180	356.
35	95.0	190	374.
36	96.8	200	392.
37	98.6	250	482.
38	100.4	300	572.
39	102.2	350	662.
40	104.0	400	752.
41	105.8	500	932.
42	107.6	600	1112.
43	109.4	700	1292.
44	111.2	800	1472.
45	113.0	900	1652.
46	114.8	1000	1832.

Fig. 25-2. Table for converting Celsius temperatures to Fahrenheit.

(Temperature, 20° C)

Gauge Number	Diameter (mm)	Cross Section (mm²)	Resistance (Ω/km)	Resistance (m/Ω)
0000	11.68	107.2	0.1608	6219
000	10.40	85.03	0.2028	4932
00	9.266	67.43	0.2557	3911
0	8.252	53.48	0.3224	3102
1	7.348	42.41	0.4066	2460
2	6.544	33.63	0.5027	1951
3	5.827	26.67	0.6465	1547
4	5.189	21.15	0.8152	1227
5	4.621	16.77	1.028	9729
6	4.115	13.30	1.296	771.5
7	3.665	10.55	1.634	611.8
8	3.264	8.366	2.061	485.2
9	2.906	6.634	2.599	384.8
10	2.588	5.261	3.277	305.1
11	2.305	4.172	4.132	242.0
12	2.053	3.309	5.211	191.9
13	1.828	2.624	6.571	152.2
14	1.628	2.081	8.258	120.7
15	1.450	1.650	10.45	95.71
16	1.291	1.309	13.17	75.90
17	1.150	1.038	16.61	60.20
18	1.024	0.8231	20.95	47.74
19	0.9116	0.6527	26.42	37.86
20	0.8118	0.5176	33.31	30.02
21	0.7230	0.4105	42.00	23.81
22	0.6438	0.3255	52.96	18.88
23	0.5733	0.2582	66.79	14.97
24	0.5106	0.2047	84.21	11.87
25	0.4547	0.1624	106.2	9.415
26	0.4049	0.1288	133.9	7.486
27	0.3606	0.1021	168.9	5.922
28	0.3211	0.08098	212.9	4.697
29	0.2859	0.06422	268.5	3.725
30	0.2546	0.05093	338.6	2.954
31	0.2268	0.04039	426.9	2.342
32	0.2019	0.03203	538.3	1.858
33	0.1798	0.02540	678.8	1.473
34	0.1606	0.02014	856.0	1.168
35	0.1426	0.01597	1079	0.9265
36	0.1270	0.01267	1361	0.7347
37	0.1131	0.01005	1716	0.5827
38	0.1007	0.007967	2164	0.4621
39	0.08996	0.006318	2729	0.3664
40	0.07987	0.005010	3441	0.2906

Fig. 25-3. Properties of copper wire.

Modern Residential Wiring

Rating	NEMA Config. and Config. No.	Page	Rating	NEMA Config. and Config. No.	Page	Rating	NEMA Config. and Config. No.	Page
2-Pole—2-Wire			**2-Pole—3-Wire—Grounding (Continued)**			**3-Pole—3-Wire—Not Grounding (Continued)**		
15A., 125V.; 10A., 250V. (Midget)	ML-1R	4	20A., 600V.	L9-20R	14	20A., 3φ, 480V.	L12-20R	22
15A., 125V.	L1-15R	5	30A., 125V.	L5-30R	15	30A., 250V.		23
20A., 250V.	L2-20R	6	30A., 250V.	L6-30R	15	30A., 125/250V.	L10-30R	24
2-Pole—3-Wire—Grounding			30A., 277V., AC	L7-30R	15	30A., 3φ, 250V.	L11-30R	24
15A., 125V. (Midget)	ML-2R	7	30A., 480V.	L8-30R	16	30A., 3φ, 480V.	L12-30R	25
15A., 125V.	L5-15R	8, 9	30A., 600V.	L9-30R	16	30A., 3φ, 600V.	L13-30R	25
15A., 250V.	L6-15R	10	**3-Pole—3-Wire—Not Grounding**			50A., 250V., DC 50A., 600V., AC		26
15A., 277V., AC	L7-15R	11	15A., 125/250 V. (Midget)	ML-3R	17	**3-Pole—4-Wire—Grounding**		
20A., 125V.	L5-20R	12	15A., 125V.; 10A., 250V. For replacement use—Not UL listed		18, 19	20A., 125/250V.	L14-20R	27
20A., 250V.	L6-20R	13	20A., 125/250V.		20, 21	20A., 3φ, 250V.	L15-20R	27
20A., 277V., AC	L7-20R	13	20A., 125/250V.	L10-20R	22	20A., 3φ, 480V.	L16-20R	27
20A., 480V.	L8-20R	14	20A., 3φ, 250V.	L11-20R	22	30A., 125/250V.	L14-30R	28

Fig. 25-4. Plug receptacle configurations.

Technical Information

Rating	NEMA Config. and Config. No.	Page	Rating	NEMA Config. and Config. No.	Page	Rating	NEMA Config. and Config. No.	Page
3-Pole—4-Wire—Grounding (Continued)			**4-Pole—4-Wire—Not Grounding (Continued)**			**POWER INTERRUPTING DEVICES**		
30A., 3φ, 250V.	L15-30R	28	30A., 3φY 277/480V.	L19-30R	35	**3-Pole—3-Wire & 2-Pole—3-Wire**		
30A., 3φ, 480V.	L16-30R	29	30A., 3φY 347/600V.	L20-30R	35	20A., 125V. AC or DC 10A., 250V. 480V., AC		40
30A., 3φ, 600V.	L17-30R	29	30A., 250V. AC or DC 30A., 600V. AC		36	**4-Pole—4-Wire & 3-Pole—4-Wire**		
3-Pole—4-Wire with Equipment Ground			**4-Pole—5-Wire—Grounding**			20A., 250V., DC 30A., 600V., AC		41
50A., 250V., DC 50A., 600V., AC		30	20A., 3φY 120/208V.	L21-20R	37	20A., 250V., DC 30A., 600V., AC		42
50A., 250V., DC 50A., 600V., AC		31	20A., 3φY 277/480V.	L22-20R	37	**4-Pole—5-Wire**		
4-Pole—4-Wire—Not Grounding			20A., 3φY 347/600V.	L23-20R	37	20A., 250V., DC 30A., 600V., AC		43
20A., 250V. AC or DC 10A., 600V. AC		32, 33	30A., 3φY 120/208V.	L21-30R	38			
20A., 3φY 120/208V.	L18-20R	34	30A., 3φY 277/480V.	L22-30R	38			
20A., 3φY 277/480V.	L19-20R	34	30A., 3φY 347/600V.	L23-30R	38			
20A., 3φY 347/600V.	L20-20R	34						
30A., 3φY 120/208 V.	L18-30R	35						

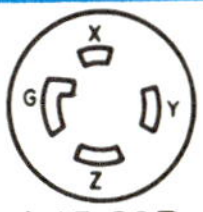

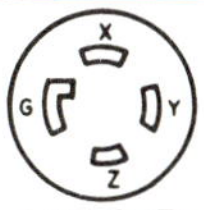

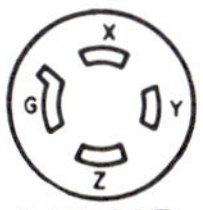

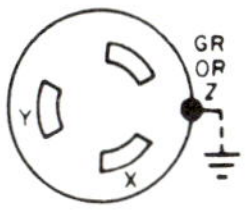

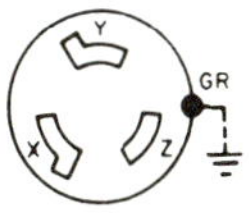

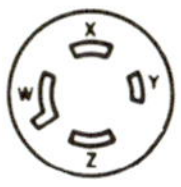

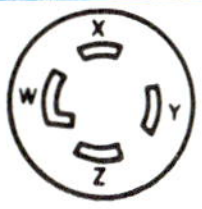

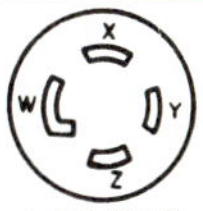

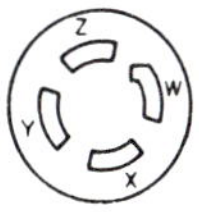

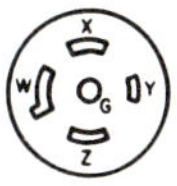

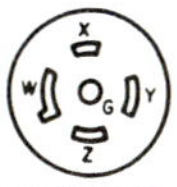

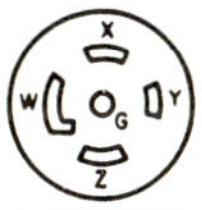

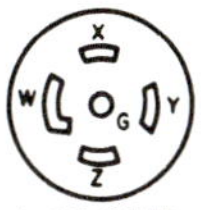

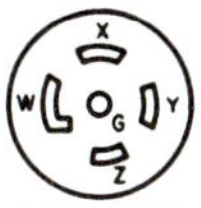

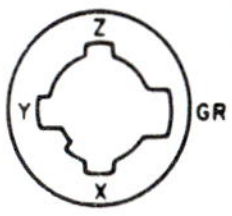

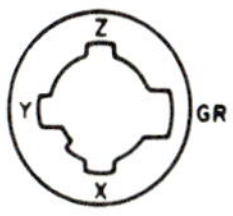

Note that "L" series of catalog numbers in this section relate directly to NEMA Configuration numbers.

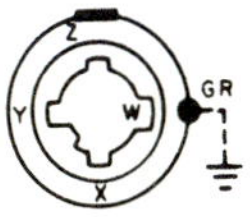

Fig. 25-4. Continued.

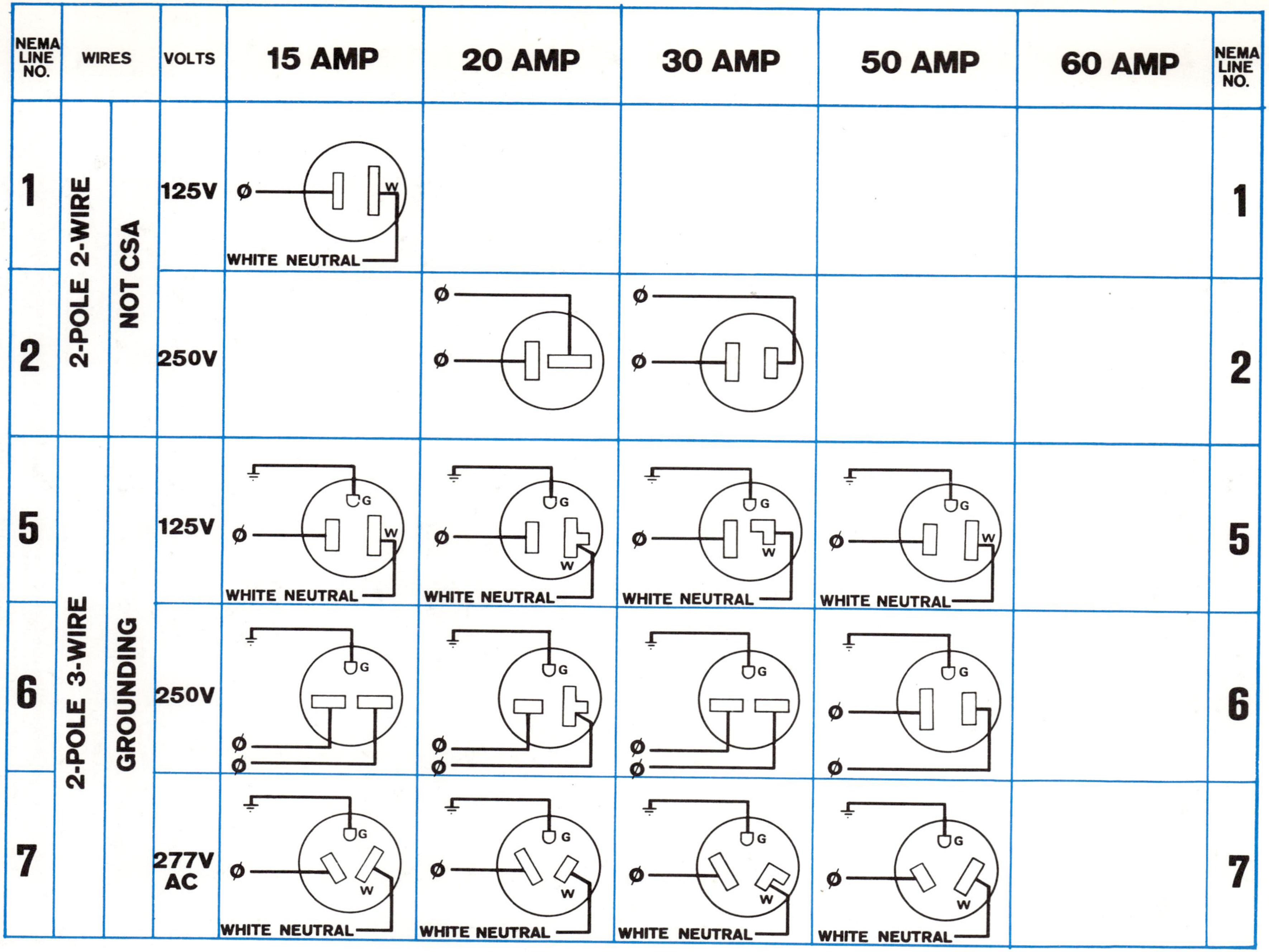

Modern Residential Wiring

NEMA LINE NO.		15 AMP	20 AMP	30 AMP	50 AMP	60 AMP
1						
2						
5						
6						
7						

NEMA LINE NO.	WIRES	VOLTS	15 AMP	20 AMP	30 AMP	50 AMP	60 AMP
1	2-POLE 2-WIRE	125V	○ WHITE NEUTRAL				
2	NOT CSA	250V		○ ○	○ ○		
5	2-POLE 3-WIRE GROUNDING	125V	G W WHITE NEUTRAL	G W WHITE NEUTRAL	G W WHITE NEUTRAL	G W WHITE NEUTRAL	
6		250V	G WHITE NEUTRAL	G WHITE NEUTRAL	G WHITE NEUTRAL	G WHITE NEUTRAL	
7		277V AC	G W WHITE NEUTRAL	G W WHITE NEUTRAL	G W WHITE NEUTRAL	G W WHITE NEUTRAL	

Fig. 25-5. Straight blade receptacle wiring configuations Letters "x," "y," and "z" refer to "hot" terminals. "W" indicates the neutral or white grounded wire terminal. "G" stands for the equipment ground terminal. (General Electric Co.)

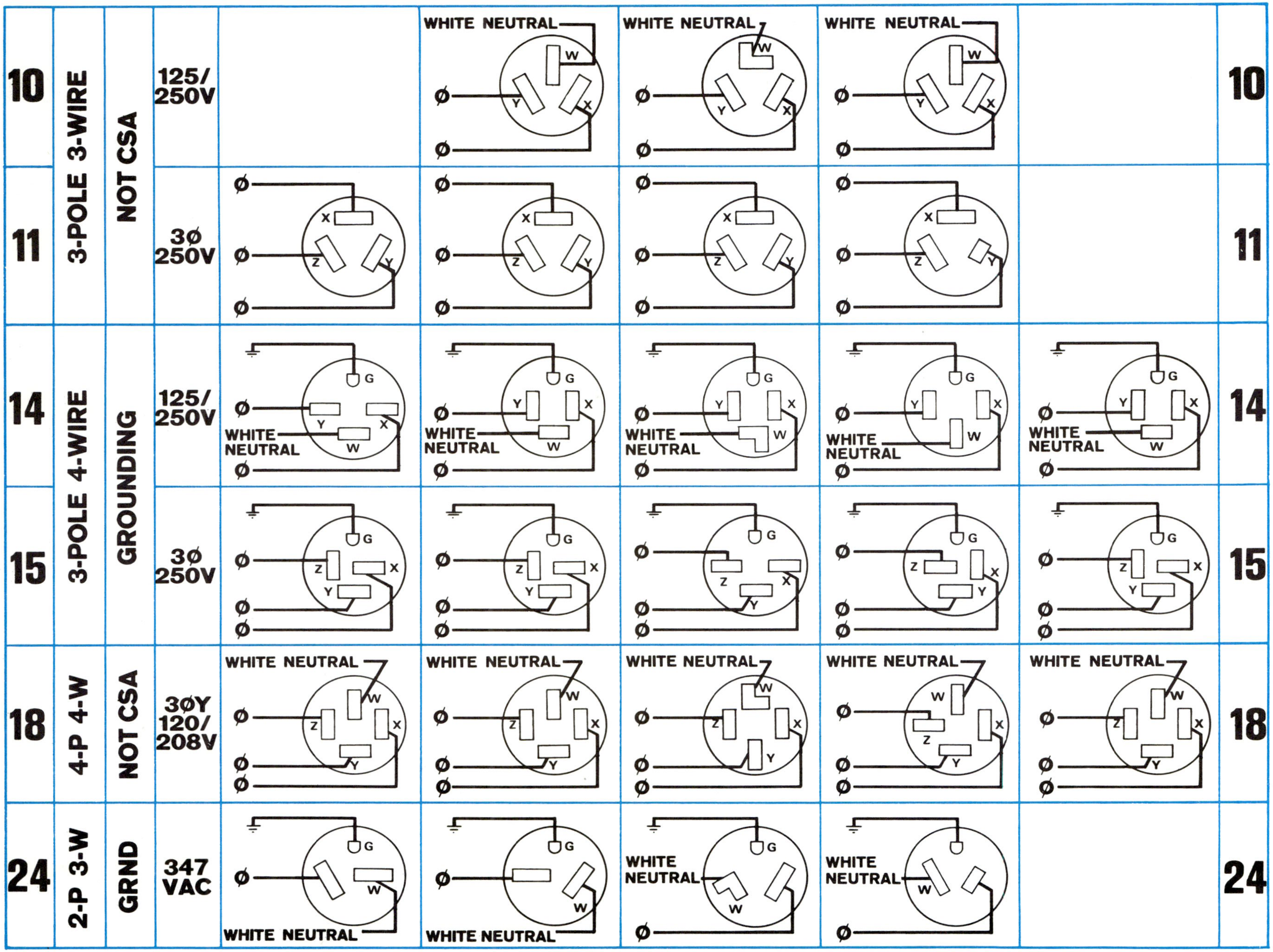

Fig. 25-5. Continued.

Modern Residential Wiring

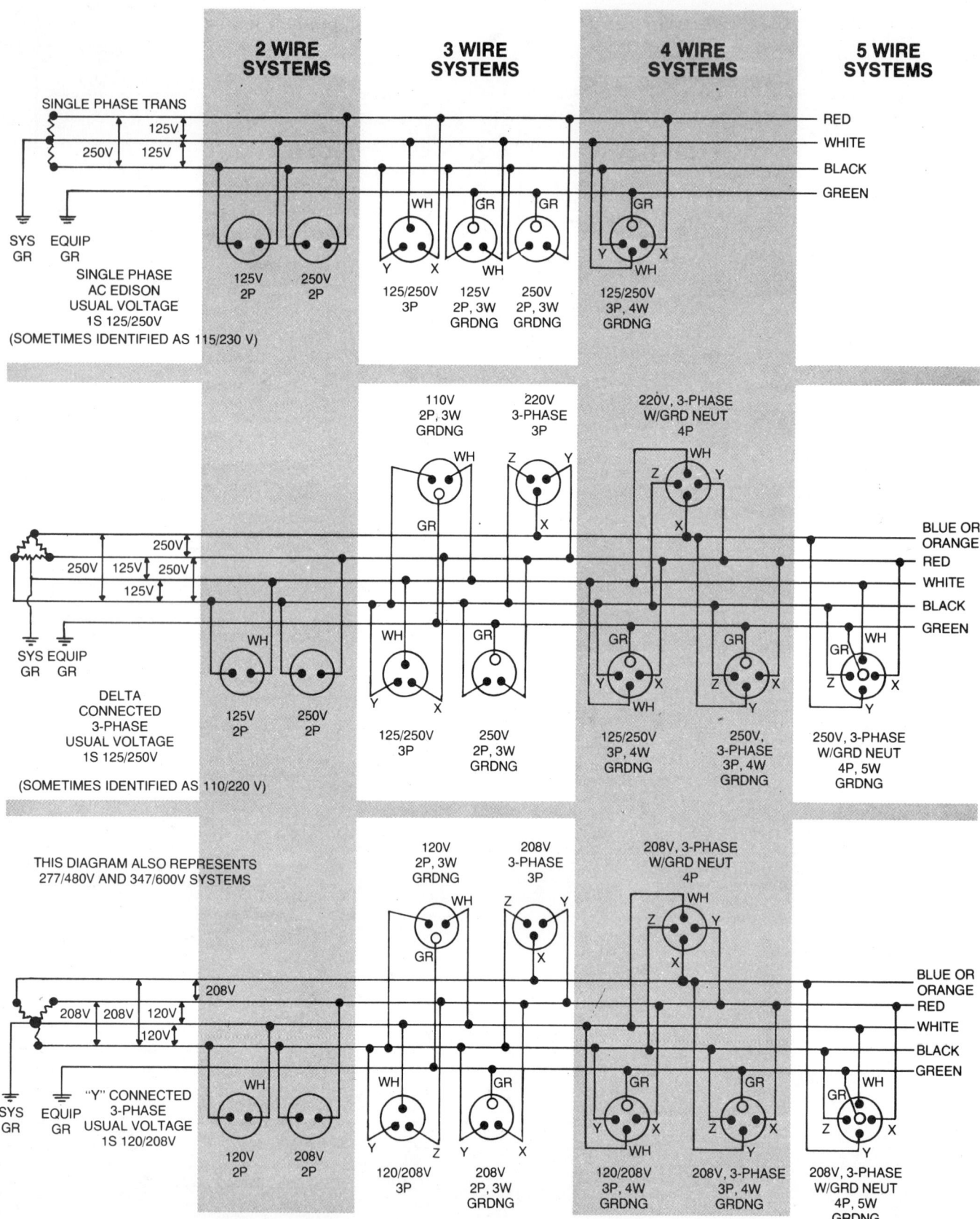

Fig. 25-6. Power circuits and terminal identification. (Slater Electric, Inc.)

Technical Information

$$\text{Area of triangle} = \frac{\text{base} \times \text{altitude}}{2}$$

$$\text{Circumference of circle} = \pi \times \text{diameter} = 2\pi \times \text{radius}$$

$$\text{Area of circle} = \pi \times \text{radius}^2 = \frac{\pi}{4} \times \text{diameter}^2$$

$$\text{Surface of sphere} = 4\pi \times \text{radius}^2 = \pi \times \text{diameter}^2$$

$$\text{Volume of sphere} = \frac{4\pi \times \text{radius}^3}{3} = \frac{\pi}{6} \times \text{diameter}^3$$

$$\left.\begin{array}{l}\text{Volume of prism}\\\text{Volume of cylinder}\end{array}\right\} = \text{area of base} \times \text{altitude}$$

$$\pi = 3\tfrac{1}{7} = 3.14$$

Fig. 25-7. Rules for computing areas and volumes of different shapes.

	Average Wattage	Est KWH Used Annually	Est KWH Used Monthly		Average Wattage	Est KWH Used Annually	Est KWH Used Monthly
FOOD PREPARATION				**LAUNDRY**			
Blender	300	1	*	Clothes Dryer	4,856	993	83
Broiler	1,140	85	7	Iron (hand)	1,100	60	5
Carving Knife	92	8	*	Washing Machine			
Coffee Maker	894	106	9	Automatic	512	103	9
Deep Fryer	1,448	83	7	Non-Automatic	286	76	6
Dishwasher	1,201	363	30	Water Heater			
Egg Cooker	516	14	1	With Laundry	4,500	6,000	500
Frying Pan	1,196	100	8	Without Laundry	4,500	4,800	400
Hot Plate	1,200	90	7				
Mixer	127	2	*	**HEALTH AND BEAUTY**			
Oven, Microwave (only)	1,450	190	16	Germicidal Lamp	20	141	12
Range—Conventional Oven	12,200	1,175	98	Hair Dryer	381	14	1
—Self-Cleaning Oven	12,200	1,205	100	Heat Lamp, (Infrared)	250	13	1
Roaster	1,333	60	5	Shaver	15	.05	*
Sandwich Grill	1,161	33	3	Sun Lamp	279	16	1
Toaster	1,146	39	3	Tooth Brush	1.1	1.0	*
Trash Compactor	400	50	4	Vibrator	40	2	*
Waffle Iron	1,200	20	2				
Waste Disposal	445	7	*	**HOME ENTERTAINMENT**			
				Radio	71	86	7
FOOD PRESERVATION				Radio/Record Player	109	109	12
Freezer (15-21 cu. ft.)				Television			
Chest type, manual defrost	—	1,320	110	Black and White			
Upright type				Tube Type	100	220	18
Manual Defrost	—	1,320	110	Solid State	45	100	8
Automatic Defrost	—	1,985	165	Color			
Refrigerator/Freezer				Tube Type	240	528	44
Manual Defrost				Solid State	145	320	27
10-15 cu. ft.	—	700	58				
Automatic Defrost				**HOUSEWARES**			
16-18 cu. ft.	—	1,795	150	Clock	2	17	1
Automatic Defrost				Floor Polisher	305	15	1
20 cu. ft. and up	—	1,895	158	Sewing Machine	75	11	1
				Vacuum Cleaner	630	45	4

Fig. 25-8. Appliance power consumption.

Chapter 26
GLOSSARY OF TERMS

ACCESSIBLE: Easily approached, removed or exposed. Not permanently concealed.

AIEE: American Institute of Electrical Engineers.

ALTERNATING CURRENT (ac): A periodic electric current that changes direction 60 times per second.

ALTERNATOR: An alternating current machine which changes mechanical energy into electrical energy.

AMBIENT TEMPERATURE: The temperature of the area which surrounds an appliance or the parts of an apparatus.

AMERICAN WIRE GAUGE: The accepted standard for sizing wire-copper wire, brass wire, German silver wire and also the thickness of sheets of these materials, (AWG).

AMMETER: Instrument which measures electric current.

AMPACITY: Current carrying capacity expressed in amperes.

AMPERE: An electron flow of one coulomb per second.

ANION: A negative ion; an ion which moves toward the anode in electrolysis.

ANSI: American National Standards Institute.

ANODE: The positive terminal, such as the plate in the electron tube or the positive post in electrolysis.

ANNUNCIATOR: Electromagnetically operated signaling device.

APPARENT POWER: Product of the effective current times the effective voltage.

APPLIANCE: A current consuming device; utilization equipment such as clothes dryer, air conditioner, food processor or cooking devices.

APPRENTICE: A beginner in a trade, usually serving several years as a helper to a journeyman.

APPROVED: Acceptable to authority or regulatory agency.

ARCING: The forming of an electric arc across contacts of a switch or at motor or generator.

ARMATURE: The revolving part in a generator or a motor. The vibrating or moving part of a relay or buzzer.

ARMOR: A wrapping of metal, used to protect conductors.

ARMOR CLAMP: A fitting used to attach armored cable to a junction box or other device.

ARMORED CABLE: A flexible metal protective covering enclosing electrical conductors.

ARRESTOR (lightning rod): A device used to protect buildings, including electrical devices, from damage by lightning.

ATOM: The smallest particle that makes up an element, of matter.

AWG: American Wire Gauge, used in sizing wire by numbers. It is the same as Brown & Sharpe Gauge.

BATTERY: Several voltaic cells connected in series or parallel; usually contained in one case.

BATTERY CAPACITY: The ability of a battery to produce a given current over a period of time measured in ampere hours. A rated 100 ampere-hour battery will theoretically produce one ampere for 100 hours or two amperes for 50 hours, etc.

BATTERY RESISTANCE: The internal resistance between plates and electrolyte in a cell or battery.

BELL TRANSFORMER: A small transformer used to change the line voltage for operation of door bells or chimes.

Glossary of Terms

BOND: The continuity of an electric connection across a joint or otherwise separated conductors.

BONDING JUMPER: A conductor between metal parts which establishes electrical continuity.

BOX: An enclosed panel for connecting conductors or mounting devices.

BRAID: An interwoven fibrous or metal covering for conductors.

BRANCH CIRCUIT: The part of a wiring system which extends beyond the control devices.

BRANCH CIRCUIT CENTER: A central supply, from which the branch circuits extend.

BRITISH THERMAL UNIT (Btu): The quantity of heat required to raise one pound of water one degree Fahrenheit.

BROWN & SHARPE: The former name of the American Wire Gauge system of designation of wire sizes. B & S Gauge.

BRUSH: A sliding contact to make connections to a rotating armature in a generator or motor.

BURIED CABLE: Cable which lies below the grade surface. Also called direct burial cable.

BUS: A conductor or common connection for two or more circuits.

BUSHING: A device used to mechanically protect and insulate conductors passing through abrasive openings.

BUILDING: A separate and distinct structure.

BX CABLE: A flexible armored electric cable; flexible metallic cover covers conductor wires.

CABINET: An enclosure used to house the branch circuit connections and protective devices.

CABLE: An arrangement of conductors in a protective covering.

CABLE JOINT: A connection of two or more lengths of cable. Often called a splice.

CALORIE: The amount of heat energy required to raise one gram of water one degree Celsius.

CANDLE: A unit of luminous intensity or brightness.

CAPACITANCE: The inherent property of an electric circuit that opposes a change in voltage. The property of a circuit whereby energy may be stored in an electrostatic field.

CAPACITOR MOTOR: A modified version of the split-phase motor, employing a capacitor in series with its starting winding, to produce a phase displacement for starting.

CARRYING CAPACITY: The ability of a conductor to carry a current, expressed in amperes. Refer to the *National Electric Code.*

CATION: A positively charged ion.

CENTER-TAP: A connection made to the center of a coil.

CELSIUS: A temperature measuring system in which 0° equals the freezing point of water and 100° equals the boiling point.

CENTIMETER: One hundredth part of a meter; .3937 inches.

CHARGE (Electric): An unequal distribution of electrons within or on an object.

CIRCUIT: A conducting path for electrons.

CIRCUIT BREAKER: A protective device, in the form of a relay, which opens the circuit in case of an overload; a device which can open or close an electrical circuit.

CIRCULAR MIL: The area of a circle .001 in. in diameter; used to express wire cross-sectional area.

CODE: See *National Electrical Code.* (NEC).

COIL: Wire arranged in a spiral so as to create a magnetic field from current passing through it.

COMMUTATOR: A group of bars providing connections between armature coils and brushes. A mechanical switch to maintain current in one direction in external circuit.

CONDENSER (Capacitor): A device which stores an electrical charge.

CONDUCTANCE: The ability of an object to permit the flow of electrons; the opposite of resistance.

CONDUCTOR: A substance which allows electrons to flow freely through it; an object having good conductance.

CONDUIT: Electrical tubing either rigid or flexible, metallic or nonmetallic.

CONDUIT BODY: A separate and accessible portion of a conduit system.

CONDUIT FITTINGS: Materials used to connect a conduit system of electrical wiring.

CONDUIT RUN: Continuous electrical duct between two points or devices.

CONNECTED LOAD: The electrical power that would be needed if every load connected to the system were drawing power at the same time.

CONTACTS: Conducting devices which may open or close to complete or interrupt a circuit.

CONTINUITY: The ability of a current to flow continuously through a length of conductor.

CONTINUOUS DUTY: The operation of a circuit or device under a constant load for an indefinite period of time.

CONTROLLER: A system of switches, relays, and instrumentation used to regulate voltage, current, speed, and other predetermined actions of an electrical machine or group of machines.

CONVECTION: Transfer of heat by air, gas, or liquid containing thermal energy.

CORD: Flexible insulated cable having small size conductors.

COULOMB: 6.28×10^{18} electrons is equivalent to 1 coulomb.

COUPLING: A threaded sleeve used to join the ends of two lengths of conduit.

CSA: Canadian Standards Association. Operates a listing service for electrical materials and equipment.

CURRENT: The rate of electron flow. A current of

1 ampere is equal to one coulomb per second.

D'ARSONVAL METER: A stationary-magnet moving coil meter.

DEAD: Functioning portions of an electrical system having no voltage or charge.

DEAD FRONT: Without exposed live parts.

DELTA CONNECTION: A method of connecting three-phase alternators and transfers so that the start-end of one winding is connected to the finish-end of the second. The circuit configuration resembles the Greek letter Δ (Delta).

DEMAND: The amount of power that a power company must supply at a given time or period.

DEMAND LOAD: Amount of power which would most probably be needed at any given time. Generally, the minimum demand load is considered to be about 35 percent of the connected load.

DIAGRAM, ELECTRICAL: Drawing which indicates parts and layout of electrical circuit.

DIRECT CURRENT (dc): A steady, nonperiodic, current which does not vary.

DROP, SERVICE: The overhead service conductors between the utility company's last pole and the customer's first point of attachment.

DROP, VOLTAGE: The electromotive force needed to cause the current to flow through a resistor.

DUCT: Tube for carrying electrical conductors.

DUPLEX RECEPTACLE: A double outlet receptacle used in house wiring. It provides outlets to connect lamps and appliances. The duplex receptacle is mounted in a metal box in the wall.

EFFECTIVE VALUE: The root-mean-square value of alternating current or voltage.

EFFICIENCY, ELECTRICAL: A percentage representing the ratio of power output to power input.

ELECTRICITY: A form of energy related to the atomic structure of matter.

ELECTRODE: A conducting substance through which electricity goes to or from an electrical device.

ELECTROLYTE: A liquid which provides good electrical conduction.

ELECTROMOTIVE FORCE (EMF): The electric pressure caused by a flow of electrons from one point to another; voltage.

ELECTROPLATING: Deposition of metallic ions on an electrode.

ELECTRON: A negative subatomic particle having a charge of 1.6×10^{-19} coulomb.

ELECTROSTATICS: The science or study of electricity at rest.

EMT: Electrical metallic tubing. Another name for thinwall conduit.

ENCLOSURE: Housing of device or apparatus protecting personnel from contact with energized parts.

ENERGY: The ability to do work. Electrical energy is measured in watt-seconds.

ENTRANCE CAP: A weatherproof, insulated cap for terminating the power line connections to a building; service head.

ENTRANCE ELL: A metal box to complete a 90° angle with conduit. A cover allows the electrician to pull wires through conduit in either direction.

ENTRANCE, SERVICE: The part of the electrical installation from the service drop to the main service panel of fuse or breaker box.

EQUIPMENT (SERVICE): All the necessary equipment which constitutes the main control and cutoff of the electrical supply to the premises.

FARAD: Unit of capacitance.

FEEDERS: A conductor or group of conductors between the service equipment and the final branch circuit overcurrent device.

FESTOON LIGHTING: Outdoor string of lights suspended between points exceeding 15 ft. (4.6 m).

FISHING: A means of pulling wires through an enclosed wall section or conduit by means of a single wire or rope.

FISH TAPE: A flexible wire that can be pushed through conduit and around bends. Used to pull wire through conduit.

FITTING: Mechanical accessory, like bushing or locknut, of a wiring system.

FIXTURE STUD: A special fitting used to connect a light fixture to an outlet box.

FOOT-CANDLE: A unit of measurement of light; the light produced on a surface from a source of one candle at a distance to one foot.

FOUR-WAY SWITCH: A switch used in conjunction with three-way switches when control is desired at three or more places.

FUSE: A protective device consisting of an element which melts when subjected to high temperature; opens an electric circuit during surge of excessive current; a current limiting device.

FUSE, PLUG: Household type fuse having threaded base or Type S base. Rated at 0 to 30 amperes at 125 volts.

FUSE, ENCLOSED CARTRIDGE TYPE: Tubular fuse with terminal at each end with a casing that insulates the fusible element and contains the arc.

FUSE, TIME-DELAY TYPE: Enclosed fuse having a time delay on overloads.

GALVANOMETER: Instrument which measures small electric currents.

GENERATOR: A rotating electric machine to provide a large source of electrical energy. A generator converts mechanical energy to electric energy.

GFCI: Ground Fault Circuit Interrupter.

GREENFIELD: A flexible metallic conduit used in the connection of electrical power to machinery and in applications which require bends of the conduit at various angles.

Glossary of Terms

GROUND (EARTH): A conductor which provides connection between a circuit and the earth.

GROUND CLAMP: A mechanical clamping device to connect a ground wire or conduit to ground.

GROUNDED: Refers to the condition of having a proper ground.

GROUND FAULT: Conducting connection, intentional or accidental, between any conductor of an electrical system and the conducting material (metal conduit, metal cabinet, etc.) which encloses the conductors as well as any conductors which are grounded or may become grounded.

GROUNDING ELECTRODE: A conductor placed in the earth, providing a connection to a circuit.

GROUNDING ELECTRODE CONDUCTOR: A conductor interconnecting the grounding electrode to the neutral bus of the service equipment.

HEATER: Device (appliances, etc.) used to raise the temperature of the surrounding area.

HENRY: Unit of inductance.

HERTZ (Hz): Term expressing frequency or cycles per second.

HICKEY: A device used to bend conduit, also, a special fitting for attaching light fixtures to outlet boxes.

HORSEPOWER: 33,000 ft.-lbs. of work per minute or 550 ft. lbs. of work per second equals one horsepower. Also 746 watts = 1 Hp.

HOT: Carrying a current; danger of shock.

IAEI: International Association of Electrical Inspectors.

IBEW: International Brotherhood of Electrical Workers.

IDENTIFIED: Refers to equipment suitable for a specific purpose, application, etc.

IES: Illuminating Engineering Society.

IMPEDANCE: The total resistance to the flow of an ac current as a result of resistance and reactance.

INCANDESCENT LAMP: A lamp which gives light due to a filament glowing at a white heat.

INDUCED CURRENT: The current that flows as the result of an induced emf.

INDUCED EMF: Voltage induced in a conductor as it moves through a magnetic field.

INDUCTANCE: The inherent property of an electrical circuit that opposes a change in current. The property of a circuit whereby energy may be stored in a magnetic field.

INDUCTION MOTOR: An ac motor operating on the principle of a rotating magnetic field produced by out-of-phase currents. The rotor has no electrical connections, but receives energy by transformer action from the field windings. Motor torque is developed by the interaction of rotor current and the rotating field.

INSULATION: Materials which are poor conductors of electricity and which are used to cover wires and components to prevent short circuits and accidental shock hazards.

INSULATORS: Substances containing very few free electrons and requiring large amounts of energy to break electrons loose from the influence of the nucleus.

INTERRUPTER: A switching device which opens and closes a circuit many times per second.

INTERRUPTING RATING: Highest current that a fuse or breaker will permit under ideal or test conditions.

ION: An atom which has lost or gained some electrons. It may be positive or negative depending on the net charge.

ISOLATED: Not easily accessible; requiring special ways of clearance to obtain entry.

JACKET: The outer nonmetallic coating or covering applied over insulated wire or cable.

JOULE: A watt-second, unit of electric power.

JOURNEYMAN: A tradesperson who has served his or her apprenticeship and is qualified to perform the skills of the trade.

JUMPER: A length of wire used to connect a portion of the circuit.

JUNCTION: A point in a circuit where the current branches out into other sections.

JUNCTION BOX: An enclosure for electrical connections.

KILO: Prefix meaning one thousand times. Ten kilovolts means ten thousand volts.

KILOVOLT-AMPERES (kVA): The product of voltage and amperage (power), multiplied by 1000.

KILOWATT: One thousand watts.

KILOWATT-HOUR: 1000 watts per hour. Common unit of measurement of electrical energy for home and industrial use. Power is priced by the kWh.

KIRCHHOFF'S CURRENT LAW: At any junction of conductors in a circuit; the algebraic sum of the current is zero.

KIRCHHOFF'S LAW OF VOLTAGES: In a simple circuit, the algebraic sum of the voltages around a circuit is equal to zero.

KNOCKOUT: Part of an electrical box which can be removed easily to allow conduit or cable connection.

LABELED: Equipment or materials to which a label or other identifying means has been attached which is acceptable to the authority having jurisdiction concerning the products use and evaluation.

LINE DROP: The voltage drop due to resistance in an electrical conductor.

LISTED: Equipment or materials included on a recognized list indicating that the product meets standards or has been tested for a specific use.

LOADS: All the devices which tax a circuit and use electrical energy.

LUGS: Terminals; terminal places on the ends of a

wire or equipment to facilitate rapid connection.

LUMEN: A unit of measurement for the flow of light.

MAGNETIC FIELD: Space around a magnet in which magnetic forces are detected.

MAIN: The main circuit which supplies all others; main disconnect.

MASTER: A tradesperson possessing the qualifications of a journeyman and also the knowledge of the physical laws affecting his/her work and installations. An expert in his/her field.

MASTER SWITCH: A switch which controls the operation of several other switches.

MEGA: Numerical prefix meaning 1,000,000.

METER, ELECTRICAL: An instrument used to measure characteristics of an electric circuit.

METER, MULTIPLIER: A resistor connected to a meter (in series) for purposes of providing greater range capacity.

MHO: Unit of conductance. The reciprocal of the ohm.

MILLIAMMETER: A meter which measures in the milliammeter range of currents.

MILLIAMPERE: One thousandth of an ampere; 1×10^{-3} amperes or .001 amperes.

MILLIVOLT: One thousandth of a volt; 1×10^{-3} volts or .001 volts.

MOTOR: A rotating device which converts electrical energy into mechanical energy.

MOTOR STARTER: A hand or automatic variable resistance box used to limit the current to a motor during the time it is increasing to rated speed. As speed increases, the resistance is decreased.

MTW: Machine tool wire.

MULTIMETER: A combination volt, ampere, and ohm meter. VOM.

NATIONAL ELECTRICAL CODE: A set of rules and regulations to be used by the electrician when installing electric wiring, appliances and machinery.

NEGATIVE: A non-positive, below zero value, an electric polarity sign indicating an abundance of electrons.

NEMA: National Electrical Manufacturers Association.

NETWORK: A series-parallel circuit.

NEUTRAL: Neither positive or negative.

NEUTRAL CONDUCTOR: The grounded wire in a two wire system; the grounded third wire in a three wire system.

NEUTRAL WIRE: The balance wire in a three wire electrical distribution system. The grounded conductor.

NFPA: National Fire Protection Association, sponsor of the *National Electrical Code.*

NOMINAL (VOLTAGE): Value assigned to a circuit or electrical system to designate its voltage (for example, 120/240, 277, 600, etc.).

NONCONDUCTOR: An insulator: material which does not conduct electricity.

OHM: A unit of electrical resistance.

OHMMETER: Electrical instrument used to measure circuit resistance.

OHM'S LAW: Electrical circuit law which states that the current is proportional to the voltage, but inversely proportional to the resistance, reactance, or impedance.

OSCILLOSCOPE: Instrument which displays ac waveforms.

OSHA: Occupational Safety and Health Act. Law passed in 1970 to protect persons in places of employment.

OUTLET: A point at which current may be supplied to a load.

OUTLET BOX: A box, used to terminate a cable or conduit. Connections are made in the box. A variety of covers and plates are available to close the box.

OVERCURRENT: Current in excess of the rated current of equipment or conductors.

OVERLOAD: An excessive demand on an electric circuit. Usually of brief duration and relatively harmless. Caused by temporary surges of current.

PERMANENT MAGNET: A magnet which retains its characteristics for a long period of time.

PLUG: The end of a conductor which is designed to be inserted into an electrical receptacle.

POLARITY: Current flow direction designation.

POSITIVE: A non-negative, greater than zero value; an area, in an electric circuit, which has a deficiency of electrons.

POTENTIAL DIFFERENCE: The amount of work per unit charge needed to move the charge between points. It is measured in volts and is equivalent to joules per coulomb.

POWER: The rate of doing work. Electric power is measured in watts or volt-amperes.

POWER FACTOR: The relationship between the true power and the apparent power of a circuit.

PRIMARY: The input windings of a transformer.

PULL BOX: A metal box at a sharp corner in a conduit, to facilitate pulling wires through the conduit.

RACEWAY: A protected runway or enclosure for conductors of electricity. A continuous channel for holding conductors or cables; conduit, conduit bodies, or cable tray.

RATING: The maximum operating characteristics (volts, amperes, watts, etc.) of a device.

REACTANCE (symbol X): The opposition to an alternating current as a result of inductance or capacitance.

RECEPTACLE: Point along an electrical circuit to which a cord plug is attached for the purpose of using the current supplied by that circuit.

RECTIFIER: A device with a high resistance on one end and a low resistance on the other.

RELAY: A device in a circuit which activates other

Glossary of Terms

parts of the circuit.

REMOTE-CONTROL CIRCUIT: Electric circuit that controls another circuit through relays.

REPULSION-START: A motor which develops starting torque by interaction of rotor currents and a single-phase stator field.

RESISTANCE: The opposition to current flow. A characteristic of electricity rated in Ohms.

RESISTOR: A component of an electric circuit which opposes current flow.

RHEOSTAT: A device having a variable resistance to current flow.

RISERS: Wires and cables which run vertically between floors of a building.

ROOT-MEAN-SQUARE: The effective value of an alternating current or voltage, RMS.

SECONDARY WINDING: The coil which receives energy from the primary winding by mutual induction and delivers energy to the load.

SERIES CIRCUIT: A circuit having a single path for current flow.

SERVICE: All the conductors and equipment for supplying electrical energy to the dwelling.

SERVICE DROP: Connecting wires from the power lines to the point of entry to building.

SERVICE ENTRANCE: The associated cables, conduit, boxes and meters used to bring electric power from main power line to a building.

SERVICE LATERAL: Underground service conductors extending from street main to service entrance conductors.

SHORT CIRCUIT: Conducting current, accidental or intentional, between any of the conductors of an electrical system. This connection may be from line to line or line to the neutral (grounded) conductor.

SHUNT: Any parallel connection in a circuit.

SINE WAVE: A wave form of a single frequency alternating current. The graphical representation of all points traced by the sine of an angle as the angle is rotated through 360 deg.

SINGLE PHASE: Only one alternating current or voltage produced or used.

SINGLE-POLE SWITCH: A switch which opens and closes one side of a circuit only.

SLIP RINGS: Metal rings connected to rotating armature windings in a generator. Brushes sliding on these rings provide connections for the external circuit.

SNAKE: A flexible wire used to push or pull wires through a conduit, a partition, or other inaccessible place.

SNM: Special cable designed for use in hazardous locations.

SOCKET: A device for electrical connection of a plug, bulb, etc.; slang electrical term.

SOLDER: An alloy of tin and lead used to bond wires electrically.

SPLICE: Connection of two or more conductors.

STATIC ELECTRICITY: Electricity at rest as opposed to an electric current.

STATOR: The stationary coils of an ac generator.

STRAND: A single uninsulated conductor.

STRANDED CONDUCTOR: A conductor of several strands of solid wire twisted together. Standard cables have 7, 19, or 37 strands.

SUBSTATION: A station in a power transmission system at which electric power is transformed to a conveniently used form. The station may consist of transformers, switches, circuit breakers, and other auxiliary equipment.

SWITCH: A device for opening or closing a connection in an electric circuit.

SYMBOL: Letter or picture representing some part of an electric circuit or an electrical device.

TAP: A method of drawing current from a conductor.

TERMINAL: Point on an electrical device where connections may be made.

THERMAL CUTOUT: Overcurrent device which protects the circuit during excessive heating conditions.

THERMOPLASTIC: Plastic material which softens when heated and hardens when cooled.

THHN: 90°C, 600 volt wire.

THREE-WAY SWITCH: A three-terminal switch in which a circuit may be switched to either of two paths.

THWN: 75°C, 600 volt wire.

TORQUE: Forces producing twisting or rotating motion. It is measured in foot-pounds.

TRANSFORMER: A regulatory device which "steps up" or "steps down" voltage or current for transfer purposes.

UNDERGROUND CABLE: Cable specially designed for below-grade installation. Type UF is an underground approved cable.

UNDERWRITERS LABORATORIES, INC.: A laboratory which tests devices and materials for compliance with the standards of construction and performance established by the Laboratory and with regards to their suitability for installation in accordance with the appropriate standards of the National Board of Fire Underwriters.

VOLT: A unit of electrical pressure (voltage).

VOLTAGE: The electromotive force or electrical pressure along the conductor of a circuit. The potential difference between portions of a conductor is a circuit.

VOLTAIC CELL: A cell produced by suspending two dissimilar elements in an acid solution. A potential difference is developed by chemical action.

VOLT-AMPERE: The unit of measurement of apparent power; also the unit of electrical power.

VOLTMETER: Instrument used to measure voltage values.

VOM: A common test instrument which combines a voltmeter, ohmmeter, and milliammeter in one case.

WATERTIGHT: Constructed so that water or moisture is excluded from an enclosure.

WATT: The unit of electrical power.

WATT-HOUR: A unit of electrical energy expenditure. It is the equivalent of 1 watt of power operating for one hour of time.

WATT-HOUR METER: A meter which indicates the instantaneous rate of power consumption of a device or circuit. Used by power companies to monitor consumers electrical consumption.

WATTMETER: An instrument used to measure electrical power.

WEATHER-PROOF: Electrical device(s) designed to be protected from the weather.

WESTERN UNION SPLICE: A standard splice made by twisting two wires together.

WIRE: A conductor, bare or insulated.

WIRING DIAGRAM: A diagram showing electrical devices by symbols and their interconnections.

WORK: The result of energy expenditure. A force applied through a distance.

WYE CONNECTION: A method of connecting three-phase alternators and transformers so that the end of each coil or winding has a common neutral point. The circuit configuration resembles the letter Y.

ZERO ADJUST: A mechanical means (dial, screw, button, etc.) for setting a meter pointer on the zero mark of the meter scale.

ZERO POTENTIAL: Zero voltage. No electrical potential.

ACKNOWLEDGMENTS

My appreciation to all those who assisted me in completion of this publication. Special thanks to the following:

My wife, Bambi, for typing the manuscript.
Bud Smith for careful editing.
Malanie Wern for many of the line illustrations.
Various industrial firms for illustrations.

Harvey N. Holzman

INDEX

A

Acceptability of equipment, 85
Ac generators, how they work, 12, 13
Adapting meters to ac measurement, 204, 205
Alternating current generator, 11, 12
Ammeter, 201, 202
Ampacity allowances of insulated copper wire, 28
Amperage, 15, 16
Appliance circuits, 107, 108
Appliance classifications, 154-157
  heaters, 155
  types of heaters, 155-157
Appliance power consumption, 257
Appliance wiring and special outlets, 153-160
  general considerations, 153, 154
Application, low-voltage circuits, 180
Armored cable, 56, 57
  connecting to boxes, 82
  installation, 80-82
  working with, 81
Attaching boxes, 70, 71
Attaching conductors to device terminals, 88, 89

B

Back-to-back bends, 75
Balancing circuit loads, 109-112
  bathroom circuits, 110, 112
  overall design, 109, 110
  special outlets, 109
Barn electrical requirements, 170, 171
Basic remote-control operation, 179, 180
Basic telephone system, 218, 219
Basic tool list, 32
Basic trigonometry, 248
Bathroom circuits, balancing circuit loads, 110, 112
Battery power, 8
Bending IMC and EMT, specs, 78
Bending metallic conduit, 72-78
  back-to-back bends, 75
  making the offset bend, 76
  90 degree bend, 73-75
  offset bend, 75, 76
  right angle bends, 72, 73
  saddle bend, 76, 77
  tools, 72, 73
Bending nonmetallic conduit, 78
Blueprints and wiring circuits, reading, 114-130
Bonding, 47
Boring and notching for conduit, 71
Box and conductor installation, 69-83
Box construction, 59, 60
  handy box, 60
  pull boxes, 60
  types and uses, 60
Box covers and accessories, 67, 68
Boxes, attaching, 70, 71
  fittings, and covers, 59-68
  ganging, 61
  nonmetallic, 62, 64
  telephone wiring, 221
Box extension rings, fittings, 64
Box fill table, 67
Box installation, 193-199
  adding a sub-panel, 195, 196
  contact power company, 194, 195
  modernizing service entrances, 194
  surface wiring, 196-199
Box mounting systems, 61, 62
Box shapes, 59
Branch circuit design, 104-107
  plug-in receptacles, 104, 105
  receptacle, 105, 106
  spacing, 106, 107
  split-wiring, 106, 107
  wall switches, 104
  wire capacity, 104
Branch circuit needs, determining, 110, 111
Branch circuit planning, lighting, 107
Branch circuits, planning, 101-113
  types, 101, 102
Breakers, 211
Building categories and service drops, 163
Building construction, electrical remodeling, 189-193

extending a circuit, 192, 193
installing cable, 189, 190
using nonmetallic boxes, 191, 192
using the fish tape, 190, 191
wall openings, 189, 190
wiring behind a baseboard, 192, 193
Bushings, 66

C

Cable and devices, low-voltage, 180-183
Cable, armored, 56, 57, 80-82
Cable ends, stripping, 81, 82
Cable, nonmetallic sheathed, 57
Cable rough-in, 80
Cable splicers, 241, 242
Calculating panelboard load, mobile
   home wiring, 176
Career categories, 240-242
Careers, cable splicers, 241, 242
  electrical, 240-245
  electricians, construction, 243
  maintenance electricians, 242
  selecting helpful courses, 243
  troubleshooters, 241
Circuit breaker, 144, 145
Circuit controllers (switches), 30
Circuit (electrical path), 9, 10
Circuit fundamentals, 20-22
  parallel circuit rules, 21
  parallel circuits, 21
  series circuit, 20, 21
  series-parallel combination circuits, 21, 22
Circuit guide, general purpose, motors and
   motor circuits, 230
Circuit loads, balancing, 109-112
Circuit protection, 28-30
  circuit controllers, 30
  motors and motor circuits, 229
  overloads and short circuits, 28
  protective device rating, 28-30
Circuits, appliance, 107, 108
  extending, 192, 193
  fixtures, and receptacles, specialized
   wiring, 215, 216
  general lighting, figuring, 108, 109
  improperly connected, 210
  low-voltage, 179-186
  number of, 107
  parallel, 9, 10
  series, 9
  types of, 9, 10
Circuits summary, 22
Circuit wiring for emergency systems, 217
Code enforcement, 16
Code rules, L-V, 184-186
Color coding, telephone wiring, 221
Commercial service entrance requirements, 147-151

four-wire, three-phase systems, 148, 149
Commercial wiring, finding loads, 161, 162
Commercial wiring, light, 161-164
Common electrical symbols, 116
Compression connectors, 91-94
Computing farm power requirements, 170
Conductor allotment, metal boxes, 67
Conductor markings, 27
Conductor properties, 251
Conductors, 212
  EMT or thin-wall conduit, 53
  intermediate metal conduit, 53
  preparation of, 86-88
  rigid conduit, 53
  service entrance, 136
  splicing, 89, 90
  stripping, 86, 87
  types, 52, 53
Conductors to device terminals,
   attaching, 88, 89
Conduit, bending metallic, 72-78
  bending nonmetallic, 78
  boring and notching, 71
  cutting, 71, 72
  flexible metal (Greenfield), 54, 55
  installing flexible metal, 79
  liquid-tight flexible metal, (LTFM), 55, 56
  making connections, 79-83
  rigid nonmetallic, 57
  supporting, 78, 79
Conduit fill, 54-58
  capacity, 58
  special notes, 58
Conduit runs, installing, 71
Conduit size, finding, 54
Conduit wire allotment, 54
Connectors and clamps, fittings, 64, 65
Continuous ground check, 209
Control methods, controller requirements, 230, 231
Controller requirements, control
   methods, 230, 231
  motors and motor circuits, 230
  motor duty, 232
  sample problem, 232
Converting fraction to percentage, 247, 248
Converting fractions to decimals and
   vice versa, 247
Cooking tops and wall mounted cooking
   units, 159, 160
Copper wire properties, 251
Copper wire, resistance of soft or
   annealed, 258
Current capacity emergency systems, 217
Current-carrying electrical conductors,
   attaching, 96, 97
Cutting and sawing tools, 33, 34
Cutting conduit, 71, 72

Index

D

Dairy barn, 168
Definitions, branch-circuit selection current, 232
  mobile home wiring, 172-175
  nameplate data, 227, 228
  rated-load current, 232
Delta four-wire systems, 150, 151
Determining loads, 103, 104
Devices and outlets, mobile home wiring, 175
Device wiring, 85-100
Diagnosing problems, 208, 209
  continuous ground check, 209
  testing fixtures, 209
  testing receptacles, 208, 209
  testing switches, 209
Dial light transformer, indoor telephone wiring, 225
Direct current generator, 13, 14
Direction of induced current, 11-13
Dishwashers, 159
Distribution panelboard, mobile home wiring, 173, 174
Distribution panel hookup, mobile home wiring, 175, 176
Drilling tools, 36, 37
Duplex receptacles, 95, 96, 120-122

E

Effects of electrical shock, 42
Electrical careers, 240-245
Electrical circuit components, 25-31
Electrical circuit theory, 19-24
Electrical codes and safety standards, 16
Electrical connections, 85
Electrical energy consumers, 30, 31
Electrical energy fundamentals, 7-18
Electrical energy, sources of, 8
Electrical meters, 200-206
Electrical remodeling, 187-199
  basic considerations, 187
  building construction, 189-193
  materials, 188, 189
  safety, 187
  special tools, 187, 188
Electrical requirements, barn, 170, 171
Electrical resistance, 19
  length and thickness, 19
  nature of the material, 19
Electrical shock, effects, 42
Electrical symbols, 116, 117
Electrical troubleshooting, 207-211
  breakers, 211
  diagnosing problems, 208, 209
  fuses, 210, 211
  overloaded neutrals, 209, 210
  unbalanced currents, 209, 210
Electric current, types of, 7
Electricians, construction, careers, 243

Electric motors, 15
Electric motors, general provisions, 160
Electric power, transmission of, 14, 15
Electromagnetic induction, 10, 11
Electron theory, 7, 8
  electron travel, 7
  equilibrium, 7, 8
  free electrons, 7
Electron travel, 7
Emergency and standby systems, 216, 217
Emergency power, connection procedures, 217, 218
Emergency power equipment, general types, 217
Emergency switching location and accessibility, 217
Emergency systems, circuit wiring, 217
  locations, 216
  maintenance and testing, 217
EMT or thin-wall conduit, 53
Energy measurement, 22-24
Energy measurement, power, 23
Equilibrium, 7, 8
Equipment acceptability, 85
Equipment grounding, 48
Equipment testing agencies, 16, 17

F

Farm equipment, wiring, 165, 166
Farm power requirements, computing, 170
Farm wiring, 165-171
  grounding, 166, 167
  locating lights, 168, 169
  power distribution, 165-171
  running cable or conduit, 168
  special devices, 167-171
Fastening tools, 35, 36
Feeder assembly, mobile home wiring, 174, 175
Fill allotment, 66, 67
Finding loads, light commercial wiring, 161, 162
Finding percentage, 247
Finishing, 85
Fish tape, building construction, 190, 191
Fittings, 63-66
  box extension rings, 64
  bushings, 66
  connectors and clamps, 64, 65
  ground clip or screw, 66
  rigid conduit bodies, 65
Fixtures, mounting, 98, 99
Fixture wiring, 96, 98
Flexible metal conduit (Greenfield), 54, 55
Formula organization, 248
Four-way switches, 125-127
Four-wire, three-phase systems, 148, 149
Frame number, motor nameplate, 228

Free electrons, 7
Fuses, 145, 210, 211

G

Galvanometer, 201
Ganging boxes, 61
Garbage disposal unit, 159
General layout, motors and motor circuits, 230
General lighting circuits, figuring, 108, 109
General lighting, finding load, 162
General purpose circuit guide, motors and
    motor circuits, 230
Generator power, 8
Generators and alternators, 11-14
Glossary of terms, 258
Ground clip or screw, 66
Ground fault circuit interrupters
    (GFCI), 48-50, 129
  how they work, 49, 50
Grounding, 43
  conductor, 142
  electrode conductor, sizing, 142
  equipment, 48
  essentials, 40-51
  farm wiring, 166, 167
  remodeled systems, 196
  system, 44-47
  theory, 43-47
Grounding the receptacle, 120, 121, 122

H

Handy box, 60
Heater circuits, 157, 158
Heater circuits, water heaters, 157, 158
Heater installation, 155
Heaters, 155
  baseboard units, 157
  floor units, 156
  types, 155-157
  wall units, 156
Heating, air conditioning and water heating,
    finding load, 162
Hog barn, sheep house, horse barn, 169
House loads, 164

I

Identification for safety, 86
Indoor telephone wiring, installing, 222-225
Induced current, direction of, 11-13
Information, careers in electrical trades, 245
Inspection, permits, and licensing, 16
Installation rules, lighting fixtures, 107
Installing boxes and conductors, 69-83
  locating boxes, 69, 70
  planning the rough-in, 69
Installing cable, building
    construction, 189, 190
Installing conduit runs, 71

Installing flexible metal conduit, 79
Installing remote control systems, 183, 184
Insulation color, 27
Insulation systems for small motors, 234
Insulators, 138, 139
Intermediate metal conduit, 53

J

Joining wire, other methods, 91
Junction boxes, 119

K

Knockouts and pryouts, 60, 61

L

Ladder safety, 41
Legally required standby systems, 217
Light commercial wiring, 161-164
Lighting, 107
Lighting, show window, finding load, 162
Lighting fixture installation rules, 107
Line installers and repairers, 241
Liquid-tight flexible metal conduit,
    (LTFMC), 55, 56
Load determination, 103, 104
Loads, 10
Locating boxes, standard box heights, 70
Locating lights, farm wiring, 168, 169
Low-voltage cable and devices, 180-183
Low-voltage circuits, 179-186
  application, 180
  power-limited, 180
  remote-control, 180
  signaling, 180
  types, 180
  wiring advantages, 180
Low-voltage switch, 181, 182
Low-voltage symbols, 184
Low-voltage wiring advantages, 180
L-V code rules, 184-186

M

Makeup of a receptacle, 95
Making conduit connections, 79-83
  pulling wire, 79
  using the fish tape, 79, 80
Making the 90 degree bend, 73-75
Main disconnect and service panel, 140
Main disconnect switch, 140
Main service entrance conductors, 164
Maintenance and testing emergency
    and standby systems, 217
Maintenance electricians, 242
Master switch, low-voltage cable and
    devices, 182, 183
Materials, electrical remodeling, 188, 189
Math review, 247-249
  basic trigonometry, 248

Index

circle area and circumference, 248
converting fraction to percentage, 247, 248
converting fractions to decimals and vice versa, 247
finding percentage, 247
formula organization, 248
rectangle, area, square area, and circumference, 248
Maximum number of outlets, 102, 103
Measuring electricity, 15, 16
 amperage, 15, 16
 resistance/ohms, 16
 voltage, 16
 wattage, 16
Measuring tools, 37
Mechanical considerations, 85
Meter care, 205
Meter design, 200, 201
Meter enclosure, 138
Meter function, 201
Meters, adapting to ac measurement, 204, 205
 ammeter, 201, 202
 electrical, 200-206
 galvanometer, 201
 multimeter, 203, 204
 ohmmeter, 202, 203
 types, 201-204
 voltmeter, 202
Miscellaneous tools, 37-39
Mobile home distribution panelboard, 172, 173
 calculating load, 176
Mobile home feeder assembly, 172, 174
Mobile home hookup, 178
Mobile home parks, 176, 177
Mobile home service equipment, 175
Mobile home service lateral, 175
Mobile home wiring, 172-178
Motor branch circuit, 230
Motor failure, bearings, 235
 causes, 234, 235
 connecting to load, 35
 moisture, 235
 mounting, 235
 overheating, 234, 235
 starting mechanism, 235
Motor feeder conductors, size and protection, 229
Motor installation tips, 234, 235
Motor nameplate, 227, 228
 frame number, 228
 proper operation factors, 228
Motors and motor circuits, 227-239
Motor service operations, 236, 237, 239
Motors, service and repair, 235-239
Mounting fixtures, 98, 99
Mounting, motor failure, 235
Multimeter, 203, 204
Multiphase systems, 146, 147

N

Nameplate data definitions, 227, 228
Nonmetallic boxes, 62, 64
Nonmetallic boxes, building construction, 191, 192
Nonmetallic cable (NM, NMC), preparing, 82, 83
 working with, 82, 83
Nonmetallic sheathed cable, 57
Number of circuits, planning branch circuits, 107

O

Obsolete wiring, 101
Occupational Safety and Health Act (OSHA), 17
Offset bend, 75, 76
Offset bend, making, 76
Ohmmeter, 202, 203
Optional standby power, 217, 218
Operation factors, proper, 228
Operation of equipment, telephone wiring, 220
Outbuildings, farm wiring, 169
Outdoor wiring procedures, 129, 130
Outlets, maximum number, 102, 103
Overall design, balancing circuit loads, 109, 110
Overcurrent protection, 144, 145
Overhead installation, mobile home wiring, 175
Overloaded neutrals and unbalanced currents, 209, 210
Overloads and short circuits, 28

P

Parallel circuit, 9, 10, 21
Parallel circuit rules, 21
Phase concept, service ratings, 146
Phone wiring plan, installing indoor telephone wiring, 222-224
Pilot switches, 125, 127, 128
Planning branch circuits, 101-113
Pliers, 35
Plug-in receptacles, 104, 105
Plug receptacle configurations, 252, 253
Polarity in electrical wiring, 119
Polarity wiring, reasons for, 119
Pole transformer, 131
Poultry house, farm wiring, 169
Power, 23
Power circuit and terminal identification, 256
Power company contact, box installations, 194, 195
Power company wires, 134, 135
Power distribution, farm wiring, 165-171
Power-limited circuits, 180
Preparation of conductors, 86-88
 stripping, 86, 87
 using stripping knife, 87
 using wire stripper, 88
Problem diagnosing, 208, 209
Protective devices, 28

Protective device rating, 28-30
Pull boxes, 60
Pull-chain fixture, 119, 120
Pulling wires, 54, 79

R

Reading blueprints and wiring circuits, 114-130
Receptacles, duplex, 95, 96
  location, 105, 106
  makeup of, 95
  making the ground, 96
  split-wired, 96-98
  wiring, 95
Refrigerator and freezers, 159
Relays, low-voltage cable and devices, 180, 181
Remote control circuits, 180
Remote control operation, basic, 179, 180
Remote control systems, installing, 183, 184
Repairing and troubleshooting, indoor telephone wiring, 225, 226
Resistance/ohms, 16
Retail store, small, finding load, 161, 162
Rigid conduit, 53
Rigid conduit bodies, 65
Rigid nonmetallic conduit, 57
Room air-conditioner units, 159
Rough-in, planning, 69
Rules for computing areas and volumes of different shapes, 257
Running cable or conduit, farm wiring, 168

S

Saddle bend, 76, 77
Saddle bend, making, 76-78
Safety and grounding essentials, 40-51
Safety, electrical remodeling, 187
  ladder, 41
  scaffolding, 41
  telephone wiring, 220
Safety considerations, electrical troubleshooting, 207
Safety during installation, 40, 41
Scaffolding safety, 41
Sealed hermetic motor, 231, 232
Selecting helpful courses, careers, 243
Selecting the proper size disconnect, 232
Series circuit, 9, 20, 21
Series-parallel combination circuits, 21, 22
Service and repair of motors, 235-239
Service clearances, 142-144
Service completion, 142, 144
Service conductors, sizing, 136
Service designation, 147, 148
Service drop mast and insulator, 137-140
Service drop, small multifamily dwelling, 162
Service drop, yard pole, 166
Service entrance, 131-151
  components, 131, 132
  conductors, 136
  modernizing, 194
  requirements, commercial, 147-151
Service grounding, 141, 142
Service head, 137
Service lateral, mobile home wiring, 175
Service load, 162
  multiple occupancies, 163, 164
  power company wires, 134, 135
Service location, 132-134
Service panel main disconnect, 140, 141
Service ratings, 145, 146
Service supplies and fittings, 144
Shock victim, helping, 42, 43
Show window lighting, finding load, 162
Signaling circuits, 180, 212
Single and three-phase electricity, 13, 14
Single-phase connections, 148, 150
Size and protection of motor feeder conductors, 229
Sizing circuit components for combination load, 232
  step-by-step method, 232, 233
Sizing grounding electrode conductor, 142
Small appliance circuits, figuring number of, 109
Small motors, insulation systems, 234
Small multifamily dwellings, 162
Soldering technique, proper, 91
Soldering tools, 37
Sources of electrical energy, 8
Sources of electrical energy, battery power, 8
Sources of electrical power, generator power, 8
Spacing, branch circuit design, 106, 107
Special devices, farm wiring, 167-171
Specialized wiring, 212-226
  additional considerations, 215
  garages and outbuildings, 215, 216
Special outlets, 109
  and appliance wiring, 153-160
  units, 160
Special tools, electrical remodeling, 187, 188
Specs for bending IMC and EMT, 78
Splicing conductors, 89, 90
  compression conductors, 91
  other methods of joining wire, 91
  proper soldering technique, 91
  Western Union splice, 90
Split-wired receptacles, 96-98, 122, 123, 125
Split-wiring, 106, 107
Standby power, optional, 217, 218
Standard box heights, 70
Standardized symbols, 116, 117
State and local codes, 16, 17
  code enforcement, 16
  equipment testing agencies, 16, 17
  inspection, permits, and licensing, 16
Straight blade electric receptacle wiring

Index

configurations, 254, 255
Striking tools, 33
Stripping cable ends, 81, 82
Stripping conductors, 86, 87
Stripping knife, using, 87
Sub-panel addition, box installations, 195, 196
Sub-panel power source, specialized wiring, 216
Supporting conduit, 78, 79
Surface wiring, box installations, 196-199
Switch wiring, 94, 95
Symbols, low-voltage, 184
System ground, how it works, 48
System grounding, 44-47

T

Tables, copper wire properties, 251
  temperature conversion, 251
  voltage drop, 250
Tap joint connector, 139
Technical information, 250-257
Telephone system, basic, 218, 219
Telephone wiring, 218-222
  boxes, 221
  color codes, 221
  indoor, concealing surface wiring, 224, 225
  indoor, dial light transformer, 225
  indoor, new construction, 223, 224
  indoor, old work, 224
  indoor, repairing and troubleshooting, 225, 226
  installing indoor, 222-225
  material and tools, 220-222
  safety, 220
  terminal blocks, 221
  tools, 222, 223
  wall plates, 221, 222
  wire types, 220, 221
Temperature conversion table, 251
Temperature, electrical resistance, 20
Temporary wiring, 41, 42
Terminal blocks, telephone wiring, 221
Testing fixtures, 209
Testing receptacles, 208, 209
Testing switches, 209
Three-way switches, 124, 125
Tools, cutting and sawing, 33, 34
  drilling, 36, 37
  electrician, 32-39
  essential, 32, 39
  fastening, 35, 36
  for bending metallic conduit, 72, 73
  measuring, 37
  miscellaneous, 37-39
  pliers, 35
  soldering, 37
  striking, 33
  telephone wiring, 222, 223
  wire and cable cutters, 34, 35

Training, informal, 243
Training, other qualifications, and advancement, 242, 244
Transformer, 212
  hazardous locations, 213, 214
  low-voltage cable and devices, 180
  operation, 146, 147
  principles, 14, 15
  rule, 147
  wiring of hazardous locations, 214, 215
Transmission and distribution electricians, 241, 242
Transmission of electric power, 14, 15
  transformer principles, 14, 15
Trigonometry, basics, 248
Troubleshooting tools, 207, 208
Types of branch circuits, 101, 102
Types of circuits, 9, 10
Types of electric current, 8
Types of meters, 201-204

U

Unbalanced currents, 209, 210
Using the fish tape, 79, 80

V

Voltage drop, motors and motor circuits, 228, 229
Voltage drop table, 250
Voltmeter, 202

W

Wall plates, telephone wiring, 221, 222
Wall openings, building construction, 189, 190
Wall switches, 104
Water heaters, 157, 158
Wattage, 16
Watt/hour meter, 205
Wattmeter and watt/hour meter, 204, 205
Weatherproof connectors, 138, 139
Western Union splice, 90
Where to locate service, 133, 134
Wire and cable cutters, 34, 35
Wire capacity, 104
Wire coverings (insulation), 26, 27
Wire sizes, 25-27
Wire sizes, voltage drop, 26
Wire stripper, using, 88
Wire types, telephone wiring, 220, 221
Wiring behind a baseboard, 192, 193
Wiring circuits, 117-130
  duplex receptacles, 120-122
  four-way switches, 125-127
  ground fault circuit interrupters, 129
  junction boxes, 119
  outdoor wiring procedures, 129, 130
  pilot switches, 125, 127, 128
  polarity in electrical wiring, 119